Lecture Notes in Economics and Mathematical Systems

Springer
Berlin
Heidelberg
New York
Barcelona
Budapest
Hong Kong
London
Milan
Paris
Tokyo

Jürg Kohlas   Paul-André Monney

A Mathematical Theory of Hints

An Approach to the Dempster-Shafer Theory of Evidence

Springer

Authors

Prof. Dr. Jürg Kohlas
Institute of Informatics
University of Fribourg
CH-1700 Fribourg, Switzerland
e-mail juerg. kohlas @ unifr.ch

Dr. Paul-André Monney
Insitute of Informatics
University of Fribourg
CH-1700 Fribourg, Switzerland

ISBN 3-540-59176-1 Springer-Verlag Berlin Heidelberg New York

Ligrary of Congress Cataloging-in-Publication Data

Kohlas, Jürg, 1939- A mathematical theory of hints: an approach to the Dempster-Shafer theory of evidence / Jürg Kohlas, Paul-André Monney. p.cm. – (Lecture notes in economics and mathematical systems; 425
ISBN 0-387-59176-1
Probabilities. 2. Mathematical statistics. 3. Evidence. I. Monney, Paul-André. II. Title. III. Series. QA273.K597 1995 519.5'4–dc20 95-12080

Printed in Germany

Typesetting: Camera ready by author
SPIN: 10486761 42/3142-543210 - Printed on acid-free paper

Preface

This book is about the theory of hints, a variant of the Dempster-Shafer theory of evidence. Since the reader may already have heard of or even be familiar with the the theory of evidence, let us begin by giving some historical notes. Evidence theory started from a course on statistical inference taught by Arthur P. Dempster at Harvard University in the late sixties. Then Dempster's ideas were reinterpreted by Glenn Shafer in his book *A Mathematical Theory of Evidence* published by Princeton University Press in 1976. Evidence is an appealing term and it is no surprise that the theory of evidence found much interest among knowledge engineers who have to model uncertain and imprecise information. After a lot of ad hoc attempts to model uncertainty in expert systems, a serious theory seemed finally available to treat this basic problem of the field. However, as the theory is not really compatible with the simple paradigms of early rule-based expert systems, misinterpretations and oversimplistic misuses of the theory caused much deception. Only slowly begins the true nature and the real meaning of evidence theory to emerge and to be understood. And hopefully this book will contribute to its better understanding as well demonstrate its usefulness and mathematical elegance.

To put the subject of this book into perspective, it is a good idea to have a closer look at Dempster's and Shafer's conceptions of the theory of evidence. It will then be easier to explain what is the essence of the theory of hints, which is our own view of the theory of evidence. Dempster originally developed a theory of lower and upper probabilities in an attempt to reconcile Bayesian statistics with Fisher's fiducial argument (Dempster, 1967, 1968). The following citation (up to symbols) is taken from Dempster (1967): "Consider a pair of spaces Ω and Θ and a multivalued mapping Γ which assigns a subset $\Gamma(\omega) \subseteq \Theta$ to every $\omega \in \Omega$. Suppose that P is a probability measure which assigns probabilities to the members of a class $\mathcal{A}$ of subsets of Ω. If P is acceptable for the probability judgements about an uncertain outcome $\omega \in \Omega$, and if this uncertain outcome ω is known to correspond to an uncertain outcome $\theta \in \Gamma(\omega)$, what probability judgements may be made about the uncertain outcome $\theta \in \Theta$?" As Dempster argues, this leads to the concept of compatible probability measures and he refers to the bounds as lower and upper probabilities.

On the other hand, Shafer states in the preface of his book: "It *(his book)* offers a reinterpretation of Dempster's work, a reinterpretation that identifies his lower probabilites as epistemic probabilities or degrees of belief, takes the rule for combining such degrees of belief as fundamental, and abandons the idea that they arise as lower bounds over classes of Bayesian probabilities." Given a piece of evidence relative to a question whose possible answers form the set Θ, Shafer postulates the existence of a function Bel on the power set of Θ satisfying some axiomatic conditions. He interprets the numerical value $Bel(H)$, a number between 0 and 1, as the degree of belief that H contains the answer and calls the function Bel a belief function. When Θ is finite he shows that belief functions are generated by so-called basic probability assignments which in turn are used to combine belief functions.

So where does the theory of hints stand with respect to these two points of view ? The point of view of hints is closer to Dempster's approach than to the one of Shafer. However it gives a new interpretation to the probabilistic structure considered in Dempster. So we keep Dempster's original structure $(\Omega, P, \Gamma, \Theta)$ but we give it the following meaning:

Suppose that a certain question, whose answer is unknown, has to be studied. The elements of Θ represent the possible answers to the question and we assume that there is exactly one correct, but unknown answer in Θ. The goal is to make assertions about the answer in the light of the available information. We assume that this information allows for several different interpretations, depending on some unknown circumstances. These interpretations are regrouped into the set Ω and there is exactly one correct interpretation. Not all interpretations are equally likely and the known probability measure P on Ω reflects our information in that respect. Furthermore, if the interpretation $\omega \in \Omega$ is the correct one, then the answer is known to be in the subset $\Gamma(\omega) \subseteq \Theta$. Such a structure $\mathcal{H} = (\Omega, P, \Gamma, \Theta)$ is called a hint. The hint $\mathcal{H}$ can then be used to judge the fact or the hypothesis, which can be true or false, that the answer is in a subset H of Θ by assinging to it a degree of support and a degree of plausibility. An interpretation $\omega \in \Omega$ supports the hypothesis H if $\Gamma(\omega) \subseteq H$ because in this case the answer is necessarily in H. The degree of support of H, denoted by $sp(H)$, is defined as the probability of all supporting interpretations of H. So $sp(H)$ can be regarded as the probability with which H can be infered or proved from the available information encoded in the hint. In a similar way, if the interpretation $\omega \in \Omega$ is correct, the subset H possibly contains the answer if $\Gamma(\omega) \cap H \neq \emptyset$. Then the degree of plausibility of H, denoted by $pl(H)$, is defined as the probability of all interpretations under which H is possible. So like in Shafer's theory it is possible to judge hypotheses about the answer, but in contrast to it this can be done only in a second stage, after the hint has been constructed. For us, the degrees of support and plausibility are concepts which are derived from the notion of a hint, and not primitive objects like in Shafer's approach. Also, we speak of degrees of support and not degrees of belief because the word belief suggests an idea of personal judgement which is absent as soon as the hint is given. Considering all possible subsets $H \subseteq \Theta$ leads to support and plausibility functions sp and pl.

Dempster mathematically defines $sp(H)$ and $pl(H)$ exactly in the same way as we did above, but the interpretation he gives to these values is different from ours. He points out that $sp(H)$ can be regarded as the minimal amount of probability which can be transfered from Ω to outcomes $\theta \in H$ and $pl(H)$ as the largest possible amount of probability which can be transfered to outcomes $\theta \in H$. Then he introduced the concept of compatible probability measures: P is compatible if $sp(H) \leq P(H) \leq pl(H)$ for all $H \subseteq \Theta$. Ever since sp and pl have often been looked at as lower and upper bounds for some unknown probability distribution by many authors. This has had unfortunate effects because it led to intriguing misinterpretations of the theory of evidence. In our opinion, the question treated by the theory of evidence is whether a hypothesis H is true or not, or more precisely, to what degree a piece of evidence supports a hypothesis H. It is quite a different question to ask what could be the *probability* of the hypothesis H. In the latter case the unknown is *not* whether H is true or not, but the value of its probability. So let us emphasize that although approaches using partially specified probabilities may be very useful in many circumstances, the theory of evidence as developed in this book as a theory of hints does *not* belong to this class of methods.

It is clear today that the theory of evidence can be given various different forms. Some of them are based on probability theory and others are axiomatic theories, a priori without reference to probability theory. The first approach tries to integrate evidence theory in the framework of classical probability theory, and the theory of hints belongs to this category, whereas the second one deliberately goes beyond classical probability. Shafer's theory of belief functions and Smets' transferable belief model (Smets, 1988) are the two best known representatives of this non-standard probability approach. Despite the differences in approach and interpretation, all of them lead essentially to mathematically equivalent theories in the finite case, which implies that they share the same basic theorems and the same computational procedures apply. However, substantial differences appear when the infinite case is considered.

This book is divided into three Parts: Part I (Chapters 1 to 8) presents the elementary theory of hints, Part II (Chapters 9 to 12) discusses some possible applications of this theory and Part III (Chapters 13 to 16) develops the rudiments of a general theory of hints.

By discussing some simple examples, Chapter 1 intuitively introduces the idea of a hint and the related concepts of support and plausibility. These concepts are then defined and studied formally in Chapters 2 and 3. Chapter 4 explains how to combine hints, an operation that is fundamental to the whole theory. Chapter 5 shows that the theory of hints provides for an adequate framework for probabilistic assumption-based reasoning, thereby establishing the link with propositional logic. The degree of support of a logical formula h is then essentially equal to the probability of the assumptions that permit to logically infer h, given the knowledge base Σ. This idea goes back to Pearl's notion of probability of provability (Pearl, 1988), an idea also present in Laskey, Lehner (1989) and Provan (1990). The assumptions can be seen as arguments in favour of hypotheses. These arguments are weighed by their likelihoods and the degree of support can then be viewed as the reliability of an associated proof system. This point of view is considered in Chapter 6. Let us note that this approach can be developed beyond the material presented in this book: The idea of arguments for hypotheses suggests that it might be possible to develop a purely qualitative theory of evidence by dropping the probabilities on the arguments. This has been done in Kohlas (1993 a) for propositional logic and was generalized to Boolean algebras and lattices in Kohlas (1994). It is even possible to go one step further by generalizing the notion of provability to arbitrary consequence relations in the sense of Tarski (Besnard, Kohlas, 1994). Of course, in a second step, it is always possible to reintroduce probabilities into these qualitative or symbolic versions of evidence theory. Chapter 7 is concerned with the analysis of families of frames of discernment. This done in the abstract setting of lattices and important notions like conditional independence of frames are discussed. Chapter 8 deals with the problem of combining hints efficiently and local propagation schemes are presented.

Chapter 9 shows how statistical inference can be developed from the point of view of the theory of hints. This leads to interesting reinterpretations of classical statistical results. Chapter 10 discusses uncertainty in dynamical systems from the perspective of hints and Chapter 11 considers diagnostic problems from an evidential point of view. The last chapter of Part II, Chapter 12, explains how the theory of hints can be used to represent and support spatial and temporal reasoning. A very general class of models based on groups is analyzed and computational methods are presented.

Chapter 13 presents a general theory of hints, where the sets of interpretations and the frames of discernment are not necessarily finite. The main result in this chapter is that every support function is generated by a generalized hint. Chapter 14 introduces the notion of inclusion between hints, as well as the the notion of canonical hint of a support function. Chapter 15 shows how to combine general hints by Dempster's rule and it is proved that inclusion of hints is preserved by this operation. Finally, the last Chapter 16 studies closed random intervals and random variables from the perspective of hints.

Part I should be read first, and then either Part II and then Part III, or Part III and then Part II. Within parts the chapters should be read in sequence.

There is no doubt that theoretical foundations, practical use in modeling and computational aspects represent some of the possible topics for future research in evidence theory. On the theoretical level, evidence theory on topological spaces is a large open field for investigation. Also, a hint can be seen as the generalization of a random variable and classical probability theory can largely be considered as a theory of random variables. However, Dempster's rule of combination of hints, which is a key concept in the whole theory of evidence, does not appear as such in probability theory. What is the role of this rule in probability theory ? Also, recently connections between evidence theory and modal logic have been put forward, but there remains a lot of open questions about the relationship between evidence theory and non-standard logics. Furthermore, it seems that evidential information is essentially relational in nature, but conditional information has been proved very important in knowledge engineering (e.g. in Bayesian networks where the knowledge is essentially encoded as conditional probability distributions). Does this aspect really restrict the applicability of evidence theory ? Finally, from the point of view of computations, there is clearly a need for efficient approximations and heuristics for evidential calculations. In order to study a question, probably not all available evidence has to be considered because the influence of far-fetched information might not be substantial. So focussing strategies to simplify the reasoning is another important topic for investigation.

Many of the developments in this book are tentative, especially in Parts II and III. This is due to the fact that the theory is still new and many important questions remain open. It is hoped that this text contributes to expand the new ideas on evidence theory and its applications.

Fribourg, December 1994
J. Kohlas, P.A. Monney

Acknowledgments

We would like to thank a number of people who have helped to bring this book into existence. We are deeply indebted to Miss Ruth Leu, who corrected a lot of english mistakes and greatly improved the style of the text. We also wish to express our appreciation to Prof. Prakash Shenoy for reading some chapters, and to Louis Cardona, Rolf Haenni, Urs Hänni and Norbert Lehmann for implementing some of the ideas contained in the book. Also, we would like to thank the Swiss National Foundation for Scientific Research and the Swiss Federal Ministry of Science and Education for their financial support (Grants Nr. 21-30186.90 and 21-32660.91). Finally, one of the authors (P.A. Monney) expresses his gratitude to the Research Commission of the University of Fribourg and the Swiss National Foundation for Scientific Research for supporting him during his twelve months stay at the University of Kansas. In that regard, he wants to thank the entire University of Kansas community for its welcome, especially Prof. Prakash Shenoy for his invitation to be part of this community.

Table of Contents

Part I

Elements of the Theory of Hints

1. Introductory Examples

1.1 Pure Arguments, Unreliable Rules and Assumption-Based Reasoning

The following scheme is perhaps one of the most common in everyday evidential situations: We know that a implies b (if it rains we get wet), but it is not sure whether a is true (whether it will rain). Let's admit that a known probability $p(a)$ for a being true is given. What can then be concluded about b ? A probabilist would apply the theorem of total probability to compute the probability of b

$$p(b) = p(b|a)p(a) + p(b|\neg a)p(\neg a). \tag{1.1}$$

The conditional probability $p(b|a)$ of b given a equals one by the fact that a implies b. The second conditional probability $p(b|\neg a)$ is not known, but must have a value somewhere between zero and one. Hence it follows from (1.1) that $p(b)$ is not smaller than $p(a)$ (if $p(b|\neg a) = 0$) and at most one (if $p(b|\neg a) = 1$) : $p(a) \leq p(b) \leq 1$.

Instead, in a different manner, we may say that there is an uncertain and not fully reliable argument a in favour of b. If a happens to be true, then b can be deduced. Thus, b is supported by the available knowledge to the degree that a is true. The probability that b can be deduced or proved, given the available information equals $p(a)$. This measures not so much a degree of belief in b, but rather a *degree of credibility* or *support* of b in the light of the available explicit knowledge:

$$cred(b) = p(a). \tag{1.2}$$

In fact the degree of belief in b may subjectively be much stronger than $p(a)$ (a pessimist is nearly sure that he will get wet, even if meteorological predictions assign a small risk to rain only). However this belief can only be supported, argued or sustained to the degree $cred(b)$ by the available evidence. Furthermore, as there is no argument in favor of the *negation* of b, we may conclude that b is entirely possible or plausible and assign it a *degree of plausibility* of one,

$$pl(b) = 1. \tag{1.3}$$

Thus, we examine both to what degree a hypothesis is *supported* and to what degree it is *contradicted* by the available evidence.

This is representative of the prefered approach taken in this book. Rather than determine an interval of possible degrees of belief, the available arguments are identified, their reliability is assessed and the probability of proof, deduction or inference of a hypothesis or its negation is computed. This has the advantage of providing explicit explanations and foundations for the credibility and plausibility of the hypothesis in the light of the available information and knowledge. Even if a numerical assessment of the reliability of arguments (i.e their probabilities) is not available or difficult to obtain, the elaboration of possible proofs or deductions is already in itself of great value for a qualitative judgment of hypotheses, their credibility and plausibility.

The scheme sketched here corresponds to Jakob Bernoulli's notion of *pure argument* as developed in his "Ars Conjectandi" (the art of conjecture), a concept nearly forgotten afterwards. Pure evidence splits into x cases which favour a hypothesis and y cases which are neutral (Hacking, 1975). Alternatively, mixed evidence decomposes into x cases that count for the hypothesis and y cases that count against it. Bernoulli himself cites the following example:

" ... A certain man has been stabbed with a sword in the midst of a rowdy mob, and it is established by the testimony of trustworthy men who were standing at a distance that the crime was committed by a man in black coat. If it is found that Gracchus and three others in the crowd were wearing tunics of that colour, this tunic is something of an argument that the murder was committed by Gracchus, but it is mixed; for in one case it proves his guilt, in three cases his innocence, according to whether the murder was perpetrated by himself or by one of the remaining three; for it is not possible that one of these perpetrated it without Gracchus being thereby supposed innocent. But if indeed in a subsequent hearing Gracchus paled, this pallor of face is a pure argument; for it proves Gracchus' guilt if it arises from a guilty conscience, but it does not, on the other hand, prove his innocence if it arises otherwise; for it could be that Gracchus pales from a different cause yet is still the murderer." (Ars Conjectandi, p. 218).

See also Shafer, 1978, who rediscovered this remarkable concept of Bernoulli and noted its relationship to his mathematical theory of evidence.

In this view, the crucial question is then, how to combine different mixed and pure arguments, a problem which Bernoulli did also grapple with without finding satisfactory answers (according to Hacking, 1975 and Shafer, 1978). Let's for the time being limit the consideration to "pure" arguments. Suppose it is known that a_1 implies b and that a_2 implies b. But neither a_1 nor a_2 are known to be true for certain. Each of the two propositions is a sufficient argument to deduce b. The degree of credibility of b equals thus the probability that at least one of the two propositions is true. In order to determine this probability, the combined sample space with the four possible combinations $(a_1, a_2), (a_1, \neg a_2), (\neg a_1, a_2)$ and $(\neg a_1, \neg a_2)$ has to be considered and a probability measure P to be defined on it. It follows then $cred(b) = P\{(a_1, a_2), (a_1, \neg a_2), (\neg a_1, a_2)\} = 1 - P(\neg a_1, \neg a_2)$. If there is sufficient reason to assume that a_1 and a_2 are independent from each other (in the stochastic sense), then

$$cred(b) = 1 - (1 - p_1)(1 - p_2) \tag{1.3}$$

where p_1 and p_2 are the probabilities for a_1 and a_2 respectively. This is one of the basic combination rules for pure evidence already developed by Bernoulli. The formula (1.3) is reminiscent of reliability theory. In fact, we have here a "proof machinery" for hypothesis b which is functioning if any one of the two "component arguments" a_1 or a_2 of the machinery is functioning. This corresponds to the well known "parallel systems" of reliability theory and (1.3) is the corresponding reliability formula. There exists also the analogy with "series systems". Suppose only a_1 and a_2 together imply b. Then, in order to prove b both a_1 and a_2 must be simultaneously true; this corresponds to a "series system" of reliability theory where a system is functioning only if all its components do. Assuming independence between a_1 and a_2, it follows that

$$cred(b) = p_1 p_2 \tag{1.4}$$

in this case.

Let's consider now still another simple case involving "mixed" arguments. This time a1 is known to imply b whereas a_2 is known to imply $\neg b$. Here we have an argument in favour of b and another one against b, and both are uncertain. However the case that both a_1 and a_2 are simultaneously true must be excluded on the ground that it would lead to a *contradiction*, since b and $\neg b$ can not be true at the same time. The sample space of possible combinations to consider is thus limited to the set $\{(a_1, \neg a_2), (\neg a_1, a_2), (\neg a_1, \neg a_2)\}$. A probability defined on this sample space permits to evaluate the degrees of credibility of b and $\neg b$: $cred(b) = P(a_1, \neg a_2), cred(\neg b) = P(\neg a_1, a_2)$. The doubt in b is defined as the probability that $\neg b$ can be proved. Since the plausibility is as much smaller as the doubt is high, it is natural to define the degree of plausibility of b as $pl(b) = 1 - cred(\neg b)$, which equals $P(\{(a_1, \neg a_2), (\neg a_1, \neg a_2)\})$. In some cases it may be reasonably assumed that a_1 and a_2 are a priori independent (before considering the contradiction implied by the knowledge that a_1 implies b and a_2 implies $\neg b$). Examples where this is clearly true will be presented in Sect. 1.3). Given the contradiction generated by the combination (a_1, a_2), the corresponding product probability on the sample space $\{(a_1, a_2), (\neg a_1, a_2), (a_1, \neg a_2), (\neg a_1, \neg a_2)\}$ must be conditioned on the fact that contradictions are not possible. This leads then to

$$\begin{aligned} cred(b) &= p_1(1 - p_2)/(1 - p_1 p_2) \\ cred(\neg b) &= (1 - p_1)p_2/(1 - p_1 p_2) \\ pl(b) &= (1 - p_2)/(1 - p_1 p_2). \end{aligned} \tag{1.5}$$

This is an instance of a more general rule proposed by Dempster (1967) which will play a fundamental role in this book.

It is also possible, that the implications are not entirely certain or reliable; it may be that x implies y in general, but that under some circumstances this relation is not valid. Such a situation could arise if a rule "if x and a then y" is given, where the condition a is uncertain. If a is true, then x indeed implies y, otherwise nothing can be said, although x and y are still compatible. Examples for such *uncertain rules* will be found in the following Sect. 1.2. Similarly, we consider rules R such as "if x_1 and x_2 and $\ldots x_n$ then y" whose validity depends on some proposition r which has a probability q. Hence r is a pure argument for the validity of the rule. A system of rules like for example

$$R_1 \text{ :if } x_1 \text{and } x_2 \text{then } x_0$$
$$R_2 \text{ :if } x_2 \text{and } x_3 \text{then } x_0$$
$$R_3 \text{ :if } x_4 \text{then } x_0$$

can be graphically represented as in Fig. 1.1. Suppose that in addition there are some pure arguments a_i which hold with probabilities p_i for propositions x_i. It is then possible to assemble arguments or proofs for each proposition x_i by traversing the network of Fig. 1.1 in a forward sweep. Proposition x_4 has the only argument a_4. Proposition x_2 has also its direct argument a_2; in addition,however, x_2 can be deduced if x_4 can be proved and rule R_3 holds. By combining the proof $\{a_4\}$ for x_4 and the argument r_3 for R_3 we get a second proof (a_4, r_3) for x_2. Propositions x_1 and x_3 have again only their direct arguments for proof. Apart from its direct argument a_0, x_0 can also be deduced if either rule R_1 or R_2 holds and their respective conditions x_1 and x_2 or x_2 and x_3 respectively can be proved. The proofs already assembled for x_1, x_2 and x_3 permit to construct further proofs for x_0 as indicated in Fig. 1.1.

In fact, this can be interpreted as follows: x_0 is proved if the Boolean expression (compare Fig. 1.1)

$$a_0 \vee (a_1 \wedge r_1 \wedge (a_2 \vee (a_4 \wedge r_3))) \vee (a_3 \wedge r_2 \wedge (a_2 \vee (a_4 \wedge r_3)))$$

has the value "true". Given the probabilities for the arguments, the probability of the value "true" for this Boolean expression has to be computed. This probability is the degree of credibility of a_0. This is a quite familiar problem in probability theory and especially in reliability theory. There exist numerous exact and approximate methods for this computation under various assumptions concerning the stochastic dependence between the arguments; there are furthermore different lower and upper bounds for this probability available. This subject will be discussed in detail in Chap. 5.

Consider another small example:

$$R_1 \text{ :if } x_1 \text{then } x_0$$
$$R_2 \text{ :if } x_2 \text{then } \neg x_0.$$

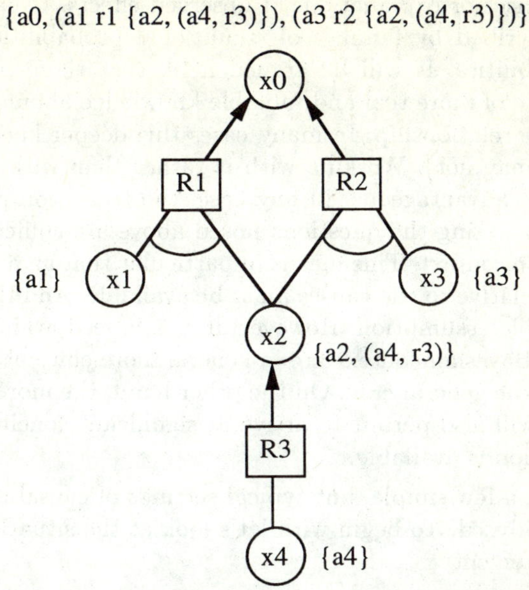

Fig. 1.1. Graphical representation of a system of uncertain rules

Then (r_1, a_1) is a proof of x_0 whereas (r_2, a_2) is a proof of $\neg x_0$. However (r_1, r_2, a_1, a_2) would lead to a contradiction, because both x_0 and $\neg x_0$ would be proved. In the theory of *assumption based reasoning* such a context is called a "no-good environment" (Laskey, Lehner, 1988). Here we shall call it a contradictory argument. In evaluating degrees of credibility of propositions like x_0 it will be necessary to elicit contradictory arguments and to take care of them in the computations. The manner of handling them will be by conditioning on non-contradictory contexts as in the example of Dempster's rule above. These first introductory examples show that the theory to be developed in this book has close connections with assumption-based truth maintenance (De Kleer, 1986), a method well-known in artificial intelligence.

1.2 Causal and Evidential Modeling

A popular way of representing knowledge is by means of causal relations. If such and such a cause is present, then it is likely, probable or possible that such and such an effect is produced. There are then two kinds of questions which are normally asked. Given some evidence about the presence of a cause, how likely is it that a specified effect is produced? Or, inversely, given that some effect is observed, how likely is it that a specified cause produced it? The first question concerns *prediction* of effects, whereas the second one looks for *diagnostics* of causes or explanations of observed effects. Usually such causal relations are described by families of conditional probabilities, often called transition probabilities. It will be argued here, that these probabilities are only the surrogate of more real and tangible knowledge about the mechanism which creates the relationship. In many cases this deeper knowledge is available (it some it may not). Working with it rather than with its surrogate is then shown to be advantageous. In any case, the transition probabilities for analyzing and answering the questions posed above are sufficient only in the so-called *Bayesian* context. This means in particular that an a priori probability distribution relative to the causes must be available. And this is very often a very questionable assumption. However, if it is indeed available, then computations in the Bayesian context are in general more efficient and the results stronger than in the general case. On the other hand, the more explicit model considered here will also permit to arrive at significant conclusions, when no a priori distribution is available.

In the sequel, a few simple, but typical settings of causal relations will be discussed and analyzed. To begin with let's look at the situation described in the following statement:

(1) "e can – under some circumstances – be produced by h; however, no other cause than h can produce e".

This means first that it is not certain that h produces e; whether this happens depends on some other circumstances. To formally describe this, one may suppose that some other assumption a must hold in order that h produces e. First a will be considered as a sufficient condition only for h producing e. Let p be the probability that a holds. Then the causal relationship (1) can formally be defined by the following two rules or implications:

$$R_1 : h \wedge a \text{ implies } e$$
$$R_2 : \neg h \text{ implies } \neg e.$$

The first rule R_1 can also be seen as an uncertain rule as considered in the previous section. The second rule is also important; it expresses the second fact contained in statement (1) that h is the only cause capable of producing e. Thus if h is not present then there is no possibility that e is produced.

This model can be used for predictive evaluations as well as for diagnostics. Suppose h is present. Then, if a is true, e must necessarily be true. If a is not true, e is still possible because a was assumed to be only sufficient, but not necessary for h producing e. Thus a is the only argument to prove e in this situation, $cred(e) = p$; however $pl(e) = 1$ because e is possible in all cases, whether a holds or not. Second, if h is absent ($\neg h$ true), then necessarily $\neg e$ (rule R_2). In this case it follows thus that $cred(e) = pl(e) = 0$. Inversely, for a diagnostic inference, suppose that e is observed. Then it follows necessarily that h holds, because $\neg h$ would imply $\neg e$. Thus e is unambiguous evidence for h, and therefore $cred(h) = pl(h) = 1$ in this case. If e is not present, there are two cases to be considered according to whether a is true or not. If a

is true, then h is excluded, because otherwise rule R_1 would imply e; hence in this case a is a proof for $\neg h$. If a is not true, $\neg h$ cannot be proved and thus h not excluded. Therefore it follows that $cred(h) = 0,\ pl(h) = 1 - p$ or $cred(\neg h) = p,\ pl(\neg h) = 1$.

Note that these conclusions have been drawn without an a priori probability distribution about h. If however such a distribution is available, then this information can be combined with the relational model to obtain more precise results. In order to be capable of combining these two items of information and knowledge, it is necessary to model the a priori distribution by a "mixed" argument in the sense of Bernoulli (Sect. 1.1). Thus let b be a mixed argument for h which holds with probability q. If b holds, then h holds and if $\neg b$ holds, then $\neg h$ holds. In the predictive analysis, using the above reflections, it follows that a and b together form a proof of e, whereas $\neg b$ alone forms a proof of $\neg e$. Supposing a and b stochastically independent, we get $cred(e) = pq,\ cred(\neg e) = 1 - q$. In the diagnostical evaluation, assuming e has been observed, $\neg b$ is in contradiction with e (see rule R_2) and must be excluded. It follows as above that $cred(h) = pl(h) = 1$. If however $\neg e$ is observed, then the combination a and b must be excluded because it would imply e and the remaining cases must be conditioned on this exclusion. Thus h is proved by $\neg a$ and b together, whereas $\neg h$ is proved by $\neg b$ alone. Because the probability of no contradiction (assuming a priori stochastic independence between a and b) equals $1 - pq$, it follows

$$\begin{aligned} cred(h) &= (1 - p2)q/(1 - pq) \\ cred(\neg h) &= (1 - q)/(1 - pq). \end{aligned} \tag{1.6}$$

Note that in this case $cred(h) + cred(\neg h) = 1$ and $cred(h) = pl(h)$, $cred(\neg h) = pl(\neg h)$; this is in fact a probability distribution.

As a small variant suppose that a is not only sufficient, but also necessary for h producing e. As a consequence, rule

$$R_3 : \ h \wedge \neg a \text{ implies } \neg e$$

has to be added to R_1 and R_2. Then, in the predictive analysis given h, $\neg a$ necessarily implies $\neg e$ and $cred(e) = p,\ cred(\neg e) = 1 - p$, which represents also a probability distribution. As before, given $\neg h$, it follows that $cred(e) = 0,\ cred(\neg e) = 1$, again a probability distribution. These are in fact transition probabilities $P(e|h) = p, P(\neg e|h) = 1 - p, P(e|\neg h) = 0, P(\neg e|\neg h) = 1$. We note them as conditional probabilities although strictly speaking they are not defined as such; this is common practice in similar cases of applied probability theory (see below).

The diagnostic analysis however does not change with respect to the analysis with R_1 and R_2 only. This implies in particular that (1.6) is still valid, when an a priori distribution P is given. But then we have in fact a Bayesian context. Defining $P(h, e) = P(e|h)P(h), P(h, \neg e) = P(\neg e|h)P(h), \ldots$, a probability is determined on the sample space $\{(h, e), (h, \neg e), (\neg h, e), (\neg h, \neg e)\}$ and with respect to this probability $P(h) = q$, $P(\neg h) = 1 - q$ is a marginal distribution and $P(e|h), \ldots$ become veritable conditional probabilities. In this context the diagnostic analysis is made using Bayes' formula to obtain

$$\begin{aligned} P(h|e) &= P(e|h)P(h)/P(e) = 1 \\ P(h|\neg e) &= P(\neg e|h)P(h)/P(\neg e) \\ &= P(\neg e|h)P(h)/(P(\neg e|h)P(h) + P(\neg e|\neg h)P(\neg h)) \\ &= (1-p)q/(1-pq). \end{aligned} \tag{1.7}$$

This corresponds exactly to (1.6) which shows that Bayesian analysis can also be very well performed in the context of reasoning with uncertain arguments. Remark furthermore that (1.6) holds even in the non-Bayesian context of the original model were a was only sufficient but not necessary for h producing e. The existence of transition probabilities of an entirely Bayesian context, is thus not even really necessary for Bayesian analysis.

The same model can now also be considered from another point of view. e can be interpreted as a *sensor* for h, which however suffers from the possibility of an *error of the first kind* :

Even if h holds it is possible that $\neg e$ is observed

(the error of the second kind will be introduced below with respect to the next model). a and $\neg a$ represent in this point of view internal states of the sensor, which decides whether the error of the first kind occurs or not. In state a the error does not occur (which means that e is observed if and only if h holds), whereas in state $\neg a$ the error does occur (which means that the observation e is sufficient but not necessary for h to hold). If e is observed, this is an unambiguous hint for h. An observation of $\neg e$ however is a hint with reliability p for $\neg h$; there remains a probability of $1 - p$ that the observation $\neg e$ does not prove $\neg h$. In a purely logical setting, note that rule R_1 can also be written as

$$R_1 : \ \neg e \wedge a \text{ implies } \neg h.$$

Rules R_2 and R_3 can be transformed into

$$R_2 : e \text{ implies } h$$
$$R_3 : e \text{ implies } \neg h.$$

Now, from this form of rules R_2 and R_3, it becomes clear that e and $\neg a$ are incompatible and rule R_3 can be replaced by

$$R_3' : \ \neg a \text{ implies } \neg e.$$

These rules, which are equivalent to the original ones, describe the sensor model and represent the evidential point of view – as opposed to the causal point of view – which will be favored in this book. Note however that the causal and evidential points of view do not represent contradictory, divergent positions, but rather complementary approaches.

There is still another view of the model. Rules R_1 and R_2 represent both possible uncertain relations between h and e. In Fig. 1.2 it is shown which pairs of h, $\neg h$ and e, $\neg e$ are compatible according to rule R_1, depending on a or $\neg a$, and which pairs are compatible under rule R_2. For both possibilities a and $\neg a$, the two rules together are represented by the intersection of the corresponding relations; the result is shown in Fig. 1.2 c. This is an entirely symmetrical picture of the relation between h and e. No connotation to cause and effect, to hypothesis and evidence or to predictive and diagnostic analysis remains. h and e play fully symmetrical roles, h may be evidence for e and e may be evidence for h. This is the view which will mainly be adopted throughout the book, especially when computational issues are under discussion. The elimination of directionality allows uniform inference procedures in knowledge bases in all directions.

Let's consider a second, slightly more general case:

(2) "e can – under some circumstances – be produced by h; however, other causes are also possible"

The model to formally represent this situation contains two independent arguments a_1 and a_2 with probabilities p_1 and p_2. a_1 is considered as a sufficient reason for h to produce e, whereas a_2 is a sufficient argument for e, independent of h. a_2 encloses all "other causes" which produce e and p_2 represents the probability that any such cause is present and produces e. The causal relation is defined in this case by the system of rules R_1 to R_3:

$$R_1 : h \wedge a_1 \text{ implies } e$$
$$R_2 : a_2 \text{ implies } e$$
$$R_3 : \neg h \wedge \neg a_2 \text{ implies } \neg e.$$

Rule R_3 is called the *accountability condition* (Pearl, 1988). It is very important, especially for diagnostic analyses. It states that the model contains explicitly all causes which may produce e, even if some of them are – implicitly accounted for – grouped into argument a_2.

Predictive and diagnostic conclusions can be drawn as in the former example. If cause h is present, then either a_1 or a_2 is sufficient to conclude e. If neither a_1 nor a_2 are present ($\neg a_1 \wedge \neg a_2$), then e is still possible, but not proved. Hence we get $cred(e) = 1 - (1 - p_1)(1 - p_2)$ and $pl(e) = 1$. If cause h is absent, then a_2 proves e, whereas $\neg a_2$ proves $\neg e$. It follows $cred(e) = pl(e) = p_2$, $cred(\neg e) = pl(\neg e) = 1 - p_2$, which represents a probability distribution. If e is observed, then $\neg a_2$ suffices to deduce h, and h is possible, but not provable if a_2 is present. Hence $cred(h) = 1 - p_2$, $pl(h) = 1$. If finally $\neg e$ is observed, then a_1 implies necessarily $\neg h$, otherwise h is possible. Thus $cred(\neg h) = p_1$ or $pl(h) = 1 - p_1$. These uncertain arguments may be combined just as before with an a priori distribution on the cause h.

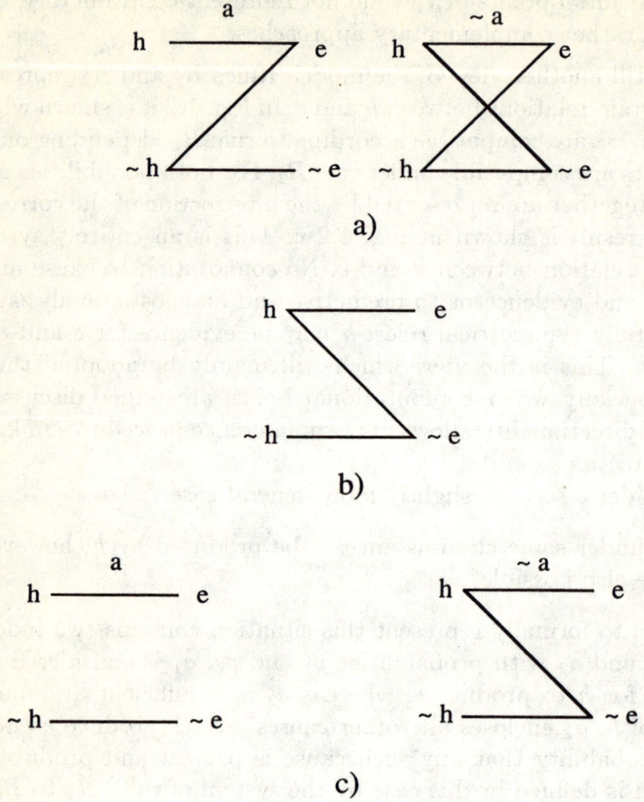

Fig. 1.2. a) Compatibility of pairs according to the rule R_1: if $h \wedge a$ then e, depending on whether a or $\neg a$. b) Compatibility of pairs according to the rule R_2: if $\neg h$ then $\neg e$. c) Compatibility of pairs under both rules

e can again be considered as a *sensor* for h. In fact, rules R_1, R_2 and R_3 can equivalently be written in the evidential form

$$R_1 : \neg e \wedge a_1 \text{ implies } \neg h$$

$$R_2 : \neg e \text{ implies } \neg a_2$$

$$R_3 : e \wedge \neg a_2 \text{ implies } h.$$

This time the observation of e is not sufficient to conclude h. There is the possibility of an *error of the second kind* :

Even if h is not present e may be observed.

a_1 and a_2 can be considered as describing internal states of the sensor. When a_2 is true, then the sensor is in the state which produces e independently whether h is present or not, whereas with $\neg a_2$ the observation e implies h. When both $\neg a_1$ and $\neg a_2$ are true, $\neg e$ can be observed whether h is present or not (error of the first kind). In addition, a_2 is in contradiction with $\neg e$. Thus an observation of e is a hint with reliability of $1 - p_2$ for h and the observation of $\neg e$ is a hint with reliability of p_1 for $\neg h$. This general sensor with two kinds of possible errors is a typical evidential model. The use and application of sensor models will be illustrated in the following section.

1.3 Pooling of Evidence, Cascaded Inference and Multiple Causes

The sensor models presented in the previous section – or some variants and extensions thereof – will be applied in this section to an example given by Pearl (1988) and analyzed there from a Bayesian point of view. Suppose, a burglar alarm system is installed in the house of Mr. Holmes. Being an electronic system there is some risk that the system fails to operate properly. In particular, the system may not react to a burglary, although it is functioning perfectly. Let a_1 denote the event that the burglar makes enough trouble for the alarm to be triggered. The probability of a_1 measures the reliability of the burglar alarm system and let $p_1 = 0.95$ denote this probability. An alarm system may however respond to other stimuli than a burglar, for example to strong vibrations as generated by earthquakes. Denote by a_2 the presence of such other causes for an alarm and suppose its probability is $p_2 = 0.01$. This probability measures the over-sensitivity of the system. Moreover, we suppose that a_1 and a_2 are independent. The causal mechanism of this burglar alarm system can then be described by the following rules:

$$R_1 : \text{if burglary } \wedge a_1 \text{ then alarm}$$

$$R_2 : \text{if } a_2 \text{ then alarm}$$

$$R_3 : \text{if } \neg \text{ burglary} \wedge \neg a_2 \text{ then } \neg \text{ alarm.}$$

This defines a sensor with the possibility of an error of both kinds. Thus, if the alarm sounds, this is evidence which proves burglary with a probability $1 - p_2 = 0.99$. Or, in other words, we get $cred(burglary) = 0.99$ and $pl(burglary) = 1$ because there is no argument in favor of the absence of a burglary. On the other hand, if we listen and hear no sound from the alarm, this observation $\neg alarm$ is a hint for $\neg burglary$ which proves this hypothesis with a probability $p_1 = 0.95$. So, an alarm is strong evidence for a burglary, whereas no alarm is rather strong evidence for the absence of a burglary.

Now, suppose that a credibility of 0.95 for the absence of a burglary when the alarm is quiet is not sufficient for Mr. Holmes' confidence. Then a second alarm system is installed. If the new system is of the same kind as the first one, then nothing is changed because in this case there is full dependence between arguments for both systems.

In contrast, if the second system is of a different kind from the first one, then let a'_1 denote the event that the burglar makes enough trouble for the alarm of the second system to be triggered. We suppose that a_1 and a'_1 are independent, a hypothesis which is not necessarily true in reality, but the model considered here is rather simple and of course more sophisticated ones can be envisaged. Moreover, the second system may react to other causes than burglary, but this time not to vibrations as for the first system, but to other independent causes as for example an animal crossing a light ray. If a'_2 denotes the presence of such extra causes which are capable to trigger the alarm of the second system, then a'_2 and a_2 are independent.

In this case, assuming for the sake of simplicity that a_1 and a'_1 have equal probability p_1, and that a_2 and a'_2 have equal probability p_2, a simultaneous alarm of both systems increases the credibility of burglary to $1 - p_2^2 = 0.9999$. This is so, because the absence of at least one of the two extra causes is sufficient to prove burglary. Also important, the redundancy of the system increases the risk of a false alarm (error of the second kind). In fact, if there is no burglary, then the absence of both extra causes a_2 and a'_2 implies that both alarms are quiet. Since the presence of the extra cause for one system and its absence for the other one is incompatible with the absence of a burglary, it follows that the credibility that both alarms are quiet equals $(1 - p_2)^2/(1 - p_2(1 - p_2))$ when there is no burglary. Hence the credibility that at least one alarm rings when there is no burglary is given by $p_2/(1 - p_2(1 - p_2)) = 0.0101$. The credibility of a false alarm when there is only one system is $p_2 = 0.01$. In the absence of a burglary, it is of course more credible to hear sound if two alarm systems are installed than if there is only one system.

Now suppose that both alarms are quiet. In this case, a combination of arguments is not contradictory when both extra causes are absent, which has a probability of $(1 - p_2)^2$. Since the combination $(\neg a_1, \neg a_2, \neg a'_1, \neg a'_2)$ does not permit to prove the presence or the absence of a burglary, whereas the other three non-contradictory combinations prove the absence of a burglary, it follows that

$$cred(\neg h) = \frac{(1 - p_2)^2 - (1 - p_2)^2(1 - p_1)^2}{(1 - p_2)^2} = 0.9975.$$

Hence the credibility of the absence of a burglary is greater with two systems than with a single system when we hear no sound. In a similar way, the situation where one alarm sounds while the other one is quiet could be analyzed.

These are examples of the *pooling of evidence* coming from different, more or less independent sources. The method relies once more on logical analysis, proving hypothesis under different assumptions and evaluating the probabilities of the proofs of hypotheses. An essential ingredient of the method is the analysis of contradictory assumptions and their exclusion as possible contexts. This is done by conditioning the probabilities of the remaining combinations of assumptions on the condition of no contradiction. This will be the very basic method studied in this book for the combination of evidence. Further examples of the application of this approach follow.

Let's consider the following story (extracted from Pearl, 1988, p. 42) :

"Mr. Holmes receives a telephone call from his neighbor Dr. Watson who states that he hears the sound of a burglar alarm from the direction of of Mr. Holmes' house. While preparing to rush home, Mr. Holmes recalls that Dr. Watson is known to be a tasteless practical joker ..."

How can the credibility of a burglary be evaluated in view of Dr. Watson's call ? Dr. Watson may be considered as a sensor with no error of the first kind, i.e. a neighbour who reliably reports an alarm, if there is one. On the other hand there is a real risk of an error of the second kind; Dr. Watson may be suspected to report an alarm only as a joke. Here it is admittedly difficult to assess the probability that Dr. Watson is making a joke. Nevertheless let's assume some probability q_1 for this event. With this model the hypothesis of an alarm is proved with probability $1 - q_1$. As an alarm proves a burglary with probability $1 - p_2 = 0.99$ it follows that $cred(burglary) = 0.99(1 - q_2)$. The credibility of the hypothesis of a burglary is reduced by a factor equal to the reliability of the reporting neighbour. This is an example of a cascaded inference, where the available evidence does not bear directly on the hypothesis of interest, but only on an intermediate event. Of course, such a chain of inference with error-prone sensors can only decrease the reliability of proofs.

Let's continue the story (Pearl, 1988) :

... Mr. Holmes recalls that Dr. Watson is known to be a tasteless practical joker, and he decides to first call another neighbor, Mrs Gibbson, who is far more reliable.

As a sensor, Mrs Gibbson does not commit errors of the second kind. However, because she does not hear very well, it is possible, that she did not hear the alarm (error of the first kind). The probability of this state is assessed to equal q_2. So, if Mrs Gibbson confirms the alarm, its credibility is one and the doubt about Dr Watson's testimony vanishes. If however Mrs Gibbson did not hear an alarm, then it is not possible that simultaneously Dr. Watson is not joking and Mrs Gibbson did hear well. This combination is to be excluded. The assumption that Dr. Watson makes no joke and Mrs Gibbson did not hear well proves an alarm, whereas the assumption that Dr. Watson makes a joke and Mrs. Gibbson did hear well proves the absence of an alarm. It follows that $cred(alarm) = (1-q_1)_/(1-(1-q_1)(1-q_2))$ and $cred(\neg alarm) = q_1(1-q_2)/(1-(1-q_1)(1-q_2))$. This is again a case of contradictory evidence which leads to support for and against a hypothesis. If for example $q_1 = 0.4$ and $q_2 = 0.2$ we obtain $cred(alarm) = 0.23$, $cred(\neg alarm) = 0.61$. From this, the credibility for a burglary may be obtained, $cred(burglary) = 0.99 \times 0.23 = 0.22$ and $cred(\neg burglary) = 0.95 \times 0.61 = 0.5795$. So the plausibility of a burglary is $pl(burglary) = 1 - cred(\neg burglary) = 0.42$. By the way, note that in all cases credibilities of complementary hypotheses add up to less than one, such that the plausibility of a hypothesis is always larger than its credibility.

The story continues (Pearl, 1988):

Immediately after his conversation with Mrs. Gibbson, as Mr. Holmes is preparing to leave his office, he recalls that his daughter is scheduled to arrive home at any minute. If greeted by an alarm sound she surely would phone him for instructions. Now he wonders whether he should wait a few more minutes in case she calls.

This assumes a further causal relation described by the rules

R_1 : if alarm then phone call
R_2 : if x then phone call
R_3 : if $\neg alarm \wedge \neg x$ then $\neg$ phone call.

Here x denotes any other reason for which she may have to call her father. The phone call can be necessarily predicted, if an alarm is proved or if x is present. Suppose a probability $q = 0.1$ for x and x is reasonably assumed independent of a. Then it follows $cred(phone\ call) = 1-(1-cred(alarm))(1-q) = 0.307$. On the other hand, $\neg$ *phone call* follows if $\neg$ *alarm* is proved and $\neg x$ takes place, $cred(\neg phone\ call) = cred(\neg alarm)(1-q) = 0.$ and thus $pl(phone\ call) = 0.41$. It is very easy, as the example shows, to obtain predictive information about unknown or future events.

The story is still not finished (adapted from Pearl, 1988, p. 49) :

As he is debating whether or not to rush home, Mr. Holmes remembers reading in the instruction manual of his alarm system that the device is sensitive to earthquakes and can be accidentally triggered by one. So he decides to phone to the earthquake observation station to inform himself whether an earthquake took place.

With this new reflection, the alternative cause a_2 for an alarm becomes identified: it represents the possibility of an earthquake. Now there is information available about the occurrence of an earthquake. Let's suppose that the earthquake observation station disposes of earthquake sensor devices with the two kinds of possible errors; say probability of 0.01 for an error of the second kind and of 0.1 for an error of the first kind. Assume first, that the observation station did not observe an earthquake. Then a_2 can be excluded with a probability of 0.9 from this information. However we also had the a priori probability of a_2 (0.01); so, clearly, the supplementary information from the observation station increases the probability of $\neg a_2$ above 0.99 (we pass the computations). This in turn reinforces an alarm as an evidence for a burglary since $cred(burglary)$ equals the probability of $\neg a_2$. As the evidence for no earthquake is already nearly one, the credibility of burglary will however not increase much.

More interesting is the case where the observation station reports an earthquake. To simplify the computation let's neglect the a priori probabilities about earthquakes. Then a_2 (earthquake) is proved with probability 0.99 by the observation of the station. Thus, the alarm becomes a hint for burglary with a reliability of only 0.01, i.e. it supports the hypothesis of a burglary only to the degree of 0.01, even if it still contains no argument against burglary.

We have now the result that $cred(burglary) = 0.01$, which practically equals zero, whereas $pl(burglary) = 1$. Nothing excludes a burglary taking place at the same time as an earthquake.

Essentially, this result means that in the presence of an earthquake, an alarm contains practically no more any information about a hypothetical burglary. The burglary as a possible cause of the alarm is *explained away by the earthquake*. In this analysis, note that at no time an a priori distribution about the relative frequencies of burglaries was used. There is no need for such an a priori distribution. This is in sharp contrast to the analysis of the same example by the usual causal probabilistic approach in a Bayesian context (compare Pearl, 1988). There the a priori distribution is an essential element of the method from the beginning and it determines the results of the calculations very strongly. This is unfortunate because among all data needed for the analysis, this one is most difficult to assess.

1.4 Evidential Reasoning in Expert Systems

One of the first practical expert systems which helped to raise interest in expert systems technology and artificial intelligence was MYCIN. This system was developed to determine the likely identity of pathogens in patients and to assist in the selection of a therapeutic regimen appropriate for treating the organisms under consideration. It is a rule based system designed for computer use of medical knowledge. The following is an example of the kind of rules used in MYCIN:

IF: 1) The stain of the organism is gram positive, and
2) The morphology of the organism is coccus, and
3) the growth confirmation of the organism is chains

THEN: There is suggestive evidence (0.7) that the identity of the organism is streptococcus

The number 0.7 indicates 70% of belief that the conclusion is valid, if the conditions hold (Shortliffe, Buchanan, 1975). Questioning of experts seems to reveal that this number is not to be considered as a conditional probability $P(h|s_1, s_2, s_3) = 0.7$. Experts are uneasy about the conclusion of this interpretation that $P(\neg h|s_1, s_2, s_3) = 0.3$. The observations s_1, s_2 and s_3 are seen *in favour* of the conclusion streptococcus; however experts do not want to use them either as evidence (to a degree of 0.3) against streptococcus (Shortliffe, Buchanan, 1975).

One way to interpret this, is that the conditions s_1, s_2, s_3 are arguments for streptococcus, although not fully conclusive ones, hence the degree 0.7. But they are no arguments against streptococcus. The arguments are not fully conclusive, because some other argument a, which is not explicitly formulated, must be present to conclude streptococcus without any reserve. Rather than include a explicitly into the rule, its probability is estimated (0.7) and considered as the degree of support or credibility with which the conclusion can be inferred from the rule. This then brings us back to the pure arguments of Sect. 1. For the sake of historical correctness let's note that initially an ad hoc formalism based on so called certainty factors was developed (Shortliffe, Buchanan, 1975; Heckermann, 1986). However, later it was remarked that the Dempster-Shafer theory could as well or better represent the kind of uncertainty modelled in MYCIN (Gordon, Shortliffe, 1984).

The conclusion of a rule represents a statement concerning a specified question. In a medical diagnosis system for example the question considered is the cause of cholestatic jaundice. The possible answers to such a question are the elements of a set Θ which is called frame of discernment. In the example the possible answers may be hepatitis (hep), cirrhosis (cirr), gallstone (gall) and pancreatic cancer (pan), such that $\Theta = \{hep, cirr, gall, pan\}$ (the example presented here is taken from Gordon, Shortliffe, 1984). A statement or proposition about the answer is represented by a subset of Θ. For example the statement "Intrahepatic Cholestasis" is defined as the subset $\{hep, cirr\}$ of Θ, "Extrahepatic Cholestasis" as $\{gall, pan\}$ (see Fig. 1.3) and the statement "not hepatitis" corresponds to the subset $\{cirr, gall, pan\}$. A statement about the answer constrains the possible answers into a corresponding subset of Θ. That is exactly the effect of information with respect to the question to be considered, namely to restrict the possible answers into smaller sets.

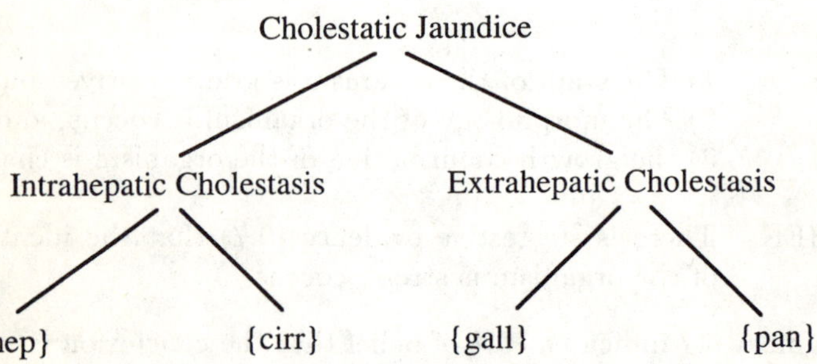

Fig. 1.3. The frame of discernment for Cholestatic Jaundice and statements about it

The rules of an expert system permit now to describe evidence relative to the questions considered in the system. Depending on conditions which must be fulfilled, statements about the answer may be confirmed with more or less certainty. It is assumed here that each rule possesses an implicit, unstated condition whose probability is given. This is then the probability that the conclusion of the rule holds, if its explicit conditions are satisfied. If the conclusion of the rule R is a statement $A \subseteq \Theta$, then its conditions are pure evidence for A and their reliability is the probability $p(R)$ of its implicit condition. Let's suppose in a simple example that the available evidence permit to apply three rules with the following conclusions:

R1: Supports or confirms "Intrahepatic Cholestasis" with a reliability $p(R_1) = 0.6$

R2: Disconfirms the diagnosis of hepatitis or equivalently confirms "not hepatitis" with a reliability of $p(R_2) = 0.7$

R3: Confirms the diagnosis of hepatitis to the degree $p(R_3) = 0.8$.

In a synthesis of these evidential elements the different combinations of arguments or valid rules may be enumerated and their combined conclusions evaluated. For example, if only R_1 is assumed to be reliable, then only "Intrahepatic Cholestasis" can be concluded, but nothing about hep or cirr. If rules R_1 and R_2 are supposed to be valid, but not rule R_3, then both "Intrahepatic Cholestasis" and "not hepatitis" follow and the diagnosis must be in the intersection of the two corresponding sets, $\{hep, cirr\} \cap \{cirr, gall, pan\} = \{cirr\}$. This assumption permits then to conclude cirr. Rules R_2 and R_3 can not both be valid simultaneously, because their conclusions are contradictory; this combination is to be excluded and the probabilities of the remaining cases are to be conditioned on the event that this combination does not occur. Table 1.1 summarizes the complete analysis. As an illustration of how the probabilities are computed, let's look at a first case, where rule R_1 only is assumed to be reliable. This means that R_2 and R_3 are not valid and – if independence of rules is assumed – the probability of this combination is $0.6 \times (1 - 0.7) \times (1 - 0.8)$. This probability has to be conditioned on non-contradictory combination of rules. The probability of this condition is $1 - (0.224 + 0.336) = 0.44$ (see the last two rows in Tab. 1.1). Thus we finally get $0.6 \times 0.3 \times 0.2/0.44 = 0.082$. The other probabilities are calculated in the same way.

Table 1.1. The combined evidence on Cholestatic Jaundice

Reliable Rules	Conclusion	Probability
∅	{hep, cirr, gall, pan}	0.055
R1	{hep, cirr}	0.082
R2	{cirr, gall, pan}	0.127
R3	{hep}	0.218
R1, R2	{cirr}	0.191
R1, R3	{hep}	0.327
R2, R3	contradiction	(0.224)
R1, R2, R3	contradiction	(0.336)

If one is interested in the hypothesis hep, then Tab. 1.1 shows that there are two cases which support this hypothesis, namely the rule R_3 alone as well as in combination with rule R_1. The total probability of these two cases is $0.218 + 0.327 = 0.545$ and thus the degree of credibility of hepatitis is $cred(hepp) = 0.545$. But there are also two cases against this hypothesis, namely rule R_2 alone and together with R_1. It follows that $cred(not\ hepatitis) = 0.127 + 0.191 = 0.318$ and the plausibility of hepatitis is only $pl(hep) = 1 - 0.318 = 0.682$. In the same way we find from Tab. 1.1 that $cred(cirr) = 0.191$, $pl(cirr) = 0.455$. The hypothesis "Intrahepatic Cholestasis" has the combined support of hepp and cirr plus its proper support by rule R_1 alone; hence $cred(\{hep, cirr\}) = 0.818$ and $pl(\{hep, cirr\}) = 1$. This shows how the credibility and plausibility of hypotheses can be evaluated in the light of the available evidence. These degrees measure the force with which the evidence supports or proves a hypothesis as well as the force with which the evidence disagrees with the hypothesis.

1.5 Measurement, Estimation and Diagnostics

Here we consider a measurement device which measures a numerical quantity T (like temperature, pressure, etc.) with some error ϵ. To simplify, we assume, to start with, that both T and the error take integer values. Furthermore, we suppose that the probability distribution of the error is known to be $p(i) = P(\epsilon = i)$ for each relative integer i. If the unknown exact value θ of the quantity is measured by a value x, then $x = \theta + \epsilon$. Then this measurement x brings some evidence for the unknown value θ of T. Namely, if we assume an error of e, then we may conclude $\theta = x - e$, and this is an argument of reliability $p(e)$. Suppose now, that n independent measurements of the same value θ have been made and that the measurement values $x_1, x_2, \ldots, x_n$ resulted. What can then be concluded about the unknown value of T ? This is of course a basic and classical question of statistics. But let's look at this question from the point of view of our general approach of reasoning with uncertain arguments. Given the measurement x_i, an argument would be to assume an error of e_i and to conclude $\theta = t = x_i - e_i$. This conclusion would have the reliability $p(e_i)$. If such arguments were selected for all n measurements, then the condition

$$x_1 - e_1 = x_2 - e_2 = \ldots = x_n - e_n (= t) \tag{1.8}$$

must be satisfied; otherwise the compound argument would be contradictory. A non-contradictory array $(e_1, e_2, ..., e_n)$ of arguments or errors such that $x_i - e_i = t$ for all $i = 1, 2, \ldots, n$ is a combined argument for the quantity T to equal t. And it is the only one for this value t. Hence the degree of credibility as well as the degree of plausibility for t equals

$$cred(t) = k \prod_{i=1}^{n} p(x_i - t) \tag{1.9}$$

where $1/k$ is the probability of no contradiction between the errors. This constant can be computed from the normalization condition

$$\sum_t cred(t) = 1. \tag{1.10}$$

Thus $cred(t)$ becomes in this case a *probability distribution*. As a function of t, $\prod_{i=1}^n p(x_i - t)$ is of course nothing else than the well known *likelihood function* in statistics for the unknown value θ. So, our method of reasoning with unreliable arguments appears to be linked to the statistical theory of likelihood functions (see however Chap. 9 for a more thorough discussion of statistical analysis). A classical estimator of θ is the value t which maximizes the likelihood function, the maximum likelihood estimator.

It is now easy to generalize these considerations to the more interesting and realistic case of continuous errors. Suppose then that the probability distribution of the error of the measurement device is defined by a density function $f(e)$. If, as above, n independent measurements are made and the measurement values $x_1, x_2, \ldots, x_n$ are obtained, then an array of measurement errors $(e_1, e_2, \ldots, e_n)$ satisfying (1.8) is still the unique argument for $t = x_i - e_i, i = 1, 2, \ldots, n$. In this case we obtain a *credibility density fcred* for the possible values of t

$$fcred(t) = k \prod_{i=1}^n f(x_i - t). \tag{1.11}$$

The normalization constant k can be determined from the normalization condition

$$\int_{-\infty}^{+\infty} fcred(t)\, dt = 1. \tag{1.12}$$

In the case of Gaussian errors with zero mean value, $f(e)$ is proportional to $exp(-e^2/(2\sigma^2))$. It follows then from (1.11) that

$$\begin{aligned} fcred(t) &= k_1\, exp\left(-\sum_{i=1}^n (x_i - t)^2/(2\sigma^2)\right) \\ &= k_2\, exp\left(-n(t-\bar{x})^2/(2\sigma^2)\right) \end{aligned} \tag{1.13}$$

with $\bar{x} = (1/n)\sum_{i=1}^n x_i$.

The credibility distribution of t is again a Gaussian distribution with mean value $\bar{x}$ and variance σ^2/n. It follows immediately that the maximum likelihood estimator $\hat{t}$ for θ is $\bar{x}$. This is of course very well known. What is new in this analysis is the credibility distribution (1.13) of the unknown value θ. In statistics one normally asks, what is a good estimator for θ and what is its error or its distribution. In the case considered here, the answer is that $\bar{x}$ is a good estimator, that it is a random variable and that its distribution is Gaussian with mean θ and variance σ^2/n. On the other hand, our analysis basically looks at hypotheses about possible values for θ, like for example that θ is in the interval $\bar{x} \pm a$, and tries to determine the credibility and plausibility of such a hypothesis. This is clearly closely related to the statistical analysis of *confidence intervals*, but we pretend that the approach described here is more direct and general. Thus, from (1.13), it follows, for example, immediately that the credibility of the hypothesis that θ is contained in the interval $\bar{x} \pm 2\sigma/\sqrt{n}$ is greater than 95%. This kind of statistical analysis will be further developed in Chap. 9.

Let's suppose now that measurement devices may become defective. We assume that there is a known, in general rather small probability α for a measurement device to be defective. If the device is working correctly, then an error distribution $p(e)$ as before is assumed to be known (we return here temporarily again to the discrete model). It becomes now necessary to describe also the behavior of the device, or at least what we know about it, when it is defective. We adopt first a very general model and we shall see that finally only very little information about this model is really needed. This is a lucky circumstance which greatly facilitates the application of the model. Thus, it will be assumed that a defective system can be in any one of a set of failure modes $W = \{\omega_1, \omega_2, \ldots, \omega_n\}$ and that the probabilities $q(\omega_j), j = 1, \ldots, n$ of these modes are known. If the device is in failure mode ω_j, then the error is only known to be in some subset $\Gamma(\omega_j)$ of the integers. This is a very general model, which includes many interesting cases. For example, if an error distribution is known even in the case of a defective device, then let W be the possible error values and set $\Gamma(\omega_j) = \{\omega_j\}$. On the other hand, if nothing is known about the errors of a defective device, let $W = \{\omega\}$ and define $\Gamma(\omega)$ as the set of all integers. This is a model of complete ignorance about the behavior of a defective measurement device.

Two questions arise when measurements are observed from possibly defective devices. The first is how to estimate the unknown quantity measured or rather – in our philosophy – how to evaluate the credibility and plausibility of different hypotheses about the unknown value. The second question is how to detect the defective devices. This will be called the problem of *diagnostics*. These questions are related to the statistical question of how to treat and detect outliers in data sets.

First the question of estimation is considered. Let there be n measurements $x_1, x_2, \ldots, x_n$ for the exact value θ of a quantity T made with independent measurement devices $i = 1, 2, \ldots, n$. Here, independence means that the devices fail independently from each other and that their measurement errors and failure modes are stochastically independent. In order to analyze the evidence about the unknown value θ contained in the measurements $x_1, x_2, \ldots, x_n$, the different possible combination of arguments must be considered. Here, arguments are whether device i is defective (di) or normally functioning (fi), the measurement errors e_i of functioning devices and the failure modes $\omega_j(i)$ of defective devices. Denote by D the set of functioning devices and by F the subset of defective devices. Such a combination of arguments is non-contradictory if

$$x_i - e_i = t \text{ for all } i \in F \text{ and for some value } t$$
$$x_i - t \in \Gamma(\omega_j(i)) \text{ for some } j \text{ and all } i \in D.$$

Such a combination of arguments is a proof of the value t for the unknown quantity θ. It is this time, however, no longer the only combination yielding the value t; other combinations of defective devices and failure modes could prove the same value t. In order to compute the degree of credibility of a given value t, the probabilities of all different combinations proving t must be added up. In order to avoid contradictions, for each defective device i, there must be a failure mode $\omega_j(i)$ such that $x_i - t \in \Gamma(\omega_j(i))$. Hence we have

$$\begin{aligned} cred(t) = k \sum \{&\alpha^{|D|}(1-\alpha)^{|F|}. \\ &(\prod_{i \in F} p(x_i - t))(\prod_{i \in D} \sum \{q(\omega_j(i)) : x_i - t \in \Gamma(\omega_j(i))\}) : \\ &D + F = \{1, \ldots, n\}, F \neq \emptyset\}. \end{aligned} \tag{1.14}$$

The case $F = \emptyset$ were all devices are defective is excluded in the sum, because we make the further assumption, that in this case there is no way to prove t. This excludes in particular the case where a defective device has a known error distribution. But even in this and similar cases, we may assume – in an approximate analysis – that the probability of all devices failing is negligible. As a fortunate consequence, it can be shown that $cred(t)$ is a probability distribution over the possible values for t, which would not necessarily have been the case otherwise.

The term $\sum \{q(\omega_j(i)) : x_i - t \in \Gamma(\omega_j(i))\}$ represents, for a defective device, the probability that the failure mode of the device is compatible with the hypothetical error $e = x_i - t$. In other words, this term equals the degree of plausibility of the error $x_i - t$ of a defective measurement device. If $pl(x_i - t)$ denotes this plausibility, then (1.14) becomes

$$\begin{aligned} cred(t) = k \sum \{&\alpha^{|D|}(1-\alpha)^{|F|}. \\ &(\prod_{i \in F} p(x_i - t))(\prod_{i \in D} pl(x_i - t)) : \\ &D + F = \{1, \ldots, n\}, F \neq \emptyset\}. \end{aligned} \tag{1.15}$$

This formula shows that it is in fact not necessary to specify the whole model of a defective measurement device. It suffices to indicate the plausibility of errors, and even for these values only relative magnitudes are necessary, because k normalizes $cred(t)$. This is very fortunate, because it is relatively easy to specify reasonable plausibility functions of errors. For example, one may assume that for a defective measurement device small errors are implausible, because small errors are by definition characteristic for a well functioning device, whereas large errors are very plausible for a defective device. Figure 1.4 shows an example of such a plausibility function.

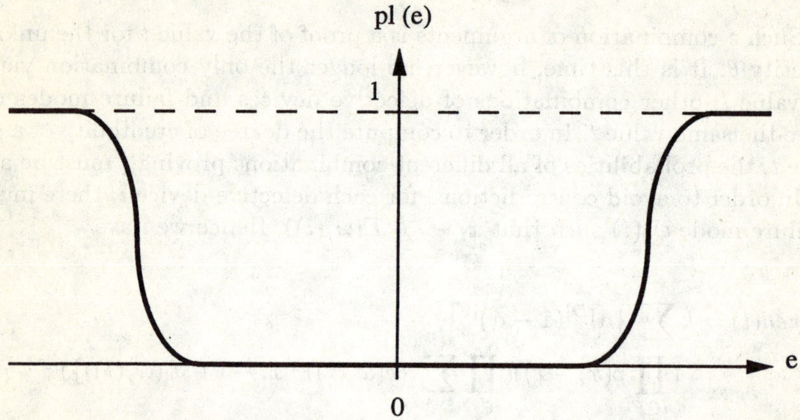

Fig. 1.4. A plausibility function for the error of a defective measurement device

Neglecting the normalization constant k in (1.15) yields a likelihood function for the unknown quantity θ, now according to a model with possible failures of the measurement device. Intuitively, this formula can also be used to get likelihood functions for continuous error models with a density function $f(e)$ instead of the discrete probabilities $p(e)$. For an illustrating example let's assume a Gaussian distribution with mean 0 and variance 1 for the errors of a properly functioning measurement device. Furthermore – in order to simplify calculations – it is assumed that the failure probability and the number of measurements are both sufficiently small, so that the probability of more than one failing measurement device can be neglected. Then, in (1.15) only terms with $|D| = 0$ and $|D| = 1$ are to be included and the likelihood function becomes

$$\begin{aligned}
&(1-\alpha)^n \prod_{i=1}^{n} exp(-(x_i - t)^2/2) \\
&+ \sum_{j=1}^{n} \alpha(1-\alpha)^{n-1} pl(x_j - t) \prod \{exp(-(x_i - t)^2/2) : i = 1, \ldots, n; i \neq j\} \\
&= (1-\alpha)^n exp(-ns^2/2) exp(-n(\bar{x} - t)^2/2) \\
&+ \sum_{j=1}^{n} \alpha(1-\alpha)^{n-1} pl(x_j - t) exp(-(n-1)s_j^2/2) exp(-(n-1)(\bar{x}_j - t)^2/2)
\end{aligned} \tag{1.16}$$

where

$$\begin{aligned}
\bar{x} &= (1/n)\sum_{i=1}^{n} x_i,\ s^2 = (1/n)\sum_{i=1}^{n}(x_i - \bar{x})^2 \\
\bar{x}_j &= (1/(n-1))\sum \{x_i : i = 1,\ldots,n; i \neq j\} \qquad (1.17) \\
s_j^2 &= (1/(n-1))\sum \{(x_i - \bar{x}_j)^2 : i = 1,\ldots,n; i \neq j\}.
\end{aligned}$$

These formulas have simple interpretations : If the measurements are all from non defective devices then s^2 as well as all s_j^2 will be in the same order of magnitude as the variance σ^2 (which is 1 in our example), because they are estimators of the variance. Because the failure probability α is small, the first term in (1.16) will largely dominate in the likelihood function and the maximum likelihood estimator will be $\bar{x}$ as usual. If however one device, say number k, is defective and the corresponding measurement value x_k is an outlier, then only s_k^2 will be in the order of σ^2, all the other values s^2 and s_j^2 for $j \neq k$ will be of the same order of magnitude, but much larger than s_k^2. Hence, even for a small failure probability α, the k-th term of the sum in (1.16) will dominate all other terms in (1.16). The maximum likelihood estimator in this case will be appropriatively $\bar{x}_k$: the outlier will be justly excluded in the estimation.

For a numerical illustration suppose three measurements $-0.6, 1.2$ and 4 are obtained and suppose $\alpha = 0.01$. The plausibility function for the errors of a defective measurement equipment is defined as $pl(\omega) = 0$ when $|\omega| < 3$, $pl(\omega) = 1$ when $|\omega| > 6$ and linearly between 0 and 1 if $3 \leq |\omega| \leq 6$. The unnormalized function (1.16) is plotted in Fig. 1.5. With a standard deviation of one, the measurement 4 can not really be qualified as an outlier. The first term of (1.16) dominates. This corresponds to maximum likelihood at 1.53, corresponding to the mean $\bar{x}$ of the three measurements. However there is a second marked local maximum around zero, more or less corresponding to the mean of the first two measurements.

If the measurement 4 is replaced by 10, a clear outlier, then (again with $\alpha = 0.01$) the likelihood function becomes the curve plotted in Fig. 1.6. It excludes the outlier very neatly and concentrates around 0.3, the mean of the first two measurements. Note that if α is set to zero, that is, if the possibility of a defective measurement device is excluded, then the curve plotted in Fig. 1.7 is obtained. It concentrates around the mean 3.53 of all three measurements. The outlier greatly distorts the support, if the possibility of a defective measurement device is not explicitly accounted for.

In order to discuss the problem of diagnosis, the detection of defective measurement devices, let's suppose that a known value (say zero) is measured by the measurement device under inspection. If the measurement value deviates greatly form zero, then this is evidence for a defect. To formalize this, we return to our general model of a defective device introduced above and assume that a measurement value x is observed. Let f and d denote the states of functioning and defectiveness respectively. The measurement x implies that

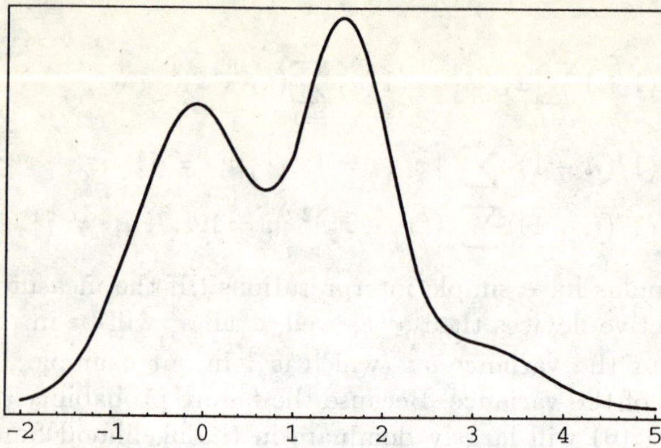

Fig. 1.5. Measurements $-0.6, 1.2, 4$ with failure probability of measurement device 1 %

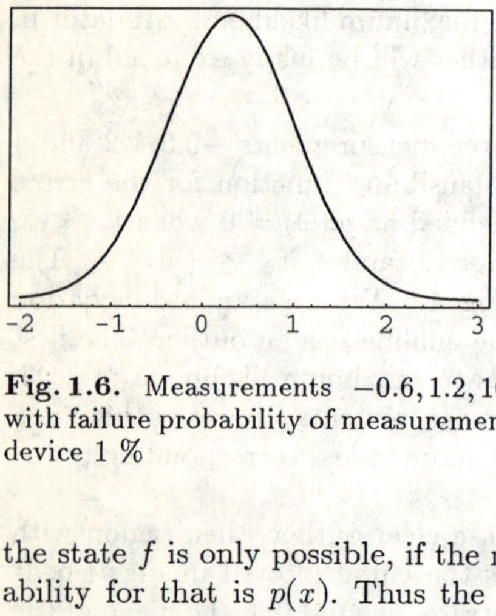

Fig. 1.6. Measurements $-0.6, 1.2, 10$ with failure probability of measurement device 1 %

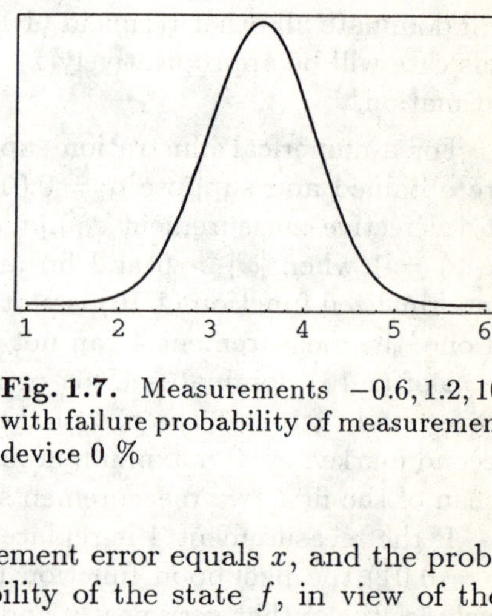

Fig. 1.7. Measurements $-0.6, 1.2, 10$ with failure probability of measurement device 0 %

the state f is only possible, if the measurement error equals x, and the probability for that is $p(x)$. Thus the plausibility of the state f, in view of the measurement x, equals $pl(f) = k\ p(x)$. The state d on the other hand is possible under all failure modes ω_j such that $x \in \Gamma(\omega_j)$. Hence the plausibility of state d in view of the measurement x equals $pl(d) = k\ \sum\{q(\omega_j) : x \in \Gamma(\omega_j)\}$. The latter sum represents once more the plausibility $pl(x)$ of an error x for a defective measurement device, hence $pl(d) = k\ pl(x)$. Note that the normalization factor k is still necessary to take into account possible contradictions of arguments.

This evidence can now be combined with the a priori probability α of a defect. This is a mixed argument in the sense of J. Bernoulli. In order to make our process of reasoning clear, let's introduce an argument a which implies d, if a is true, and assume that a has the probability α. Now, in order to prove d, using the evidence from the measurement x and from the a priori probability, both a and a failure mode ω_j, such that $x \in \Gamma(\omega_j)$ must be true. The probability of such a proof is $k\ \alpha\ pl(x)$, where k accounts once more for the possibility of contradictory arguments. On the other hand, in order to prove f, $\neg a$ must hold and the error of the functioning device must be x. This proof of f has the probability $k\ (a - \alpha)\ p(x)$. Because only d or f can be proved, these two credibilities defined by these probabilities of proofs must add up to one,

$$k\ (\alpha pl(x) + (1 - \alpha)p(x)) = 1 \tag{1.18}$$

which permits to determine k. It follows from the measurement x and the a priori probability of a defective device that the degrees of credibility of the two states are:

$$\begin{aligned} cred(f) &= \frac{(1-\alpha)p(x)}{\alpha pl(x) + (1-\alpha)p(x)} \\ cred(d) &= \frac{\alpha pl(x)}{\alpha pl(x) + (1-\alpha)p(x)} . \end{aligned} \tag{1.19}$$

It turns out that, just as for the estimation problem, it is here again sufficient to know the plausibilities of errors of a defective measurement device and not the whole error model. The decision for or against a defect depends on the ratio $(1 - \alpha)p(x)/(\alpha pl(x))$. Assuming a probability $p(x)$ decreasing with $|x|$ and a plausibility function $pl(x)$ as in Fig. 1.4., for a small error x, $p(x)$ will be large with respect to $pl(x)$ (which may be zero) and $cred(f)$ will be near one, $cred(d)$ near zero. One may safely assume that the device is working properly. If however the error x is large, $p(x)$ will become very small, even in comparison to α, and $pl(x)$ will be near one. Then the credibility of f will become very small and the credibility of d nearly one. In this case one has to assume that the device is defective.

1.6 Imprecise Measurements

Still in the context of measurement of numerical quantities we want to introduce an alternative measurement model. This time the measurement error is not assumed to be stochastic, but rather simply unknown although bounded. In the simplest case the error is known to lie within some finite interval $[a, b]$, where in most cases a will be negative and b positive. The width of the interval is a measure of the imprecision of the measurement. It will, however, be assumed that the precision of the measurement device used is not necessarily known. That is, several intervals $[a(\alpha), b(\alpha)]$ are possible error ranges, depending on an unknown precision index α. Two basic assumptions will be made about these precision indices:

(1) The precision is non-decreasing with increasing α. This means that if $\alpha' \leq \alpha''$ then $[a(\alpha'), b(\alpha')] \supseteq [a(\alpha''), b(\alpha'')]$. Furthermore, the precision

index is assumed to vary between 0 and 1, 0 being the most imprecise case and 1 representing the most precise case.

(2) A uniform probability distribution is assumed over the range of α from 0 to 1.

Assumption (2) is not as special as it may seem at first sight. If this is the knowledge we have about the possible errors of a measurement device and if we use this device to measure some unknown numerical quantity θ, a measurement value x represents some evidence about θ. In fact, assuming a precision determined by an index α, the interval $\theta + [a(\alpha), b(\alpha)]$ represents the range of possible measurement values. Inversely, again for this precision index, a measurement value of x implies that $\theta \in x - [a(\alpha), b(\alpha)]$. Let's fix a value of t and ask, to what degree is this a possible value for θ, given the measurement x. The condition for t to be possible, is that $t \in x - [a(\alpha), b(\alpha)]$. Denote by $\alpha(t)$ the supremum of all values of α for which t is possible, $\alpha(t) = sup\ \{\alpha : t \in x - [a(\alpha), b(\alpha)]\}$. Then $\alpha(t)$ is the probability that t is a possible value for θ, given the measurement x. This determines the plausibility of a value t

$$pl(t) = \alpha(t). \tag{1.20}$$

If $pl(t)$ is a continuous function, then the equation $\alpha = pl(t)$ has two solutions. These solutions define an interval $[x - b(\alpha), x - a(\alpha)]$ which is called an α-cut (see Fig. 1.8). Thus α-cuts are obtained by cutting a plausibility function at level α. If $pl(t)$ is not continuous, then the notion of an α-cut is defined in an analogous way (Dubois, Prade, 1985).

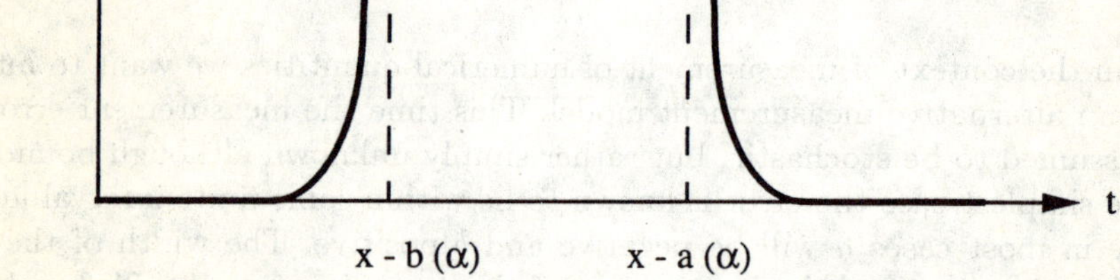

Fig. 1.8. The α-cuts of the plausibility function

One may also ask, what is the plausibility, that θ is within some interval $[a, b]$. An α is a precision index for which it is not excluded that $\theta \in [a, b]$ if and only if $[a, b] \cap x - [a(\alpha), b(\alpha)]$ is not empty. As above, let $\alpha[a, b]$ be the supremum of all values of α for which this condition holds. Then $pl([a, b]) = \alpha[a, b]$. On the other hand, let α such that $[a,b] \cap x - [a(\alpha), b(\alpha)]$ is not empty. Then there exists $t \in [a,b]$ such that $pl(t) = \alpha(t) \geq \alpha$ and hence $sup \{pl(t) : t \in [a,b]\} \geq \alpha$, which shows that $sup \{pl(t) : t \in [a,b]\} \geq \alpha[a,b]$. Inversely, since $pl(t) \leq pl([a,b])$ for all $t \in [a,b]$, it follows that $sup \{pl(t) : t \in [a,b]\} \leq pl([a,b])$ and thus

$$pl([a,b]) = sup \{pl(t) : t \in [a,b]\}. \tag{1.21}$$

Suppose now, that the same measurement device is used to measure two lengths θ_1 and θ_2 and that the measurements are x_1 and x_2. What can then be said about the total length $\theta = \theta_1 + \theta_2$? For a given precision index α it follows necessarily, that $\theta \in (x_1 - [a(\alpha), b(\alpha)] + (x_2 - [a(\alpha), b(\alpha)] = (x_1 + x_2) - [2a(\alpha), 2b(\alpha)]$. This permits to compute the plausibility function of θ in view of the two measurements x_1 and x_2 and the relation $\theta = \theta_1 + \theta_2$. The method consists of adding the α-cuts of the respective plausibility functions for all $\alpha \in [0,1]$. For an example suppose that the interval $[a(\alpha), b(\alpha)]$ shrinks to the point $\{0\}$ and that for $\alpha = 0$ the interval is $[-1, +1]$, whereas between these extremes a linear interpolation takes place (see Fig. 1.9).

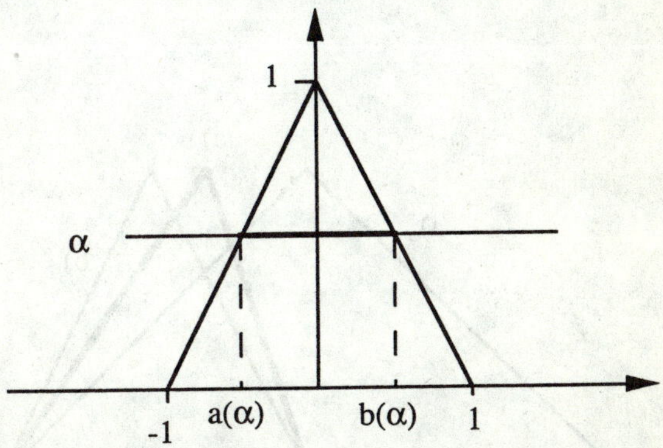

Fig. 1.9. The intervals $[a(\alpha), b(\alpha)]$

The addition of two such triangular plausibility functions gives again a triangular plausibility function but with the base doubled. Suppose, furthermore, that the values measured for θ_1 and θ_2 are $x_1 = 1.8$ and $x_2 = 2.2$ respectively. This gives then the plausibility function $pl'(t)$ for θ as drawn in Fig. 1.10 a). Furthermore, assume that the value of θ is also directly measured by the same measurement device with unknown precision. The measurement value x yields a plausibility function $pl''(t)$ for θ defined by $x - [a(\alpha), b(\alpha)]$. In the example, if $x = 5$ then $pl''(t)$ is as shown in Fig. 1.10 b). The question is now, how to combine the two evidences regarding θ resulting from the two measurements. The answer should be clear by now: If α is the correct precision index, then θ must be in the intersection of the two α-cuts $[a'(\alpha), b'(\alpha)]$ and $[a''(\alpha), b''(\alpha)]$ of the two plausibility functions. It is easy to see that the combined plausibility function so far is then determined by $pl(t) = min\ \{pl'(t), pl''(t)\}$. However, there may be values of α for which the above intersection of the α-cuts are empty. But these are clearly precision indices which are to be excluded, because they lead to an impossible conclusion. Let α^* be the supremum of all α for which the intersection is not empty. The uniform distribution of α is then to be conditioned on the event that $\alpha \leq \alpha^*$; i.e. the new distribution is uniform over the interval from 0 to α^*. It is convenient to stretch α again to the interval from 0 to 1 by posing $\alpha' = \alpha/\alpha^*$. The distribution of α' is then of course still uniform, but on the interval from 0 to 1. This amounts only to renormalize $pl(t)$ such that its maximum value becomes again 1. Thus

$$pl(t) = k \min \{pl'(t), pl''(t)\} \tag{1.22}$$

with

$$k = 1/\sup_t \{\min \{pl'(t), pl''(t)\}\}$$

becomes the final plausibility function combining the evidence from the two measurements. This is shown in Fig. 1.10 c) for our example.

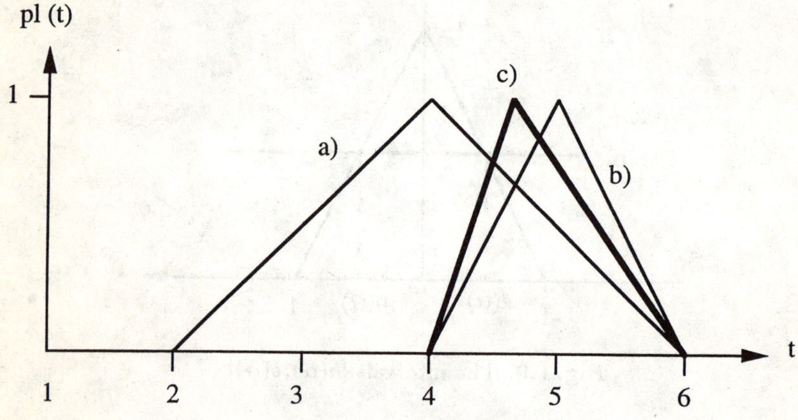

Fig. 1.10. Plausibility functions. a) of t in view of the measurements x_1 and x_2. b) of t in view of the measurement x. c) of t in view of the measurements x_1, x_2 and x

This model of knowledge about measurement errors is close to the approach of fuzzy set theory or possibility theory to describe vague information (see e.g. Dubois, Prade, 1985).

1.7 Conclusion

The examples of this chapter were selected to illustrate a uniform and new approach to the treatment of a variety of different problems. The essence of the method is to model uncertainty by uncertain arguments with known probabilities or probability distributions and known consequences. In order to analyze certain hypotheses, proofs for these hypotheses or their negations are constructed using the available arguments. This is the *logical* aspect of the approach. Using the probabilities of the arguments the probability that the proofs actually work or the reliability of the proofs are computed. This permits to evaluate the degree of credibility of a hypothesis and also its degree of plausibility. The latter measures to what extent there is no evidence or proof against the hypothesis. Credibility and plausibility are in general different. In extreme cases there may be no credibility in a hypothesis, no way to prove it with the available evidence; but still the hypothesis may be fully plausible.

The procedures, methods and approaches developed here for some simple examples deserve more complete and more systematic study. This is the goal of this book, and the systematic construction of a corresponding theory starts in the next chapter. The examples considered here will reappear to illustrate notions and concepts introduced in the course of development. Furthermore, new applications will emerge and be studied in order to illustrate the range and significance of the concept of logical reasoning with uncertain arguments.

2. The Mathematical Concept of a Hint

2.1 Questions and Answers

Information is always relative to one or several questions which may be put forward. Thus, in order to model information, it is necessary to represent questions and possible answers to these questions formally. Once a precise question is fixed, the set of possible answers to the question can be specified. Let's denote this set Θ. The generic elements of Θ will be denoted by symbols like $\theta, \theta', \theta'', \theta_i \ldots$ etc. Each element θ of Θ represents a possible answer to the given question. Unless explicitly said, Θ is throughout this book assumed to be complete in the sense that the true (but unknown) answer is exactly one of the elements of Θ. This is a *closed world assumption.* An open world assumption which allows for the possibility that the true answer could be outside Θ would also be possible, see Smets (1988).

The kind and nature of the possible answers to a question depend partly on the question and is partly a modeling decision. The possible answers to the question "who is the murderer ?" form a list of suspects, the possible answers to the question "what is the colour of the murderer's car ?" a list of colours. The possible answers to the question "what will the temperature be tomorrow ?" can be of qualitative nature such as "very low, low, medium, high, very high" or quantitatively determined as integers or even real numbers. In this paper, Θ is thus either a finite set or an infinite set like for example $\mathcal{R}, \mathcal{R}^2$ or $\mathcal{R}^n$ or a subset of one of them. It is part of the model building process to determine and fix Θ.

During a reasoning process it may become necessary to refine the possible answers. For example, whereas at the beginning of the argument it seemed sufficient to give a qualitative answer to the question about tomorrow's temperature, the need may arise to consider more precise answers such as degree of centigrade as integers or even more precisely including fractions of degrees. Or, on the contrary, the precision of the possible answers may be relaxed to cruder statements. Also compound questions such as "what will tomorrow's temperature and humidity be ?" or "what is the type and the colour of the murderer's car ?" may at some stage of the discussion be reduced to simpler, one-dimensional questions focusing only on the temperature or on the colour of the car. The inverse process may also become necessary when considering humidity besides temperature, type of the car besides its colour, etc.

Associated with these shifts between relevant questions is always a shift with respect to the sets Θ of possible answers reflecting the more or less crude or fine discernment of possibilities. At any time of the discourse a fixed Θ will be considered, which is called the (actual) *frame of discernment.* It becomes thus necessary to consider dynamically changing frames of discernment. This will be discussed in Chap. 6, whereas for the moment Θwill be considered as a fixed set.

It may also become necessary or useful to consider statements which are cruder than single answers. For example, one might be interested in the statement "the temperature will be below zero degree centigrade" or "the murderer was left-handed". Such propositions are represented by subsets of the corresponding frame of discernment, for example the subset of all suspects who are left-handed. This presupposes that the frame Θ is sufficiently fine to discern the proposition. If Θ contains only qualitative indications of tomorrow's possible temperature, then it is probably not fine enough to discern the above proposition and must eventually be refined. Relative to a fixed frame Θ, the propositions discerned by it (i.e. its power set $\mathcal{P}(\Theta)$) form a Boolean algebra permitting, as usual, to represent logical connectives by the corresponding set operations. Also, as usual, the empty subset represents the *impossible proposition* and Θ the trivially *true proposition.*

Information pertaining to a question must now be specified and described with respect to the corresponding frame of discernment. How this can be done will be discussed in detail in the following section.

2.2 Describing Hints

A hint is considered here as a piece or body of information relative to a single or compound question which is in general *imprecise* in the sense that it does not point to a precise answer. Rather it points to a whole range of possible answers. A hint is also often *uncertain* in the sense that it is not entirely certain which is the correct interpretation of the information conveyed by the hint. There may also be internal *contradictions* within a hint since under different possible interpretations contradictory answers may arise. Finally there can exist external contradictions between different hints relative to the same question. The aim of this section is to develop a mathematical model for hints which accounts for these elements. It takes A. Dempster's (1967) multivalued mappings as its starting point and develops into similar lines as in G. Shafer's *"Mathematical Theory of Evidence"* (1976 a). There exist however subtle differences with respect to the latter theory. These differences will be discussed in this chapter and also in the second part of the book dedicated to the general theory of hints.

Hints representing information about the answer to a fixed question whose frame of discernment is Θ are now formally introduced. However, let's stress already here, that the same hint may also very well carry information relative to other questions (see Chap. 7). This is an important remark, which however will be neglected for the time being. First of all, it is assumed that a hint allows for a number of possible interpretations and it is unknown, which is the correct one. Let $\Omega = \{\omega_1, \omega_2, \ldots, \omega_m\}$ represent the set of possible *interpretations* of the information contained in the hint. As we shall see in the examples, this generic term of interpretation can take various specialized forms like the internal states of a technical device or the possible scenarios about the evolution of a process. One of these possible interpretations must be the correct one, but it is unknown which one. In this first part of the book, only hints with a *finite* number of possible interpretations will be considered. It is very well possible and also both useful and interesting to study hints with an infinite number of interpretations, for example continuously varying interpretations. This generalization is the subject of part II of the book.

Each interpretation – if assumed to be the correct one – permits to restrict the possible answers to a nonempty subset $\Gamma(\omega_i)$ of Θ. This means that if ω_i proves to be the correct interpretation of the hint, then the true answer must be within the subset $\Gamma(\omega_i)$. Thus, even the knowledge of the correct interpretation of a hint does in general not permit to fix the true answer precisely , but only to restrict the possibilities to a subset of Θ. This represents the *imprecision* of a hint. Γ is a multivalued mapping from Ω into Θ or a mapping from Ω to the power set $\mathcal{P}(\Theta)$. The sets $\Gamma(\omega_i)$ are called the *focal sets* of the hint. They are not necessarily all different, two different interpretations may very well have the same focal set. Two interpretations of a hint are (internally and partially) *contradictory* if their associated focal sets are not nested, i.e. if each of the two focal sets contains possibilities which are not present in the other one. They are fully contradictory if their focal sets do not intersect.

Now, interpretations are not all equally likely. Thus let $p_i > 0,\ i = 1, \ldots, m$ be the probability that interpretation ω_i is the correct one and assume that

$$\sum_{i=1}^{m} p_i = 1 \tag{2.1}$$

because one of the possible interpretations must be the correct one. This represents the *uncertainty* of the information conveyed by the hint.

A hint h relative to a frame Θ is thus a quadruple $(\Omega, P, \Gamma, \Theta)$ where Ω is the set of possible interpretations, P a probability measure over this (finite) set – defined by its discrete density p – and Γ a multivalued mapping from Ω to Θ. The core of a hint is the union of all its focal sets and thus the true answer to the question considered must be within the core.

To illustrate this concept let's consider a few very simple examples:

Example 2.1 (Gas Tank). Consider the question whether the gas tank of your car is empty or not. Thus let $\Theta = \{e, f\}$ (e for empty, f for full). Suppose you look at your gas gauge, which indeed reads empty. This information is a hint, which has two possible interpretations: first of course you assume that your gauge is functioning correctly (state (interpretation) ω_1: gas gauge ok), in which case we may indeed infer the answer e, i.e. $\Gamma(\omega_1) = \{e\}$. But it is not excluded that the gas gauge has broken down (state ω_2: gas gauge down). Under this interpretation nothing at all can be inferred about the question, $\Gamma(\omega_2) = \Theta$, since both e and f remain perfectly possible under this interpretation. Note carefully that the alternative interpretation ω_2 does not allow to infer the complementary answer of the first one. The probability of the first interpretation represents the reliability of the gas gauge, i.e. the probability that, at an arbitrary time, the gas gauge is functioning. The probability of the second interpretation is simply the complement to 1 of the first one.

Example 2.2 (Strike on the Metro). Suppose you take the night train to Paris on Sunday evening. In the train you find a piece of a newspaper dating from Sunday announcing a strike on the Paris metro for Monday. Unfortunately the date of the newspaper is illegible, so that you are not sure whether indeed the newspaper dates from this Sunday or whether it is an old one. But the question arises whether you have to cope with a strike on the metro upon your arrival in Paris on Monday morning (possibility s) or whether on the contrary there will be no strike (possibility $\neg s$) and the metro will operate normally. Thus $\Theta = \{s, \neg s\}$. The newspaper you find is a hint for you. If it dates from this Sunday (interpretation ω_1), then surely you must be prepared to take a taxi, $\Gamma(\omega_1) = \{s\}$. If, however, you assume that the newspaper is an old one (interpretation ω_2), then nothing can be said about the question of a strike because you may still be confronted with a strike and thus $\Gamma(\omega_2) = \Theta$. The probabilities you assign to the two possible interpretations depend on the appearance of the piece of newspaper. If it is clean and fresh, then you may give a larger probability to ω_1 than if it is crumpled and dirty.

You might also wish to take into account other information. For example it is possible that you know (maybe from the same newspaper) that, on the average, there are x strike days a month on the metro in autumn. This is a hint of statistical nature. Suppose that the night you travel is in autumn. If you are unlucky enough for your Monday to fall on a strike day (possibility ω_1), then $\Gamma(\omega_1) = \{s\}$. Otherwise, if you assume that your Monday is not a strike day (possibility ω_2), then $\Gamma(\omega_2) = \{\neg s\}$. The probabilities of the two possibilities are estimated by the statistical information as $x/30$ and $(30 - x)/30$ respectively. Then, of course, the question arises as to how to combine these two hints to get an overall appreciation of your risk regarding the strike. This question will be treated in Chap. 4.

Example 2.3 (Duration of an activity). A project is usually composed of many activities which must satisfy certain precedence conditions. Typically, an activity cannot begin before some other activities are terminated. The question is then to determine the duration of the project. In this type of problems – called project scheduling problems – the duration of activities is traditionally supposed to be known. But in reality, we only dispose of hints regarding these durations. In fact, the duration may depend for example on the realization of a particular scenario about future meteorological conditions. For example, the possible scenarios may be ω_1 = good weather, ω_2 = cloudy, and ω_3 = bad weather. If durations are expressed in days, then we may take Θ as the non-negative integers and for a particular activity $\Gamma(\omega_1)$ could be the integer interval between 5 and 7, $\Gamma(\omega_2)$ between 6 and 9 and $\Gamma(\omega_3)$ between 8 and 12. If probabilities are given to the scenarios, then we have a hint for the duration of the activity.

Example 2.4 (Decision analysis). Suppose you have to choose one action among a whole family of actions which could be taken. Each action is rated according to a certain number of criteria by a mark between 1 and 10. The higher the mark, the better the action. With respect to a particular criteria, it is quite frequent that the rating of an action depends on the political or the economical situation which will prevail at the moment the action will be undertaken. Many scenarios can be envisaged, and we may well assume that for each of them only a range of possible ratings can be given. If this is completed by the addition of probabilities on the scenarios, we then have a hint regarding the evaluation of an action. Then, what is the definition of a best action in this type of model with hints, and how can we find it? This problem has not a unique solution, one of them is presented in Schaller (1991).

Example 2.5 (Amount of a check). As an investigator in a corruption scandal, you find a check which was given to a certain person as a bribe. Unfortunately, the amount of the check is no longer legible, the only thing you can be sure of is that it is a four digit number. In view of the context of this affair, you are able to assess probabilities for the value of the first digit on the left. But if this digit is known, then the exact amount of the check is certain to be between two bounds which can be calculated. This defines a hint regarding the amount of the check.

Example 2.6 (Error of measurement). Suppose that the error e liable to be commited by a certain type of measurement device increases with age. For each possible age a in Ω, the absolute value of the measurement error will not exceed a certain known limit $l(a)$, i.e. $e \in [-l(a), l(a)]$. Now we dispose of an instance of such a device and we ignore its exact age : we can only assess a probability $p(a)$ to each possible age $a \in \Omega$. Then $\mathcal{H} = (\Omega, p, \Gamma, \mathcal{R})$ with $\Gamma(a) = [-l(a), l(a)]$ is a hint for the actual error commited by the device. The age sensitivity implies that the focal sets $\Gamma(a)$, $a \in \Omega$ are nested.

There are a few special and important classes of hints:

(1) First of all, a hint whose focal sets all equal Θ represents total ignorance relative to the question represented by Θ (remember that the hint may carry information relative to other questions). In fact, under no interpretation is it possible to restrict even slightly the set containing the answer.

Such a hint is called *vacuous*. For example, if we are fully ignorant about the answer, then this fact might be considered as an interpretation and its focal set is of course Θ itself. Note how perfect full ignorance can be represented by hints and compare this to the problem of representing ignorance if you restrict yourself to treat uncertainty in the framework of ordinary probability theory. Any mathematical theory of uncertainty should at least permit a neat representation of ignorance.

(2) Second, a hint which has at most one focal set F distinct from Θ is called *simple*. This means that all interpretations which indeed carry some information relative to Θ restrict the answer to be within the same subset F. A "pure argument" in the sense of Bernoulli is an example of a simple hint (see Chap. 1).

(3) A hint which assures that the answer is within a subset F under all interpretations is called *deterministic* because we know for sure that F contains the answer. All hints with only one possible interpretation are deterministic.

(4) If the focal sets of a hint are nested, i.e. if they can be numbered such that $F_1 \subseteq F_2 \subseteq \ldots$, then the hint is called *consonant*. In this case all possible interpretations point into the same direction, although with varying precision. But there is no internal contradiction between the interpretations of a consonant hint unlike in general hints. The hint described in example 2.6 is consonant.

(5) If all focal sets are singletons, then the hint is called *precise*. In this case Γ is in fact a random variable and there is full contradiction between interpretations pointing to different singletons. Hence all (discrete) random variables can be viewed as precise hints. In this sense, the theory of hints contains probability theory (as a theory of random variables) as a special case.

(6) If the focal sets of a hint are disjoint, then the hint is called *Bayesian*. Any precise hint is Bayesian since two different singletons are necessarily disjoint. For example, a "mixed argument" in the sense of Bernoulli is a precise hint (see Chap. 1). The hint given in example 2.5 is also Bayesian.

2.3 Modeling with Hints

The notion of hint – at least in its general definition (see Part III) – can be used as a modeling tool whenever we dispose of partial or full information regarding the answer to a particular, well defined question. For example, a "pure argument" in favour of a proposition A discerned by the frame Θ can be described as a simple hint. In fact, all possible cases to be considered in the argument form the set of interpretations Ω, and if $\Gamma(\omega) = A$ or $\Gamma(\omega) = \Theta$, depending on whether ω is in favour of A or ω is neutral respectively, then we have an application Γ from Ω to the power set of Θ. If the probability $p(\omega)$ is known for all $\omega \in \Omega$, then $\mathcal{H} = (\Omega, p, \Gamma, \Theta)$ is a hint representing the pure argument. Similarly, defining in addition $\Gamma(\omega) = A^c$ for each case which speaks against A permits to represent a "mixed argument" as a hint.

A reasoning activity usually involves many propositions $P_i, i \in I$ which can be true or false. For each proposition, let $\Theta_i = \{true, false\}$ be its frame of discernment. Now, information is typically pertaining to a group of propositions $P_j, j \in J, J \subseteq I$. A simple example will be given below. The special case where J contains a single element is of course not excluded. The frame of discernment is then $\Theta_J = \prod_{j \in J} \Theta_j$ because the information concerns the simultaneous truth value of each proposition in the group. This knowledge is represented by a hint $\mathcal{H} = (\Omega, P, \Gamma, \Theta_J)$. Since knowing that the answer is in the focal set $\Gamma(\omega))$ is exactly the same as knowing that the answer is in $\Gamma(\omega) \times \Theta_{I-J}$, nothing is changed if we express $\mathcal{H}$ with respect to the overall frame $\Theta = \Theta_I$. This has the advantage that information pertaining to different groups of propositions is represented with respect to a common frame of discernement. It is on this frame that hints can be combined (see Chap. 4). The frame Θ generated by a family of propositions is a special case of general product space structures, which in turn are special cases of families of compatible frames of discernment. This subject will be discussed in detail in Chap. 7.

Propositions $P_i, i \in I$ can be used to build so-called logical formulas from the logical connectives $\neg, \wedge, lor, \Rightarrow, \Leftrightarrow$. Here are the precise rules for constructing formulas

(1) Any proposition standing alone is a formula
(2) Given any formula X and Y already constructed, the expressions $(X \wedge Y), (X \vee Y), (X \Rightarrow Y)$ or $(X \Leftrightarrow Y)$ are again formulas, and so is the expression $\neg X$.

If Θ is the frame generated by the family of propositions P_i, then let $f(\theta)$ represent the truth value of the formula f under the instanciation $\theta \in \Theta$ of the propositions $P_i, i \in I$. To each formula f a subset $T(f)$ of Θ corresponds :

$$T(f) = \{\theta \in \Theta : f(\theta) = true\}. \tag{2.2}$$

This subset of Θ represents the fact that the formula f is true. For example, if $I = \{1, 2\}$ and f is the formula $P_1 \Rightarrow P_2$, then $T(f) = \Theta - \{(true, false)\}$ represents the fact that this implication is true.

The multi-proposition framework is particularly well suited for representing uncertain information of the form

$$\text{If } P_1 \wedge \ldots \wedge P_n \wedge a \text{ then } Q. \tag{2.3}$$

Here a is an *implicit* assumption which may be valid or not. The probability $p(a)$ expresses the likelihood that this assumption is verified. In this case, the implication or formula $f = (P_1 \wedge \ldots \wedge P_n \Rightarrow Q)$ is true and is thus represented by its corresponding subset $T(f)$ in Θ. If the assumption a is not valid, then the implication $P_1 \wedge \ldots \wedge P_n \Rightarrow Q$ may be true or false, which is in fact a vacuous information. This shows that the uncertain rule (2.3) can be represented by the simple hint $\mathcal{H} = (\Omega, p, \Gamma, \Theta)$ with $\Omega = \{a, \neg a\}$ and $\Gamma(a) = T(f)$, $\Gamma(\neg a) = \Theta$.

To give a concrete example, consider the sensor model with two possible kinds of error described in Sect. 1.2. The causal relations are defined by the following rules R1 to R3 :

$$R_1 : h \wedge a_1 \text{ implies } e$$

$$R_2 : a_2 \text{ implies } e.$$

$$R_3 : \neg h \wedge \neg a_2 \text{ implies } \neg e.$$

Here a_1 is considered as a sufficient reason for h to produce e, whereas a_2 represents the presence of another cause which produces e. The frame of discernment to be considered is $\Theta = \Theta_h \times \Theta_e$ where $\Theta_h = \{h, \neg h\}$ accounts for the presence or the absence of the cause h and $\Theta_e = \{e, \neg e\}$ accounts for the presence or absence of the effect e. Rule R_1 can be represented by the hint $\mathcal{H}_1 = (\Omega_1, P_1, \Gamma_1, \Theta)$ with $\Omega_1 = \{a_1, \neg a_1\}$, $\Gamma_1(a_1) = \Theta - \{(h, \neg e)\}$, $\Gamma_1(\neg a_1) = \Theta$ and a known probability $p_1(a_1)$. Since the same element a_2 appears in both rules R_2 and R_3, they must be considered simultaneously. These rules define a second hint $\mathcal{H}_2 = (\Omega_2, P_2, \Gamma_2, \Theta)$ with $\Omega_2 = \{a_2, \neg a_2\}$, $\Gamma_2(a_2) = \{(h, e), (\neg h, e)\}$, $\Gamma_2(\neg a_2) = \Theta - \{(\neg h, e)\}$ and a known probability $p_2(a_2)$. These two hints represent the functioning mechanism of the sensor. This type of information is sometimes called the structural knowledge. Besides, there is also information coming from observation : we observe the presence of the effect (possibility e) or we observe its absence (possibility $\neg e$). This kind of knowledge is called factual. For example, observing the presence of the effect is represented by the deterministic hint $\mathcal{H}_e$ having a single interpretation with associated focal set $\{(h, e), (\neg h, e)\} \subseteq \Theta$. In measurement models, a measurement value x is also a factual information which can be represented by a deterministic hint.

In the Cholestatic Jaundice example described in Sec. 1.4, the frame of discernment is $\Theta = \{hep, cirr, gall, pan\}$. Each of the three unreliable rules which can be applied, using the available evidence, defines a simple hint on Θ. For example, R_1 is represented by the hint $\mathcal{H}(R_1) = (\Omega, P, \Gamma, \Theta)$, where $\Omega = \{R_1, \neg R_1\}$ expressing the fact that the rule is reliable or not, $\Gamma(R_1) = \{hep, cirr\}$, $\Gamma(\neg R_1) = \Theta$ and $p(R_1)$ is the probability that the rule is reliable.

In the imprecise measurement model of Sect. 1.6, the error is known to lie within some finite interval $[a(\alpha), b(\alpha)]$ depending on the precision level α of the device. This level is an unknown parameter in the interval $[0, 1]$. If we consider the possible levels of precision as the set of interpretations Ω – assuming there is a finite number of them – then $\mathcal{H} = (\Omega, P, \Gamma, \mathcal{R})$ with $\Gamma(\alpha) = [a(\alpha), b(\alpha)]$ is a consonant hint for the measurement error produced by the device, provided that P is a known probability measure on Ω. When Ω is the continuous interval $[0, 1]$ and P is the uniform distribution over this interval, then we have a continuous consonant model. Hints with an infinite number of interpretations will be discussed in Part III of the book.

3. Support, Credibility, Plausibility and Possibility

3.1 Judging Hypotheses in the Light of Hints

Often one is not so much interested in the precise answer to the question considered, but rather in some proposition discerned by the frame Θ. Thus, let H be a subset of Θ representing such a proposition. Then the question is: To what extent does a hint relative to Θ prove the proposition, or, in other words, how strong does the hint support the hypothesis that H contains the answer. Let $\mathcal{H} = (\Omega, P, \Gamma, \Theta)$ be a hint for the question considered. If ω is the correct interpretation of $\mathcal{H}$, then the answer necessarily lies in $\Gamma(\omega)$. So, if $\Gamma(\omega) \subseteq H$, then the answer is a fortiori also in H. Hence, each interpretation ω such that $\Gamma(\omega) \subseteq H$ speaks in favour, supports or proves the hypothesis H. Therefore, H can be infered if and only if the correct interpretation happens to be in the set

$$u(H) = \{\omega \in \Omega : \Gamma(\omega) \subseteq H\}. \tag{3.1}$$

Since the correct interpretation ω is unknown, it is uncertain whether H can be infered from the hint. However, the probability of the event (3.1), $P(u(H))$, measures the likelihood that H may indeed be infered. Therefore, it is natural to define the *degree of credibility* or the *degree of support* of H by

$$sp(H) = P(u(H)). \tag{3.2}$$

This value measures to what extent the hypothesis H is proved by the hint. The terminology used here does not correspond exactly to the one introduced by Shafer (1976 a). There, the term support is used only when simple hints are combined. Shafer calls $sp(H)$ a degree of belief and he denotes it by $Bel(H)$. However, we claim that a hint supports a hypothesis to a certain degree which is measured by $sp(H)$, whereas one's (subjective) belief in H may well be larger than this value. It seems thus reasonable and necessary to distinguish between support and belief. It is important to note that the theory developed in this book concerns support and not belief.

We may also ask to what extent the hypothesis H appears as possible or plausible in view of the hint. Since the hypothesis H remains possible under an interpretation ω if and only if $\Gamma(\omega) \cap H$ is not empty, it is natural to define the *degree of plausibility* of H by

$$pl(H) = P(v(H)) \tag{3.3}$$

with

$$v(H) = \{\omega \in \Omega : \Gamma(\omega) \cap H \neq \emptyset\}. \tag{3.4}$$

We can also forumulate the following theorem.

Theorem 3.1. *If H, H' and H'' are subsets of Θ, then*

(1) $u(\emptyset) = v(\emptyset) = \emptyset,\ u(\Theta) = v(\Theta) = \Omega$
(2) $v(H) = u(H^c)^c,\ u(H) = v(H^c)^c$
(3) $u(H) \subseteq v(H)$
(4) If $H' \subseteq H''$ *then* $u(H') \subseteq u(H'')$ *and* $v(H') \subseteq v(H'')$.

The proof of this theorem, like that of all theorems in this chapter, can be found in the mathematical appendix at the end of the chapter. This theorem shows that the support and plausibility of the empty hypothesis is zero, whereas these are equal to one for the hypothesis Θ. This theorem implies also that the plausibility can be computed from the degree of support. In fact,

$$pl(H) = 1 - sp(H^c) \tag{3.5}$$

because of theorem 3.1 (2). Thus a hypothesis appears as less plausible as its negation is more credible. The following theorem is a direct consequence of theorem 3.1.

Theorem 3.2. *Let H, H' and H'' be subsets of Θ.*

(1) $sp(\emptyset) = pl(\emptyset) = 0,\ sp(\Theta) = pl(\Theta) = 1$
(2) $sp(H) = 1 - pl(H^c),\ pl(H) = 1 - sp(H^c)$
(3) $sp(H) \leq pl(H)$
(4) If $H' \subseteq H''$ *then* $sp(H') \subseteq sp(H'')$ *and* $pl(H') \subseteq pl(H'')$.

Other properties of the mappings u and v from $\mathcal{P}(\Theta)$ to $\mathcal{P}(\Omega)$ are given in the next theorem.

Theorem 3.3. *Let $\mathcal{F} \subseteq \mathcal{P}(\Theta)$ be an arbitrary family of subsets of Θ.*

(1) $u(\cap \{H : H \in \mathcal{F}\} = \cap \{u(H) : H \in \mathcal{F}\}$
(2) $v(\cup \{H : H \in \mathcal{F}\} = \cup \{v(H) : H \in \mathcal{F}\}$
(3) $u(\cup \{H : H \in \mathcal{F}\} \supseteq \cup \{u(H) : H \in \mathcal{F}\}$
(4) $v(\cap \{H : H \in \mathcal{F}\} \subseteq \cap \{v(H) : H \in \mathcal{F}\}$

In view of (1) and (2), the mappings u and v are called $\cap$-homomorphism and $\cup$-homomorphism respectively.

In the gas tank example 2.1, let p_1 denote the probability that the gas gauge is functioning correctly, i.e. its reliability. Then the degree of support that the tank is empty is p_1 and its degree of plausibility is 1 because no interpretation speaks in favour of the hypothesis that the tank is not empty. In the example 2.3, let p_i denote the probability of the interpretation $\omega_i, i = 1, 2, 3$. Then the credibility that the activity will last 5 or less days is zero whereas its plausibility is $1 - (p_2 + p_3)$. The credibility that the duration of the activity is between 6 and 10 days equals p_2, whereas its plausibility is 1.

3.2 Elementary Properties of Support and Plausibility Functions

Of course, the degrees of support $sp(H)$ and of plausibility $pl(H)$ can be defined for any subset H of Θ. Thus sp is a function from $\mathcal{P}(\Theta)$ to $[0, 1]$ and is called a *support function* or a *credibility function* and similarly the function $pl : \cap P(\Theta) \rightarrow [0, 1]$ is called a *plausibility function*. So, to each hint $\mathcal{H} = (\Omega, P, \Gamma, \Theta)$ corresponds exactly one support function and one credibility function, both defined on $\mathcal{P}(\Theta)$. Let's collect a few important basic results about support and plausibility functions.

Theorem 3.4.

(1) If $H_1 \supseteq H_2 \supseteq \ldots$ is a decreasing sequence of subsets of Θ, then

$$sp(\cap_{i=1}^{\infty} H_i) = \lim_{i \to \infty} sp(H_i). \tag{3.6}$$

(2) If $H_1 \subseteq H_2 \supseteq \ldots$ is an increasing sequence of subsets of Θ, then

$$pl(\cup_{i=1}^{\infty} H_i) = \lim_{i \to \infty} pl(H_i). \tag{3.7}$$

This theorem asserts that the set functions sp and pl are *continuous*. The following theorem states a very basic property of support functions and plausibility functions.

Theorem 3.5.

(1) A support function sp is monotone of order ∞, i.e.

$$sp(H) \geq \sum \{(-1)^{|I|+1} sp(\cap_{i \in I} H_i) : \emptyset \neq I \subseteq \{1, \ldots, n\}\} \tag{3.8}$$

for all $n \geq 1$ and all subsets H, H_i of Θ such that $H \supseteq H_i$ for all $i = 1, \ldots, n$.

(2) A plausibility function pl is alternating of order ∞, i.e.

$$pl(H) \leq \sum \{(-1)^{|I|+1} pl(\cup_{i \in I} H_i) : \emptyset \neq I \subseteq \{1, \ldots, n\}\} \tag{3.9}$$

for all $n \geq 1$ and all subsets H, H_i of Θ such that $H \subseteq H_i$ for all $i = 1, \ldots, n$.

This theorem was proved by Shafer (1976 a) for finite frames of discernment Θ, but it extends easily also to infinite frames. Together with theorem 3.2 (1), it shows that a support function is a *Choquet capacity* of infinite order (Choquet, 1953). One may ask whether any function $f : \mathcal{P}(\Theta) \to [0,1]$ satisfying (3.7) and $f(\emptyset) = 0$, $f(\Theta) = 1$ (i.e. a Choquet capacity of infinite order) is the support function of some hint. Since it is known that a Choquet capacity of infinite order defined on an infinite set is not necessarily continuous, the answer is no (see however theorem 3.10 in Sect. 3.3). The notion of hint must be extended in order that any infinite order Choquet capacity is the support function of some hint. This general theory will be presented in Part II of the book. Note that Shafer (1979) defines belief functions as Choquet capacities of infinite order without a reference to the notion of a hint.

Under our general hypothesis that there is only a finite number of interpretations in Ω, we can formulate the following simple result:

Theorem 3.6. *If H is any subset of Θ, then*

(1) $pl(H) = max\ \{pl(I) : I \text{ finite}, I \subseteq H\}$
(2) $sp(H) = min\ \{sp(C) : C \text{ cofinite}, C \supseteq H\}$.

The plausibility of any hypothesis is fully determined by the plausibilities of its finite subsets. For consonant hints, the whole plausibility function, and therefore also the support function, are already determined by the plausibilities of the singletons (the degree of plausibility of such hypotheses $pl(\{\theta\})$ will be abbreviated by $pl(\theta)$) :

Theorem 3.7. *Let pl and sp be the respective plausibility and support functions of a consonant hint. If H, A and B are subsets of Θ, then*

(1) $pl(H) = max\ \{pl(\theta) : \theta \in H\}$
(2) $pl(A \cup B) = max\ \{pl(A), pl(B)\}$
(3) $sp(A \cap B) = min\ \{sp(A), sp(B)\}$.

These formulas are reminiscent of rules advocated in possibility theory (Dubois, Prade, 1985). In fact, a set function satisfying (2) is called a possibility measure by these authors, and if it satisfies (3) it is called a necessity measure. In this way, consonant hints provide a new semantics to possibility theory. This has already been noticed by Shafer (1987).

Note that equality holds in (3.8) when $H = \cup_{i=1}^{n} H_i$ and sp is replaced by a probability measure, as is well known from probability theory. Equality holds also in (3.9) when $H = \cap_{i=1}^{n} H_i$ and pl is replaced by a probability measure. Together with theorems 3.2 and 3.4, this permits to consider support and plausibility functions as generalizations of probability measures.

The following theorem asserts that the support and plausibility functions of a precise hint are in fact probability measures.

Theorem 3.8. *Let sp and pl be the support and plausibility functions of a precise hint. Then*

(1) $sp(H) = pl(H)$ *for all* $H \subseteq \Theta$
(2) sp is a probability measure on $\mathcal{P}(\Theta)$, *i.e.*

(i) $sp(H) \geq 0$ *for all* $H \subseteq \Theta$
(ii) $sp(\Theta) = 1$
(iii) sp is σ*-additive, i.e. if* $A_1, A_2, \ldots$ *is a finite or infinite sequence of mutually disjoint subsets of* Θ, *then*

$$sp(\cup_{i=1}^{\infty} A_i) = \sum_{i=1}^{\infty} sp(A_i). \tag{3.10}$$

In fact, a precise hint $\mathcal{H} = (\Omega, P, \Gamma, \Theta)$ can be seen as a random variable

$$(\Omega, \mathcal{P}(\Omega), P) \xrightarrow{\Gamma} (\Theta, \mathcal{P}(\Theta)) \tag{3.11}$$

when the singletons $\Gamma(\omega) \subseteq \Theta$ are no longer considered as sets but as elements of Θ. In this way, $\Gamma^{-1}(H) = u(H)$ and hence the support function sp is the image probability of P by the mapping Γ. So, formally, a precise hint is a random variable, but the semantics is different. From the point of view of hints, the set Ω is not necessarily a sample space but merely represents a set of possible, mutually exclusive and exhaustive interpretations to be considered. The next example illustrates this remark.

Example 3.1. Suppose that the question considered is what is the result of flipping a coin. Then of course $\Theta = \{heads, tails\}$. Suppose there are several people who tell us the result; the problem is only that these people may be liars who tell exactly the contrary of the truth. So, somebody asserting that the coin turned "heads" defines a hint with two possible interpretations ω_1 and ω_2. If he tells the truth (interpretation ω_1), then $\Gamma(\omega_1) = \{heads\}$ and if he is a liar (interpretation ω_2), then $\Gamma(\omega_2) = \{tails\}$. Let p be the probability of a person being a liar. This defines a precise hint for the result of flipping the coin. A similar hint is defined for every person saying that the coin turned "tails". Note carefully that all these hints do not inform us about the random variable (i.e. the probability model) describing the random outcome of flipping that coin (i.e. for example whether the coin is fair or not). They inform us only on the result of the one particular coin flipping ! This remark is very important and its misunderstanding is the source of many misinterpretations of the theory of evidence.

3.3 Equivalence of Hints

Let $\mathcal{H} = (\Omega, P, \Gamma, \Theta)$ be a hint on the frame of discernment Θ. Since Γ is a mapping from Ω to $\mathcal{P}(\Theta)$, we can define

$$\Gamma^{-1}(A) = \{\omega \in \Omega : \Gamma(\omega) = A\} \tag{3.12}$$

for all $A \subseteq \Theta$ ($G^{-1}(A)$ may be empty). The application $m : \mathcal{P}(\Theta) \to [0,1]$ given by

$$m(A) = P(\Gamma^{-1}(A)) \tag{3.13}$$

is called the *basic probability assignment* (bpa) of the hint (Shafer, 1976 a). If the addition of sets is assumed to mean their disjoint union and if $\mathcal{F} \subseteq \mathcal{P}(\Theta)$ is the finite family of focal sets of $\mathcal{H}$, then

$$u(H) = \sum \{\Gamma^{-1}(F) : F \in \mathcal{F}, F \subseteq H\} \tag{3.14}$$

and hence the right-hand side of (3.14) forms a partition of $u(H)$ according to interpretations having the same focal set. Similarly, we have

$$v(H) = \sum \{\Gamma^{-1}(F) : F \in \mathcal{F}, F \cap H \neq \emptyset\}. \tag{3.15}$$

But this implies that the degrees of credibility and plausibility can be expressed in terms of the function m:

$$\begin{aligned} sp(H) &= \sum \{m(F) : F \in \mathcal{F}, F \subseteq H\} \\ pl(H) &= \sum \{m(F) : F \in \mathcal{F}, F \cap H \neq \emptyset\}. \end{aligned} \tag{3.16}$$

Since $\mathcal{F}$ is a finite family and $m(F) = 0$ whenever $F \notin \mathcal{F}$, these equations can also be written

$$\begin{aligned} sp(H) &= \sum \{m(A) : A \subseteq H\} \\ pl(H) &= \sum \{m(A) : A \cap H \neq \emptyset\} \end{aligned} \tag{3.17}$$

for all $H \subseteq \Theta$.

For example, if m is the bpa of a simple hint with focal sets $\{F, \Theta\}$, then

$$sp(H) = \begin{cases} 1 & \text{if } H = \Theta \\ m(F) & \text{if } F \subseteq H \neq \Theta \\ 0 & \text{otherwise.} \end{cases}$$

$$pl(H) = \begin{cases} 1 & \text{if } H \cap F \neq \emptyset \\ m(\Theta) & \text{if } H \cap F = \emptyset \neq H \\ 0 & \text{otherwise.} \end{cases}$$

Let m be the bpa of a consonant hint with focal sets $F_1 \subseteq F_2 \subseteq \ldots \subseteq F_m$ and let F_0 be the empty set. For a given hypothesis H, let j denote the greatest index i such that $F_i \subseteq H$ ($j = 0$ if F_1 is not included in H). Then we have

$$sp(H) = \sum_{i=0}^{j} m(F_i). \tag{3.18}$$

Similarly, if F_{m+1} is defined to be Θ when $F_m \neq \Theta$, then there is a smallest index k such that $F_i \cap H \neq \emptyset$ ($k = m+1$ if $F_m \cap H = \emptyset$). Then we have

$$pl(H) = \sum_{i=k}^{m+1} m(F_i). \tag{3.19}$$

Suppose now that m is the bpa of a precise hint. Then $m(H)$ is different from zero only for the finite set $\mathcal{F}$ of focal sets which are all singletons. Hence, according to (3.16), m can be considered as a probability density on this finite set. This is also true for sp and pl because $m(\theta) = pl(\theta) = sp(\theta)$ for all $\{\theta\}$ in $\mathcal{F}$.

More generally, the equality $sp(\Theta) = 1$ implies that

$$\sum_{A \subseteq \Theta} m(A) = 1 \tag{3.20}$$

because $m(A) = 0$ for non focal sets A ($m(A)$ is positive only for a finite number of subsets $A \subseteq \Theta$). Since $m(\emptyset) = 0$, this shows that m is a discrete probability density on the power set $\mathcal{P}(\Theta)$. Hence a hint generates a non-empty *random subset* of Θ. As already mentioned, this random subset becomes a random variable when the hint is precise.

In consequence of (3.17), two hints having the same bpa will have identical support and plausibility functions.For this reason, two hints $\mathcal{H}_1$ and $\mathcal{H}_2$ having the same bpa will be called equivalent and we write $\mathcal{H}_1 \simeq \mathcal{H}_2$ (the relation is reflexive, symmetric and transitive). Two hints are equal when the four components which they are made of, are identical. In addition, if id is the identity mapping, then $\mathcal{H}_c = (\mathcal{F}, m, id, \Theta)$ is a particular representative of the equivalence class $[\mathcal{H}]$ of the hint $\mathcal{H}$. This hint $\mathcal{H}_c$ is called the *canonical representative* of the class. Note that this hint is completely determined by the bpa m defining the equivalence class because $\mathcal{F} = \{A \subseteq \Theta : m(A) > 0\}$. If Θ is a finite frame, then the bpa of $\mathcal{H}$ – and hence its canonical representative $\mathcal{H}_c$ – can be computed directly from its support function sp. The following theorem gives the explicit formula.

Theorem 3.9. *Let m be the bpa of a hint $\mathcal{H} = (\Omega, P, \Gamma, \Theta)$ on the finite frame Θ. If sp is the support function of $\mathcal{H}$, then*

$$m(A) = \sum_{B \subseteq A} (-1)^{|A-B|} sp(B) \tag{3.21}$$

for all $A \subseteq \Theta$.

Since the support function of a precise hint can be considered as a probability density, the next theorem generalizes the result from probability theory asserting that any probability measure is the image probability of a random variable.

Theorem 3.10. *Every Choquet capacity of infinite order on a finite set Θ is the support function of a hint.*

We have already mentioned that this theorem is not valid for general frames Θ unless the notion of hint is generalized (see Part III).

Theorem 3.11. *Suppose that Θ is finite and let $\mathcal{G}$ denote the family of hints having the infinite order Choquet capacity φ on Θ as support function. Then $\mathcal{G}$ is an equivalence class whose canonical representative is $(\mathcal{F}, m, id, \Theta)$ with*

$$m(A) = \sum_{B \subseteq A} (-1)^{|A-B|} \varphi(B) \tag{3.22}$$

for all $A \subseteq \Theta$ and $\mathcal{F} = \{A \subseteq \Theta : m(A) > 0\}$.

According to this theorem, we can speak of the hint $(\mathcal{F}, m, id, \Theta)$ as the canonical representative of the class generated by φ, or short the canonical representative of φ. In addition, two hints with the same support function on a finite frame are equivalent.

The elements Ω, P and Γ allow for another interpretation of support functions. In fact, let sp be the support function of the hint $\mathcal{H} = (\Omega, P, \Gamma, \Theta)$ with $\Omega = \{\omega_1, \ldots, \omega_m\}$. If $n = 2^m$, then the elements $M_1, \ldots, M_n$ of $\mathcal{P}(\Omega)$ correspond to the so-called portions of belief or the so-called probability masses introduced by Shafer (1976 a), the whole belief being represented by Ω. Furthermore, for any $A \in \mathcal{P}(\Theta)$, the set $u(A) \in \mathcal{P}(\Omega)$ is thought of as the portion of belief committed to A. According to theorem 3.3 (1), the mapping $u : \mathcal{P}(\Theta) \to \mathcal{P}(\Omega)$ is an $\cap$-homomorphism and so the portion of belief committed to $A \cap B$ corresponds precisely to the intersection of the portions committed to A and to B respectively, which is quite reasonable. For any $M \in \mathcal{P}(\Omega)$, Shafer calls $P(M)$ the measure of the probability mass M. And since the belief of a proposition H in $\mathcal{P}(\Theta)$ is defined to be the measure of the portion of belief committed to H, we have

$$Bel(H) = P(u(H)) = sp(H) \tag{3.23}$$

and hence the support function sp is now interpreted as a belief function.

Following Shafer, the $\cap$-homomorphism u lends itself to a geometric intuition. Think that the masses M in $\mathcal{P}(\Omega)$ are spread over the set Θ. But instead of distributing these masses in a fixed way, require that a mass M be constrained to remain inside the set $A \subseteq \Theta$ if, and only if, $M \subseteq u(A)$. In this case, we write $M \; ct \; A$. Therefore, for any subset $A \subseteq \Theta$, $\Gamma^{-1}(A)$ is the mass that is confined to A, but not to any proper subset of A. In other words, $\Gamma^{-1}(A)$ is the mass that is confined to A, but can move freely to every point of A. More generally, we have the following result:

Theorem 3.12. *For each $M \in \mathcal{P}(\Omega)$, the set $\Gamma(M) = \cup_{\omega \in M}\Gamma(\omega)$ is the smallest subset of Θ in which M is constrained to move, i.e.*

$$\Gamma(M) = \cap \{A : M \text{ ct } A\}. \tag{3.24}$$

This means that M is confined to $\Gamma(M)$ but can move freely in it. If $\mathcal{M}_1 = \{M \in \mathcal{P}(\Omega) : \Gamma(M)$ is a focal set included in $H\}$, then $sp(H) = \sum \{P(M) : M \in \mathcal{M}_1\}$ and hence the support of H is the total measure of the masses that can move freely in focal sets included in H. Similarly, if $\mathcal{M}_2 = \{M \in \mathcal{P}(\Omega) : \Gamma(M)$ is a focal set cutting $H\}$ then $pl(H) = \sum \{P(M) : M \in \mathcal{M}_2\}$ and thus the plausibility of H is the total measure of the masses that can move freely in focal sets cutting H.

Theorem 3.12 will be used to justify the name given to an important function to be defined in the next section. To conclude this discussion, let's stress again that the theory of hints is merely a theory of support rather than a theory of belief. For example, this is the reason why we are going to characterize Choquet capacities by extending the notion of focal set and not by attempting to interpret them in terms of belief like Shafer did (see Chap. 13). It is in this respect that our theory differs essentially from that of Shafer. Since a hint incorporates a multivalued mapping Γ, we may say that the theory of hints is somewhat closer to Dempster's theory of upper and lower probabilities than to Shafer's theory of belief functions (see Dempster, 1967)

3.4 Commonality Functions

Let $\mathcal{H} = (\Omega, P, \Gamma, \Theta)$ be a hint with support function sp, plausibility function pl and basic probability assignment m. For $H \subseteq \Theta$, let $w(H) = \{\omega \in \Omega : H \subseteq \Gamma(\omega)\}$. Then the quantity

$$q(H) = P(w(H)) \tag{3.25}$$

is called the *commonality number* of H and the function $q : \mathcal{P}(\Theta) \rightarrow [0, 1]$ is the *commonality function* of the hint $\mathcal{H}$. If $\mathcal{M} = \{M \in \mathcal{P}(\Omega) : \Gamma(M)$ is a focal set and $H \subseteq \Gamma(M)\}$, then H is a common domain of movement of the masses in $\mathcal{M}$ and $q(H)$ is the total measure of these masses, i.e.

$$q(H) = \sum \{P(M) : M \in \mathcal{M}\} = \sum \{m(A) : H \subseteq A\} \tag{3.26}$$

where the last sum is finite because $m(A)$ is zero whenever A is not a focal set. This observation explains why q is called the commonality function. Although the commonality numbers are not of direct interest like the degrees of support and plausibility, they provide an important technical aid to study hints. This was already remarked by Dempster (1967) who introduced them but it is

Shafer who first called these quantities commonality numbers. There is a result similar to theorem 3.6 (1).

Theorem 3.13. *Let $\mathcal{H} = (\Omega, P, \Gamma, \Theta)$ be a hint with commonality function q. If H is any subset of Θ, then*

$$q(H) = min\ \{q(I) : I \text{ finite},\ I \subseteq H\}. \tag{3.27}$$

The following theorem shows how to compute the plausibility of finite sets and the support of cofinite sets from the commonality function. This theorem has been proved by Shafer for the case where Θ is finite.

Theorem 3.14. *Let q be the commonality function, sp the support function and pl the plausibility function of a hint $\mathcal{H}$. If I is any non-empty finite subset of Θ and C any cofinite subset of Θ, then*

$$(1)\quad pl(I) = \sum \{(-1)^{|J|+1} q(J) : \emptyset \neq J \subseteq I\} \tag{3.28}$$

$$(2)\quad sp(C) = \sum \{(-1)^{|J|} q(J) : J \subseteq C^c\}. \tag{3.29}$$

Note that by theorems 3.6 and 3.13 the complete support and plausibility functions are determined by the commonality numbers of finite subsets of Θ. In view of theorem 3.13, the next result shows that the support function and the plausibility function each completely determine the commonality function.

Theorem 3.15. *Let q denote the commonality function, sp the support function and pl the plausibility function of a hint $\mathcal{H}$. Then for any non-empty finite subset I of Θ we have*

$$(1)\quad q(I) = \sum \{(-1)^{|J|} sp(J^c) : J \subseteq I\} \tag{3.30}$$

$$(2)\quad q(I) = \sum \{(-1)^{|J|+1} pl(J) : J \subseteq I\} \tag{3.31}$$

In particular, $q(\theta) = pl(\theta)$ for all $\theta \in \Theta$.

Now let's see what the commonality function of particular classes of hints looks like. First, let $\mathcal{H} = (\Omega, P, \Gamma, \Theta)$ be a simple hint with $\Omega = \Omega_1 + \Omega_2$ and $\Gamma(\omega) = F$ for all $\omega \in \Omega_1$ and $\Gamma(\omega) = \Theta$ for all $\omega \in \Omega_2$. Then, for all $H \subseteq \Theta$, we have

$$q(H) = \begin{cases} 1 & \text{if } H \subseteq F \\ 1 - m(F) & \text{otherwise.} \end{cases} \tag{3.32}$$

Suppose now that $\mathcal{H}$ is a consonant hint with focal sets $F_1 \subseteq F_2 \subseteq \ldots \subseteq F_m$. It is convenient to define $F_0 = \emptyset$, and if $F_m \neq \Theta$ let $F_{m+1} = \Theta$. If $H \subseteq \Theta$, then H is always included in at least one F_i and hence

$$k = min \ \{i : H \subseteq F_i\} \tag{3.33}$$

is well defined. This implies that

$$q(H) = \sum_{i=k}^{m+1} m(F_i). \tag{3.34}$$

Furthermore, we have a result similar to theorem 3.7.

Theorem 3.16. *Let q be the commonality function of a consonant hint. If H, A and B are subsets of Θ, then*

$$(1) \quad q(H) = min \ \{q(\theta) : \theta \in H\} \tag{3.35}$$

$$(2) \quad q(A \cup B) = min \ \{q(A), q(B)\}. \tag{3.36}$$

Next let $\mathcal{H}$ be a Bayesian hint with $\Omega = \Omega_1 + \ldots + \Omega_n$ such that $\Gamma(\omega) = F_i, i = 1, \ldots, n$ for $\omega \in \Omega_i$ and the different focal sets F_i are mutually disjoint. This implies that for any non-empty set $H \subseteq \Theta$ there is at most one F_i such that $H \subseteq F_i$ and therefore

$$q(H) = \begin{cases} 1 & \text{if } H = \emptyset \\ m(F_i) & \text{if if } \emptyset \neq H \subseteq F_i \\ 0 & \text{otherwise.} \end{cases}$$

This result takes the following special form when the Bayesian hint is precise:

$$q(H) = \begin{cases} 1 & \text{if } H = \emptyset \\ m(\theta) & \text{if if } H = \{\theta\} \\ 0 & \text{otherwise.} \end{cases}$$

So the commonality number of any subset with more than one element is zero.

To conclude this section, let's mention that we will be in a position to better appreciate the importance and applicability of the commonality function when we discuss the combination of hints. This is precisely the subject of the next chapter.

A3 Appendix: Proofs of Theorems in Chapter 3

Proof of Theorem 3.1

Assertion (1) is trivial. By definition,

$$\begin{aligned} v(H)^c &= \{\omega \in \Omega : \Gamma(\omega) \cap H = \emptyset\} \\ &= \{\omega \in \Omega : \Gamma(\omega) \subseteq H^c\} = u(H^c). \end{aligned} \tag{A3.1}$$

The equality $u(H) = v(H^c)^c$ follows from the application of the equality $v(H) = u(H^c)^c$ to the hypothesis H^c. Assertions (3) and (4) are direct consequences of the definition of u and v. □

Proof of Theorem 3.3

First we prove assertion (1). If $\omega \in u(\cap\{H : H \in \mathcal{F}\})$, then $\Gamma(\omega) \subseteq \cap\{H : H \in \mathcal{F})$ and thus $\Gamma(\omega) \subseteq H$ for all $H \in \mathcal{F}$, which shows that $\omega \in \cap\{u(H) : H \in \mathcal{F})$. Conversely, if $\omega \in \cap\{u(H) : H \in \mathcal{F}\}$, then $\Gamma(\omega) \subseteq H$ for all $H \in \mathcal{F}$ and thus $\Gamma(\omega) \subseteq \cap\{H : H \in \mathcal{F}\}$, which shows that $\omega \in u(\cap\{H; H \in \mathcal{F}\})$.

Now we prove assertion (3). If $\omega \in \cup\{u(H) : H \in \mathcal{F}\}$, then $\Gamma(\omega) \subseteq H$ for some $H \in \mathcal{F}$ and thus $\Gamma(\omega) \subseteq \cup\{H : H \in \mathcal{F}\}$, which shows that $\omega \in u(\cup\{H : H \in \mathcal{F}\})$. According to theorem 3.1 (2) and the first assertion, we have

$$\begin{aligned} v(\cup\{H : H \in \mathcal{F}\}) &= u(\cap\{H^c : H \in \mathcal{F}\})^c = (\cap\{u(H^c) : H \in \mathcal{F}\})^c \\ &= \cup\{u(H^c)^c : H \in \mathcal{F}\} = \cup\{v(H) : H \in \mathcal{F}\} \end{aligned} \tag{A3.2}$$

and (2) is proved. Finally, theorem 3.1 (2) and assertion (3) imply

$$\begin{aligned} v(\cap\{H : H \in \mathcal{F}\}) &= u(\cup\{H^c : H \in \mathcal{F}\})^c \subseteq (\cup\{u(H^c) : H \in \mathcal{F}\})^c \\ &= \cap\{u(H^c)^c : H \in \mathcal{F}\} = \cap\{v(H) : H \in \mathcal{F}\} \end{aligned} \tag{A3.3}$$

and (4) is proved. □

Proof of theorem 3.4

Assertion (1): According to theorem 3.3 (1) and theorem 3.1 (4) we have

$$\begin{aligned} sp(\cap_{i=1}^{\infty} H_i) &= P(u(\cap_{i=1}^{\infty} H_i)) = P(\cap_{i=1}^{\infty} u(H_i)) \\ &= \lim_{i\to\infty} P(u(H_i)) = \lim_{i\to\infty} sp(H_i) \end{aligned} \tag{A3.4}$$

by the lower continuity property of the probability measure P (since P is discrete on the finite set Ω, the limit is attained for a finite n).

Assertion (2): According to theorem 3.3 (2) and theorem 3.1 (4), we have

$$\begin{aligned} pl(\cup_{i=1}^{\infty} H_i) &= P(v(\cup_{i=1}^{\infty} H_i)) = P(\cap_{i=1}^{\infty} v(H_i)) \\ &= \lim_{i\to\infty} P(v(H_i)) = \lim_{i\to\infty} pl(H_i) \end{aligned} \tag{A3.5}$$

by the upper continuity property of the probability measure P. □

Proof of theorem 3.5

Assertion (1): Using the well-known inclusion-exclusion formula (Feller, 1968) from probability theory together with theorem 3.2 (4) and theorem 3.3 (3) and (1), we have

$$
\begin{aligned}
sp(H) \geq sp(\cup_{i=1}^{n} H_i) &= P(u(\cup_{i=1}^{n} H_i)) \geq P(\cup_{i=1}^{n} u(H_i)) \\
&= \sum\{(-1)^{|I|+1} P(\cap_{i\in I} u(H_i)) : \emptyset \neq I \subseteq \{1,\ldots,n\}\} \qquad (A3.6) \\
&= \sum\{(-1)^{|I|+1} sp(\cap_{i\in I} H_i) : \emptyset \neq I \subseteq \{1,\ldots,n\}\}.
\end{aligned}
$$

Assertion (2): Similarly, using theorem 3.2 (4) and theorem 3.3 (4) and (2), we have

$$
\begin{aligned}
pl(H) \leq pl(\cap_{i=1}^{n} H_i) &= P(v(\cap_{i=1}^{n} H_i)) \leq P(\cap_{i=1}^{n} v(H_i)) \\
&\leq \sum\{(-1)^{|I|+1} P(\cup_{i\in I} v(H_i) : \emptyset \neq I \subseteq \{1,\ldots,n\}\} \qquad (A3.7) \\
&= \sum\{(-1)^{|I|+1} pl(\cup_{i\in I} H_i) : \emptyset \neq I \subseteq \{1,\ldots,n\}\}.
\end{aligned}
$$

□

Proof of theorem 3.6

Assertion (1): According to theorem 3.2 (4), we have

$$pl(H) \geq pl(I) \qquad (A3.8)$$

for all finite subset $I \subseteq H$. Now we show that $pl(H) = pl(H')$ for some finite subset $H' \subseteq \Theta$ such that $H' \subseteq H$, which implies (1). For each $\omega \in v(H)$, arbitrarily select one element $x(\omega)$ in $\Gamma(\omega) \cap H$ and let H' be this set of elements, i.e.

$$H' = \{x(\omega) \in \Theta : \omega \in v(H)\}. \qquad (A3.9)$$

Then of course H' is finite because Ω is finite, and $H' \subseteq H$. But $v(H') \subseteq v(H)$ and $\omega \in v(H)$ implies that $x(\omega)$ exists and lies in $H' \cap \Gamma(\omega)$, which shows that $\omega \in v(H')$. Thus $v(H') = v(H)$ and $pl(H') = pl(H)$.

Assertion (2): We can write

$$
\begin{aligned}
sp(H) &= 1 - pl(H^c) \\
&= 1 - max\{pl(I) : I \ finite,\ I \subseteq H^c\} \\
&= min\ \{1 - pl(I) : I \ finite,\ I \subseteq H^c\} \qquad (A3.10) \\
&= min\ \{sp(I^c) : I \ finite,\ H \subseteq I^c\} \\
&= min\ \{sp(C) : C \ cofinite,\ H \subseteq C\}.
\end{aligned}
$$

□

Proof of theorem 3.7

Assertion (1): This assertion is clearly satisfied when H is empty. Since $pl(H) \geq sup\{pl(\theta) : \theta \in H\}$ by theorem 3.2 (4), it remains to prove that there exists $\theta \in H$ such that $pl(H) = pl(\theta)$. If $v(H) = \emptyset$, then $pl(H) = 0$ and $pl(\theta) = 0$ for any $\theta \in H$. Otherwise, if $v(H) \neq \emptyset$, let $F = \cap\{\Gamma(\omega) : \omega \in v(H)\}$. Since the hint is consonant, there is $\omega' \in v(H)$ such that $F = \Gamma(\omega')$.

Arbitrarily select one element $\theta \in \Gamma(\omega') \cap H$. Now we show that $v(\theta) = v(H)$, which implies that $pl(\theta) = pl(H)$. In fact, if $\omega \in v(H)$, then $\theta \in \Gamma(\omega)$ by construction. But $\theta \in H$ and so $\theta \in \Gamma(\omega) \cap H$ which proves that $\omega \in v(H)$. On the other hand, if $\omega_i \in v(H)$, then θ is in

$$F = \cap\{\Gamma(\omega) : \omega \in v(H)\} \subseteq \Gamma(\omega_i) \tag{A3.11}$$

and hence $\theta \in \Gamma(\omega_i)$, which shows that $\omega_i \in v(\theta)$.

Assertion (2): The inequality $pl(A \cup B) \geq max\{pl(A), pl(B)\}$ holds by theorem 3.2 (4). According to the first assertion, there is $\theta \in A \cup B$ such that $pl(A \cup B) = pl(\theta)$. If θ happens to be within A, then

$$\begin{aligned} pl(A \cup B) &= pl(\theta) \leq max\{pl(\theta) : \theta \in A\} \\ &= pl(A) \leq max\{pl(A), pl(B)\}. \end{aligned} \tag{A3.12}$$

If θ is in B, then we can also prove similarly that $pl(A \cup B) \leq max\{pl(A), pl(B)\}$. Since θ is either in A or B, the assertion is proved.

Assertion (3): We have

$$\begin{aligned} sp(A \cap B) &= 1 - pl(A^c \cup B^c) = 1 - max\{pl(A^c), pl(B^c)\} \\ &= min\{1 - pl(A^c), 1 - pl(B^c)\} = min\{sp(A), sp(B)\}. \end{aligned} \tag{A3.13}$$

□

Proof of theorem 3.8

Assertion (1): Since focal sets are singletons, $u(H) = v(H)$ and hence $sp(H) = pl(H)$ for all $H \subseteq \Theta$.

Assertion (2): Relation (i) is trivial and (ii) holds according to theorem 3.2 (1). We can show (3.10) for pl instead of sp according to the first assertion of the theorem. Theorem 3.3 (2) and the σ-additivity of probability measures imply that

$$\begin{aligned} pl(\cup_{i=1}^{\infty} A_i) &= P(v(\cup_{i=1}^{\infty} A_i)) = P(\cup_{i=1}^{\infty} v(A_i)) \\ &= \sum_{i=1}^{\infty} P(v(A_i)) = \sum_{i=1}^{\infty} pl(A_i) \end{aligned} \tag{A3.14}$$

because the $v(A_i), i = 1, 2, \ldots$ are disjoint. The proof is similar when there is only a finite number of sets A_i. □

Remark.

The proofs of theorems 3.9 and 3.10 are borrowed and adapted from Shafer, 1976 a. The next three lemmas are well known and will be useful to prove these theorems (see Shafer, 1976 a).

Lemma A3-1. *If A is a finite set, then*

$$\sum_{B \subseteq A} (-1)^{|B|} = \begin{cases} 1 & \textit{if } A = \emptyset \\ 0 & \textit{otherwise.} \end{cases} \tag{A3.15}$$

Proof of Lemma A3-1

By the binomial theorem,

$$\binom{n}{0} - \binom{n}{1} + \binom{n}{2} - + \ldots + (-1)^n \binom{n}{n} = (1-1)^n = 0 \tag{A3.16}$$

whenever n is a positive integer. It follows that whenever $A = \{\theta_1, \ldots, \theta_n\}$ is a finite non-empty set,

$$\begin{aligned} \sum_{B \subseteq A} (-1)^{|B|} &= (-1)^{|\emptyset|} + \sum_i (-1)^{|\{\theta_i\}|} + \sum_{i<j} (-1)^{|\{\theta_i, \theta_j\}|} + \ldots + (-1)^{|A|} \\ &= \binom{n}{0} - \binom{n}{1} + \binom{n}{2} - + \ldots + (-1)^n \binom{n}{n} = 0. \end{aligned} \tag{A3.17}$$

And of course, when A is empty,

$$\sum_{B \subseteq A} (-1)^{|B|} = (-1)^{|A|} = (-1)^0 = 1. \tag{A3.18}$$

□

Lemma A3-2. *If A is a finite set and $B \subseteq A$, then*

$$\sum_{B \subseteq C \subseteq A} (-1)^{|C|} = \begin{cases} (-1)^{|A|} & \textit{if } A = B \\ 0 & \textit{otherwise.} \end{cases} \tag{A3.19}$$

Proof of Lemma A3-2

This lemma is seen to follow from lemma A3-1 once it is noticed that

$$\sum_{B \subseteq C \subseteq A} (-1)^{|C|} = \sum_{D \subseteq (A-B)} (-1)^{|B \cup D|} = (-1)^{|B|} \sum_{D \subseteq (A-B)} (-1)^{|D|} \tag{A3.20}$$

□

According to Shafer (1976 a), the following lemma is an example of a general combinatorial phenomena that G.C. Rota (1964) christened Möbius inversion. See also page 15 of Hall's Combinatorial Theory (1967).

Lemma A3-3. *Suppose that Θ is a finite set and f and g are functions on $\mathcal{P}(\Theta)$. Then*

$$f(A) = \sum_{B \subseteq A} g(B) \tag{A3.21}$$

for all $A \subseteq \Theta$ if and only if

$$g(A) = \sum_{B \subseteq A} (-1)^{|A-B|} f(B) \tag{A3.22}$$

for all $A \subseteq \Theta$.

Proof of Lemma A3-3

Both implications follow by simple calculations using lemma A3-2. (i). If (A3.21) is satisfied for all $A \subseteq \Theta$, then

$$\begin{aligned}
\sum_{B \subseteq A} (-1)^{|A-B|} f(B) &= (-1)^{|A|} \sum_{B \subseteq A} (-1)^{|B|} f(B) \\
&= (-1)^{|A|} \sum_{B \subseteq A} (-1)^{|B|} \sum_{C \subseteq B} g(C) \\
&= (-1)^{|A|} \sum_{C \subseteq A} g(C) \sum_{C \subseteq B \subseteq A} (-1)^{|B|} \\
&= (-1)^{|A|} g(A) (-1)^{|A|} = g(A).
\end{aligned} \tag{A3.23}$$

On the other hand, if (A3.22) is satisfied for all $A \subseteq \Theta$, then

$$\begin{aligned}
\sum_{B \subseteq A} g(B) &= \sum_{B \subseteq A} \sum_{C \subseteq B} (-1)^{|B-C|} f(C) \\
&= \sum_{C \subseteq A} (-1)^{|C|} f(C) \sum_{C \subseteq B \subseteq A} (-1)^{|B|} \\
&= (-1)^{|A|} f(A) (-1)^{|A|} = f(A).
\end{aligned} \tag{A3.24}$$

□

Proof of theorem 3.9

According to (3.17) we have

$$sp(H) = \sum_{F \subseteq H} m(F) \tag{A3.25}$$

Since sp and m are both functions on $\mathcal{P}(\Theta)$, lemma A3-3 implies that

$$m(A) = \sum_{B \subseteq A} (-1)^{|A-B|} sp(B) \tag{A3.26}$$

for all $A \subseteq \Theta$. □

Proof of theorem 3.10

Let $\varphi : \mathcal{P}(\Theta) \rightarrow [0,1]$ be a Choquet capacity of infinite order, i.e. φ is monotone of order infinity and $\varphi(\emptyset) = 0, \varphi(\Theta) = 1$. Define the function $p : \mathcal{P}(\Theta) \rightarrow \mathcal{R}$ by

$$p(A) = \sum_{B \subseteq A} (-1)^{|A-B|} \varphi(B). \tag{A3.27}$$

Then we show that p is a probability density on $\mathcal{P}(\Theta)$, i.e. $p(\emptyset) = 0, p(A) \geq 0$ for all $A \in \mathcal{P}(\Theta)$ and $\sum_{A \subseteq \Theta} = 1$. By lemma A3-3 above, we can write

$$\varphi(A) = \sum_{B \subseteq A} p(B) \tag{A3.28}$$

for all $A \subseteq \Theta$. This implies $p(\emptyset) = 0$ and $\sum_{A \subseteq \Theta} p(A) = 1$. To see that $p(A) \geq 0$ whenever $A \neq \emptyset$, let $A = \{\theta_1, \ldots, \theta_n\}$, where $n \geq 1$ and the θ_i are distinct. Also set $A_i = A - \{\theta_i\}$, so that $A_1, \ldots, A_n$ are precisely the subsets of A that omit exactly one element of A. Then every proper subset B of A can be uniquely expressed as an intersection of the A_i; indeed, if $A - B = \{\theta_{i1}, \ldots, \theta_{ik}\}$, then $B = A_{i1} \cap \ldots \cap A_{ik}$. Thus

$$\begin{aligned} p(A) &= \sum_{B \subseteq A} (-1)^{|A-B|} \varphi(B) \\ &= \varphi(A) + \sum \{(-1)^{|I|} \varphi(\cap_{i \in I} A_i) : \emptyset \neq I \subseteq \{1, \ldots, n\}\} \\ &= \varphi(A) - \sum \{(-1)^{|I|+1} \varphi(\cap_{i \in I} A_i) : \emptyset \neq I \subseteq \{1, \ldots, n\}\}. \end{aligned} \tag{A3.29}$$

This is non-negative because φ is monotone of infinite order. This permits to define the hint $\mathcal{H} = (\mathcal{F}, P, id, \Theta)$ where $\mathcal{F} = \{A \subseteq \Theta : p(A) > 0\}$ and $P(X) = \sum_{A \in X} p(A)$ for all $X \subseteq \mathcal{F}$. If m and sp are the bpa and the support function of this hint, then $m(A) = p(A)$ for all $A \subseteq \Theta$ by definition and so

$$sp(H) = \sum_{B \subseteq H} m(B) = \sum_{B \subseteq H} p(B) = \varphi(H) \tag{A3.30}$$

by equation (A3.28). Hence $\varphi = sp$ and φ is the support function of the hint $\mathcal{H}$. □

Proof of theorem 3.11

According to theorem 3.10 there exists a hint $\mathcal{H}$ in $\mathcal{G}$ and we show that $\mathcal{G}$ is the equivalence class $[\mathcal{H}]$ of the hint $\mathcal{H}$. Let $\mathcal{H}'$ be a hint in $\mathcal{G}$. According to theorem 3.9 and using self-explanatory notations we can write

$$\begin{aligned} m_{\mathcal{H}'}(A) &= \sum_{B \subseteq A} (-1)^{|A-B|} sp_{\mathcal{H}'}(B) \\ &= \sum_{B \subseteq A} (-1)^{|A-B|} \varphi(B) \\ &= \sum_{B \subseteq A} (-1)^{|A-B|} sp_{\mathcal{H}}(B) = m_{\mathcal{H}}(A) \end{aligned} \tag{A3.31}$$

and hence $\mathcal{H}' \in [\mathcal{H}]$.

Conversely, if $\mathcal{H}'$ is in $[\mathcal{H}]$, then $m_{\mathcal{H}'} = m_{\mathcal{H}}$ which implies that $sp_{\mathcal{H}'} = sp_{\mathcal{H}}$ by equation (3.16). Since $sp_{\mathcal{H}} = \varphi$ we have $sp_{\mathcal{H}'} = \varphi$ and hence $\mathcal{H}'$ is in $\mathcal{G}$, which proves that $\mathcal{G} = [\mathcal{H}]$.

To prove that $(\mathcal{F}, m, id, \Theta)$ is the canonical representative of the class $[\mathcal{G}]$, it must be shown that m is the bpa characterizing this class. Since $\mathcal{H} \in \mathcal{G}$, it is sufficient to verify that $m(A) = m_{\mathcal{H}}(A)$ for all $A \subseteq \Theta$, which is true because

$$m(A) = \sum_{B \subseteq A} (-1)^{|A-B|} \varphi(B) = \sum_{B \subseteq A} (-1)^{|A-B|} sp_{\mathcal{H}}(B) = m_{\mathcal{H}}(A). \quad (A3.32)$$

□

Proof of theorem 3.12

Suppose that $M\ ct\ A$. Then $M \subseteq u(A)$ and thus $\Gamma(\omega) \subseteq A$ for all $\omega \in M$ and therefore $\Gamma(M) \subseteq A$. This implies that $\Gamma(M) \subseteq \cap\{A : M\ ct\ A\}$. On the other hand, $M\ ct\ \Gamma(M)$ because $\Gamma(\omega) \subseteq \Gamma(M)$ for all $\omega \in M$. This shows that $\cap\{A : M\ ct\ A\} \subseteq \Gamma(M)$ and the theorem is proved. □

Proof of theorem 3.13

First, note that the assertion is clearly proved when H is empty and of course $q(\emptyset) = 1$. So suppose that H is not empty. Since $A \subseteq B$ implies $q(A) \geq q(B)$, we have

$$q(H) \leq q(I) \quad (A3.33)$$

for all finite subsets $I \subseteq H$. Now we show that there always exists a finite set $X \subseteq H$ such that $w(H) = w(X)$, which proves the theorem. If $w(H) = \Omega$, then arbitrarily select an element $x \in H$. If $X = \{x\}$, then of course $w(H) = w(X)$. If $w(H) \neq \Omega$, then for all $\omega \in \Omega - w(H)$ there exists an element $x(\omega) \in H$ such that $x(\omega) \notin \Gamma(\omega)$. Define $X = \{x(\omega) : \omega \in \Omega - w(H)\}$. Then $X \subseteq H$ and $X \not\subseteq \Gamma(\omega)$ for all $\omega \in \Omega - w(H)$ because $x(\omega) \notin \Gamma(\omega)$. This implies that $w(X) = w(H)$ and the theorem is proved. □

Proof of theorem 3.14

First we prove the equality (1). Using lemma A3-1 we find that

$$\begin{aligned}
&\sum\{(-1)^{|J|+1} q(J) : \emptyset \neq J \subseteq I\} = \sum_{J \subseteq I} (-1)^{|J|+1} q(J) - (-1) \\
&= 1 - \sum_{J \subseteq I} (-1)^{|J|} q(J) = 1 - \sum_{J \subseteq I} \{(-1)^{|J|} \cdot \sum_{J \subseteq K} m(K)\} \\
&= 1 - \sum_{K \subseteq \Theta} \{m(K) \cdot \sum_{J \subseteq K \cap I} (-1)^{|J|}\} = 1 - \sum_{K \cap I = \emptyset} m(K) \\
&= 1 - \sum_{K \subseteq I^c} m(K) = 1 - sp(I^c) = pl(I).
\end{aligned} \quad (A3.34)$$

Now we prove (2). If $C = \Theta$, the assertion clearly holds. So let $C \neq \Theta$. Using the first assertion we have

$$\begin{aligned} sp(C) &= 1 - pl(C^c) = 1 - \sum_{\emptyset \neq J \subseteq C^c} (-1)^{|J|+1} q(J) \\ &= q(\emptyset) + \sum_{\emptyset \neq J \subseteq C^c} (-1)^{|J|} q(J) = \sum_{J \subseteq C^c} (-1)^{|J|} q(J). \end{aligned} \tag{A3.35}$$

□

In order to prove theorem 3.15, we need two more lemmas.

Lemma A3-4. *Let f and g be real functions on $\mathcal{P}(\Theta)$. Then*

$$f(A) = \sum_{B \subseteq A} (-1)^{|B|+1} g(B) \tag{A3.36}$$

for all finite subsets $A \subseteq \Theta$ if and only if

$$g(A) = \sum_{B \subseteq A} (-1)^{|B|+1} f(B) \tag{A3.37}$$

for all finite subsets $A \subseteq \Theta$.

Proof of lemma A3-4

If (A3.36) holds, then by lemma A3-2 we have for any finite subset A of Θ

$$\begin{aligned} \sum_{B \subseteq A} (-1)^{|B|+1} f(B) &= \sum_{B \subseteq A} \{(-1)^{|B|+1} \cdot \sum_{C \subseteq B} (-1)^{|C|+1} g(C)\} \\ &= \sum_{C \subseteq A} (-1)^{|C|} g(C) \cdot \{ \sum_{C \subseteq B \subseteq A} (-1)^{|B|}\} = g(A). \end{aligned} \tag{A3.38}$$

The other implication is proved by inverting the roles of f and g in the above development. □

Lemma A3-5. *Let f and g be real functions on $\mathcal{P}(\Theta)$. Then*

$$f(A) = \sum_{B \subseteq A^c} (-1)^{|B|} g(B) \tag{A3.39}$$

for all cofinite subsets $A \subseteq \Theta$ if and only if

$$g(A) = \sum_{B \subseteq A} (-1)^{|B|} f(B^c) \tag{A3.40}$$

for all finite subsets $A \subseteq \Theta$.

Proof of lemma A3-5

Let h denote the function on $\mathcal{P}(\Theta)$ given by $h(A) = -f(A^c)$. (i). Suppose that (A3.39) holds. Then for any finite set $A \subseteq \Theta$ we have

$$h(A) = -f(A^c) = \sum_{B \subseteq A} (-1)^{|B|+1} g(B). \tag{A3.41}$$

By lemma A3-4, for any subset $A \subseteq \Theta$, we have :

$$g(A) = \sum_{B \subseteq A} (-1)^{|B|+1} h(B) = \sum_{B \subseteq A} (-1)^{|B|} f(B^c) \tag{A3.42}$$

and (A3.40) is proved.

(ii). Suppose that (A3.40) holds and let A be a finite subset of Θ. Then

$$g(A) = \sum_{B \subseteq A} (-1)^{|B|} f(B^c) = \sum_{B \subseteq A} (-1)^{|B|+1} h(B) \tag{A3.43}$$

and by lemma A3-4 the following equations are equivalent for any finite subset A of Θ:

$$\begin{aligned} h(A) &= \sum_{B \subseteq A} (-1)^{|B|+1} g(B) \\ -f(A^c) &= \sum_{B \subseteq A} (-1)^{|B|+1} g(B) \\ f(A^c) &= \sum_{B \subseteq A} (-1)^{|B|} g(B). \end{aligned} \tag{A3.44}$$

□

Proof of theorem 3.15

(1): Theorem 3.13 and lemma A3-5 imply that

$$q(I) = \sum_{J \subseteq I} (-1)^{|J|} sp(J^c) \tag{A3.45}$$

for all finite subsets $I \subseteq \Theta$. (2): Since I is a finite non-empty subset of Θ, equation (3.30) and lemma A3-1 imply that

$$\begin{aligned} q(I) &= \sum_{J \subseteq I} (-1)^{|J|} sp(J^c) = \sum_{J \subseteq I} (-1)^{|J|} (1 - pl(J)) \\ &= \sum_{J \subseteq I} \{(-1)^{|J|} + (-1)^{|J|+1} pl(J)\} \\ &= \sum_{J \subseteq I} (-1)^{|J|+1} pl(J) \end{aligned} \tag{A3.46}$$

and (3.31) is proved. □

Proof of theorem 3.16

(1). The assertion is clearly true when H is empty because by convention $\emptyset \in \emptyset$. Since $A \subseteq B$ implies $q(A) \geq q(B)$, we have

$$q(H) \leq q(\theta) \tag{A3.47}$$

for all $\theta \in H$. If k is defined as in (3.33), then there exists $x \in H$ such that $x \notin F_{k-1}$. But the consonance of the hint implies that $x \notin F_i$ for all $i \leq k-1$. This shows that $w(H) = w(x)$ and hence $q(H) = q(x)$, which proves the assertion.

(ii). We obviously have $q(A \cup B) \leq min\ \{q(A), q(B)\}$ because q is a decreasing function. According to the first assertion, there is $\theta \in A \cup B$ such that $q(A \cup B) = q(\theta)$. If θ happens to be within A, then

$$q(A \cup B) = q(\theta) \geq min\ \{q(\theta) : \theta \in A\} = q(A) \geq min\{q(A), q(B)\} \tag{A3.48}$$

If θ is in B, then we can prove in a similar way that $q(A \cup B) \geq min\ (q(A), q(B))$. Since θ is either in A or B, the assertion is proved. □

4. Combining Hints

4.1 Dempster's Rule

If there are several hints relative to the same question and hence relative to the same frame of discernment Θ, then the problem of combining the information contained in each hint arises. This is the very basic operation to be defined in the theory of hints. Its intuitive idea is very simple and logically follows from the semantics given to the four elements defining a hint. Recall that for each hint to be combined there is exactly one correct interpretation and the answer to the question considered necessarily lies in its corresponding focal set. The vector formed by these correct interpretations is called the correct *combined* interpretation. Whereas there are several vectors of interpretations (combined interpretations) which are obtained by selecting an interpretation for each hint, there is exactly one *correct* combined interpretation. Then, of course, the answer to the question considered must be in the *intersection* of the focal sets associated with the individual interpretations forming this combined interpretation. Therefore, a combined interpretation leading to an empty intersection is certainly not correct and it is called *contradictory.* Contradictory interpretations represent a new information for the determination of the correct combined interpretation. This information must be integrated in the original knowledge available for this question, knowledge which is described by the probability that an arbitrary combined interpretation is correct. Of course, the integration is realized by *conditioning* this probability on non-contradictory interpretations.

Let's formalize these ideas. Let $\mathcal{H}_i = (\Omega_i, P_i, \Gamma_i, \Theta)$, $i = 1, \ldots, n$ be a family of hints relative to a common frame of discernment Θ. Then $\Omega_1 \times \cdots \times \Omega_n$ represents the set of all *combined interpretations*. The hints are called *independent* if, before conditioning, the probability that a combined interpretation $(\omega_1, \ldots, \omega_n), \omega_i \in \Omega_i$ is correct is given by the product $\prod_{i=1}^n p_i(\omega_i)$. This is the case when the correct interpretations of the hints to be combined (considered as random variables) are stochastically independent. If the hints $\mathcal{H}_1, \ldots \mathcal{H}_n$ are independent, then

$$k = \sum \{\prod_{i=1}^{n} p_i(\omega_i) : \cap_{i=1}^{n} \Gamma_i(\omega_i) = \emptyset\} \tag{4.1}$$

is the probability that a combined interpretation is contradictory. Since we have to condition on the fact that the correct combined interpretation is not condradictory, let Ω denote the set of non-contradictory interpretations. Ω is the set of possible combined interpretations. Define

$$p(\omega) = \Big(\prod_{i=1}^{n} p_i(\omega_i)\Big)/(1-k) \tag{4.2}$$

for all $\omega = (\omega_i, \ldots, \omega_n) \in \Omega$. Note that k must be different from 1 when each Ω_i is finite because there exists at least one possible combined interpretation, namely the correct one. If k is nevertheless equal to 1, then this indicates that the hypotheses underlying the definition of a hint have not been respected and the hints cannot be combined. For each $\omega = (\omega_1, \ldots, \omega_n) \in \Omega$, let

$$\Gamma(\omega) = \cap_{i=1}^{n} \Gamma_i(\omega_i) \tag{4.3}$$

represent the non-empty focal set associated to ω. If P is the probability measure defined by (4.2), then $\mathcal{H} = (\Omega, P, \Gamma, \Theta)$ is a new hint relative to Θ. It is called the *combination* of $\mathcal{H}_1, \ldots, \mathcal{H}_n$ and is denoted by

$$\mathcal{H}_1 \oplus \cdots \oplus \mathcal{H}_n. \tag{4.4}$$

This procedure to combine independent hints is called *Dempster's rule of combination* because it was introduced by A.P. Dempster (1967) in the context of multivalued mappings. However, note that it can be generalized to combine dependent hints: simply replace the product probability measure on $\Omega_1 \times \cdots \times \Omega_n$ by a probability measure reflecting the stochastic dependencies between the correct interpretations of the hints to be combined. Dempster's rule can also be formulated in terms of the bpas of the hints involved as the following theorem shows.

Theorem 4.1. *Let $\mathcal{H}_1, \ldots, \mathcal{H}_n$ be a family of hints relative to the frame Θ. If m_i is the bpa of the hint $\mathcal{H}_i$ and m is the bpa of the combined hint $\mathcal{H} = \oplus_{i=1}^{n} \mathcal{H}_i$, then*

$$m(A) = \sum \{\prod_{i=1}^{n} m_i(A_i) : A_i \subseteq \Theta, \cap_{i=1}^{n} A_i = A\}/(1-k) \tag{4.5}$$

for all non-empty subsets A of Θ and

$$k = \sum \{\prod_{i=1}^{n} m_i(A_i) : A_i \subseteq \Theta, \cap_{i=1}^{n} A_i = \emptyset\}. \tag{4.6}$$

The two sums in (4.5) and (4.6) are finite because $m_i(A_i)$ is different from zero only when A_i is a focal set of $\mathcal{H}_i$ and there is only a finite number of them. The following three elementary results about Dempster's rule are immediate consequences of this theorem.

Corollary 4.1. *The Dempster's rule of combination is a class operation. In other words, if $\mathcal{H}_1 \simeq \mathcal{H}_1'$ and $\mathcal{H}_2 \simeq \mathcal{H}_2'$, then $\mathcal{H}_1 \oplus \mathcal{H}_2 \simeq \mathcal{H}_1' \oplus \mathcal{H}_2'$.*

Corollary 4.2. *The combination of hints by Dempster's rule is both commutative and associative. In other words, if $\mathcal{H}_i, i = 1, \ldots, 3$ are three hints relative to the same frame, then $\mathcal{H}_1 \oplus \mathcal{H}_2 \simeq \mathcal{H}_2 \oplus \mathcal{H}_1$ and $\mathcal{H}_1 \oplus (\mathcal{H}_2 \oplus \mathcal{H}_3) \simeq (\mathcal{H}_1 \oplus \mathcal{H}_2) \oplus \mathcal{H}_3$.*

This implies that the sequence used to combine several hints is irrelevant, a result which is expected since usually there is no natural order among the hints: they all just represent a piece of information about the question under investigation.

Corollary 4.3. *A vacuous hint is a neutral element with respect to Dempster's rule of combination. In other words, if $\mathcal{H}_0$ is a vacuous hint with respect to the frame Θ and $\mathcal{H}$ is an arbitrary hint relative to Θ, then $\mathcal{H}_0 \oplus \mathcal{H} \simeq \mathcal{H}$.*

Let $\mathcal{H}_i = (\Omega_i, P_i, \Gamma_i, \Theta), i = 1, 2$ be two hints relative to the same frame Θ and let m_i, sp_i, pl_i denote their bpa, support function and plausibility function respectively. If m, sp and pl represent these functions for the combined hint $\mathcal{H}_1 \oplus \mathcal{H}_2$, then we have the following result:

Theorem 4.2. *There exists a constant c such that*

$$sp(H) = c \cdot \sum_{\omega_1 \in \Omega_1} p_1(\omega_1) sp_2(H \cup \Gamma_1(\omega_1)^c) - (c-1) \tag{4.7}$$

$$pl(H) = c \cdot \sum_{\omega_1 \in \Omega_1} p_1(\omega_1) pl_2(H \cap \Gamma_1(\omega_1)) \tag{4.8}$$

for all $H \subseteq \Theta$. This constant c equals $(1-k)^{-1}$ where k is the probability of contradictory interpretations given in (4.6). Moreover, we have

$$1 - k = \sum_{\omega_1 \in \Omega_1} p_1(\omega_1) pl_2(\Gamma_1(\omega_1)). \tag{4.9}$$

Since the Dempster's rule is commutative, symmetric formulas are obtained by exchanging the subscripts 1 and 2 in the formulas (4.7), (4.8) and (4.9).

All examples from chapters 1 and 2 involving combination of information can be analyzed from the Dempster's rule point of view. Three of them will now be presented in more details.

Example 4.1. Gas tank, continued. This is a continuation of example 2.1. Besides the hint coming from the gas gauge, we dispose of a second hint from the mileage counter. Knowing the car's consumption, you can compute the distance d you can drive with a completely full tank. On the other hand, each time your car is refueled, the driver usually resets the mileage counter to zero but sometimes he forgets to do it. Suppose that your mileage counter is currently indicating a larger distance than d. This information can be represented by a hint with two different interpretations. In the first interpretation ω'_1, it is supposed that the mileage counter was indeed reset to zero at the last tank filling, whereas the interpretation ω'_2 assumes that this is not the case. This implies that the focal sets corresponding to ω'_1 and ω'_2 are $\{e\}$ and Θ respectively. Suppose that on the average seven times out of ten the mileage counter is indeed reset to zero at tank filling. If the reliability of the gauge is known to be 0.99 (i.e. 99% of the time the gauge is functioning correctly), then the two hints coming from the gas gauge and the mileage counter can be combined by Dempster's rule.

The four combined interpretations $(\omega_1, \omega'_1), (\omega_1, \omega'_2), (\omega_2, \omega'_1), (\omega_2, \omega'_2)$ are not contradictory and their respective probabilities are 0.693, 0.297, 0.007 and 0.003. The first three interpretations support the hypothesis $\{e\}$ and therefore its degree of credibility equals 0.997. The combined evidence very strongly supports the hypothesis that the gas tank is empty. Furthermore, since there is no interpretation supporting the hypothesis $\{f\}$, its degree of credibility is 0 and the plausibility of $\{e\}$ is therefore equal to 1. Finally, the plausibility of a non-empty tank is very small, namely 0.003.

In another situation, suppose that the mileage counter indicates a distance which is below the limit d. This information can again be represented by a hint having two different interpretations. The first interpretation ω'_1 assumes that the counter correctly indicates the driven distance since the last tank filling. In this case, the tank is necessarily not empty and the corresponding focal set is therefore equal to $\{f\}$. In the second interpretation, it is assumed that the counter does not indicate the correct driven distance since the last tank filling because someone reset it afterwards or for any other reason. The focal set associated to this interpretation is of course the whole frame Θ. If we suppose that the probability of each interpretation is 0.5, then we have a hint which can be combined with the hint coming from the gas gauge.

Since the combined interpretation (ω_1, ω'_1) having probability 0.495 is contradictory, the probability of the other three interpretations must be divided by 0.505 corresponding to a conditioning on non-contradictory interpretations. Table 4.1 shows the interpretations, the probabilities and the focal sets of the combined hint. The degree of support and the degree of plausibility of hypotheses can easily be computed from Tab. 4.1. In particular, the degree of support that the tank is empty equals 0.9802, which is still quite a large value in spite of the fact that the evidence coming from the mileage counter contradicts the evidence from the gas gauge (recall that this degree of support is 0.997 when both hints speak in favour of an empty tank, i.e. when the gas gauge reads empty and the counter indicates a distance larger than d). Similarly, we see that the degrees of support and plausibility practically do no change when the counter indicates a value which is either larger or smaller than d.

Example 4.2. Strike on the Metro, continued. This is a continuation of the example 2.2. There are two hints, one coming from the newspaper and another from the statistical information on the frequency of strikes on the Paris Metro

Table 4.1. The combined hint when the mileage counter indicates a distance smaller than d

Interpretation	Focal set	Probability
$(\omega 1, \omega 2')$	{e}	0.9802
$(\omega 2, \omega 1')$	{f}	0.0099
$(\omega 2, \omega 2')$	Θ	0.0099

in autumn. Let $\Omega = \{\omega_1, \omega_2\}$ and $\Omega' = \{\omega'_1, \omega'_2\}$ be the set of interpretations of the first and second hint respectively. Let ω_1 represent the possibility that the newspaper is from this Sunday and suppose its probability equals 0.8. Similarly, let ω'_1 represent the possibility that the Monday you arrive in Paris is indeed a strike day and assume that, on the average, there are three strike days a month on the Metro in autumn. This implies that the probability of ω'_1 is 0.1. The interpretations ω_2 and ω'_2 representing the alternative interpretation for each hint have probabilities 0.2 and 0.9 respectively.

The combined interpretation (ω_1, ω'_2) is contradictory and its probability equals $0.8 \cdot 0.9 = 0.72$. This indicates that the two hints are rather contradictory. The remaining possible interpretations of the combined hint are thus $\{(\omega_1, \omega'_1), (\omega_2, \omega'_1), (\omega_2, \omega'_2)\}$ and their probabilities are 0.286, 0.071 and 0.643 respectively. Furthermore, the focal sets associated with the first two interpretations are both equal to $\{s\}$, whereas the focal set of the last interpretation is $\{\neg s\}$. This defines the combined hint which can be used to compute the degrees of support and plausibility of hypotheses as shown in Chap. 3. The degree of support of $\{s\}$, which is also its degree of plausibility, is equal to 0.357. Similarly, the degree of support of $\{\neg s\}$, which is also its degree of plausibility, is given by 0.643. You are rather inclined to be optimistic and hope that there is no strike, although you are not too sure about it.

Example 4.3. Sensor model. Consider again the sensor model with two possible kinds of error described in Sect. 1.2. We have seen in Sect. 2.3 that this model can be described by two hints $\mathcal{H}_1$ and $\mathcal{H}_2$ on the frame $\Theta = \Theta_h \times \Theta_e$ where $\Theta_h = \{h, \neg h\}$ and $\Theta_e = \{e, \neg e\}$. These two hints have to be combined in order to get an overall hint $\mathcal{H}'$ representing the functioning mechanism of the sensor. It can easily be verified that the focal sets associated with the interpretations (a_1, a_2) and $(\neg a_1, a_2)$ are both equal to $\{(h, e), (\neg h, e)\}$, whereas the focal sets associated with $(a_1, \neg a_2)$ and $(\neg a_1, \neg a_2)$ are $\{(h, e), (\neg h, \neg e)\}$ and $\Theta - \{(\neg h, e)\}$ respectively.

This hint represents the structural description of the sensor. But there may also be the evidential information coming from the observation e, information which is represented by the deterministic hint $\mathcal{H}_e$. These two kinds of information must be combined to obtain a global image of the knowledge available. So we combine $\mathcal{H}'$ with $\mathcal{H}_e$ by Dempster's rule. Since the single focal set of $\mathcal{H}_e$ is $\{(h,e),(\neg h,e)\}$, it is easy to see (using theorem 4.1) that $\mathcal{H}' \oplus \mathcal{H}_e$ is equivalent to a hint with focal sets $\{(h,e),(\neg h,e)\}$, $\{(h,e)\}$ and bpa $m(\{(h,e),(\neg h,e)\}) = p_2(a_2), m(\{(h,e)\}) = 1 - p_2(a_2)$. This implies of course that the support of h is $1 - p_2(a_2)$, which is the same result as obtained in Sect. 1.2.

Now suppose that we observe $\neg e$ instead of e. This information is represented by the deterministic hint $\mathcal{H}_{\neg e}$ which must be combined with $\mathcal{H}'$. Again using theorem 4.1, we can see that the combined hint has focal sets $\{(\neg h, \neg e)\}$ and $\{(h,\neg e),(\neg h,\neg e)\}$ with bpa $p_1(a_1)$ and $p_1(\neg a_1)$ respectively. This implies that the support of $\neg h$ is $p_1(a_1)$, which is again the same result as obtained in Sect. 1.2.

This last example illustrates the fact that on some occasions it is necessary to extend a focal set from its original and natural frame of discernment to another one. This is the case for the hint representing the observation e. The natural frame of discernment for this information is of course Θ_e, but in order to be combined with the structural knowledge it was necessary to extend it to the frame Θ. In this small example the extension procedure does not require a big effort, but when there are many hints defined on different frames it becomes an important problem. Since quite often information is organized into small pieces pertaining to different frames of discernment, there is an obvious need for a more elaborated theory for combining hints. This kind of questions will be discussed extensively in chapters 7 and 8.

Before we pass to the relation between Dempster's rule and commonality functions, it is worthwhile to examine the combination of a few special cases of hints.

(1) Homogeneous hints. Consider n simple hints $\mathcal{H}_i = (\Omega_i, P_i, \Gamma_i, \Theta)$ with $\Omega_i = \Omega'_i + \Omega''_i$ and

$$\Gamma_i(\omega_i) = \begin{cases} F & \text{if } \omega_i \in \Omega'_i \\ \Theta & \text{if } \omega_i \in \Omega''_i. \end{cases} \tag{4.10}$$

If m_i is the bpa of $\mathcal{H}_i$, then let $p_i = m_i(F)$. There is no contradictory combined interpretation when we combine these hints. Moreover, if Γ is the multivalued mapping of the combined hint, then

$$\Gamma(\omega) = \begin{cases} \Theta & \text{if } \omega \in \prod_{i=1}^n \Omega''_i \\ F & \text{otherwise,} \end{cases} \tag{4.11}$$

which implies that the combined hint is also simple. If m is its bpa, then

$$m(F) = 1 - \prod_{i=1}^{n} (1 - p_i). \tag{4.12}$$

This implies that the support of a hypothesis containing F converges to 1 as more and more independent simple hints focussing on F become available. In the theory of reliability of systems, equation (4.7) represents the reliability of a so-called parallel system: it is sufficient that at least one argument (a simple hint) holds in order to prove a hypothesis H containing F. This connection of the theory of hints with the theory of reliability of complex systems consisting of many components will be pursued in the next chapter.

(2) Heterogeneous hints. This time, let $\mathcal{H}_1$ and $\mathcal{H}_2$ be two simple hints whose focal sets are F_1, Θ and F_2, Θ respectively and suppose that $F_1 \cap F_2 \neq \emptyset$. Then the focal sets of $\mathcal{H}_1 \oplus \mathcal{H}_2$ are $F_1 \cap F_2, F_1, F_2, \Theta$ and their bpas are $p_1p_2, p_1(1-p_2), (1-p_1)p_2$ and $(1-p_1)(1-p_2)$ respectively ($p_i = m_i(F_i)$). There is still no contradiction between the two hints but the combination is no longer simple. Moreover, the support of a hypothesis different from Θ and containing F_1 and F_2 is given by $1-(1-p_1)(1-p_2)$. Hence the support of a hypothesis supported by both hints is larger with the combined hint than with a single hint.

(3) Conflicting hints. Consider once more two simple hints $\mathcal{H}_1$ and $\mathcal{H}_2$ as in (2) above, but assume this time that $F_1 \cap F_2 = \emptyset$. The probability of contradiction is $k = p_1p_2$. The combined hint $\mathcal{H}_1 \oplus \mathcal{H}_2$ has focal sets F_1, F_2, Θ and their bpas are $p_1(1-p_2)/(1-p_1p_2), p_2(1-p_1)/(1-p_1p_2)$ and $(1-p_1)(1-p_2)/(1-p_1p_2)$ respectively. It can easily be verified that the degree of support of a hypothesis is always smaller in the case of conflicting hints than in the case of heterogeneous hints. Moreover, if H is a hypothesis not supported by $\mathcal{H}_2$ (i.e. $H \not\supseteq F_2$, then the support of H with the combined hint is smaller than with $\mathcal{H}_1$ alone ! Symmetrically, this is also true for a hypothesis not supported by $\mathcal{H}_1$. This is explained by the presence of contradictions between the two hints. The second hint sheds some doubt on the validity or reliability of the first one. This aspect of the combination of hints reveals the important feature of non-monotonicity in reasoning with conflicting evidence. This property makes the mathematics of hints a bit harder, but it reflects also a very frequently encountered phenomenon in everyday reasoning.

4.2 Dempster's Rule and Commonality Functions

The commonality functions introduced in Sect. 3.4 are particularly well suited for representing the combination of hints. The following theorem gives conditions under which it is possible to combine hints (i.e. when the probability of contradiction k is different from 1). This result uses the notion of the *core of a hint*, a subset of Θ defined as the union of its focal sets.

Theorem 4.3. *Let $\mathcal{H}_1, \ldots, \mathcal{H}_n$ be hints on the same frame Θ and denote their cores by $\mathcal{C}_1, \ldots, \mathcal{C}_n$ and their commonality functions by $q_1, \ldots, q_n$ respectively. Then the following conditions are equivalent:*

(1) The hints $\mathcal{H}_1, \ldots, \mathcal{H}_n$ cannot be combined, i.e. $k = 1$

(2) $\cap_{i=1}^n \mathcal{C}_i = \emptyset$
(3) $\prod_{i=1}^n q_i(A) = 0$ *for all non-empty subsets* $A \subseteq \Theta$.

It should be clear that hints whose cores do not intersect cannot be combined because, according to our general assumptions about the definition of a hint, the answer must necessarily lie in the core of a hint and hence, if we combine several hints, in the intersection of their cores. Throughout this book and unless the contrary is explicitely asserted, we'll always assume that the hints we combine can indeed be combined. The following important theorem shows how to compute the commonality function of the hint resulting from the combination of several hints.

Theorem 4.4. *Let* $\mathcal{H}_1, \ldots, \mathcal{H}_n$ *be hints with commonality functions* $q_1, \ldots, q_n$ *respectively. If* q *denotes the commonality function of the combined hint* $\oplus_{i=1}^n \mathcal{H}_i$*, then*

$$q(A) = \Big(\prod_{i=1}^n q_i(A)\Big)/(1-k) \tag{4.13}$$

for all non-empty subsets $A \subseteq \Theta$ *and* k *is given by (4.6) if* m_i *is the bpa of* $\mathcal{H}_i$.

This theorem shows that commonality functions essentially multiply when hints are combined. If we think of a commonality function as a transformation of a hint (or of its bpa), then this result means that the transformation of the combination of hints equals the multiplication of their transformations. This is reminiscent of the Fourier transform which replaces convolution by multiplication. Thoma (1989) shows that we can indeed interpret the commonality function of a hint as a Fourier transform or as the characteristic function of its basic probability assignment in a precise mathematical sense (albeit he speaks of belief functions instead of hints). Thereby, the Fourier transform is based on the semi-group $(\mathcal{P}(\Theta), \cap)$ rather than on the usual group $(\mathbf{R}, +)$. The following theorem gives a formula for computing the denominator $1-k$ in (4.13) directly in terms of the commonality functions of the hints to be combined when Θ is finite.

Theorem 4.5. *If* Θ *is finite, then the normalization constant* $(1-k)$ *in (4.13) is given by*

$$\sum \{(-1)^{|B|+1} \prod_{i-1}^n q_i(B) : \emptyset \neq B \subseteq \Theta\}. \tag{4.14}$$

4.3 Conditioning of Hints

An important special case of combination of hints arises when we learn that the answer definitely lies in a subset B of Θ. This information is of course represented by the deterministic hint having a single interpretation whose corresponding focal set is B. This hint will also be denoted by B. Incorporating this piece of evidence into the current state of knowledge represented by a hint $\mathcal{H}$ on Θ requires the computing of the combined hint $\mathcal{H} \oplus B$. This combination is called *conditioning* because we update our knowledge to the new information that the answer is in B. The new hint $\mathcal{H} \oplus B$ is called a *conditional hint.*

Theorem 4.6. *If pl is the plausibility function of the hint* $\mathcal{H} = (\Omega, P, \Gamma, \Theta)$, *then* $\mathcal{H} \oplus B \simeq (\Omega', P', \Gamma', \Theta)$ *where* $\Omega' = v(B)$, $p'(\omega) = p(\omega)/pl(B)$ *and* $\Gamma'(\omega) = \Gamma(\omega) \cap B$ *for all* $\omega \in \Omega'$.

The following theorem of Shafer (1976 a) gives formulas for the support and plausibility functions of the conditional hint $\mathcal{H} \oplus B$.

Theorem 4.7. *Let sp and pl be the support and plausibility functions of* $\mathcal{H}$. *If* $sp(\cdot|B)$ *and* $pl(\cdot|B)$ *denote the respective support and plausibility functions of the hint* $\mathcal{H} \oplus B$, *then*

$$\begin{aligned} sp(H|B) &= \frac{sp(H \cup B^c) - sp(B^c)}{1 - sp(B^c)} \\ pl(H|B) &= pl(H \cap B)/pl(B) \end{aligned} \tag{4.15}$$

for all $H \subseteq \Theta$.

In particular, note the similarity of the conditional plausibility with the usual conditional probability (see also the following section 4.4). Conditional information is a very common and very natural way of representing and expressing knowledge.

4.4 Combining with Precise Hints

In Chap. 2 we defined a hint $\mathcal{H} = (\Omega, P, \Gamma, \Theta)$ to be precise when all its focal sets are singleton subsets of Θ. Such hints are called precise because the uncertainty is concentrated in the unknown correct interpretation only, but not in the associated focal set which is precise. According to theorem 3.8, the support and plausibility functions of a precise hint coincide and represent indeed a probability measure on $\mathcal{P}(\Theta)$. When Ω is finite, this measure has a *discrete probability density* given by the restriction to singletons of the basic probability assignment m of the hint. For the sake of simplicity of notations, we will write this density $m(\theta)$ instead of $m(\{\theta\})$ for all $\theta \in \Theta$. By the way, note that if C is the core of $\mathcal{H}$ then the canonical representative of $\mathcal{H}$ is the hint

$\mathcal{H}_c = (\mathcal{C}, m, id, \Theta)$. Precise hints impose themselves strongly when combined with other hints as the following theorem of Shafer shows.

Theorem 4.8. *Let $\mathcal{H}_0$ be a precise hint on Θ and $\mathcal{H}_1$ an arbitrary hint on Θ. If m_0 is the bpa of $\mathcal{H}_0$ and pl_1 is the plausibility function of $\mathcal{H}_1$, then $\mathcal{H}_0 \oplus \mathcal{H}_1$ is a precise hint with probability density*

$$m(\theta) = c\, m_0(\theta) pl_1(\theta) \tag{4.16}$$

where

$$c = \Big(\sum_{\theta\in\Theta} m_0(\theta) pl_1(\theta)\Big)^{-1}. \tag{4.17}$$

This theorem is illustrated by example 4.1. For an arbitrary positive constant α, define $l_1(\theta) = \alpha\, pl_1(\theta)$ for all $\theta \in \Theta$. The $l_1(\theta)$ are called *relative plausibilities of singletons*. Theorem 4.8 has the following direct corollary.

Corollary 4.4. *Let $\mathcal{H}_0$ be a precise hint on Θ and $\mathcal{H}_1$ an arbitrary hint on Θ. If m_0 is the bpa of $\mathcal{H}_0$ and $l_1(\theta), \theta \in \Theta$ relative plausibilities of singletons of hint $\mathcal{H}_1$, then $\mathcal{H}_0 \oplus \mathcal{H}_1$ is a precise hint with probability density*

$$m(\theta) = c\, m_0(\theta) l_1(\theta) \tag{4.18}$$

where

$$c = \Big(\sum_{\theta\in\Theta} m_0(\theta) l_1(\theta)\Big)^{-1}. \tag{4.19}$$

In order to combine a hint with a precise hint, it is therefore sufficient to know relative plausibilities of singletons of the former hint (remember that the bpa m completely defines the equivalence class of the combined hint). This is a very limited information which considerably simplifies the combination of hints in the presence of precise hints. Therefore, working with precise hints makes many facets of the evidence contained in the other hints irrelevant. This is a fortunate result because, in many circumstances, it is difficult to define the full hint whereas relative plausibilities of singletons may be easier to specify (compare Sect. 1.5).

Theorem 4.9. *If $\mathcal{H}_1, \ldots, \mathcal{H}_n$ are arbitrary hints on Θ with $l_1(\theta), \ldots, l_n(\theta)$ as respective relative plausibilities of singletons, then the values*

$$l(\theta) = \prod_{i=1}^{n} l_i(\theta) \tag{4.20}$$

are relative plausibilities of singletons of the hint $\mathcal{H} = \oplus_{i=1}^{n} \mathcal{H}_i$.

This theorem clearly allows for an extension of corollary 4.4 (Shafer, 1976 a).

Corollary 4.5. *Let $\mathcal{H}_0$ be a precise hint on Θ and $\mathcal{H}_1, \ldots, \mathcal{H}_n$ arbitrary hints on Θ with $l_1(\theta), \ldots, l_n(\theta)$ as relative plausibilities of singletons. If m_0 is the bpa of $\mathcal{H}_0$, then the combined hint $\mathcal{H} = \oplus_{i=0}^{n} \mathcal{H}_i$ is precise and its probability density is given by*

$$m(\theta) = c\, m_0(\theta) \prod_{i=1}^{n} l_i(\theta) \tag{4.21}$$

where

$$c = \Big(\sum_{\theta \in \Theta} m_0(\theta) \prod_{i=1}^{n} l_i(\theta) \Big)^{-1}. \tag{4.22}$$

As an application of theorem 4.8, consider the combination of a precise hint $\mathcal{H}_0$ with a deterministic hint $\mathcal{H}_1$ having a unique focal set B. This is the conditioning of $\mathcal{H}_0$ with respect to B. According to theorem 4.8, $\mathcal{H}_0 \oplus \mathcal{H}_1$ is a precise hint with probability density

$$m(\theta) = \begin{cases} m_0(\theta)/P_0(B) & \text{if } \theta \in B \\ 0 & \text{otherwise,} \end{cases} \tag{4.23}$$

where P_0 is the probability measure on $\mathcal{P}(\Theta)$ induced by the discrete probability density m_0 (P_0 is indeed nothing but the support and plausibility function of $\mathcal{H}_0$). If P is the probability measure on $\mathcal{P}(\Theta)$ induced by the density m, then this implies that

$$P(A) = P_0(A \cap B)/P_0(B) \tag{4.24}$$

for any subset A of Θ. Note that this result can also be deduced directly by applying theorem 4.7 to the precise hint $\mathcal{H}_0$: in this case $sp(\cdot|B) = pl(\cdot|B) = P$ and $sp = pl = P_0$. Thus conditional hints generalize traditional conditional probabilities since the right-hand side of (4.24) is the usual definition of the *conditional probability* of A given B with respect to P_0. Note however that in the context of hints this is not a definition but a result derived from Dempster's rule because P is the probability measure on $\mathcal{P}(\Theta)$ induced by the precise conditional hint $\mathcal{H}_0 \oplus \mathcal{H}_1$.

Since, for precise hints, the bpa of a singleton equals its plausibility, corollary 4.5 implies the following corollary.

Corollary 4.6. *If $\mathcal{H}_1, \ldots, \mathcal{H}_n$ are precise hints on Θ with probability densities $m_1, \ldots, m_n$ respectively, then $\oplus_{i=1}^{n} \mathcal{H}_i$ is again precise and has probability density*

$$m(\theta) = c \prod_{i=1}^{n} m_i(\theta) \tag{4.25}$$

where

$$c = \Big(\sum_{\theta \in \Theta} \prod_{i=1}^{n} m_i(\theta) \Big)^{-1}. \tag{4.26}$$

In particular, if all hints have identical probability density $m_0(\theta)$, then the combined hint has density

$$m(\theta) = \frac{m_0(\theta)^n}{\sum_{\theta \in \Theta} m_0(\theta)^n}. \tag{4.27}$$

Therefore, if $m_0(\theta)$ is a function with a unique maximum at the value θ_0, then $m_0(\theta_0)$ converges to 1 for increasing n, whereas $m(\theta)$ converges to 0 for all other values θ different from θ_0. In other words, the combined hint asymptotically approaches a deterministic precise hint pointing to the most plausible value θ_0 of each individual hint. This means that at the end of the process, we know for certain that the correct answer is θ_0, which is a very reasonable result. As an illustration, we continue example 3.1.

Example 4.4. This is a continuation of example 3.1. Suppose that n persons assert that the coin indeed turned heads. This is represented by n independent precise hints having the same probability density m_0. If $p < 0.5$ is the probability that a person is a liar telling exactly the opposite of the truth, then $m_0(heads) = 1 - p$ and $m_0(tails) = p$. Therefore, if m is the probability density of the combination of these hints, then

$$m(heads) = \frac{(1-p)^n}{(1-p)^n + p^n}. \tag{4.28}$$

This value approaches 1 as n tends to infinity, such that it becomes more and more credible that the coin in reality turned heads as the number of testimonies increases. This is a case of concordant testimonies. If $p > 0.5$, then it is similarly almost certain that the coin turned tails. Now suppose that n_1 persons pretend heads whereas n_2 persons pretend tails, $n_2 = n_1 + k$ for some $k \geq 0$. In this situation, we obtain the following result.

Theorem 4.10. *If m is the probability density resulting from the combination of all heads and tails testimonies, then*

$$m(tails) = \frac{(1-p)^k}{(1-p)^k + p^k}. \tag{4.29}$$

These are simple cases of analysis of testimonies, a classical subject of investigation in the eighteenth century probability theory (Shafer, 1978). Whereas most authors treated the problem in the nowadays classical framework of probability theory, Lambert (1764) already solved it in the spirit of evidence theory, essentially using Dempster's rule of combination.

4.5 Consonant Combination

Let $\mathcal{H}_1$ and $\mathcal{H}_2$ be two consonant hints. Then the focal sets of $\mathcal{H}_1 \oplus \mathcal{H}_2$ are clearly no longer nested. The combination of independent consonant hints is therefore a non-consonant hint.

However, the combination of *dependent* consonant hints preserves consonance. By the word dependent, we mean here that the hints are fully dependent, which means that all hints have the same and identical set of interpretations and therefore also the same correct interpretation. We suppose also that the probability of an interpretation is the same for all hints, only the focal sets of the interpretations are different when we pass from one hint to another. More concretely, let $\Omega = \{\omega_1, \ldots, \omega_m\}$ be a set of interpretations and let $\mathcal{H}_i = (\Omega, P, \Gamma_i, \Theta), i = 1, \ldots, n$ be a family of dependent consonant hints such that $\Gamma_i(\omega_1) \subseteq \Gamma_i(\omega_2) \subseteq \ldots \subseteq \Gamma_i(\omega_m)$ for all $i = 1, \ldots n$. Following the idea which permitted to define the Dempster's rule for combining independent hints, if the interpretation $\omega \in \Omega$ is correct, then the answer must be in

$$\Gamma(\omega) = \cap_{i=1}^{n} \Gamma_i(\omega). \tag{4.30}$$

Thus we have $\Gamma(\omega_1) \subseteq \Gamma(\omega_2) \subseteq \ldots \subseteq \Gamma(\omega_m)$ and let h denote the largest index j such that $\Gamma(\omega_j) = \emptyset$ (let $h = 0$ if $\Gamma(\omega_1)$ is not empty). If $h = m$ the hints cannot be combined. Otherwise, if $h < m$, only the interpretations in $\Omega' = \{\omega_{h+1}, \ldots, \omega_m\}$ remain possible because an interpretation in $\Omega - \Omega'$ cannot be correct since it leads to an empty intersection. The probability of a possible interpretation is thus given by the conditional probabilities $p'(\omega_j) = p(\omega_j)/(1-k)$ where $k = \sum_{j=1}^{h} p(\omega_j)$ (p is the discrete density of P). The combined hint $(\Omega', P', \Gamma, \Theta)$ is thus still consonant. This special case of Dempster's rule involving consonant dependent hints can be expressed directly in terms of the plausibility functions of the hints to be combined. As a matter of fact, it is sufficient to consider the plausibilities of singletons as the following theorem shows (remember that for consonant hints the plausibility of singletons completely determines the plausibility function, see theorem 3.7).

Theorem 4.11. *Let $\mathcal{H}_i = (\Omega, P, \Gamma_i, \Theta), i = 1, \ldots, n$ be a family of consonant dependent hints with $\Omega = \{\omega_1, \ldots, \omega_m\}, p(\omega) > 0$ for all $\omega \in \Omega$ and such that $\Gamma_i(\omega_1) \subseteq \Gamma_i(\omega_2) \subseteq \ldots \subseteq \Gamma_i(\omega_m)$ for all $i = 1, \ldots, n$. If $pl_1, \ldots, pl_n$ are their respective plausibility functions, then they can be combined if and only if*

$$1 - k = \max_{\theta \in \Theta} \min\,(pl_1(\theta), \ldots, pl_n(\theta)) > 0. \tag{4.31}$$

In this case, the combined hint is still consonant and if pl denotes its plausibility function, then

$$pl(\theta) = \min\,(pl_1(\theta), \ldots, pl_n(\theta))/(1-k) \tag{4.32}$$

for all $\theta \in \Theta$.

This type of combination arises in particular when we have to combine several hints expressing the errors of a measurement device working with an unknown but fixed level of precision. Refer to Sect. 1.6 about imprecise measurements for a complete description of this kind of model. Note in particular that Eq. (1.22) corresponds exactly to (4.32). Moreover, equation (4.32) corresponds essentially to the rule of combination for possibility measures or the intersection of fuzzy sets (Dubois, Prade, 1985). The connections between the theory of support functions – or belief functions as Shafer calls them – and the theory of possibility functions have been studied by Shafer (1987) and Dubois, Prade (1987, 1990).

A4 Appendix: Proofs of Theorems in Chapter 4

Proof of Theorem 4.1

In order to prove this theorem, we have to show that the right-hand side of (4.5) equals zero when A is not a focal set of $\mathcal{H}$ and equals $P(\Gamma^{-1}(A))$ when A is a focal set of $\mathcal{H}$. First suppose that A is not a focal set of $\mathcal{H}$. Then, for any family $A_i, i = 1, \ldots, n$ of subsets of Θ such that $\cap_{i=1}^n A_i = A$, there must exist an index i_0 such that A_{i_0} is not a focal set of $\mathcal{H}_{i_0}$. This implies that $m_{i_0}(A_{i_0}) = 0$ and hence $\prod_{i=1}^n m_i(A_i) = 0$, which shows that the right-hand side of (4.5) is zero.

Now suppose that A is a focal set of $\mathcal{H}$. Then we may consider that in the rigth-hand side of (4.5) the sum extends over families $F = (A_1, \ldots, A_n)$ such that $\cap_{i=1}^n A_i = A$ and A_i is a focal set of $\mathcal{H}_i$. Furthermore, if $\mathcal{A}$ denotes the set of all these families and if the addition of sets is supposed to represent their disjoint union, then

$$\Gamma^{-1}(A) = \sum \{\prod_{i=1}^{n} \Gamma_i^{-1}(A_i) : F \in \mathcal{A}\}. \tag{A4.1}$$

Similarly, if $\mathcal{A}_0$ is the set of all families of focal sets $F = (A_1, \ldots, A_n)$ such that $\cap_{i=1}^n A_i = \emptyset$, then

$$\{\omega \in \prod_{i=1}^{n} \Omega_i : \cap_{i=1}^n \Gamma_i(\omega_i) = \emptyset\} = \sum \{\prod_{i=1}^{n} \Gamma_i^{-1}(A_i) : F \in \mathcal{A}_0\} \tag{A4.2}$$

and we can assume that the sum in (4.6) extends over all families $(A_i) \subset \mathcal{A}_0$.

On the other hand, let P' denote the product probability of the probabilities P_i for $i = 1, \ldots, n$. Furthermore, if $\mathcal{B}$ is either the set $\mathcal{A}$ or the set $\mathcal{A}_0$, then define

$$\tau(\mathcal{B}) = \sum \{\prod_{i=1}^{n} \Gamma_i^{-1}(A_i) : F \in \mathcal{B}\}. \tag{A4.3}$$

Therefore, by definition of P we have

$$m(A) = P(\Gamma^{-1}(A)) = \frac{P'(\tau(\mathcal{A}))}{1 - P'(\tau(\mathcal{A}_0))}, \tag{A4.4}$$

which proves the theorem because $P_i(\Gamma_i^{-1}(A_i)) = m_i(A_i)$. □

Proof of Theorem 4.2

Formula (4.9) is an immediate consequence of the definition of k given in formula (4.1) and of the definition of the degree of plausibility:

$$\begin{aligned} 1 - k &= \sum \{p_1(\omega_1)p_2(\omega_2) : \Gamma_1(\omega_1) \cap \Gamma_2(\omega_2) \neq \emptyset\} \\ &= \sum_{\omega_1 \in \Omega_1} p_1(\omega_1) \cdot \sum \{p_2(\omega_2) : \Gamma_1(\omega_1) \cap \Gamma_2(\omega_2) \neq \emptyset\} \\ &= \sum_{\omega_1 \in \Omega_1} p_1(\omega_1) pl_2(\Gamma_1(\omega_1)). \end{aligned} \tag{A4.5}$$

Now we prove formula (4.7). According to theorem 4.1, for all non-empty subsets $A \subseteq \Theta$, we have

$$m(A) = (1-k)^{-1} \cdot \sum_{A_1 \subseteq \Theta} m_1(A_1) \cdot \sum \{m_2(A_2) : A_1 \cap A_2 = A\} \tag{A4.6}$$

where k is given by (4.6). This implies that

$$\begin{aligned} sp(H) &= \sum_{D \subseteq H} m(D) = \sum \{m(D) : \emptyset \neq D \subseteq H\} \\ &= (1-k)^{-1} \cdot \sum_{\emptyset \neq D \subseteq H} \Big(\sum_{A_1 \subseteq \Theta} m_1(A_1) \cdot \sum \{m_2(A_2) : A_1 \cap A_2 = D\} \Big) \\ &= (1-k)^{-1} \cdot \sum_{A_1 \subseteq \Theta} m_1(A_1) \cdot \sum \{m_2(A_2) : \emptyset \neq A_1 \cap A_2 \subseteq H\} \\ &= (1-k)^{-1} \cdot \sum_{A_1 \subseteq \Theta} m_1(A_1) \cdot \sum \{m_2(A_2) : A_2 \subseteq H \cup A_1^c, A_2 \not\subseteq A_1^c\}, \end{aligned} \tag{A4.7}$$

and therefore

$$\begin{aligned} sp(H) &= (1-k)^{-1} \cdot \sum_{A_1 \subseteq \Theta} m_1(A_1) \cdot \Big(\sum_{A_2 \subseteq H \cup A_1^c} m_2(A_2) - \sum_{A_2 \subseteq A_1^c} m_2(A_2) \Big) \\ &= (1-k)^{-1} \cdot \sum_{A_1 \subseteq \Theta} m_1(A_1)(sp_2(H \cup A_1^c) - sp_2(A_1^c)) \\ &= (1-k)^{-1} \cdot \sum_{\omega_1 \in \Omega_1} p_1(\omega_1)(sp_2(H \cup \Gamma_1(\omega_1)^c) - sp_2(\Gamma_1(\omega_1)^c)) \end{aligned} \tag{A4.8}$$

because $m_1(A_1) = 0$ when A_1 is not a focal set of $\mathcal{H}_1$. If we define $c = (1-k)^{-1}$, then we can write

$$\begin{aligned} sp(H) &= c \cdot \sum_{\omega_1} p_1(\omega_1) sp_2(H \cup \Gamma_1(\omega_1)^c) - c \cdot \sum_{\omega_1} p_1(\omega_1)(1 - pl_2(\Gamma_1(\omega_1))) \\ &= c \cdot \sum_{\omega_1} p_1(\omega_1) sp_2(H \cup \Gamma_1(\omega_1)^c) - c(1 - \sum_{\omega_1} p_1(\omega_1) pl_2(\Gamma_1(\omega_1))) \\ &= c \cdot \sum_{\omega_1} p_1(\omega_1) sp_2(H \cup \Gamma_1(\omega_1)^c) - c(1 - (1 - k)) \\ &= c \cdot \sum_{\omega_1} p_1(\omega_1) sp_2(H \cup \Gamma_1(\omega_1)^c) - c + 1 \end{aligned} \tag{A4.9}$$

and formula (4.7) is proved. Now we prove formula (4.8):

$$\begin{aligned} pl(H) = 1 - sp(H^c) &= 1 - c \cdot \sum_{\omega_1} p_1(\omega_1) sp_2(H^c \cup \Gamma_1(\omega_1)^c) + (c - 1) \\ &= 1 - c \cdot \sum_{\omega_1} p_1(\omega_1)(1 - pl_2(H \cap \Gamma_1(\omega_1))) + (c - 1) \\ &= 1 - c(1 - \sum_{\omega_1} p_1(\omega_1) pl_2(H \cap \Gamma_1(\omega_1))) + (c - 1) \\ &= 1 - c + c \cdot \sum_{\omega_1} p_1(\omega_1) pl_2(H \cap \Gamma_1(\omega_1)) + c - 1 \\ &= c \cdot \sum_{\omega_1} p_1(\omega_1) pl_2(H \cap \Gamma_1(\omega_1)). \end{aligned} \tag{A4.10}$$

□

Proof of Theorem 4.3

First we prove that (1) is equivalent to (2). If $k = 1$, then $\cap_{i=1}^n \Gamma_i(\omega_i) = \emptyset$ for all combined interpretation $\omega = (\omega_1, \ldots, \omega_n)$ and hence

$$\cup_\omega \cap_{i=1}^n \Gamma_i(\omega_i) = \cap_{i=1}^n \cup_{\omega_i} \Gamma_i(\omega_i) = \cap_{i=1}^n C_i \tag{A4.11}$$

is also empty. If $\cap_{i=1}^n C_i$ is empty, then by (A4.11) it follows that $\cap_{i=1}^n \Gamma_i(\omega_i)$ is also empty for all ω and hence $k = 1$.

Now we prove that (2) is equivalent to (3). Let A be a non-empty subset of Θ and suppose that $\cap_{i=1}^n C_i$ is empty. Then there must exist an index i_0 such that $A \not\subseteq C_{i_0}$. But this implies that $q_{i_0}(A) = \emptyset$ and hence $\prod_{i=1}^n q_i(A) = 0$. Now we show that if $\cap_{i=1}^n C_i$ is not empty, then there is a non-empty subset $A \subseteq \Theta$ such that $\prod_{i=1}^n q_i(A) > 0$. Arbitrarily select an element $\theta \in \cap_{i=1}^n C_i$ and define $A = \{\theta\}$. Then by construction $q_i(A) > 0$ for all $i = 1, \ldots, n$ and hence $\prod_{i=1}^n q_i(A) > 0$. □

Proof of Theorem 4.4 (Shafer, 1976 a)

If m denotes the bpa of the combined hint, then theorem 4.1 yields

$$\begin{aligned} q(A) &= \sum_{A \subseteq B} m(B) \\ &= (1-k)^{-1} \sum_{A \subseteq B} \sum \{\prod_{i=1}^{n} m_i(B_i) : \cap_{i=1}^{n} B_i = B\} \\ &= (1-k)^{-1} \sum \{\prod_{i=1}^{n} m_i(B_i) : A \subseteq \cap_{i=1}^{n} B_i\} \qquad (A4.12) \\ &= (1-k)^{-1} \prod_{i=1}^{n} \sum_{A \subseteq B_i} m_i(B_i) \\ &= (1-k)^{-1} \prod_{i=1}^{n} q_i(A) \end{aligned}$$

and the theorem is proved. □

Proof of Theorem 4.5 (Shafer, 1976 a)

Theorem 3.14 (2) implies that

$$\sum_{B \subseteq \Theta} (-1)^{|B|} q(B) = 0 \qquad (A4.13)$$

which implies of course that

$$\sum \{(-1)^{|B|+1} q(B) : \emptyset \neq B \subseteq \Theta\} = 1. \qquad (A4.14)$$

Hence by theorem 4.4 we have

$$(1-k)^{-1} \sum \{(-1)^{|B|+1} \prod_{i=1}^{n} q_i(B) : \emptyset \neq B \subseteq \Theta\} = 1 \qquad (A4.15)$$

and the theorem is proved. □

Proof of Theorem 4.6

If $\bar{\omega}$ denotes the interpretation of B, then the combined contradictory interpretations of $\mathcal{H} \oplus B$ are

$$\{(\omega, \bar{\omega}) : \Gamma(\omega) \cap B = \emptyset\} = v(B)^c \times \{\bar{\omega}\} \qquad (A4.16)$$

and thus the interpretations of $\mathcal{H} \oplus B$ are $v(B) \times \{\bar{\omega}\}$. Since $P(v(B)) = pl(B)$ it is clear that the hint $(\Omega', P', \Gamma', \Theta)$ defined in the theorem is equivalent to $\mathcal{H} \oplus B$. □

Proof of Theorem 4.7

Let m_1 denote the bpa of the hint $\mathcal{H} \oplus B = (\Omega', P', \Gamma', \Theta)$ (cf. theorem 4.6). If m is the bpa of $\mathcal{H}$, then theorem 4.2 implies that $1 - k = pl(B)$. Again by theorem 4.2 we have

$$\begin{aligned} sp(H|B) &= \frac{sp(H \cup B^c)}{pl(B)} - \Big(\frac{1}{pl(B)} - 1\Big) \\ &= \frac{sp(H \cup B^c)}{pl(B)} - \frac{1 - pl(B)}{pl(B)} \\ &= \frac{sp(H \cup B^c) - sp(B^c)}{1 - sp(B^c)} \end{aligned} \tag{A4.17}$$

and

$$pl(H|B) = \frac{pl(H \cap B)}{pl(B)}. \tag{A4.18}$$

□

Proof of Theorem 4.8

To prove this theorem, we need the following lemma.

Lemma A4-1 *If q is the commonality function and m the bpa of a precise hint, then $m(\theta) = q(\theta)$ for all $\theta \in \Theta$.*

In other words, this lemma asserts that the probability density of a precise hint is nothing but its commonality function restricted to singletons.

Proof of Lemma A4-1

If pl is the plausibility function of the hint, then by definition we have $q(\theta) = pl(\theta)$ for all $\theta \in \Theta$. Since $pl(\theta) = m(\theta)$ for precise hints, it follows that $m(\theta) = q(\theta)$ and the lemma is proved. □

First we prove that $\mathcal{H}_0 \oplus \mathcal{H}_1$ is precise. In fact, any combined interpretation (ω_0, ω_1) is either contradictory or points to $\Gamma_0(\omega_0)$ depending on whether the singleton $\Gamma_0(\omega_0)$ is included in $\Gamma_1(\omega_1)$ or not. Hence $\mathcal{H}_0 \oplus \mathcal{H}_1$ is again a precise hint.

Let q_0, q_1 and q denote the commonality functions of $\mathcal{H}_0, \mathcal{H}_1$ and $\mathcal{H}_0 \oplus \mathcal{H}_1$ respectively. If $\theta \in \Theta$, then by theorem 4.4 we can write $q(\theta) = \alpha\, q_0(\theta) q_1(\theta)$ for some constant α. But this equation can be written in a different manner. In fact, by definition, the commonality function and the plausibility function coincide on singletons: $q_1(\theta) = pl_1(\theta)$ for all $\theta \in \Theta$. On the other hand, lemma A4-1 implies that $q(\theta) = m(\theta)$ and $q_0(\theta) = m_0(\theta)$. Therefore $m(\theta) = \alpha\, m_0(\theta) pl_1(\theta)$ and since $m(\theta), \theta \in \Theta$ must be a probability density we have

$$\alpha = \Big(\sum_{\theta \in \Theta} m_0(\theta) pl_1(\theta)\Big)^{-1} \tag{A4.19}$$

and the theorem is proved. □

Proof of Theorem 4.9

Let q and pl denote the respective commonality and support function of $\mathcal{H}$. According to theorem 4.4 we have

$$q(\theta) = \alpha \prod_{i=1}^{n} q_i(\theta) \tag{A4.20}$$

for some positive constant α. But theorem 3.15 (2) implies that $q(\theta) = pl(\theta)$ for all $\theta \in \Theta$. Hence

$$\alpha^{-1}\ pl(\theta) = \prod_{i=1}^{n} q_i(\theta) \tag{A4.21}$$

and the theorem is proved. □

Proof of Theorem 4.10

If p is the probability that a person is a liar and if m_1 denotes the probability density of the hint coming from somebody asserting heads, then $m_1(heads) = 1 - p$ and $m_1(tails) = p$. Similarly, for someone asserting tails we have $m_2(heads) = p$ and $m_2(tails) = 1 - p$. If m' denotes the probability density resulting from the combination of all heads testimonies, then (4.27) implies that

$$m'(tails) = \frac{p^{n_1}}{p^{n_1} + (1-p)^{n_1}} \tag{A4.22}$$

$$m'(heads) = \frac{(1-p)^{n_1}}{p^{n_1} + (1-p)^{n_1}} \tag{A4.23}$$

Similarly, if m'' denotes the probability density resulting from the combination of all tails testimonies, then

$$m''(tails) = \frac{(1-p)^{n_2}}{p^{n_2} + (1-p)^{n_2}} \tag{A4.24}$$

$$m''(heads) = \frac{p^{n_2}}{p^{n_2} + (1-p)^{n_2}} \tag{A4.25}$$

Since Dempster's rule is associative, corollary 4.6 implies that

$$m(tails) = c \cdot m'(tails) m''(tails)$$

where

$$c = \frac{(p^{n_1} + (1-p)^{n_1})(p^{n_2} + (1-p)^{n_2})}{p^{n_1}(1-p)^{n_2} + (1-p)^{n_1} p^{n_2}}. \tag{A4.26}$$

Therefore,

$$
\begin{aligned}
m(tails) &= \frac{p^{n_1}(1-p)^{n_2}}{p^{n_1}(1-p)^{n_2}+(1-p)^{n_1}p^{n_2}} \\
&= \frac{p^{n_1}(1-p)^{n_1}(1-p)^k}{p^{n_1}(1-p)^{n_1}(1-p)^k+(1-p)^{n_1}p^{n_1}p^k} \\
&= \frac{(1-p)^k}{(1-p)^k+p^k}.
\end{aligned}
\tag{A4.27}
$$

□

Proof of Theorem 4.11

First note that the core $\mathcal{C}_i$ of $\mathcal{H}_i$ is simply $\Gamma_i(\omega_m)$ for all $i = 1, \ldots, n$. Then the hints are combinable if and only if $\cap_{i=1}^n \Gamma_i(\omega_m)$ is not empty. But this last condition is equivalent to the following four assertions:

(1) there exists $\theta \in \Theta$ such that $\theta \in \Gamma_i(\omega_m)$ for all $i = 1, \ldots, n$
(2) there exists $\theta \in \Theta$ such that $pl_i(\theta) > 0$ for all $i = 1, \ldots, n$
(3) there exists $\theta \in \Theta$ such that $\min \{pl_1(\theta), \ldots, pl_n(\theta)\} > 0$
(4) $\max_{\theta\in\Theta} \{\min (pl_1(\theta), \ldots, pl_n(\theta)\} > 0$.

This proves the first part of the theorem.

Now we suppose that the hints can be combined. Then $\Gamma(\omega_1) \subseteq \Gamma(\omega_2) \subseteq \ldots, \subseteq \Gamma(\omega_m)$ and $\Gamma(\omega_m) = \cap_{i=1}^n \Gamma_i(\omega_m)$ is not empty since the hints are combinable. This shows that the combined hint is also consonant.

Note that Eq. (4.32) holds when θ is not in the core $\mathcal{C} = \cap_{i=1}^n \Gamma_i(\omega_m)$ of the combined hint because in this case $pl(\theta) = 0$ and $\theta \in \cup_{i=1}^n \mathcal{C}_i^c$ implies $\min (pl_1(\theta), \ldots, pl_n(\theta)) = 0$.

So let $\theta \in \mathcal{C}$ and let $j(\theta)$ denote the smallest index j for which $\theta \in \Gamma(\omega_j)$. Furthermore,

$$g(\theta) = \sum_{j \geq j(\theta)} p(\omega_j) \tag{A4.28}$$

denotes the plausibility of θ before conditioning. If $j(\theta) = 1$, then $g(\theta) = pl_i(\theta) = 1$ for all i and thus $g(\theta) = \min (pl_1(\theta), \ldots, pl_n(\theta))$. If $1 < j(\theta) \leq m$, then $\theta \in \Gamma_i(\omega_{j(\theta)})$ for all i, but $\theta \notin \Gamma_i(\omega_{j(\theta)-1})$ for at least one i. This implies $g(\theta) = pl_i(\theta)$ for such an i and hence

$$g(\theta) = \min (pl_1(\theta), \ldots, pl_n(\theta)). \tag{A4.29}$$

To prove (4.32), it remains to show that the probability of no contradiction

$$1 - k = \sum \{p(\omega_j) : \Gamma(\omega_j) \neq \emptyset\} \tag{A4.30}$$

equals

$$\max_{\theta\in\Theta} \{\min (pl_1(\theta), \ldots, pl_n(\theta)\}. \tag{A4.31}$$

If $\Gamma(\omega_l) \neq \emptyset$, then $\Gamma(\omega_j) \neq \emptyset$ for all $j \geq l$ and thus

$$\sum_{j=l}^{m} p(\omega_j) = \min\ (pl_1(\theta), \ldots, pl_n(\theta)) \tag{A4.32}$$

for all $\theta \in \Gamma(\omega_l) - \Gamma(\omega_{l-1})$. Let h denote the largest index j such that $\Gamma(\omega_j) = \emptyset$ (let $h = 0$ if $\Gamma(\omega_1)$ is not empty). If $\theta' \in \Gamma(\omega_{h+1})$, then (A4.32) implies that

$$\begin{aligned}\sum_{j=h+1}^{m} p(\omega_j) &= \min\ (pl_1(\theta'), \ldots, pl_n(\theta')) \\ &\leq \max_{\theta \in \Theta} \{\min\ (pl_1(\theta), \ldots, pl_n(\theta))\}.\end{aligned} \tag{A4.33}$$

On the other hand, the frame Θ decomposes into the disjoint union

$$\cup_{l=h+1}^{m} \Gamma(\omega_l) - \Gamma(\omega_{l-1})\ \cup\ \Gamma(\omega_m)^c. \tag{A4.34}$$

This implies that

$$\sum_{j=h+1}^{m} p(\omega_j) \geq \min\ (pl_1(\theta), \ldots, pl_n(\theta)) \tag{A4.35}$$

for all $\theta \in \Theta$ because if $\theta \in \Gamma(\omega_l) - \Gamma(\omega_{l-1})$ then $h + 1 \leq l$ and hence

$$\sum_{j=h+1}^{m} p(\omega_j) \geq \sum_{j=l}^{m} p(\omega_j) = \min\ (pl_1(\theta), \ldots, pl_n(\theta)) \tag{A4.36}$$

by Eq. (A4.32), and if $\theta \in \Gamma(\omega_m)^c$ then $\min\ (pl_1(\Theta), \ldots, pl_n(\theta)) = 0$. Thus finally

$$1 - k = \sum_{j=h+1}^{m} p(\omega_j) = \max_{\theta \in \Theta} \{\min\ (pl_1(\theta), \ldots, pl_n(\theta))\} \tag{A4.37}$$

and the theorem is proved. □

5. Probabilistic Assumption-Based Reasoning

5.1 Representing Uncertain Knowledge by Propositional Systems

In Chap. 1, sections 1.1 to 1.3 it has been shown that it is possible to represent uncertain information and uncertain knowledge by uncertain rules or - alternatively - by rules containing uncertain or only probable assumptions. Propositional systems containing uncertain assumptions provide a concise language for describing uncertain information and knowledge. They allow for efficient inference mechanisms based on well known methods from logic and from reliability theory. That is why they are of great interest. In this chapter the relevant methods and techniques will be developed and illustrated. It will be shown how propositional systems with uncertain assumptions fit into the framework of the theory of hints. It turns out that the point of view developed in this chapter is especially well suited to illustrate the basic philosophy of the theory of hints as a logic of reliable reasoning with unreliable arguments.

In this introductory section the problem to be studied will be developed informally. As an example let's summarize here the alarm story introduced in Sect. 1.3. To start with, let's define the propositions which occur in the story:

a : the alarm is ringing
b : there is a burglary
e : an earthquake has occurred
w : there is a phone call from Mr. Watson
g : there is a phone call from Mrs. Gibbson
d : there is a phone call from Mr. Holmes' daughter
c : there is confirmation of an earthquake.

In addition, a number of extra propositions, called *assumptions*, are needed:

a_1 : the alarm system is functioning
a_2 : an extra cause for the alarm is present
a_3 : Mr. Watson is in joking mood
a_4 : Mrs. Gibbson is deaf
a_5 : Mr. Holmes' daughter is at home
a_6 : the earthquake sensor is sensitive.

The difference between assumptions and the above ordinary propositions involved in the story is that although it is not known which assumptions are true or false, we are willing and ready to specify probabilities for them. In contrast, no probability information about the other propositions is assumed.

Given these definitions, the alarm system and the behavior of Mr. Watson, Mrs Gibbson and Mr. Holmes' daughter can now be described by simple inference rules:

$$
\begin{array}{llll}
R_1: & b \wedge a_1 \to a & R_8: & a \wedge \neg a_4 \to g \\
R_2: & e \wedge a_1 \to a & R_9: & a_4 \to \neg g \\
R_3: & a_2 \to a & R_{10}: & \neg a \to \neg g \\
R_4: & \neg b \wedge \neg e \wedge \neg a_2 \to \neg a & R_{11}: & a \wedge a_5 \to d \\
R_5: & a \to w & R_{12}: & \neg a_5 \to \neg d \\
R_6: & a_3 \to w & R_{13}: & e \wedge a_6 \to c \\
R_7: & \neg a \wedge \neg a_3 \to \neg w & R_{14}: & \neg e \to \neg c
\end{array}
$$

In order to formalize such models, introduce a set P of propositional symbols $p_1, p_2, \ldots, p_n$ and a set A of symbols $a_1, a_2, \ldots, a_s$ for the assumptions. The assumptions a_i have probabilities q_i assigned to them and these probabilities will be assumed to be mutually independent; q_i is the probability that the assumption a_i is true. A literal x is then any symbol out of $P \cup A$ or a negation ($\neg$) of such a symbol. Rules as formulated in the above example are to be considered as material implications in the sense of propositional calculus. They have the form $x_1 \wedge x_2 \wedge \ldots \wedge x_r \to x$, where the x_i and x are literals. A rule is uncertain if at least one of the x_i is an assumption; often rules contain at most one assumption, as in the example above. However, this will not be generally assumed. $x_1, \ldots x_r$ are called the conditions and x the conclusion of the rule. This terminology however assigns a direction to a rule which is not really important and is in fact misleading. Implications are equivalent to clauses of the form $\neg x_1 \vee \neg x_2 \vee \ldots \vee \neg x_r \vee x$. This is a symmetric notation which introduces no irrelevant direction. In this notation the above model reads as follows:

$$
\begin{array}{llll}
R_1: & \neg b \vee \neg a_1 \vee a & R_8: & \neg a \vee a_4 \vee g \\
R_2: & \neg e \vee \neg a_1 \vee a & R_9: & \neg a_4 \vee \neg g \\
R_3: & \neg a_2 \vee a & R_{10}: & a \vee \neg g \\
R_4: & b \vee e \vee a_2 \vee \neg a & R_{11}: & \neg a \vee \neg a_5 \vee d \\
R_5: & \neg a \vee w & R_{12}: & a_5 \vee \neg d \\
R_6: & \neg a_3 \vee w & R_{13}: & \neg e \vee \neg a_6 \vee c \\
R_7: & a \vee a_3 \vee \neg w & R_{14}: & e \vee \neg c
\end{array}
$$

It is customary in computer science to regard rules as directed knowledge, which allows to infer conclusions from conditions. Also, for convenience or algorithmic simplicity, often only Horn clauses, clauses with at most one non-negated literal, are admitted. Note however that already the simple example given above does not belong to this restricted category. It is thus important to treat the general case. But then it is really more natural and also more relevant to consider clauses such as the above ones as logical relations which constrain the common truth values of the propositions linked in a clause. For example the clause $e \vee \neg c$ constrains e to be true if c is true (which means that $\neg c$ is false). Therefore, a rule such as "if $\neg e$ then $\neg c$" can also be used for inferences in the opposite direction ("if c then e").

In general let Σ_K denote the set of clauses describing the logical relations between the basic propositions and the assumptions of the system. This set can be considered as a description of the available general knowledge about the system. This knowledge is uncertain as its chunks may be valid or not, depending on whether certain assumptions do or do not hold. In a given situation, further facts may be added to Σ_K. In the example above, suppose that at a given time Mr. Watson phoned (w), whereas there is no phone (yet) from Mrs. Gibbson ($\neg g$). Such facts induce simply a set Σ_F of further clauses which can be added to Σ_K in order to complete the description of all available information. So, suppose $\Sigma = \Sigma_K + \Sigma_F$ represents all the available information at a given moment or in a given situation. Using this information, it will be interesting to study the credibility or plausibility of some hypotheses h like for example that there is a burglary (proposition b).

Whether or not h can be deduced from Σ will depend on the validity of certain assumptions. There will be assumptions whose validity permits the inference of h, there will also be assumptions which allow to infer $\neg h$. It will be interesting to determine such assumptions because they represent arguments in favor and against h respectively. Even more important will be to compute their respective probabilities, because they represent essentially the degrees of support of h and $\neg h$ in the light of the available knowledge Σ. These are informally the problems to be tackled in this chapter.

Both the formal definition as well as the solution of these problems depend on propositional calculus. Therefore, the following section 5.2 is devoted to a short and concise introduction into the necessary elements of propositional calculus. In Sect. 5.3 propositional systems with uncertain assumptions will be put in the perspective of hints. In section 5.4 the notion of arguments for a hypothesis as evoked above will be stated more precisely in the framework of propositional calculus and a few theorems characterizing these arguments – called quasi-supports – will be given. Section 5.5 then presents a computational method based on resolution for actually determining the arguments for a hypothesis from the knowledge base Σ. Since this method is not sufficiently efficient for large knowledge bases, Sect. 5.6 presents another technique based on works of Siegel (1987) and Inoue (1991 a,b and 1992). A set of assumptions supporting a hypothesis is like a set of components which are sufficient for the good functioning of a technical system. Degrees of support for a hypothesis correspond therefore to reliabilities of machineries, here of logical proof machineries. Techniques from reliability theory can in fact be applied to our problem of computing degrees of support. This allows us to speak of the reliability of inference with probable arguments or assumptions. Section 5.7 discusses a few basic methods for the computation of degrees of support.

Let's point out, that the kind of systems considered in this chapter can also be treated by the methods presented in Chap. 7 and 8.

5.2 Propositional Logic and Boolean Algebra

Propositional logic is concerned with declarative sentences as those introduced at the beginning of the previous section. Such sentences or propositions are denoted by symbols like $p, q, \ldots$ or $p_1, p_2, \ldots$ etc. These symbols are called *atoms* or atomic formulas. Propositional logic has, as any logic, first of all a syntactic part, which specifies how to build compound propositions by using logical connectives. Here the four connectives $\neg$ (not), $\wedge$ (and), $\vee$ (or) and $\rightarrow$ (if... then ...) will be used. The syntactic rules for building formulas of propositional logic are as follows:

(1) An atom is a formula
(2) If f is a formula, then $\neg f$ is a formula
(3) If f and g are formulas, then $(f \wedge g)$, $(f \vee g)$ and $f \rightarrow g$ are formulas
(4) All formulas are generated by applying the above rules.

Parentheses will be omitted if not necessary. Thus $f \vee g$ is the same as $(f \vee g)$. Furthermore parentheses can be eliminated by assigning increasing priority to the propositional connectives as follows: $\rightarrow, \wedge, \vee, \neg$. Thus $p \rightarrow q \wedge r$ means $(p \rightarrow (q \wedge r))$ and $p \rightarrow q \wedge \neg r \vee s$ is $(p \rightarrow (q \wedge ((\neg r) \vee s)))$. In our applications there will always be a finite set P of atoms (an alphabet) over which formulas will be developed. The set of formulas over P will be denoted by $\mathcal{L}_P$ and is called the *propositional language* over P.

Formulas get meaning by assigning truth values t (true) or f (false) to them. This is done in the way indicated in Tab. 5.1 for the logical connectives. The assignment of truth values to the atoms of a formula is called an *interpretation* of the formula. Using the basic assignments in Tab. 5.1, the truth value of any interpretation of a formula can easily be computed. A formula f is said to be true under an interpretation, if it is evaluated to t, otherwise f is said to be false under the interpretation.

Table 5.1. Logical connectives

f	g	¬f	f ∧ g	f ∨ g	f → g
t	t	f	t	t	t
t	f	f	f	t	f
f	t	t	f	t	t
f	f	t	f	f	t

A formula with n different atoms has 2^n different interpretations. If the formula is true under all its interpretations it is called a *valid formula* or a *tautology*. If, on the contrary, a formula is true under no interpretation, it is said to be *inconsistent* or *unsatisfiable*. Most formulas are neither tautologies nor inconsistent. If such a formula is taken as a statement (implying that the formula is true), it places a constraint on the set of possible interpretations (namely restricting the possible interpretations to those under which the formula is true). In this way, a formula represents a piece of information. If $\Sigma = \{f_1, f_2, \ldots, f_r\}$ represents a number of clauses, as in the example of the previous section, then stating these r clauses is equivalent to stating the formula $f_1 \land f_2 \land \ldots, \land f_r$. In this way, a given knowledge or information is encoded.

Two formulas f and g are equivalent, denoted by $f = g$, if and only if the truth values of f and g are the same under every interpretation. Thus for example $f \to g = \neg f \lor g$ (verify using Tab. 5.1). More generally $(f_1 \land f_2 \land \ldots \land f_n \to g) = (\neg f_1 \lor \neg f_2 \lor \ldots \lor \neg f_n \lor g)$. More equivalences are given in Tab. 5.2. An algebraic system which satisfies the laws given in Tab. 5.2 is called a Boolean algebra; propositional calculus is thus an instance of a Boolean algebra. Note that equivalent formulas, as statements, refer to exactly the same information; they represent different forms of expressing the same information or knowledge. There are however important special equivalent forms into which any formula can be transformed and which are for one reason or another especially well suited to represent information. One such reason might be computational convenience, another simplicity of expression or representation. Examples of such special forms are the conjunctive and the disjunctive normal forms.

Table 5.2. $\top$ denotes the tautology, $\bot$ the inconsistency

$f \land f = f, f \lor f = f$ (idempotent)
$f \land g = g \land f, f \lor g = g \lor f$ (commutative)
$f \land (g \land h) = (f \land g) \land h, f \lor (g \lor h) = (f \lor g) \lor h$ (associative)
$f \land (f \lor g) = f, f \lor (f \land g) = f$ (absorption)
$f \land (g \lor (f \land h)) = (f \land g) \lor (f \land h)$ (modularity)
$f \lor (g \land (f \lor h)) = (f \lor g) \land (f \lor h)$
$f \land (g \lor h) = (f \land g) \lor (f \land h), f \lor (g \land h) = (f \lor g) \land (f \lor h)$ (distributive)
$f \land \bot = \bot, f \lor \bot = f, f \land \top = f, f \lor \top = \top$ (universal bounds)
$f \land \neg f = \bot, f \lor \neg f = \top$ (complements)
$\neg(\neg f)) = f$ (involution)
$\neg(f \land g) = \neg f \lor \neg g, \neg(f \lor g) = \neg f \land \neg g$ (de Morgan)

A *literal* x is an atom or the negation of an atom. A disjunction of literals $x_1 \lor x_2 \lor \ldots \lor x_n$ is called a *clause*. A formula f is called a *conjunctive normal form* (CNF), if it has the form $f_1 \land f_2 \land \ldots \land f_n, n \geq 1$, where each f_i is a clause. Remark that the chunks of knowledge in the knowledge base Σ were written as clauses and the set Σ of clauses corresponds in fact to a representation of the knowledge in conjunctive normal form. This will be the general form into which knowledge is assumed to be coded in this chapter. Similarly, a formula f is said to be a *disjunctive normal form* (DNF) if it has the form $f_1 \vee f_2 \vee \ldots \vee f_n, n \geq 1$, where each f_i is a conjunction $x_1 \wedge x_2 \wedge \ldots, \wedge x_r$ of literals. Any formula can be transformed into an equivalent disjunctive or conjunctive normal form by repeatedly using equivalences from Tab. 5.2 (see e.g. Chang, Lee, 1973).

Further notions of propositional calculus will be needed. Given formulas $f_1, f_2, \ldots, f_n$ a formula g is said to be a *logical consequence* of $f_1, f_2, \ldots, f_n$, written $\{f_1, f_2, \ldots, f_n\} \models g$, if and only if for any interpretation for which all formulas $f_1, f_2, \ldots, f_n$ are true, g is also true. Equivalently, g is a logical consequence of $f_1, f_2, \ldots, f_n$ if and only if $(f_1 \wedge f_2 \wedge \ldots \wedge f_n) \rightarrow g$ is a valid formula (Chang, Lee, 1973).

It has been said above, that any formula f has an equivalent representation in a disjunctive normal form and also in a conjunctive normal form. This raises the question of a simplest representation of a formula f. By this, one may refer to an equivalent disjunctive normal form, where the sum of the numbers of literals r_i in the conjunctive terms f_i becomes minimal. Such a form is called *minimal* among all disjunctive normal forms. One may also refer to an equivalent disjunctive normal form with a minimal number of conjunctive terms. Such a form is called *shortest* if it contains the smallest total number of literals among all equivalent forms with the minimal number of conjunctive terms. These forms stand for economic representations of a formula or a statement. The question arises however, how such representations can be found. Note in passing that by de Morgan laws the negation of a conjunctive normal form is transformed into a disjunctive normal form and vice versa. Hence, simplest conjunctive normal forms for a formula f may be found by looking for simplest disjunctive normal forms for $\neg f$. It is thus sufficient to discuss either the case of disjunctive or of conjunctive normal forms.

A conjunction of literals c is called an *implicant* of a formula f, if $c \models f$. An implicant c for f is called a *prime implicant*, if no proper subconjunction of c is an implicant for f. A disjunction d of literals, i.e. a clause, is called an *implicate* of a formula f, if $f \models d$. An implicate d of f is called a *prime implicate* of f, if no proper subclause of d is also an implicate of f. Note that if c is an implicant for f, then $\neg c$ is an implicate of $\neg f$; and if d is an implicate of f, then $\neg d$ is an implicant of $\neg f$. The following theorem shows the importance of these notions for disjunctive and conjunctive normal forms.

Theorem 5.1.

(1) If $f = f_1 \vee f_2 \vee \ldots \vee f_n$ is a disjunctive form for f, then every conjunction $f_i, i = 1, \ldots, n$ is an implicant of f.

(2) If $f = f_1 \wedge f_2 \wedge \ldots \wedge f_n$ is a conjunctive normal form for f, then every disjunction $f_i, i = 1, \ldots, n$ is an implicate of f.

The prime implicants or the prime implicates of a formula f provide for unique representations of f as the following theorem shows.

Theorem 5.2. *If $f_1, f_2, \ldots, f_r$ are all the prime implicants (implicates) of a formula f, then $f = f_1 \vee f_2 \vee, \ldots \vee f_r (f = f_1 \wedge f_2 \wedge \ldots \wedge f_r)$.*

However, in general it is not necessary to include all prime implicants in the disjunctive normal form to represent f; some of them are redundant. A disjunctive normal form $f = f_1 \vee f_2 \vee \ldots \vee f_r$ is called *irredundant* with respect to f, if (1) all terms f_i of the disjunction are prime implicants of f and (2) no term f_i in the disjunction can be deleted without destroying the equivalence to f. A similar definition can be given for conjunctive normal forms. In general there will be several different irredundant representations of f. In particular the minimal and the shortest disjunctive normal forms are among them:

Theorem 5.3. *The minimal and the shortest disjunctive or conjunctive normal forms of a formula f are irredundant with respect to f.*

The problem of determining minimal and shortest disjunctive or conjunctive normal forms is thus reduced to :

(1) Find all prime implicants or implicates of the formula f.
(2) Generate from these all irredundant forms.
(3) Choose from them those particular formulas which satisfy the chosen criterion of simplicity: minimal or shortest.

In Sect. 5.4, the relevance of these concepts for describing supporting assumptions of a hypothesis h with respect to a propositional model will be elucidated. In Sect. 5.5, methods for searching prime implicants and implicates will be presented.

The propositional calculus over a given alphabet $P = \{p_1, p_2, \ldots, p_n\}$ can also be looked at from another perspective. For convenience let's convert the truth values t and f into the numerical values 1 and 0 respectively. Any possible interpretation of the n atoms is then given by an n-tupel out of the set $B_n = \{0, 1\}^n$, the n-dimensional Boolean cube. Any n-tupel $x \in B_n$ determines an interpretation for a formula $f \in \mathcal{L}_P$. Let $f(x)$ be the truth value of the formula f under the interpretation x. f becomes thus an application from B_n to $\{0, 1\}$, that is, a so called *Boolean function.* To any formula $f \in \mathcal{L}_P$, the set $N(f)$ of all n-tupels $x \in B_n$ for which $f(x) = 1$, i.e. the set of all interpretations under which f is true, is assigned. This application enjoys the properties assembled in the following theorem.

Theorem 5.4. *Let P be an alphabet of n atoms $p_1, p_2, \ldots, p_n$ and let $f, g \in \mathcal{L}_P$. Then*

(1) $f = g$ if and only if $N(f) = N(g)$.
(2) $N(\neg f) = N(f)^c = B_n - N(f)$.

(3) $N(f \wedge g) = N(f) \cap N(g)$.
(4) $N(f \vee g) = N(f) \cup N(g)$.
(5) $f \models g$ *if and only if* $N(f) \subseteq N(g)$.

This links propositional calculus to the algebra of sets, more particularly to the algebra of subsets of Boolean cubes. Theorem 5.4 shows that the Boolean algebra of equivalent formulas over a finite alphabet is isomorphic to the Boolean algebra of subsets of the corresponding Boolean cube. Negation, conjunction and disjunction correspond to complementation, intersection and union under this isomorphism. This point of view becomes especially important for probabilistic considerations (Sect. 5.7).

The notions of implicants and implicates have their correspondents in the isomorphic subset system of B_n. A subset N of B_n is called an *interval of rank* r if there exists a conjunction c of r different literals such that $N = N(c)$. Any interval of rank r corresponds to a $(n-r)$-dimensional subcube of B_n. A subset N is called a *cointerval of rank* r if there exists a disjunction d with r different literals such that $N = N(d)$. A cointerval is the complement of an interval, $N = N(d) = N(\neg d)^c$. An interval $N(c)$ is called *permissible* with respect to a subset N of B_n if $N(c)$ is contained in N. A permissible interval $N(c)$ for N is called maximal if no other permissible interval of N contains $N(c)$. Similarly, a cointerval $N(d)$ is called permissible for a subset N if $N(d)$ contains N. A permissible cointerval $N(d)$ for N is called minimal if no other permissible cointerval for N is contained in $N(d)$. The following theorem summarizes the relations between implicant, implicates and permissible intervals, cointervals.

Theorem 5.5.

(i) If $f = f_1 \vee f_2 \vee \ldots \vee f_n$ *is a disjunctive normal form for* f*, then*
(1) Every $N(f_i)$ *is a permissible interval for* $N(f)$.
(2) f_i *is a prime implicant for* f *if and only if* $N(f_i)$ *is a maximal permissible interval of* $N(f)$.

(ii) If $f = f_1 \wedge f_2 \wedge \ldots \wedge f_n$ *is a conjunctive normal form for* f*, then*
(1) Every $N(f_i)$ *is a permissible cointerval for* $N(f)$.
(2) f_i *is a prime implicate for* f *if and only if* $N(f_i)$ *is a maximal permissible cointerval of* $N(f)$.

Theorem 5.2 has its counterpart with maximal permissible intervals and minimal permissible cointervals.

Theorem 5.6. *If* $N_1, N_2, \ldots, N_r$ *are all the maximal permissible intervals (minimal permissible cointervals) of a set* N*, then* $N = N_1 \cup \ldots \cup N_r$ $(N = N_1 \cap \ldots \cap N_r)$.

An important special case arises if the Boolean function $f(x)$ associated with a formula f is monotone: A Boolean function φ is *properly monotone*, if $\varphi(v') \leq \varphi(v'')$, whenever $v' < v''$ (i.e. $v'_i \leq v''_i$ for all i and $v'_i < v''_i$ for at least one i). A Boolean function φ is called *non-monotone*, if it is (i) not properly monotone and (ii) there exists no sequence of transformations of the form $y_i = \neg x_i$ such that φ becomes properly monotone. Note that for example $x_1 \wedge \neg x_2$ is not properly monotone, but using $y_2 = \neg x_2$ it is transformed into $x_1 \wedge y_2$ and this function is properly monotone. A Boolean function which is not non-monotone is called *monotone*; by a suitable sequence of transformations of the form $y_i = \neg x_i$ it can be made properly monotone.

The general results presented above specialize for the case of properly monotone Boolean functions as follows:

Theorem 5.7. *The Boolean function $f(x)$ of a formula f is properly monotone if and only if the prime implicants of f contain no negated atoms.*

Furthermore, the notions of minimal and shortest disjunctive normal forms coincide for formulas f with monotone Boolean functions $f(x)$ and this form is the only irredundant disjunctive representation of a monotone Boolean function φ.

Theorem 5.8. *A formula f with a monotone Boolean function $f(x)$ has only one irredundant disjunctive normal form, namely the form containing all prime implicants.*

There are models which give rise to monotone Boolean functions and these models are somewhat simpler to treat in view of the last two results. However, in general the Boolean functions arising in the context of the reasoning with unreliable arguments will not be monotone.

5.3 Propositional System With Uncertain Assumptions and Hints

To start with let us define the elements of the formal system considered henceforth in this chapter. $P = \{p_1, p_2, \ldots, p_n\}$ and $A = \{a_1, a_2, \ldots, a_s\}$ are two sets of propositions, the elements of the latter are called assumptions, $N = P \cup A$ is the entire set of symbols considered. Like in Sect. 5.1, we assume that an assumption a_i is true with a known probability $q_i, i = 1, \ldots, s$. We also assume that the assumptions are independent. If Σ is a set of formulas, then $\sim \Sigma$ denotes the set obtained by taking the negation of each element in Σ. Thus, $P^{\pm} = P \cup \sim P$ and $A^{\pm} = A \cup \sim A$ are the sets of all literals over P and A respectively. $\Sigma = \{\xi_1, \xi_2, \ldots, \xi_r\}$ is a set of clauses over literals in $N^{\pm} = N \cup \sim N$. These clauses form the knowledge base and let $\xi = \xi_1 \wedge \xi_2 \wedge \ldots \wedge \xi_r$. These clauses represent statements about the information or knowledge available; that is, they restrict the possible common truth values of the propositions involved. Let h be a formula from $\mathcal{L}_N$ representing a hypothesis which may be true or not. The problem considered here is to judge whether or not h could possibly be true in view of the knowledge available. This can be done by looking for what assumptions have to be assumed in order to infer h from Σ or, alternatively, what assumptions have to be assumed in order to infer $\neg h$ and therefore to refute h. These ideas are formalized using the theory of hints.

Since the theory of hints is designed to answer questions, the question to be considered in this case is "What are the common truth values of the propositions in N ? " The frame of discernment for this question is the Boolean cube B_{n+s} because there is exactly one vector θ in B_{n+s} corresponding to the correct truth values of the propositions in N. Each clause ξ_i from the knowledge base Σ represents a piece of information because if ξ_i is true, then θ is necessarily in $N(\xi_i) \subseteq B_{n+s}$. Since each ξ_i in Σ is true, the knowledge base Σ implies that θ is in $\cap_{i=1}^{r} N(\xi_i) = N(\xi_1 \wedge \ldots \wedge \xi_r) = N(\xi)$. The knowledge base Σ is therefore represented by the subset $N(\xi)$ in B_{n+s}.

Similarly, fixing the truth values of the assumptions $a_1, \ldots, a_s$ corresponds to a vector x in B_s. Note that B_s can be seen as a subset of B_{n+s}. Now, if we define

$$\alpha_i = \begin{cases} a_i & \text{if } x_i = 1 \\ \neg a_i & \text{if } x_i = 0 \end{cases}$$

then each α_i is true under x and hence θ is in $N(\alpha_i)$ for all $i = 1, \ldots, s$. Therefore, if $conj(x) = \alpha_1 \wedge \ldots \wedge \alpha_s$, then θ is in $\cap_{i=1}^{s} N(\alpha_i) = N(conj(x))$. This means that assuming the "truth values" x for the assumptions in A implies that θ is in $N(conj(x)) \subseteq B_{n+s}$. Combining this information with the knowledge base represented by $N(\xi)$ shows that θ is in $N(conj(x)) \cap N(\xi)$. Since we know that the probability of a_i being true equals q_i and the assumptions are independent, the probability that x is the correct vector is

$$P'(x) = \prod_{i=1}^{s} q_i^{x_i}(1-q_i)^{1-x_i}. \tag{5.1}$$

Remark that the set

$$\Gamma(x) = N(conj(x) \wedge \xi) \tag{5.2}$$

may very well be empty for some vectors x in B_s because all we know is that if x is the correct vector, then θ is in $\Gamma(x)$. But if x is not the correct vector, then $\Gamma(x)$ may or may not be empty. For example, if $P = \{p_1\}$, $A = \{a_1, a_2\}$ and $\Sigma = \{\neg a_1 \vee p_1, \neg p_1\}$, then a_1 is necessarily false according to Σ and hence $\Gamma(1,1) = N(a_1 \wedge a_2 \wedge \xi) = \emptyset$. Furthermore, among the vectors $x' = (0,0)$ and $x'' = (0,1)$, one is true and the other is false whereas both $\Gamma(x')$ and $\Gamma(x'')$ are not empty. This shows that if we learn that $\Gamma(x)$ is empty, then we can deduce that x is certainly not the correct vector. This implies that we have to *condition* the a priori probability of configurations P' on the event $\Omega = \{x \in B_s : \Gamma(x) \neq \emptyset\}$, resulting in a new probability space (Ω, P) where $P(x) = P'(x)/P'(\Omega)$. The elements of $B_s - \Omega$ are called *contradictory configurations*; they are eliminated from B_s. This name for the elements of $B_s - \Omega$ is motivated as follows: if we assume that $x \in B_s - \Omega$ is the correct configuration, then this leads to a contradiction with the accepted knowledge Σ because $N(conj(x)) \cap N(\xi) = \emptyset$, i.e. $conj(x) \wedge \xi$ is inconsistent. This permits to define the hint

$$\mathcal{H} = (\Omega, P, \Gamma, B_{n+s}). \tag{5.3}$$

Now, consider a formula $h \in \mathcal{L}_N$ representing a hypothesis which may be true or false. Notice that h is true if and only if h is true under the interpretation θ, i.e. $\theta \in N(h)$. Therefore it is interesting to look at the set

$$u(h) = \{x \in \Omega : \Gamma(x) \subseteq N(h)\} \tag{5.4}$$

because if x is the correct configuration and $x \in u(h)$, then h is necessarily true. Since x is the correct configuration with probability $P(x)$, it is natural to define the *degree of support* of h in the light of the hint $\mathcal{H}$ by

$$sp(h) = P(u(h)). \tag{5.5}$$

The rest of the chapter is concerned with the problem of finding efficient procedures for computing such degrees of support.

Now, to compute $sp(h)$, it turns out to be useful to consider a slight generalization of the notion of hint. Indeed, before conditioning, our knowledge is represented by the so-called *quasi-hint*

$$\mathcal{H}' = (B_s, P', \Gamma, B_{n+s}) \tag{5.6}$$

where this time $\Gamma(x)$ is allowed to be empty. From a quasi-hint, conditioning always permits deducing a hint which can be used to compute degrees of support. A quasi-hint allows for the computation of the so-called *degree of quasi-support* of h:

$$sp'(h) = P'(u'(h)) \tag{5.7}$$

where

$$u'(h) = \{x \in B_s : \Gamma(x) \subseteq N(h)\}. \tag{5.8}$$

Note that degrees of support must always be computed with respect to hints and not quasi-hints because contradictory configurations clearly do not contribute to the support or credibility of a hypothesis h, although $\emptyset = \Gamma(x) \subseteq N(h)$ for such configurations (Kohlas, 1993; Kohlas, Monney, 1990, 1994).

Furthermore, note that $\Omega = B_s - u'(\bot)$ and hence $P'(\Omega) = 1 - P'(u'(\bot))$. Together with $u(h) = u'(h) - u'(\bot)$ and $u'(\bot) \subseteq u'(h)$, this implies the following theorem:

Theorem 5.9 *For any formula $h \in \mathcal{L}_N$, we have*

$$sp(h) = \frac{sp'(h) - sp'(\bot)}{1 - sp'(\bot)}. \tag{5.9}$$

According to this result, in order to compute the degree of support of h, we have to compute the degree of quasi-support of h and $\bot$. For this purpose, we will take advantage of some results presented in Sect. 5.2.

5.4 Support of Hypotheses: a Logical View

In this section we are going to present a method of finding a more economic representation of $u'(h)$ for an arbitrary formula $h \in \mathcal{L}_N$ (h may very well also be the inconsistency $\perp$). In fact, for large sets of assumptions A, it is very difficult to compute $u'(h)$ explicitely because B_s has 2^s elements which must all be tested as to whether they belong to $u'(h)$. Moreover, we don't really have to compute $u'(h)$, but rather its probability $P'(u'(h))$.

Consider the formula

$$\varphi_h = \vee \{conj(x) : x \in u'(h)\} \tag{5.10}$$

in $\mathcal{L}_A$, which is a sublanguage of $\mathcal{L}_N$. We have the following result:

Theorem 5.10 *For all $h \in \mathcal{L}_N$, we have $u'(h) = N(\varphi_h)$.*

Note that since $u'(\perp) \subseteq u'(h)$, this result implies that $N(\varphi_\perp) \subseteq N(\varphi_h)$ and hence $\varphi_\perp \models \varphi_h$. Now, if we can find a disjunctive normal form representation $\varphi_h = f_1 \vee f_2 \vee \ldots \vee f_k$, then by theorem 5.4 we have the following more economic representation of $u'(h)$:

$$u'(h) = \cup_{i=1}^{k} N(f_i). \tag{5.11}$$

As we will see in Sect. 5.7, this type of representation of $u'(h)$ is very useful to compute $P'(u'(h)) = sp'(h)$, especially when the sets $N(f_i)$ are disjoint, which is the case when the conjunctions f_i are pairwise inconsistent.

Now let's look at the problem of finding a disjunctive normal form representation of φ_h. Denote by $\mathcal{C}_A$ the set of all conjunctions of zero, one or more literals in $A^\pm$ (a conjunction with zero literal is the tautology $\top$ which is always true). We say that a conjunction a in $\mathcal{C}_A$ is a *quasi-support* of h if $a \wedge \xi \models h$. Note that $\perp$ is not in $\mathcal{C}_A$ and hence cannot be a quasi-support of h. However, each $a_i \wedge \neg a_i, i = 1, \ldots s$ is in $\mathcal{C}_A$ and is in fact a quasi-support of h for any formula h in $\mathcal{L}_N$, including the inconsistency $\perp$.

Theorem 5.11 *A conjunction $a \in \mathcal{C}_A$ is a quasi-support of h if and only if $\xi \models h \vee \neg a$.*

A quasi-support of $\perp$ is called a *contradiction* and we have the following theorem. It is important to note that a contradiction is a quasi-support of any formula h in $\mathcal{L}_N$.

Theorem 5.12 *Let a be a conjunction in $\mathcal{C}_A$. The following three statements are equivalent:*

(1) a is a contradiction
(2) $\xi \models \neg a$
(3) $a \wedge \xi$ is not satisfiable, i.e. $a \wedge \xi = \perp$.

The following fundamental theorem establishes the link between quasi-supports of h and the logical formula φ_h.

Theorem 5.13 *A conjunction $a \in \mathcal{C}_A$ is an implicant of φ_h if and only if a is a quasi-support of h.*

Since theorem 5.13 is valid in particular when h is the inconsistency $\perp$, we have the following result.

Corollary 5.1 *A conjunction $a \in \mathcal{C}_A$ is an implicant of $\varphi_\perp$ if and only if a is a contradiction.*

A quasi-support of h is called *minimal* if no proper subconjunction of a is also a quasi-support of h. A contradiction a is called *minimal* if no proper subconjunction of a is also a contradiction. But beware, for any formula h in $\mathcal{L}_N$, a minimal contradiction is a quasi-support of h which is minimal within the restricted set of quasi-supports of h consisting of contradictions, but not necessarily within the whole set of quasi-supports of h. Therefore, there might exist minimal contradictions which are not minimal quasi-supports of h ! Of course, a minimal quasi-support of h which is a contradiction is necessarily a minimal contradiction.

The following two results follow directly from theorem 5.13 and corollary 5.1 respectively.

Corollary 5.2 *A conjunction $a \in \mathcal{C}_A$ is a prime implicant of φ_h if and only if a is a minimal quasi-support of h.*

Corollary 5.3 *A conjunction $a \in \mathcal{C}_A$ is a prime implicant of $\varphi_\perp$ if and only if a is a minimal contradiction.*

Again, although each implicant of $\varphi_\perp$ is also an implicant of φ_h, note that a prime implicant of $\varphi_\perp$ is not necessarily a prime implicant of φ_h because a minimal contradiction is not necessarily a minimal quasi-support of h.

Now, by theorem 5.2, if $f_1, \ldots, f_r$ are all the minimal quasi-supports of h, then $\varphi_h = f_1 \vee \ldots \vee f_r$ and we have found a disjunctive normal form representation of φ_h. Then by theorem 5.10 we have

$$u'(h) = \cup_{i=1}^{r} N(f_i). \tag{5.12}$$

Similarly, if $g_1, \ldots, g_s$ are the minimal contradictions, then $\varphi_\perp = g_1 \vee \ldots \vee g_s$ and hence

$$u'(\perp) = \cup_{i=1}^{s} N(g_i). \tag{5.13}$$

Formulas (5.12) and (5.13) can then be used to compute $sp'(h)$ and $sp'(\perp)$, which in turn can be used to compute the exact degree of support of h in the light of the available knowledge represented by the hint $\mathcal{H}$ (see (5.9)). Thus our problem is reduced to the determination of all minimal quasi-supports of h and all minimal contradictions.

Note that $\top \in \mathcal{C}_A$ is a quasi-support of h if and only if $\Sigma \models h$. In this case, $\top$ is the only minimal quasi-support of h and hence $\varphi_h = \top$, which implies that $u'(h) = N(\varphi_h) = B_s$ and so $sp'(h) = 1$ and $sp(h) = 1$.

In relation with their "Assumption Based Truth Maintenance Systems (ATMS)", Reiter, De Kleer (1987) called a contradiction a "no-good" and a quasi-support which is not a contradiction a "label" if the hypothesis h is a single literal. For our purposes however, both the contradictions and the quasi-supports and not the labels are needed. De Kleer (1986 a) originally assumed Horn clauses in Σ, that is clauses with at most one non-negated atom. They also give a number of theorems which may be adapted to compute minimal quasi-supports and minimal contradictions (see also Laskey, Lehner, 1989 and Provan, 1990). These results are given below.

Let us assume for the moment that h is a *clause*, that is a disjunction of literals from $N^{\pm}$. This assumption will be dropped again below. Then, for any quasi-support a of h, $\neg a \vee h$ is also a clause which is an implicate of ξ. In fact, by adapting theorems of Reiter, De Kleer (1987), it follows that there are even closer links between the minimal quasi-supports of h and the prime implicates of ξ.

Let's introduce some notation and conventions. First of all, clauses or conjunctions are sometimes conveniently considered as the sets of the literals they contain. Then for example $x \in c$ means that the literal x appears in c, and $c \cap c'$ denotes the set of literals contained both in c and c', but $c \cap c'$ is neither a clause nor a conjunction. A clause or conjunction f is said to *subsume* a clause or a conjunction g, if $f \supseteq g$. If f is a clause which subsumes g, then $g \models f$; if f is a conjunction which subsumes g, then $f \models g$. If f and g are two clauses, $f - g$ denotes the subclause of f which is obtained when all literals present both in f and g are eliminated from f. The set of implicates of a set of clauses Σ is denoted by $Th(\Sigma)$. For an arbitrary set of clauses S, the subset of clauses which subsume no other clause of the set, that is, which are minimal within the set, is denoted by μS. Then the set of prime implicates of ξ, i.e. the prime implicates of Σ, is $PI(\Sigma) = \mu Th(\Sigma)$.

Now, let $MQS(\Sigma, h)$ denote the set of minimal quasi-supports of h with respect to Σ and $MC(\Sigma)$ the set of minimal contradictions with respect to Σ. Let $\mathcal{C}'_A$ denote the set of all clauses composed of zero, one or more literals in $A^{\pm}$ (a clause with zero literal is the inconsistency $\perp$ which is always false). Furthermore, note that $\mathcal{C}'_A = \sim \mathcal{C}_A$.

Theorem 5.14 *If h is a clause, then*

$$MQS(\Sigma, h) = \sim \mu\{f - h \in \mathcal{C}'_A : f \in PI(\Sigma)\}. \qquad (5.14)$$

Corollary 5.4

$$MC(\Sigma) = \sim \{f \in \mathcal{C}'_A : f \in PI(\Sigma)\}. \qquad (5.15)$$

The problem of finding minimal quasi-supports of h or minimal contradictions is thus related to the problem of finding prime implicates of the conjunctive normal form $\xi = \xi_1 \wedge \ldots \wedge \xi_r$. The method of computing them, based on theorem 5.14 and corollary 5.4, depends on an algorithm for finding all prime implicates of ξ and then extracting from them the minimal contradictions and the minimal quasi-supports. A method for this task will be described in the next section. This is, however, a brute-force approach, by which often far too many unnecessary prime implicates are computed. More tightly controlled methods are possible if we adapt methods pointed out by Inoue (1991 a and b, 1992). These methods make use of concepts and procedures introduced by Bossu, Siegel (1985) and Siegel (1987).

These methods depend on the notions of production field and characteristic clauses. A *production field* P is simply a non-empty set of clauses of the language $\mathcal{L}_N$. In fact, the production field which will interest us most is the set of clauses in $\mathcal{C}'_A$. This production field will be denoted by P_A. A production field is called *stable* if any subsumed clause of a clause in P is also in P, i.e. if $c \in P$ and $c' \subseteq c$, then $c' \in P$. Clearly, the empty clause $\perp$ belongs to any stable production field and P_A is stable. $Th(\Sigma) \cap P$ represents all implicates of Σ which belong to the production field P. The minimal elements of this set are called the *characteristic clauses* of Σ with respect to P:

$$Carc(\Sigma, P) = \mu(Th(\Sigma) \cap P). \tag{5.16}$$

Similarly, we can define characteristic clauses for any collection S of formulas in $\mathcal{L}_N$:

$$Carc(S, P) = \mu(Th(S) \cap P). \tag{5.17}$$

Theorem 5.15 *If P is stable production field, then $Carc(S, P) = PI(S) \cap P$.*

Let f be an arbitrary formula in $\mathcal{L}_N$, possibly the inconsistency $\perp$. Let $MQS(\Sigma, f)$ denote the minimal quasi-supports of f with respect to Σ. Given a stable production field P, we denote the minimal quasi-supports of f belonging to $\sim P$ by

$$MQS(\Sigma, f, P) = MQS(\Sigma, f) \cap \sim P. \tag{5.18}$$

In particular, since $MC(\Sigma) = MQS(\Sigma, \perp)$, we have

$$MQS(\Sigma, \perp, P) = MC(\Sigma) \cap \sim P. \tag{5.19}$$

Theorem 5.16 *Let Σ be a set of clauses in $\mathcal{L}_N$, f an arbitrary formula in $\mathcal{L}_N$ and P a stable production field. Then the set of minimal quasi-supports of f with respect to Σ which belong to $\sim P$ is*

$$MQS(\Sigma, f, P) = \sim Carc(\Sigma \cup \{\neg f\}, P). \tag{5.20}$$

Corollary 5.5 *If P is a stable production field, then*

$$MC(\Sigma) \cap \sim P = \sim Carc(\Sigma, P). \tag{5.21}$$

In consequence, we can prove the following result.

Theorem 5.17 *For an arbitrary formula f in $\mathcal{L}_N$, we have*

$$\begin{aligned} MC(\Sigma) &= \sim Carc(\Sigma, P_A) \\ MQS(\Sigma, f) &= \sim Carc(\Sigma \cup \{\neg f\}, P_A). \end{aligned} \tag{5.22}$$

In view of the last theorem, characteristic clauses are of interest for finding minimal contradictions and minimal quasi-supports. In fact, we may more generally be interested in minimal contradictions and minimal quasi-supports belonging to a more restricted (negated) production field $\sim P$ than $\sim P_A$. For example, the production field of clauses over $A^{\pm}$ with length at most k, i.e. minimal contradictions and minimal quasi-supports which contain not too many assumptions. This is also a stable production field. Theorem 5.16 and corollary 5.5 show that these can be found by computing characteristic clauses. Note that in theorems 5.16 and 5.17 f is an arbitrary formula in $\mathcal{L}_N$ and not necessarily a clause like in theorem 5.14. In view of these theorems, we are interested in methods to compute characteristic clauses relative to a stable production field P, which may be P_A for example. This can be much more feasible than computing all prime implicates of ξ and applying theorem 5.14 and corollary 5.4 when f is a clause. Such methods will be presented in Sect. 5.6.

5.5 Computing Quasi-Supports and Contradictions by Resolution

In this section, methods for computing minimal quasi-supports and minimal contradictions with respect to a knowledge base Σ will be presented. According to theorem 5.14 and corollary 5.4 the minimal quasi-supports of clauses and the minimal contradictions can be derived from the prime implicates of Σ. Therefore, we show how all prime implicates of a conjunctive normal form can be found by the well-known technique of *resolution*. This is a naive brute-force approach which may demand too much computing time (see also Kean, Ksiknis, 1990). For this reason, a more controlled approach based on production systems introduced by Siegel (1987) will be described. Together with theorem 5.17 and a few theorems due to Inoue (1991 a and b, 1992), this also allows the computing of minimal contradictions and minimal quasi-supports for arbitrary hypotheses h this time, not only for clauses. Based on these methods, different "interpretative" and "compiled" system architectures for treating assumption-based reasoning are discussed. Several examples serve to illustrate the methods.

As in the general model of the previous section, a set of propositions $P = \{p_1, \ldots, p_n\}$ and a set of assumptions $A = \{a_1, \ldots, a_s\}$ form the basis of the model. $\Sigma = \{\xi_1, \ldots, \xi_r\}$ is a set of clauses in $\mathcal{L}_N$ representing the knowledge base. How can the prime implicates of Σ be found ? The solution to this problem is based on the well-known resolution principle due to Robinson (1965). Let f and g be two clauses containing exactly one pair of literals with opposite signs, such that $f = f' \vee p$ and $g = g' \vee \neg p$. Then the *resolvent* $\rho(f, g)$ of these two clauses is the clause $f' \vee g'$, where it is understood that identical literals in $f' \vee g'$ are regrouped such that any literal appears only once.

Theorem 5.18 *The resolvent of f and g is a logical consequence of $f \wedge g$.*

Now, the following theorem gives a method for finding all prime implicates of a conjunctive normal form.

Theorem 5.19 *Given a satisfiable formula f in conjunctive normal form, the complete set of prime implicates of f can be found by repeatedly performing the following operations on f as long as it is possible:*
(1) If a disjunctive term t of f subsumes another term s of f drop t.
(2) If for two terms s and t of f a resolution is possible and if the resolvent $\rho(s, t)$ subsumes no term already in f, then add $\rho(s, t)$ to f.

The clauses of the resulting conjunctive form upon termination of this procedure are the prime implicates of f. Let us illustrate this procedure by a few examples.

Example 5.1 To start with, consider the sensor model introduced in Sect. 1.2 which is defined by the following statements:

$R_1 : \neg h \vee \neg a_1 \vee e$
$R_2 : \neg a_2 \vee e$
$R_3 : h \vee a_2 \vee \neg e.$

Add to these relations the fact that the evidence e has been observed:

$F_1 : e$

The clauses R_1, R_2, R_3 and F_1 form the system Σ. Apply the procedure defined in theorem 5.19. Note that R_1 and R_2 both subsume F_1 so that these two terms can be dropped. Next, R_3 and F_1 resolve into $h \vee a_2$ and now R_3 may also be dropped because it subsumes this resolvent. Here the procedure stops and the following two prime implicates remain

$e, \; h \vee a_2.$

By theorem 5.14 the hypothesis h has one minimal quasi-support, namely $\neg a_2$. Also, by corollary 5.4, there is no contradiction.

Example 5.2 Consider a simplified version of the model assembled in Sect. 5.1:

$$\begin{array}{llll} R_1: & \neg b \vee \neg a_1 \vee a & R_6: & a \vee a_3 \vee \neg w \\ R_2: & \neg a_2 \vee a & R_7: & \neg a \vee a_4 \vee g \\ R_3: & b \vee a_2 \vee \neg a & R_8: & \neg a_4 \vee \neg g \\ R_4: & \neg a \vee w & R_9: & a \vee \neg g \\ R_5: & \neg a_3 \vee w & & \end{array}$$

Add the two facts

$$F_1 : w \qquad F_2 : g.$$

Now, R_4, R_5 subsume w and R_7 subsumes g; so according to the procedure of theorem 5.19 these terms may be dropped. Next resolve R_8 and R_9 with g and obtain $\neg a_4$ and a. Now, R_8 and R_9 subsume each of these resolvents and they can be dropped; R_1, R_2 and R_6 subsume the new term a and can therefore also be dropped. R_3 can be resolved with a adding the resolvent $b \vee a_2$ such that now R_3 can be dropped. At this stage the procedure stops, leaving the following complete set of prime implicates:

$$w, g, a, \neg a_4, b \vee a_2.$$

Thus, it becomes clear by theorem 5.14 and corollary 5.4 that a_4 is the only minimal contradiction and $\{a_4, \neg a_2\}$ is the set of minimal quasi-supports of b, the hypothesis of a burglary.

Suppose now, in another situation, that the facts are

$$F_1 : w \qquad F_2 : \neg g.$$

Here R_4, R_5 can again be dropped because they subsume w; R_8, R_9 can be dropped because they subsume $\neg g$. Next, resolve R_6 with w and R_7 with $\neg g$ and drop these two rules because they subsume the resolvents $a \vee a_3$ and $\neg a \vee a_4$ respectively. These two resolvents may be resolved themselves to $a_3 \vee a_4$. Furthermore, the second of the two resolvents can be resolved with R_1 and R_2 to give $\neg b \vee \neg a_1 \vee a_4$ and $\neg a_2 \vee a_4$ respectively and the first resolvent can be resolved with R_3 to give $b \vee a_2 \vee a_3$. So far, in addition to R_1, R_2 and R_3, the following terms remain:

$$\begin{array}{l} \neg a_2 \vee a_4, a_3 \vee a_4 \\ w, \neg g \\ a \vee a_3, \neg a \vee a_4 \\ b \vee a_2 \vee a_3, \neg b \vee \neg a_1 \vee a_4. \end{array}$$

Here no further resolvents can be added because they all subsume terms which are already present. For example, $\rho(\neg a_2 \vee a_4, b \vee a_2 \vee a_3)$ subsumes $a_3 \vee a_4$; the same is true for $\rho(b \vee a_2 \vee a_3, \neg b \vee \neg a_1 \vee a_4)$, etc. Also, no term of the above list subsumes another one. So this is, by theorem 5.19, the complete list of prime implicates of the corresponding system of clauses. Using theorems 5.14 and corollary 5.4, it follows that $a_2 \wedge \neg a_4$ and $\neg a_3 \wedge \neg a_4$ are the minimal contradictions and that a_2 and $\neg a_3$ are the minimal quasi-supports of the hypothesis of an alarm, whereas the only minimal quasi-support of its negation is $\neg a_4$. The minimal quasi-supports of a burglary are $\{a_2 \wedge \neg a_4, \neg a_3 \wedge \neg a_4, \neg a_2 \wedge \neg a_3\}$, whereas the minimal quasi-supports of its negation are $\{a_2 \wedge \neg a_4, \neg a_3 \wedge \neg a_4, a_1 \wedge \neg a_4\}$.

Example 5.3 Barnett's system. Consider a finite frame of discernment Θ and suppose that for all $\theta_i \in \Theta$ there is a simple hint focussing on $\{\theta_i\}$ and a simple hint focussing on $\{\theta_i\}^c$. Such systems have been studied by Barnett (1981). If θ_i denotes the proposition "θ_i is the answer", then this system can be transformed into the propositional model defined by the following clauses:

(1) $\theta_1 \vee \ldots \vee \theta_n$
(2) $\neg\theta_i \vee \neg\theta_j$ for all $i \neq j$
(3) $\neg a_i \vee \theta_i$ for all i
(4) $\neg b_i \vee \neg\theta_i$ for all i

The clause (1) says that one of the elements of Θ must be the correct answer and (2) states that only one of the elements of Θ can be the answer. Clauses (3) and (4) translate the simple hints into the propositional framework. The assumption a_i corresponds to the interpretation of the simple hint which focuses on the focal set $\{\theta_i\}$, a_i implies θ_i. Similarly, b_i implies $\neg\theta_i$. It is assumed here that there is exactly one simple hint with focal set $\{\theta_i\}$ and exactly one simple hint with focal set $\{\theta_i\}^c$ for all i. This is not really a loss of generality. If any hint is absent, then simply set the corresponding assumption a_i or b_i to false. If there is more than one hint with the same focal set, then they can be combined beforehand by Dempster's rule such that there remains only one hint focussing on the same set.

The resolution method allows one easily to find the prime implicates of the system Σ defined by (1) to (4) above. So, resolving the clauses (2) with those of (3), those of (4) with the clause (1) and the clauses (3) with (4) gives

(5) $\neg a_i \vee \neg\theta_j$ for all $i \neq j$
(6) $\theta_1 \vee \ldots \vee \neg b_i \vee \ldots \vee \theta_n$ for all i
(7) $\neg a_i \vee \neg b_i$ for all i.

No term can be dropped so far. Resolving the clauses of (3) with the new ones of (5) leads to

(8) $\neg a_i \vee \neg a_j$ for all $i \neq j$.

Furthermore, we may resolve the clauses of (4) with those of (6) to obtain eventually

(9) $\vee\{\neg b_j : \theta_j \in K^c\} \;\vee\; \vee\{\theta_i : \theta_i \in K\}$ for every proper subset K of Θ (including the empty set).

This permits to remove the clauses in (6) because they are contained in (9). None of the remaining terms subsumes another one and no further resolutions are possible. So (1) to (9) minus (6) are all prime implicates of the system (1) to (4).

From (7), (8) and (9), corollary 5.4 shows that the minimal contradictions are

$a_i \wedge b_i$ for all i, $a_i \wedge a_j$ for all $i \neq j$ and $b_1 \wedge \ldots \wedge b_n$

If H is a subset of Θ with $|H^c| \geq 2$ and $h = \vee\{\theta_i : \theta_i \in H\}$ is the clause expressing "the answer is in H", then by theorem 5.14 this hypothesis has the following minimal quasi-supports:

$$\begin{aligned} &\{a_i : \theta_i \in H\} \cup \{a_i \wedge b_i : \theta_i \in H^c\} \cup \\ &\{a_i \wedge a_j : \theta_i \neq \theta_j \text{ both in } H^c\} \cup \{\wedge\{b_j : \theta_j \in H^c\}\}. \end{aligned} \tag{5.23}$$

Note that the tautology $\top$ is the only minimal quasi-support of $H = \Theta$ and if $H = \Theta - \{\theta_k\}$ then its minimal quasi-supports are $\{a_i : \theta_i \neq \theta_k\} \cup \{b_k\}$. According to corollary 5.3, the prime implicants of $\varphi_\perp$ are precisely the minimal contradictions. Since these contain no negated symbols, theorem 5.7 implies that the Boolean function $\varphi_\perp(x)$ associated with the formula $\varphi_\perp$ is properly monotone. Also, using corollary 5.2, we can prove that the Boolean function $\varphi_h(x)$ associated with the fomula φ_h is properly monotone. This considerably simplifies the computation of $sp(h)$ (Sect. 5.7).

Example 5.4 Diagnostics with multiple faults. Let $\Theta = \{\theta_1, \ldots, \theta_n\}$ be a set of possible faults of a technical systems or sicknesses of a biological system. It is assumed that symptoms permit to locate faults with a certain reliability in certain subsets of Θ and that certain tests permit to exclude faults within certain subsets, again with more or less reliability. However it is admitted, that possibly no fault at all is present or that on the contrary possibly several faults are present simultaneously. The model presented below represents this situation in a very simple way and permits very efficient inferences. Compare this to other approaches, such as for example Bayesian models (Peng, Reggia, 1990), where it is, in general, difficult to account for the possibility of multiple faults, or to logic-based models (for example Reiter, 1987), which are computationally rather heavy.

Let $\mathcal{E} = \{E_1, \ldots, E_m\}$ be a family of subsets of Θ. It will be assumed that all evidence relates to subsets E in $\mathcal{E}$ either in a positive way, indicating a fault in E, or in a negative way, indicating that there should be no fault in E. It is supposed (without loss of generality) that all single element sets $\{\theta_i\}$ belong to $\mathcal{E}$ and that Θ belongs to $\mathcal{E}$. We associate with this family of subsets a family P of propositions h_i, where h_i states that "there is a fault in subset E_i" such that $\neg h_i$ states that "there is no fault in subset E_i". A subset E_j of $\mathcal{E}$ is called an immediate antecedent of a subset E_i of $\mathcal{E}$ if $E_j \subset E_i$ but there is no other subset E in $\mathcal{E}$ such that $E_j \subset E \subset E_i$. Figure 5.1 shows a family of subsets $\mathcal{E}$ and there is an edge between E_j and E_i if E_j is an immediate antecedent of E_i.

The edges in Fig. 5.1 represent inclusions which are translated into the implications $(h_j \rightarrow h_i) = (\neg h_j \vee h_i)$ for all E_i in $\mathcal{E}$ and for any immediate antecedent E_j of E_i. These implications express the relation that a fault in

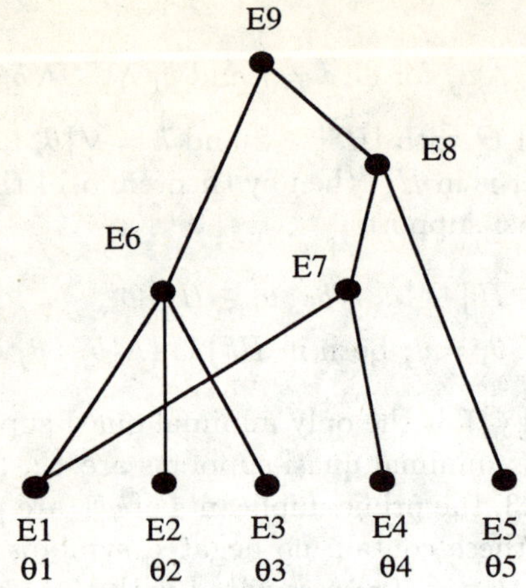

Fig. 5.1. A family of subsets

E_j means also a fault in E_i or, inversely, no fault in E_i implies no fault in E_j. So far the propositional model contains the following clauses

(1) $\neg h_j \vee h_i$, for all E_i in $\mathcal{E}$ and for all immediate antecedents E_j of E_i.

Symptoms and tests add evidence for the presence or absence of faults which will be represented by implications of the form $(a_i \rightarrow h_i) = (\neg a_i \vee h_i)$ or $(b_i \rightarrow \neg h_i) = (\neg b_i \vee \neg h_i)$. The propositions a_i and b_i are assumptions under which a given evidence for or against a fault in set E_i is valid or reliable. This represents the fact that symptoms and tests are, in general, not infallible evidence for faults. This adds the clauses

(2) $\neg a_i \vee h_i$
(3) $\neg b_i \vee \neg h_i$

to the system. We may assume a clause (2) and (3) for any E_i of $\mathcal{E}$; if there is no evidence for or against a fault in E_i, the corresponding assumptions a_i or b_i are simply set to "false" or the corresponding probability is set to zero.

First (2) and (3) may be resolved to give

(4) $\neg a_i \vee \neg b_i$ for all $E_i \in \mathcal{E}$.

Then, some clauses in (1) can be resolved among themselves to give

(5) $\neg h_j \vee h_i$ for all $E_j \subset E_i$ in $\mathcal{E}$

Then, performing resolutions between clauses (5) and (2), and between (5) and (3), and then again performing resolutions between these resulting clauses and (5) shows that we obtain the following set of clauses. In order to describe them, we need to introduce some notation. For a set E in $\mathcal{E}$, let $\mathcal{U}(E)$ denote the sets in $\mathcal{E}$ which contain E:

$$\mathcal{U}(E) = \{E_i \in \mathcal{E} : E \subseteq E_i\}. \tag{5.24}$$

Similarly, let $\mathcal{L}(E)$ denote the sets in $\mathcal{E}$ which are contained in $\mathcal{E}$:

$$\mathcal{L}(E) = \{E_i \in \mathcal{E} : E_i \subseteq E\}. \tag{5.25}$$

Then, incorporating the clauses (2) and (3), we obtain the clauses

$$(6) \quad \cup_{k=1}^{m} \{\neg a_k \vee h_i : E_i \in \mathcal{U}(E_k)\} \tag{5.26}$$

and

$$(7) \quad \cup_{k=1}^{m} \{\neg b_k \vee \neg h_i : E_i \in \mathcal{L}(E_k)\}. \tag{5.27}$$

Resolving clauses (6) and (7) and incorporating (4) gives

(8) $\neg a_i \vee \neg b_j$ for all $E_i \subseteq E_j$ in $\mathcal{E}$.

The clauses (5) to (8) are all the prime implicates of the system (1) to (3) according to theorem 5.19. By corollary (5.4), we have the minimal contradictions

(9) $a_i \wedge b_j$ for all $E_i \subseteq E_j$ in $\mathcal{E}$

This is of course what we expect, an evidence for a fault in a subset E_i of another set E_j contradicts an evidence against a fault in E_j. By theorem 5.14, a hypothesis h_i has the minimal quasi-supports

$$\{a_k : E_k \in \mathcal{L}(E_i)\} \cup \{a_k \wedge b_l : E_k \subseteq E_l, E_k \notin \mathcal{L}(E_i)\}.$$

The second set in the above union accounts for the contradictions, whereas the first set is very intuitive: an evidence for a fault in a subset of E_i is an evidence for a fault in E_i. Similarly, we can find that the minimal quasi-supports of $\neg h_i$ are

$$\{b_k : E_k \in \mathcal{U}(E_i)\} \cup \{b_k \wedge a_l : E_l \subseteq E_k, E_k \notin \mathcal{U}(E_i)\}.$$

This is again very intuitive, an evidence against a fault in a superset of E_i is an evidence against a fault in E_i. Finally, let's mention that diagnostic problems are considered in more details in Chap. 11. This completes the example 5.4

In the example 5.4, all clauses of the system (1) to (3) contain at most one non-negated literal. Such clauses are called *Horn clauses*; they correspond to implications of the form $\alpha_a \wedge \ldots \wedge \alpha_m \rightarrow \alpha_{m+1}$. The resolution method corresponds here to the replacement in an implication of a condition α_i by the conditions of another implication which has α_i as its consequence. This allows for a simple procedure for propagating assumptions forward through implications as examplified in the system above. The procedure may be generalized to arbitrary Horn clauses (see De Kleer, 1986 a and b).

One may add the information that there is at least one fault in the system. This means adding the clause

$\vee\{h_i : E_i = \{\theta_i\}\}$.

Note that this is no longer a Horn clause. Alternatively, we may add the information that there is at most one fault in the system:

$$\neg(h_i \wedge h_j) = \neg h_i \vee \neg h_j \text{ for all } i \neq j \text{ such that } E_i = \{\theta_i\}, E_j = \{\theta_j\}.$$

Or one may add both conditions, specifying that there is exactly one fault in the system. Barnett's system is a special case of such a system. The same case, but assuming that the family $\mathcal{E}$ forms a hierarchy, i.e. either $E_i \subset E_j$ or $E_j \subset E_i$ or $E_i \cap E_j = \emptyset$ whenever E_i and E_j belong to $\mathcal{E}$, has been considered by Shafer, Logan (1987). They propose an efficient algorithm for computing degrees of support based on an alternative method like those discussed in Chap. 8.

5.6 Skipped Ordered Resolution

Example 5.2 indicates that the resolution procedure may become prohibitively long if applied to a large knowledge base Σ. Obviously, the procedure defined in theorem 5.19 has to be specified more precisely, because in general several resolutions will be possible and the sequence in which resolvents are formed may be decisive for the computational effort needed to generate the prime implicates. More efficient procedures have been proposed by Inoue (1991 a, b and 1992), based upon work of Siegel (1987).

By theorem 5.17 it is sufficient to compute the set $Carc(\Sigma, P_A)$ of characteristic clauses with respect to the production field P_A in order to obtain the minimal contradictions with respect to Σ. Also, by the same result, $Carc(\Sigma \cup \{\neg f\}, P_A)$ permits one to obtain all minimal quasi-supports of an arbitrary hypothesis f in $\mathcal{L}_N$ (not necessarily a clause). If we don't want to compute all minimal contradictions and minimal quasi-supports, but only those which belong to some (negated) stable production field $\sim P$, theorem 5.16 and corollary 5.5 show that we also have to compute characteristic clauses. The problem is therefore to compute these sets. Its solution is given by a skipped, ordered, linear resolution procedure proposed by Siegel (1987). This procedure works with *structured clauses*, that is pairs (p, q^*), where p is an ordinary clause and q^* a sequence $s_1, s_2, \ldots, s_t$ of non-empty sequences s_i of literals, called a tree (Siegel, 1987). The sequences $s_i = (l_1, l_2, \ldots, l_r), r \geq 1$, are called branches and the first element l_1 of a branch is called its leaf.

We define two operations with structured clauses:

(1) **Skip**: If $(p, (ls)q^*)$ is a structured clause with a branch (l, s) with leaf l and a tree q^*, define

$$\sigma(p, (ls)s_2 \ldots s_t) = (l \vee p, s_2 \ldots s_t). \tag{5.28}$$

(2) **Resolve**: If $(p, (ls)q^*)$ is a structured clause with a branch (ls) with leaf l and a tree q^*, and if there is a clause ξ in Σ which satisfies the following conditions on *non-repetition*:

(2.1) ξ contains the literal $\neg l$
(2.2) no literal of ξ is contained in the branch (ls)
(2.3) no literal of ξ appears negated as a leaf in q^*
(2.4) no literal of ξ appears negated in p,

then

$$\rho((p,(ls)q^*),\xi) = (p,(l'_i ls)(l'_2 ls)\dots(l'_t ls)q^*) \tag{5.29}$$

where the literals l'_i are those literals of ξ which are *not removable*, that is satisfy the following conditions:

(2.2') $\neg l'_i$ does not appear in the branch (ls)
(2.3') l'_i does not appear as a leaf in q^*
(2.4') l'_i does not appear in p.

Note that the literal $\neg l$ of ξ is removable (it does not satisfy (2.2') and does therefore not appear among the l'_i). If all literals of ξ are removable, then the tree on the left in (5.29) reduces to q^*. The order in which the literals l'_i are arranged in (5.29) is irrelevant for the correctness of the algorithm which will be introduced below, but may influence its performance. Remark furthermore that the conditions of non-repetition are symmetric to the conditions of non-removability: if a literal is considered as repeated, then its negation is removable.

Given a set of clauses Σ, a *production* of a clause f from Σ is a sequence of structured clauses (p_i, q_i^*). Each of those structured clauses is called a step. This sequence must satisfy the following conditions:

(1) The first step is $(p_1, q_1^*) = (\bot, (l_1)(l_2)\dots(l_m))$, where the l_i are the literals of a clause ξ of Σ arranged in an arbitrary order.
(2) The last step is $(p_n, q_n^*) = (f, \emptyset)$, where $\emptyset$ denotes here the empty tree.
(3) (p_{i+1}, q_{i+1}^*) equals either $\sigma(p_i, q_i^*)$ or $\rho((p_i, q_i^*), \xi)$ for some ξ in Σ.

The order, in which the literals of ξ are arranged at the first step of the production is irrelevant for the correctness of the algorithm, it may however influence the performance of the algorithm. In fact, these productions do not really describe an algorithm, because at any step a number of choices is possible and in order to compute all productions, all choices have to be developed. But in any case at each step p_i never decreases. This implies that in a production any p_i subsumes any p_j with $j < i$. Hence, once a clause p_i is outside a *stable* production field P, all subsequent clauses p_j of all possible further steps and in particular the final clauses s, remain outside P. Therefore, if one wants to produce only clauses s which belong to a stable production field P, as will be the case below, then the productions may be cut at skip-steps, if $l \vee p_i$ does not belong to P.

Furthermore, because below we want to produce only minimal clauses, once a clause p_i is produced which subsumes a clause already produced, we may also cut the production. This is also because from p_i we would only produce clauses which subsume a clause already produced. So, there are important cuts which permit limiting the necessary productions.

Now, these productions are designed to generate implicates of Σ, and the following theorem tells us that this goal is achieved:

Theorem 5.20

(1) Finiteness: A finite set of clauses has a finite set of productions.
(2) Soundness: If a clause s is derived using a production from Σ, then $s \in Th(\Sigma)$
(3) Completeness: If $t \in Th(\Sigma)$, then there exists a production from Σ whose final clause s is subsumed by t.
(4) Completeness-bis: If c is a clause and $t \in Th(\Sigma \cup \{c\}) - Th(\Sigma)$, then there exists a production from $\Sigma \cup \{c\}$, starting with c, whose final clause s is subsumed by t.

Before applying this result to the generation of minimal quasi-supports and minimal contradictions, let's illustrate the production with two small examples.

Example 5.5 Let Σ consist of the following three clauses

(1) $\neg c \vee \neg a$
(2) $\neg c \vee \neg b$
(3) $a \vee b$

The following sequence is a production of $\neg c$ from Σ starting with $a \vee b$:

(1) $(\bot, (a), (b))$ first step
(2) $(\bot, (\neg c\ a)(b))$ resolve with clause (1)
(3) $(\neg c, (b))$ skip
(4) $(\neg c, \emptyset)$ resolve with clause (2).

So $\neg c$ is an implicate of the three clauses (1) to (3) above.

Example 5.6 This time Σ consists of the four clauses

(1) $\neg a \vee \neg b \vee c$
(2) $\neg x \vee \neg b \vee a$
(3) $y \vee b \vee c$
(4) $\neg c$

Here is a production of $\neg x \vee y$ from Σ starting with $\neg c$:

(1) $(\bot, (\neg c))$ first step
(2) $(\bot, (\neg a \neg c)(\neg b \neg c))$ resolve with clause (1)
(3) $(\bot, (\neg x \neg a \neg c)(\neg b \neg c))$ resolve with clause (2)
(4) $(\neg x, (\neg b \neg c))$ skip
(5) $(\neg x, (y \neg b \neg c))$ resolve with clause (3)

(6) $(\neg x \vee y, \emptyset)$ skip.

This proves that $\neg x \vee y$ is an implicate of the clauses (1) to (4) above.

Now, note that if the productions are cut every time a p_i is produced which is outside a stable production field P, theorem 5.20 remains valid with $Th(\Sigma)$ replaced by $Th(\Sigma) \cap P$. For example the completeness-bis property reads: if $t \in Th(\Sigma \cup \{c\}) \cap P - Th(\Sigma) \cap P$, then there exists a production from $\Sigma \cup \{c\}$, starting with c, whose final clause s is subsumed by t (and hence in P). Now, given a set of clauses Σ, a clause c and a stable production field P, denote by $Prod(\Sigma, c, P)$ the set of all clauses $s \in P$ which are produced from Σ, starting with c, and which do not subsume any such clause:

$$Prod(\Sigma, c, P) = \mu\{s \in P : s \text{ can be produced from } \Sigma \text{ starting with } c\}.$$

These sets are fundamental for the computational procedures proposed below. They must be determined by productions starting with the clause c nondeterministically, in order to enumerate enough produced clauses $s \in P$ to obtain all minimal ones. For this the cuts discussed above can be used to limit the computations. This is the basic computational module.

For an arbitrary formula $f \in \mathcal{L}_N$ and a set of clauses Σ in $\mathcal{L}_N$, define the *new characteristic clauses* of f with respect to Σ as those characteristic clauses of $\Sigma \cup \{f\}$ which are not characteristic clauses of Σ:

$$Newcarc(\Sigma, f, P) = Carc(\Sigma \cup \{f\}, P) - Carc(\Sigma, P). \tag{5.30}$$

The next theorem is a theoretical result which is very useful in the proofs of forthcoming theorems but which cannot be applied directly for computations. It shows that $Newcarc$ can be obtained from $Prod$.

Theorem 5.21 *If c is a clause and P is a stable production field, then*

$$Newcarc(\Sigma, c, P) = Prod(\Sigma, c, P) - Th(\Sigma) \cap P. \tag{5.31}$$

Now, the characteristic clauses can be generated incrementally. Note that if $p \vee \neg p$ belongs to a stable production field and neither p nor $\neg p$ are implicates of Σ and $\perp$ is not an implicate of Σ (which occurs if and only if ξ is satisfiable. This condition is satisfied because Σ represents accepted knowledge), then $p \vee \neg p$ belongs to $\mu(Th(\Sigma) \cap P) = Carc(\Sigma, P)$.

Theorem 5.22 *If c is a clause and P is a stable production field, then*

$$\begin{aligned} Carc(\emptyset, P) &= \{p \vee \neg p : p \in N, p \vee \neg p \in P\} \\ Carc(\Sigma \cup \{c\}, P) &= \mu(Carc(\Sigma, P) \cup Prod(\Sigma, c, P)). \end{aligned} \tag{5.32}$$

Again, the basic computational module in this incremental formula is the one which computes $Prod$.

Now, let's apply these results to the computation of minimal contradictions and minimal quasi-supports of hypotheses. According to the remarks in Sect. 5.1, the information contained in the set of clauses Σ consists on the one hand of the relatively stable knowledge base Σ_K and on the other hand of the varying facts Σ_F, which may change from one case to the other. This has to be taken into account if efficient computational procedures are to be designed.

The first task consists in the determination of the minimal contradictions with respect to Σ. Remember that these minimal contradictions are essentially given by $Carc(\Sigma, P_A)$ (theorem 5.17). We concentrate here on the case of the full production field P_A, admitting all minimal contradictions and quasi-supports. Theorem 5.22 may be applied to compute $Carc(\Sigma, P_A)$. For this purpose, the clauses of Σ are ordered into a sequence $\xi_1, \xi_2, \ldots, \xi_r$, where it is convenient to take first the s clauses $\xi_1, \ldots, \xi_s$ of Σ_K. Using theorem 5.22, the minimal contradictions $Carc(\Sigma_K, P_A)$ relative to the knowledge base Σ_K can be computed. In many practical cases, $Carc(\Sigma_K, P_A)$ will only contain the trivial clauses $a \vee \neg a$ (representing contradictions $a \wedge \neg a$) for all assumptions $a \in A$. Then the knowledge base is in itself consistent, which may even be a requirement for an acceptable knowledge base. Anyway, this computation can be considered as a *compilation* of the knowledge base.

The facts will often also arrive sequentially or can in any case be arranged into an arbitrary sequence. Then the set of minimal contradictions $MC(\Sigma)$ can be updated sequentially using theorem 5.22. According to (5.32), for any new clause ξ_{r+1} arriving, this means essentially determining $Prod(\Sigma, \xi_{r+1}, P_A)$, i.e. all produced clauses starting with the clause ξ_{r+1}, which do not subsume a clause already produced. By (5.32), we can cut computations if we arrive at a p_i which subsumes a clause already present in $Carc(\Sigma, P_A)$.

At this stage queries can be accepted. Let h be an arbitrary formula in $\mathcal{L}_N$ representing a hypothesis which is queried. By theorem 5.17, the minimal quasi-supports of h are equal to $\sim Carc(\Sigma \cup \{\neg h\}, P_A)$. If $\neg h$ is a clause, then this set can be computed using theorem 5.22:

$$Carc(\Sigma \cup \{\neg h\}, P_A) = \mu(Carc(\Sigma, P) \cup Prod(\Sigma, \neg h, P_A)). \tag{5.33}$$

and again we can cut computations in finding $Prod(\Sigma, \neg h, P_A)$ because $Carc(\Sigma, P)$ is already known.

If $\neg h$ is not a clause, then $\neg h$ must first be transformed into a conjunctive normal form, preferably a simple one (minimal or shortest). If $\neg h = h_1 \wedge \ldots \wedge h_t$ and $\mathcal{F} = \{h_1, \ldots, h_t\}$, then $Carc(\Sigma \cup \{\neg h\}, P_A) = Carc(\Sigma \cup \mathcal{F}, P_A)$ and we can again use theorem 5.22 to compute this set. Note that the basic computational module is again *Prod*, and cuts in computations are possible when we consider sequentially the clauses in $\mathcal{F}$.

The approach described so far can be qualified as an *interpretative* approach, because any query is treated on request. In fact it is a combined compiled-interpretative approach, because the minimal contradictions relative to the knowledge base Σ_K are computed once and for all and the minimal contradictions relative to the information incorporating additional facts are updated as the facts arrive, and the queries are treated as they arrive (only the minimal contradictions with respect to Σ need not be recomputed).

If there are many different queries to be treated, a fully *compiled* approach based on theorem 5.14 and corollary 5.4 might be preferable. These results presuppose the knowledge of $PI(\Sigma)$ to determine the minimal contradictions and the minimal quasi-supports for a *clause* h. Note that $PI(\Sigma) = \mu Th(\Sigma) = Carc(\Sigma, P_N)$, that is the set of characteristic clauses of Σ with respect to the stable production field P_N consisting of all clauses over $N^{\pm}$. The latter can be computed using theorem 5.22. $PI(\Sigma)$ could also be computed using the following theorem which is a slight modification of theorem 5.22.

Theorem 5.23 *If c is a clause, then*

$$\begin{aligned} PI(\emptyset) &= \{p \vee \neg p : p \in N\} \\ PI(\Sigma \cup \{c\}) &= \mu(PI(\Sigma) \cup Prod(PI(\Sigma), c, P_N)). \end{aligned} \tag{5.34}$$

So, the whole information contained in Σ is compiled into and replaced by its prime implicates $PI(\Sigma)$. Given the decomposition $\Sigma = \Sigma_K + \Sigma_F$, the compilation can again be partitioned into two phases, namely first the compilation of Σ_K and then the updating as additional facts arrive. The minimal contradictions are then obtained from corollary 5.4. Any query which is a clause can be immediately treated by using theorem 5.14. The disadvantage of this compiled approach is the large cardinality of the set $PI(\Sigma)$ which means a big compilation effort already for the knowledge base as well as for the updating phase. This presupposes also a convenient data base structure for easy retrieval of elements of $PI(\Sigma)$.

This section has proposed several approaches to obtain minimal contradictions and minimal quasi supports. They are all complete in the sense that every time all of them are computed. This may be prohibitive, even with the improved procedures of the second part of this section. It may also be unnecessary. From a practical point of view, the analysis may focus on the most important arguments in favour and against a hypothesis (see Ngair, Provan, 1993). This means focussing the effort on a part of the knowledge base only. The notion of importance of assumptions is related both to structural and probabilistic considerations as has been shown in reliability theory (see for example Barlow, Proschan, 1975). How this idea is to be incorporated into the search for quasi-supports and contradictions is still largely an open question for research.

5.7 Computing Degrees of Support

Once the minimal contradictions and the minimal quasi-supports of a formula h in a propositional system have been determined, the degree of support of h can be computed. How exactly this can be done is the subject of this section.

Consider a propositional system such as introduced at the beginning of Sect. 5.3 and let h be a hypothesis with minimal quasi-supports $f_1, \ldots, f_r$ and minimal contradictions $c_1, \ldots, c_s$. Then by theorem 5.2 we have $\varphi_h = f_1 \vee \ldots \vee f_r$ and $\varphi_\perp = c_1 \vee \ldots \vee c_s$ and hence by theorem 5.10 $u'(h) = N(\varphi_h)$ and $u'(\perp) = N(\varphi_\perp)$. Then by theorem 5.9 we just have to compute $sp'(h) = P'(N(\varphi_h))$ and $sp'(\perp) = P'(N(\varphi_\perp))$ to find the degree of support of h:

$$sp(h) = \frac{sp'(h) - sp'(\perp)}{1 - sp'(\perp)}. \tag{5.35}$$

The problem is thus to compute $P'(N(\varphi_h))$ and $P'(N(\varphi_\perp))$. Since these two tasks are of the same kind, it is sufficient to consider techniques of computing $P'(N(\varphi_h))$ only. Because the logical formula φ_h defines a Boolean function on B_s, φ_h can be considered as a binary random variable whose expected value $E(\varphi_h)$ is precisely $sp'(h) = P'(N(\varphi_h))$. We want to point out here that the problem of computing the expected value $E(\varphi)$ of a Boolean function φ is a common problem in reliability theory; however the Boolean functions considered there are mainly properly monotone, which is usually not the case for φ_h and $\varphi_\perp$. But many methods for (properly) monotone Boolean functions can be generalized to arbitrary Boolean functions. We can therefore take advantage of methods from reliability theory to solve our problems.

First, remark that the probability of any conjunction $d = \wedge\{l_j : j \in J\}$ of literals l_j in $A^\pm$ can easily be computed. In fact, if we define

$$x_l = \begin{cases} 1 & \text{if } l_j = a_j \\ 0 & \text{if } l_j = \neg a_j, \end{cases} \tag{5.36}$$

then

$$P'(N(d)) = \prod_{j \in J} q_j^{x_j} (1 - q_j)^{1 - x_j}. \tag{5.37}$$

On the other hand, we have $P'(N(\varphi_h)) = P'(\cup_{i=1}^r N(f_i))$ and the following theorem shows how the latter can be computed.

Theorem 5.24 *(Inclusion-exclusion formula).*

We have

$$P'(\cup_{i=1}^r N(f_i)) = \sum_{k=1}^r (-1)^{k+1} S_k \tag{5.38}$$

with

$$S_k = \sum \{P'(\cap_{i \in I} N(f_i)) : I \subseteq \{1, \ldots, r\}, |I| = k\}. \tag{5.39}$$

Note that the probabilities of the intersections

$$P'(\cap_{i\in I}N(f_i)) = P'(N(\wedge_{i\in I}f_i))$$

are easy to compute because $\wedge_{i\in I}f_i$ is again a conjunction, namely the union of all literals contained in the conjunctions $f_i, i \in I$. It is not excluded that two conjunctions f_i and f_k are contradictory and then $\wedge_{i\in I}f_i = \perp$, which implies that the corresponding probability is zero. This method to compute $P'(N(\varphi_h))$ is called the *inclusion-exclusion method.* Unfortunately, the sum in (5.38) contains many terms which make computations tedious and even impossible for large r. Sometimes, the bounds provided by the next theorem allow one to stop computations with an acceptable approximation at a relatively small value of h. Note however that these bounds are not necessarily monotone in h. For properly monotone Boolean functions there exist better bounds (see for example Kohlas, 1987); for non-monotone Boolean functions less bounds are known today.

Theorem 5.25 *If S_k is defined like in (5.39) and l is a positive integer such that $2l+1 \leq r$, then $sp'(h) \leq S_1$ and*

$$\sum_{k=1}^{2l}(-1)^{k+1}S_k \leq sp'(h) \leq \sum_{k=1}^{2l+1}(-1)^{k+1}S_k. \tag{5.40}$$

Example 5.7 Consider the case of example 5.2. There are two minimal contradictions $a_2 \wedge \neg a_4$ and $\neg a_3 \wedge \neg a_4$ such that $N(\varphi_\perp) = N(a_2\neg a_4) \cup N(\neg a_3 \neg a_4)$ (we drop $\wedge$ to simplify notations). Applying theorem 5.24 we obtain

$$\begin{aligned} P'(N(\varphi_\perp)) &= P'(N(a_2\neg a_4)) + P'(N(\neg a_3\neg a_4)) \\ &\quad - P'(N(a_2\neg a_4)\cap N(\neg a_3\neg a_4)) \\ &= q_2(1-q_4) + (1-q_3)(1-q_4) - q_2(1-q_3)(1-q_4) \\ &= (1-q_4)(q_2 + (1-q_3)(1-q_2)). \end{aligned} \tag{5.41}$$

On the other hand, the minimal quasi-supports of a burglary b are $a_2\neg a_4$, $\neg a_3\neg a_4$, $\neg a_2\neg a_3$, so that $\varphi_b = \varphi_\perp \vee a$ with $a = \neg a_2 \neg a_3$. This implies that

$$P'(N(\varphi_b)) = P'(N(\varphi_\perp)) + P'(N(a)) - P'(N(\varphi_\perp \wedge a)) \tag{5.42}$$

and since $\varphi_\perp \wedge a = \neg a_2\neg a_3\neg a_4$, we obtain

$$\begin{aligned} P'(N(\varphi_b)) &= P'(N(\varphi_\perp)) + (1-q_2)(1-q_3) - (1-q_2)(1-q_3)(1-q_4) \\ &= P'(N(\varphi_\perp)) + q_4(1-q_2)(1-q_3). \end{aligned} \tag{5.43}$$

Therefore, the degree of support of the hypothesis of a burglary equals

$$sp(b) = \frac{q_4(1-q_2)(1-q_3)}{1-(1-q_4)(q_2+(1-q_3)(1-q_2))}. \tag{5.44}$$

Similarly, the degree of support of other hypotheses can be computed using the results of example 5.2

For hypotheses h with many minimal quasi-supports (r is large), the inclusion-exclusion method may fail to be practicable. In such cases, so-called *disjoint decomposition algorithms* may help. These methods depend on our ability to find a decomposition of $N(\varphi_h)$ into *disjoint* subsets $N(d_k)$ with $d_k \in \mathcal{L}_A$:

$$N(\varphi_h) = \sum_{k=1}^{m} N(d_k). \tag{5.45}$$

If we have such a decomposition, then

$$P'(N(\varphi_h)) = \sum_{k=1}^{m} P'(N(d_k)). \tag{5.46}$$

So, the $N(d_k)$ should be such that

(1) $P'(N(d_k))$ is easy to compute
(2) m is as small as possible.

For example, the first requirement is fulfilled when the d_k are conjunctions. The $N(d_k)$ are disjoint if and only if the d_k are pairwise inconsistent, i.e. $d_k \wedge d_l = \perp$ whenever $k \neq l$. We say that two formulas in $\mathcal{L}_A$ are disjoint if their conjunction is equivalent to $\perp$. Furthermore, a disjunction of disjoint formulas will be written as a sum. From the disjunctive normal form representation $\varphi_h = f_1 \vee \ldots f_r$, a representation by disjoint terms can easily be obtained (for simplicity, $\wedge$ is written as a product):

$$\varphi_h = f_1 + (\neg f_1)f_2 + (\neg f_1)(\neg f_2)f_3 + \ldots + (\neg f_1)\ldots(\neg f_{r-1})f_r. \tag{5.47}$$

Note however that these terms are no longer conjunctions and their probabilities are not easily computed. In fact the disjunctions $\neg f_j$ must be developed using the distributive law in order to obtain conjunctions. This will, in general, give far too many terms. A more intelligent way of obtaining a disjoint sum representation of φ_h has been proposed by Abraham (1979), but for (properly) monotone Boolean functions only. But the method can easily be adapted to general Boolean functions. The method makes use of the following two results:

Theorem 5.26 *Two conjunctions c' and c'' are disjoint if and only if there exists a literal x in c' which appears negated ($\neg x$) in c''.*

Theorem 5.27 *Suppose that c' and c'' are two conjunctions which are not disjoint. Let $X' = \{x_1, \ldots, x_s\}$ be the set of all literals contained in c', but not in c''.*

(1) If X' is empty, then $c' \vee c'' = c'$.
(2) If X' is not empty, then

$$c' \vee c'' = c' + (\neg x_1)c'' + x_1(\neg x_2)c'' + \ldots + x_1 \ldots x_{s-1}(\neg x_s)c''. \quad (5.48)$$

Based on these results the following algorithm can be defined:

For $j = 1$ to r
define $P_{0j} = \{f_j\}$
for $i = 1$ to $j - 1$
    for all $d_k \in P_{i-1j}$
        if d_k and f_i are disjoint put d_k into P_{ij}
        else define $X' = \{x_1, \ldots, x_s\}$ to be the set of all literals in f_i not contained in d_k
        if X' is empty drop d_k
        else add $(\neg x_1)d_k, x_1(\neg x_2)d_k, \ldots, x_1 \ldots x_{s-1}(\neg x_s)d_k$ to P_{ij}.

The next theorem completes the method and proves at the same time its correctness:

Theorem 5.28 *Let $P_{j-1j}, j = 1, \ldots, r$ be the sets obtained in the j-th loop of the above algorithm. Then*

$$\begin{aligned}\varphi_h &= f_1 \vee f_2 \vee \ldots \vee f_r \\ &= \sum_{j=1}^{r} \sum \{d_k \in P_{j-1j}\}.\end{aligned} \quad (5.49)$$

So the above algorithm constructs in fact a disjoint sum representation of φ_h by conjunctions. It follows then from theorem 5.4 that

$$P'(N(\varphi_h)) = \sum_{j=1}^{r} \sum \{P'(N(d_k)) : d_k \in P_{j-1j}\} \quad (5.50)$$

and the probabilities $P'(N(d_k))$ are easily computed because d_k are all conjunctions.

Example 5.8 Look again at example 5.7 and let's apply the above algorithm to $\varphi_\perp = a_2\neg a_4 \vee \neg a_3 \neg a_4$. For $j = 1$ we obtain

$$P_{01} = \{a_2 \neg a_4\}.$$

For $j = 2$ the algorithm yields

$$P_{02} = \{\neg a_3 \neg a_4\}, \; P_{12} = \{\neg a_2 \neg a_3 \neg a_4\}.$$

It finally follows therefore (see theorem 5.28)

$$\varphi_\perp = a_2 \neg a_4 + \neg a_2 \neg a_3 \neg a_4$$

and

$$P'(N(\varphi_\perp)) = q_2(1 - q_4) + (1 - q_2)(1 - q_3)(1 - q_4).$$

Example 5.9 A more interesting example is provided by the Boolean function

$$\varphi = a_1a_2a_3(\neg a_4) \vee a_5a_6(\neg a_7)a_8. \tag{5.51}$$

Here we obtain

$$P_{01} = \{a_1a_2a_3(\neg a_4)\}$$
$$P_{02} = \{a_5a_6(\neg a_7)a_8\}$$

$$\begin{aligned} P_{12} = \{&(\neg a_1)a_5a_6(\neg a_7)a_8, a_1(\neg a_2)a_5a_6(\neg a_7)a_8, a_1a_2(\neg a_3)a_5a_6(\neg a_7)a_8, \\ &a_1a_2a_3a_4a_5a_6(\neg a_7)a_8\} \end{aligned} \tag{5.52}$$

and hence

$$\begin{aligned} \varphi =& a_1a_2a_3(\neg a_4) + (\neg a_1)a_5a_6(\neg a_7)a_8 + a_1(\neg a_2)a_5a_6(\neg a_7)a_8 + \\ & a_1a_2(\neg a_3)a_5a_6(\neg a_7)a_8 + a_1a_2a_3a_4a_5a_6(\neg a_7)a_8. \end{aligned} \tag{5.53}$$

This implies that

$$\begin{aligned} P'(N(\varphi)) = q_1q_2q_3(1-q_4) + \Big((1-q_1) + q_1(1-q_2)\Big)\Big(q_5q_6q_8(1-q_7)\Big) \\ + q_1q_2q_5q_6q_8(1-q_7)\Big((1-q_3) + q_3q_4\Big). \end{aligned} \tag{5.54}$$

Example 5.10 (Kohlas, 1991 a) We continue the analysis of Barnett's system, example 5.3 in Sect. 5.5. For this system, we have

$$\varphi_\perp = \vee_{i=1}^n (a_i \wedge b_i) \vee_{i\neq j} (a_i \wedge a_j) \vee (\wedge_{i=1}^n b_i). \tag{5.55}$$

Instead of $\varphi_\perp$, it turns out to be advantageous to consider $\neg\varphi_\perp$. From (5.55), we can deduce that

$$\neg\varphi_\perp = \Big[\wedge_{i=1}^n \neg a_i \wedge \neg(\wedge_{i=1}^n b_i)\Big] \vee (\vee_{i=1}^n (a_i \wedge \neg b_i \wedge_{j\neq i} \neg a_j)). \tag{5.56}$$

Note that the $(n+1)$ terms in this disjunction are already pairwise inconsistent. Let α_i and β_i denote the probabilities of the assumptions a_i and b_i respectively. Then, using (5.56), the probability of $N(\neg\varphi_\perp)$ can easily be computed:

$$\begin{aligned} 1 - sp'(\perp) &= P'(N(\neg\varphi_\perp)) \\ &= \Big[\prod_{i=1}^n (1-\alpha_i)\Big]\Big[1 - \prod_{i=1}^n \beta_i\Big] + \sum_{i=1}^n (\alpha_i((1-\beta_i)\prod_{j\neq i}(1-\alpha_j)) \\ &= \Big[\prod_{i=1}^n (1-\alpha_i)\Big]\Big[1 + (\sum_{i=1}^n \frac{\alpha_i(1-\beta_i)}{1-\alpha_i}) - \prod_{i=1}^n \beta_i\Big]. \end{aligned} \tag{5.57}$$

Now, given a hypothesis $h = \vee\{\theta_i : i \in H\}$, define $f = \vee_{i\in H} a_i$ and

$$g = \vee_{i\in H^c}(a_i \wedge b_i) \vee_{i\neq j\in H^c} (a_i \wedge a_j) \vee (\wedge_{i\in H^c} b_j). \tag{5.58}$$

Then equation (5.23) in example 5.3 shows that $\varphi_h = f \vee g$ and hence

$$P'(N(\varphi_h)) = P'(N(f)) + P'(N(g)) - P'(N(f \wedge g)). \tag{5.59}$$

But since the assumptions present in f correspond to indices in H whereas those present in g correspond to indices in H^c , it follows that $P'(N(f \wedge g)) = P'(N(f))P'(N(g))$ and hence

$$sp'(h) = 1 - (1 - P'(N(f)))(1 - P'(N(g))). \tag{5.60}$$

But we have clearly

$$P'(N(f)) = 1 - \prod_{i \in H}(1 - \alpha_i). \tag{5.61}$$

On the other hand, remark that g is defined exactly like $\varphi_\perp$, the only difference is that the reference set is H^c instead of $\{1, \ldots, n\}$. Therefore, by analogy to (5.57), we have

$$1 - P'(N(g)) = \Big[\prod_{i \in H^c}(1 - \alpha_i)\Big]\Big[1 + (\sum_{i \in H^c} \frac{\alpha_i(1 - \beta_i)}{1 - \alpha_i}) - \prod_{i \in H^c} \beta_i\Big]. \tag{5.62}$$

Now, by (5.60), (5.61) and (5.62), we can write

$$1 - sp'(h) = \Big[\prod_{i=1}^{n}(1 - \alpha_i)\Big]\Big[1 + (\sum_{i \in H^c} \frac{\alpha_i(1 - \beta_i)}{1 - \alpha_i}) - \prod_{i \in H^c} \beta_i\Big]. \tag{5.63}$$

But we have

$$\begin{aligned} pl(\neg h) &= 1 - sp(h) \\ &= \frac{1 - sp(\perp) - sp'(h) + sp'(\perp)}{1 - sp'(\perp)} \\ &= \frac{1 - sp'(h)}{1 - sp'(\perp)}. \end{aligned} \tag{5.64}$$

Then, it follows by (5.57) and (5.63) that

$$pl(\neg h) = K^{-1}\Big[1 + (\sum_{i \in H^c} \frac{\alpha_i(1 - \beta_i)}{1 - \alpha_i}) - \prod_{i \in H^c} \beta_i\Big] \tag{5.65}$$

where

$$K = 1 + (\sum_{i=1}^{n} \frac{\alpha_i(1 - \beta_i)}{1 - \alpha_i}) - \prod_{i=1}^{n} \beta_i. \tag{5.66}$$

From these last two equations, the degree of plausibility of h can be obtained by replacing H^c by H in (5.65). From these results, the degree of support of h or $\neg h$ can easily be computed. Note that if $H = \Theta$ then $pl(\neg h) = 0$, which is a special case of (5.65). Similarly, if $H = \Theta - \{\theta_k\}$, then the minimal quasi-supports found in example 5.4 imply that

$$\begin{aligned} 1 - sp'(h) &= (1-\beta_k)\prod_{i\neq k}(1-\alpha_i) \\ &= \Big(1+\frac{\alpha_k(1-\beta_k)}{1-\alpha_k}\Big)\prod_{i=1}^{n}(1-\alpha_i). \end{aligned} \tag{5.67}$$

which is a special case of (5.63) and hence (5.65) is also valid for this case. This shows that (5.65) is valid for any hypothesis $H \subseteq \Theta$.

For a small number r of terms in the original disjunctive normal form this algorithm may well generate more terms than the inclusion-exclusion method; for large r however, it is the contrary in general.

Another decomposition method will now be presented. This is a new method developped by R. Bertschy which generalizes to arbitrary Boolean functions the algorithm of Heidtman (1989) for monotone systems. When applied to monotone functions, this new method and the method of Heidtman give the same results. Compared to the Abraham algorithm, it is considerably more efficient and very often it gives a significantly smaller number of disjoint terms.

The idea of the method is to decompose the Boolean function

$$f = f_1 \vee \ldots \vee f_r, \tag{5.68}$$

where each f_j is a conjunction of literals over the atoms $\{x_1, \ldots, x_n\}$, into disjoint terms of the form

$$d = \left(\wedge_{k\in I_0} x_k^{e_k^0}\right) \wedge_{s=1}^{r} \left(\neg \wedge_{k\in I_s} x_k^{e_k^s}\right) \tag{5.69}$$

where $e_k^l \in \{-1, 1\}$ and x_k^1 means x_k and x_k^{-1} means $\neg x_k$. A term of the form (5.69) is called a *mix-product* if $I_u \cap I_v = \emptyset$ for any two different u and v in $\{0, 1, 2, \ldots, r\}$. Note that in a mix-product the set I_0 can be empty and r can be equal to zero in which case the mix-product is a regular conjunction of literals. So conjunctions of literals are mix-products. The algorithm yields a decomposition of f into disjoint mix-products.

The method goes as follows. First we define a procedure called *decompose* which takes as inputs a family of disjoint mix-products $m_1, \ldots, m_r$ and a family of conjunctions $c_1, \ldots, c_s$ and returns a collection of disjoint mix-products $d_1, \ldots, d_k$ such that

$$\sum_{i=1}^{r} m_i \wedge (\neg \vee_{j=1}^{s} c_j) = \sum_{i=1}^{k} d_i. \tag{5.70}$$

This function *decompose* can then be used to find a decomposition of f. Indeed, since the equality

$$f = \sum_{i=1}^{r} f_i \wedge (\neg \vee_{k=1}^{i-1} f_k) \tag{5.71}$$

always holds, applying *decompose* with arguments $m_1 = f_i$ and

$$c_1 = f_1, \ldots, c_{i-1} = f_{i-1}$$

for every $i = 1, \ldots, r$ gives a collection of disjoint mix-products $d_1, \ldots, d_l$ such that

$$f = \sum_{t=1}^{l} d_t. \tag{5.72}$$

In order to define the procedure *decompose* we need a procedure called *find-disjoint* whose arguments are a mix-product m and a conjunction c and returns a family of disjoint mix-products $q_1, \ldots, q_n$ such that

$$m \wedge \neg c = \sum_{i=1}^{n} q_i. \tag{5.73}$$

Now we describe the procedure *decompose.* First note that

$$\Big(\sum_{i=1}^{r} m_i\Big) \wedge \Big(\neg \vee_{j=1}^{s} c_j\Big) = \sum_{i=1}^{r} \Big(m_i \wedge (\neg \vee_{j=1}^{s} c_j)\Big). \tag{5.74}$$

Since the terms $m_i \wedge (\neg \vee_{j=1}^{s} c_j)$ are disjoint, for the procedure *decompose* to find a decomposition of

$$\Big(\sum_{i=1}^{r} m_i\Big) \wedge \Big(\neg \vee_{j=1}^{s} c_j\Big) \tag{5.75}$$

into disjoint mix-products it is sufficient that it finds one for every term

$$m_i \wedge (\neg \vee_{j=1}^{s} c_j). \tag{5.76}$$

But

$$\begin{aligned} m_i \wedge (\neg \vee_{j=1}^{s} c_j) &= m_i \wedge (\wedge_{j=1}^{s} \neg c_j) = (m_i \wedge \neg c_s) \wedge (\wedge_{j=1}^{s-1} \neg c_j) \\ &= (m_i \wedge \neg c_s) \wedge (\neg \vee_{j=1}^{s-1} c_j) \end{aligned} \tag{5.77}$$

and the procedure *find-disjoint* can then be invoked to obtain a collection of disjoint mix-products $q_{i_\nu}, \nu = 1, \ldots, n_i$ such that

$$m_i \wedge \neg c_s = \sum_{\nu=1}^{n_i} q_{i_\nu}. \tag{5.78}$$

Then

$$m_i \wedge (\neg \vee_{j=1}^{s} c_j) = \Big(\sum_{\nu=1}^{n_i} q_{i_\nu}\Big) \wedge \Big(\neg \vee_{j=1}^{s-1} c_j\Big) \tag{5.79}$$

and a recursive call to the procedure *decompose* with arguments $q_{i_\nu}, \nu = 1, \ldots, n_i$ and $c_1, \ldots, c_{s-1}$ then gives finally a decomposition of $m_i \wedge (\neg \vee_{j=1}^{s} c_j)$

into disjoint mix-products. This properly defines the procedure *decompose* because the number of conjunctions in the recursive call is 1 less than in the original argument and the basis of the recursion is reached when there is no conjunction at all as second argument, in which case the decomposition is simply $m_1, \ldots, m_r$. Also note that if there is no mix-product at all as first argument then *decompose* is defined to return the empty set.

Now we define the procedure *find-disjoint* whose arguments are a mix-product

$$m = \left(\wedge_{k \in I_0} x_k^{e_k^m}\right) \wedge_{s=1}^{r} \left(\neg \wedge_{k \in I_s} x_k^{e_k^m}\right) \tag{5.80}$$

and a conjunction of literals

$$c = \wedge_{k \in I_c} x_k^{e_k^c} \tag{5.81}$$

and returns a family of disjoint mix-products $q_1, \ldots, q_n$ such that $m \wedge \neg c = \sum_{i=1}^{n} q_i$.

First let's introduce some notations. For a mix-product m and a conjunction c let $j(m,c)$ denote the number of sets $I_s, 1 \leq s \leq r$ such that $I_s \cap I_c \neq \emptyset$. If $j(m,c) = 0$ then define

$$\begin{array}{lll} S' = I_c \cap I_0 & S_1' = I_c - I_0 & S_2' = I_0 - I_c \\ a' = \wedge_{k \in S_1'} x_k^{e_k^c} & b' = \wedge_{k \in S'} x_k^{e_k^c} & k' = \wedge_{k \in S_2'} x_k^{e_k^m} \end{array}$$

and

$$m' = \wedge_{s=1}^{r} (\neg \wedge_{k \in I_s} x_k^{e_k^m}) \tag{5.82}$$

and if $j(m,c) > 0$ then for an index $s_0 \in \{1, \ldots, r\}$ such that $I_{s_0} \cap I_c \neq \emptyset$ define

$$\begin{array}{lll} S = I_{s_0} \cap I_c & S_1 = \{k \in S : e_k^c = e_k^m\} & S_2 = S - S_1 \\ S_3 = I_{s_0} - S & S_4 = I_c - S & \end{array}$$

and for $i = 1, 2, 3$

$$m_i = \wedge_{k \in S_i} x_k^{e_k^m} \tag{5.83}$$

$$m'' = \left(\wedge_{k \in I_0} x_k^{e_k^m}\right) \wedge_{s \neq s_0} \left(\neg \wedge_{k \in I_s} x_k^{e_k^m}\right) \tag{5.84}$$

$$m_2^* = \wedge_{k \in S_2} x_k^{-e_k^m}, \; c' = \wedge_{k \in S_4} x_k^{e_k^c} \tag{5.85}$$

$$a = \neg(m_1 m_2 m_3) m'' (\neg m_1 m_2^*), \; b = \neg(m_1 m_2 m_3) m'' (\neg c'). \tag{5.86}$$

Furthermore, if $S_2 \neq \emptyset$ and for $k_0 \in S_2$ define

$$\begin{array}{lll} z = x_{k_0}^{e_{k_0}^m} & v = \wedge\{x_k^{e_k^m} : k \in S_2 - \{k_0\}\} & v^* = \wedge\{x_k^{-e_k^m} : k \in S_2 - \{k_0\}\} \\ d_1 = z \wedge \neg v & d_2 = \neg z \wedge \neg v^* & d_3 = \neg v^* \wedge \neg v. \end{array}$$

Now let's define the procedure *find-disjoint* as follows (the set R is the output of the procedure):

procedure find-disjoint (m, c);
if $\exists k_0 \in I_0 \cap I_c$ such that $e_{k_0}^m = -e_{k_0}^c$

or ($\exists s_0 \in \{1, \ldots, r\}$ such that $I_{s_0} \subseteq I_c$ and $e_k^m = e_k^c \; \forall k \in I_{s_0}$)
then $R = \{m\}$
else
  if $j(m, c) = 0$ then if $S_1' \neq \emptyset$ then $R = \{m \wedge \neg a'\}$ else $R = \emptyset$
  else $R = R_1 + R_2$ where
    $R_1 = R_1' + R_2' + R_3'$ where
      $R_1' = \{\neg m_1 m''\}$ if $S_1 \neq \emptyset$
      $R_1' = \emptyset$ if $S_1 = \emptyset$,
      $R_2' = \{d_1 m_1 m'', d_2 m_1 m''\}$ if $|S_2| \geq 2$
      $R_2' = \emptyset$ if $|S_2| < 2$
      $R_3' = \{m_1 m_2 \neg m_3 m''\}$ if $S_2 \neq \emptyset$ and $S_3 \neq \emptyset$
      $R_3' = \emptyset$ if $S_2 \cup S_3 = \emptyset$
    $R_2 =$ *find-disjoint* (d, c') where
      $d = m_1 \neg m_3 m''$ if $S_2 = \emptyset$
      $d = m_1 m_2^* m''$ if $S_2 \neq \emptyset$

Theorem 5.29 *The procedure find-disjoint terminates and its output R is a set of disjoint mix-products* $\{q_1, \ldots, q_n\}$ *such that*

$$m \wedge \neg c = \sum_{i=1}^{n} q_i. \tag{5.87}$$

Note that computing the probability of a mix-product is very simple:

$$P(\wedge_{k \in I_0} x_k^{e_k} \wedge_{s=1}^{r} (\neg \wedge_{k \in I_s} x_k^{e_k})) = P(\wedge_{k \in I_0} x_k^{e_k}) \prod_{s=1}^{r} P(\neg \wedge_{k \in I_s} x_k^{e_k})$$

$$= \prod_{k \in I_0} P(x_k^{e_k}) \prod_{s=1}^{r} (1 - \prod_{k \in I_s} P(x_k^{e_k}))$$

and

$$P(x_k^{e_k}) = \begin{cases} p_k & \text{if } e_k = 1 \\ q_k = 1 - p_k & \text{if } e_k = -1. \end{cases}$$

This decomposition method has been implemented by R. Bertschy and its results have been compared with those found by Abraham's method. This new method is more efficient and the number of disjoint terms is much smaller.

All these methods have exponential complexity, the problem is NP-hard, as is well known (see for example Ball, 1986). These general methods may, however, in many cases be replaced by more specific methods taking advantage of special structures of the Boolean function. Such a method is *modularization* combined with *factorization*. Its principles will be presented in the next chapter.

A5 Appendix: Proofs of Theorems in Chapter 5

Proof of Theorem 5.1

(1). For any interpretation for which the conjunction f_i is true, f is also true. (2). For any interpretation for which f is true, f_i must also be true. □

Proof of Theorem 5.2

By definition, for an interpretation for which a prime implicant f_i is true, f is also true. Inversely, consider an interpretation x for which f is true. If $p_1, \ldots, p_n$ are the atoms appearing in f, then define a conjunction $c = \wedge_{i=1}^{n} y_i$ with $y_i = p_i$ if $p_i(x) = t, y_i = \neg p_i$ if $p_i(x) = f$. c is an implicant for f because x is the only interpretation under which c is true. But clearly, every implicant must contain at least one subconjunction which is a prime implicant. This applies to c and the prime implicant f_i contained in c must also be true under the interpretation x. Hence $f_1 \vee \ldots \vee f_r$ is true under x.

The second part of the theorem follows from de Morgan laws because the prime implicates of f are obtained by negating the prime implicants of $\neg f$. □

Proof of Theorem 5.3

If a form is not irredundant, there is either a term which is not prime or one or several terms can be eliminated from the form without destroying the equivalence with f. In the first case, the form can be neither minimal nor shortest. In the second case, the resulting form contains both less literals and less terms which means also that the original form can neither be minimal nor shortest. □

Proof of Theorem 5.4

(1) follows from the definition of equivalence. (2). If x is an interpretation for which $\neg f$ is true, then x is not in $N(f)$, hence in $N(f)^c$ and vice versa. (3). If x is an interpretation for which both f and g are true, then $x \in N(f)$ and $x \in N(g)$, hence $x \in N(f) \cap N(g)$ and vice versa. (4) If x is an interpretation for which f or g or both are true, then $x \in N(f)$ or $x \in N(g)$, hence $x \in N(f) \cup N(g)$ and vice versa. (5) is trivial. □

Proof of Theorem 5.5

(i) (1). $N(f_i)$ is an interval because f_i is a conjunction and $N(f_i) \subseteq N(f)$ according to theorem 5.4 (5). (2). Suppose that f_i is a prime implicant of f and that $N(f_i)$ is not maximal permissible. Then there exists another interval $N(c)$ with $N(f_i) \subset N(c) \subseteq N(f)$ and by theorem 5.4 (5) c is an implicant of f and also a subconjunction of f_i. But this contradicts the assumption that f_i is a prime implicant. Inversely, if f_i is an implicant, but not prime, then there is a subconjunction c of f_i which is also an implicant of f and then $N(f_i) \subset N(c) \subseteq N(f)$ and $N(f_i)$ is not maximal.

(ii) is proved similarly. □

Proof of Theorem 5.6

Let f be any formula such that $N = N(f)$ and suppose f has the prime implicants $f_i, i = 1, \ldots, n$. Then by theorem 5.5 (i) (2),

$$\{N(f_1), \ldots, N(f_n)\} \subseteq \{N_1, \ldots, N_r\} \tag{A5.1}$$

because $N = N(f)$. Then by theorem 5.2 and theorem 5.4 (4), we have

$$N = N(f) = \cup_{i=1}^{n} N(f_i) \subseteq \cup_{i=1}^{r} N_i \tag{A5.2}$$

and hence $N = N_1 \cup \ldots \cup N_r$ since $N_i \subseteq N$ for all i. The second part of the theorem follows similarly. □

Proof of Theorem 5.7

(a). Suppose that $f(x)$ is properly monotone. Let the conjunction

$$c = x_1 \wedge \ldots \wedge x_n \wedge (\neg y_1) \wedge \ldots \wedge (\neg y_m) \tag{A5.3}$$

be an implicant of f, x_i represent unnegated propositions and y_j negated propositions. Then we have $f(x) = 1$ whenever $x_i = 1$ and $y_j = 0$. As $f(x)$ is properly monotone we still have $f(x) = 1$ whenever $x_i = 1, i = 1, \ldots n$, independently of the values of the other variables. Therefore the conjunction $c' = x_1 \wedge \ldots \wedge x_n$ is also an implicant for f. Hence no implicant for f with negated propositions can be prime.

(b) The inverse is evident. □

Proof of Theorem 5.8

Apply first the necessary transformations $p_i \to \neg p'_i$ corresponding to the transformations $y_i = \neg x_i$ to get a formula f' with a properly monotone function $f'(x)$. It is then sufficient to show that the disjunctive normal form containing all prime implicants of f' is irredundant.

So, let f be a form with a properly monotone Boolean function $f(x)$. Consider any prime implicant f_i of f and suppose the conjunction f_i contains the subset $J \subseteq P$ of the propositions. Let x_J denote the subvector of x corresponding to the propositions in J and x_{N-J} the subvector corresponding to the propositions not in J. Put $x_J = 1$ and $x_{N-J} = 0$. For this vector x we have $f_i(x) = 1$ and $f(x) = 1$. If f_j is another prime implicant, it must contain at least one proposition from $N - J$. But this implies $f_j(x) = 0$ and it follows that no prime implicant is redundant. This proves that the disjunctive normal form containing all prime implicants is irredundant. □

Proof of Theorem 5.9

$$
\begin{aligned}
sp(h) &= P(u(h)) = P(u'(h) - u'(\bot)) \\
&= \sum \{P(x) : x \in u'(h) - u'(\bot)\} \\
&= \sum \{P'(x|u'(\bot)^c) : x \in u'(h) - u'(\bot)\} \\
&= \sum \{\frac{P'(x)}{1 - P'(u'(\bot))} : x \in u'(h) - u'(\bot)\} \\
&= \frac{1}{1 - P'(u'(\bot))} \sum \{P'(x) : x \in u'(h) - u'(\bot)\} \\
&= \frac{1}{1 - P'(u'(\bot))} \Big(\sum_{x \in u'(h)} P'(x) - \sum_{x \in u'(\bot)} P'(x) \Big) \\
&= \frac{P'(u'(h)) - P'(u'(\bot))}{1 - P'(u'(\bot))} = \frac{sp'(h) - sp'(\bot)}{1 - sp'(\bot)}.
\end{aligned}
$$

□

Proof of Theorem 5.10

We implicitly assume that N of a formula in $\mathcal{L}_A$ is a subset of B_s. Then

$$
\begin{aligned}
N(\varphi_h) &= \cup \{N(conj(x)) : x \in u'(h)\} \\
&= \cup \{x : x \in u'(h)\} = u'(h).
\end{aligned} \tag{A5.4}
$$

□

Proof of Theorem 5.11

Since $A \subseteq B$ if and only if $A \cap B^c = \emptyset$, theorem 5.4 implies that the following statements are equivalent:

$$
\begin{aligned}
& a \wedge \xi \models h \\
& N(a \wedge \xi) \subseteq N(h) \\
& N(a) \cap N(\xi) \subseteq N(h) \\
& N(a) \cap N(\xi) \cap N(h)^c = \emptyset \\
& N(\neg a)^c \cap N(\xi) \cap N(h)^c = \emptyset \ . \\
& (N(\neg a) \cup N(h))^c \cap N(\xi) = \emptyset \\
& N(\xi) \subseteq N(\neg a) \cup N(h) \\
& N(\xi) \subseteq N(\neg a \vee h) \\
& \xi \models \neg a \vee h
\end{aligned}
$$

□

Proof of Theorem 5.12

The equivalence of (1) and (2) is a direct consequence of theorem 5.11. Furthermore, $a \wedge \xi \models \bot$ if and only if $N(a \wedge \xi) = \emptyset$, which holds if and only if $a \wedge \xi$ is not satisfiable. □

Proof of Theorem 5.13

Suppose that the conjunction a is an implicant of φ_h, i.e. $N(a) \subseteq N(\varphi_h)$. We must prove that $N(a \wedge \xi) \subseteq N(h)$. So, let $z \in N(a \wedge \xi)$. Then $\xi(z) = 1$ and $a(z) = 1$ and hence $\varphi_h(z) = 1$ because $a \models \varphi_h$. By definition of φ_h, there is an $x \in u'(h)$ such that $conj(x)(z) = 1$ and hence $(conj(x) \wedge \xi)(z) = 1$. But $x \in u'(h)$ implies that $N(conj(x) \wedge \xi) \subseteq N(h)$. Since $z \in N(conj(x) \wedge \xi)$ it follows that $z \in N(h)$.

Conversely, suppose that a is a quasi-support of h. We must prove that $a \models \varphi_h$, i.e. $N(a) \subseteq N(\varphi_h)$. Let $z \in N(a)$. Then a is a subconjunction of $conj(z)$ and thus $N(conj(z)) \subseteq N(a)$. This implies that $N(\neg a) \subseteq N(\neg conj(z))$ and hence

$$N(h \vee \neg a) = N(h) \cup N(\neg a) \subseteq N(h) \cup N(\neg conj(z)) = N(h \vee \neg conj(z)). \quad (A5.5)$$

But a is a quasi-support of h and hence $N(\xi) \subseteq N(h \vee \neg a)$, which proves that $N(\xi) \subseteq N(h \vee \neg conj(z))$, i.e. $\xi \models h \vee \neg conj(z)$ and $conj(z)$ is a quasi-support of h. This implies that $conj(z) \wedge \xi \models h$ and hence $z \in u'(h)$ by definition. Then $conj(z) \in \{conj(x) : x \in u'(h)\}$ and hence

$$\Big(\vee \{conj(x) : x \in u'(h)\}\Big)(z) = 1 \quad (A5.6)$$

because $conj(z)(z) = 1$. This proves that $\varphi_h(z) = 1$ and so $z \in N(\varphi_h)$. □

The following lemma is needed in the proof of theorem 5.14.

Lemma A5.1 *Let h be a clause and a a minimal quasi-support of h. Then there exists a prime implicate f of Σ such that $\neg a = f - h$.*

Proof of Lemma A5.1

Because a is a minimal quasi-support of h, it follows that $\Sigma \models \neg a \vee h$, and hence $\neg a \cap h = \emptyset$. There is a prime implicate f of Σ such that $f \subseteq \neg a \vee h$, say $f = \neg a' \vee h'$, where $\neg a' \subseteq \neg a$, $h' \subseteq h$.

Suppose now that $\neg a'$ is a proper subset of $\neg a$. From $f \subseteq \neg a' \vee h$ it follows that $\Sigma \models \neg a' \vee h$, which implies that a is not a minimal quasi-support, contrary to the assumption. Hence $a' = a$, $f = \neg a \vee h'$ and therefore $\neg a = f - h$. □

Proof of Theorem 5.14

Let's define the following set:

$$\Delta(\Sigma, h) = \{f - h \in \mathcal{C}'_A : f \in PI(\Sigma)\}, \quad (A5.7)$$

so that we have to prove that $MQS(\Sigma, h) = \sim \mu\Delta(\Sigma, h)$. First we show that all minimal quasi-supports of h are in $\sim \mu\Delta(\Sigma, h)$. In fact, if a is a minimal quasi support of h, then by lemma A5.1 $a \in \sim \Delta(\Sigma, h)$. It remains to show that no proper subconjunction of a is also in $\sim \Delta(\Sigma, h)$. Suppose on the contrary that there is some subconjunction a' of a in $\sim \Delta(\Sigma, h)$. From $\neg a' = f - h$ for some prime implicate f of Σ it follows that $\neg a' \vee h = (f-h) \cup h \supseteq f$ and this implies $\Sigma \models \neg c' \vee h$. Thus a' is a quasi-support for h and a is not minimal, contrary to the assumption. It follows that there can be no subconjunction of a minimal quasi support in $\sim \Delta(\Sigma, h)$ and hence $a \in \sim \mu\Delta(\Sigma, h)$.

Inversely, let's show that $a \in \sim \mu\Delta(\Sigma, h)$ implies that a is a minimal quasi-support of h. First, we prove that $\Sigma \models \neg a \vee h$. From $\neg a = f - h$ it follows that $\neg a \vee h = (f-h) \cup h \supseteq f$. Because f is an implicate of Σ, so is $\neg a \vee h$. Therefore, a is a quasi-support of h.

Finally, let's show that a is a minimal quasi-support. Suppose on the contrary that there is a proper subconjunction a' of a which is a minimal quasi support of h. By lemma A5.1 there is a prime implicate f' of Σ such that $\neg a' = f' - h$. This shows that $a' \in \sim \Delta(\Sigma, h)$ and therefore a is not in $\sim \mu\Delta(\Sigma, h)$. But this is a contradiction and a must therefore be a minimal quasi-support of h. □

Proof of Corollary 5.4

This follows immediately from the application of theorem 5.14 to the empty clause $h = \perp$. □

Proof of Theorem 5.15

Let $x \in Carc(S, P)$. Then $x \in P$ and $S \models x$. Suppose that x is not a prime implicate of S, i.e. there is $x' \subset x$ such that $S \models x'$. Since P is stable, x' is in P and hence $x' \in Th(S) \cap P$, which contradicts the minimality of x. This shows that $x \in PI(S) \cap P$. Conversely, let $x \in PI(S) \cap P$. Then clearly $x \in Th(S) \cap P$. Suppose that x is not minimal in $Th(S) \cap P$. Then there is $x' \subset x$ with $x' \in Th(S) \cap P$ and hence $S \models x'$ which is a contradiction to the fact that $x \in PI(S)$. This proves that $x \in \mu(Th(S) \cap P)$. □

Proof of Theorem 5.16

Because P is stable, in view of theorem 5.15, we must prove that

$$MQS(\Sigma, f) \cap \sim P = \sim \{PI(\Sigma \cup \{\neg f\}) \cap P\}. \qquad (A5.8)$$

Let $c \in MQS(\Sigma, f) \cap \sim P$. Then $\Sigma \models \neg c \vee f$, which is equivalent to $\Sigma \cup \{\neg f\} \models \neg c$, and $\neg c \in P$, which shows that $\neg c$ is an implicate of $\Sigma \cup \{\neg f\}$ in P. In fact, $\neg c$ is a prime implicate. On the contrary, suppose that there is $\neg c'$ such that $\neg c' \subset \neg c$ and $\Sigma \cup \{\neg f\} \models \neg c'$ or $\Sigma \models \neg c' \vee f$. Then c' is a quasi-support of f, which contradicts the minimality of c. Hence $\neg c \in PI(\Sigma \cup \{\neg f\}) \cap P$.

Conversely, suppose that $c \in \sim \{PI(\Sigma \cup \{\neg f\}) \cap P\}$. Then $\Sigma \cup \{\neg f\} \models \neg c$ and $\neg c \in P$. This shows that c is a quasi-support of f which is in $\sim P$. In fact, c is a minimal quasi-support. On the contrary, suppose that there is c' such that $c' \subset c$ and $\Sigma \models \neg c' \vee f$. Then $\Sigma \cup \{\neg f\} \models \neg c'$ and $\neg c'$ is an implicate of $\Sigma \cup \{\neg f\}$, which contradicts the fact that $\neg c \in PI(\Sigma \cup \{\neg f\}$. Hence $c \in MQS(\Sigma, f) \cap \sim P$. □

Proof of Corollary 5.5 This result follows from theorem 5.16 because of Eq. (5.19) and

$$\begin{aligned} Carc(\Sigma \cup \{\neg \perp\}, P) &= \mu(Th(\Sigma \cup \{\neg \perp\} \cap P) \\ &= \mu(Th(\Sigma) \cap P) = Carc(\Sigma, P) \end{aligned} \tag{A5.9}$$

because $Th(\Sigma \cup \{\neg \perp\}) = Th(\Sigma)$. □

Proof of Theorem 5.17

To prove this theorem, we consider the stable production field $P_A = \mathcal{C}'_A$. Since $\sim P_A = \mathcal{C}_A$ and $MC(\Sigma) \subseteq \mathcal{C}_A$, corollary 5.5 implies that $MC(\Sigma) = \sim Carc(\Sigma, P_A)$. Furthermore, since $MQS(\Sigma, f) \subseteq \mathcal{C}_A$ and $\mathcal{C}_A = \sim P_A$, equation (5.18) implies that $MQS(\Sigma, f) = MQS(\Sigma, f, P_A)$. Then theorem 5.16 implies that $MQS(\Sigma, f) =\sim Carc(\Sigma \cup \{\neg f\}, P_A)$. □

Proof of Theorem 5.18

Let $f = f' \vee p$ and $g = g' \vee (\neg p)$. Then under any interpretation for which $f \wedge g$ is true, either f' or g' is true and hence $f' \vee g' = \rho(f, g)$ is true. □

Proof of Theorem 5.19

The proof is adapted from Birkhoff, Bartee (1970). Note that the two operations of the theorem preserve equivalence to f according to theorem 5.19. Furthermore, due to the finite number of symbols involved, the procedure terminates after a finite number of operations.

Suppose that at the end of the procedure $f = g_1 \wedge \ldots \wedge g_r$. We show first that any prime implicate of f is among the elements of $\{g_1, \ldots, g_r\}$. In fact, if on the contrary there is a prime implicate g which is not contained in $\{g_1, \ldots, g_r\}$, then we show that a resolution can be formed between two terms of f which contradicts the assumption that the procedure is terminated.

By assumption g differs from all terms of f. Because g is prime, it subsumes no term of f. Add literals to g as long as possible without subsuming a term of f, until a clause g' of maximal length is obtained which subsumes g, but no term of f. Any proposition which is present as a literal in the original form of f must appear as a literal in any conjunctive normal form representation of f. Thus, if g' also contains all those propositions, it subsumes a term of any normal conjunctive form, because g' is still an implicate of f. But this is excluded, hence there is a proposition p in the original form of f such that neither p nor $\neg p$ is in g'.

Form then the terms $g' \vee p$ and $g' \vee (\neg p)$ both of which subsume some terms of f, say g_1 and g_2, and both subsume g. g_1 and g_2 contain p and $\neg p$ respectively, but otherwise they contain no pair of literals of opposite sign. Also g_1 and g_2 are not simply equal to p and $\neg p$, because otherwise $g_1 \wedge g_2 = \perp$, whereas $f \neq \perp$ because f is satisfiable. $\rho(g_1, g_2)$ will be subsumed by g', which subsumes no term in f, so $\rho(g_1, g_2)$ subsumes no term already present in f.

Secondly, we prove that all terms in $\{g_1, \ldots, g_r\}$ are prime implicates of f. On the contrary, if g is a term present in f which is not a prime implicate of f, then two cases are possible: Either (a) g subsumes a prime implicate of f already present in f, in which case operation (1) can be executed, or (b) g subsumes a prime implicate not present in f, in which case a resolution can be made and added to f according to the first part of the proof. In both cases the procedure is not terminated, contrary to the assumption. □

Proof of Theorem 5.20

For the rather lengthy proof of this theorem we refer to Siegel (1987). □

For the proof of theorem 5.21 the following two lemmas are needed.

Lemma A5.2 *If P is any production field and f is an arbitrary formula in $\mathcal{L}_N$, then*

$$Newcarc(\Sigma, f, P) = \mu(Th(\Sigma \cup \{f\}) \cap P - Th(\Sigma) \cap P). \qquad (A5.10)$$

Proof of Lemma A5.2

Let $A = Th(\Sigma \cup \{f\}) \cap P$ and $B = Th(\Sigma) \cap P$. We have to prove that $\mu(A - B) = \mu A - \mu B$.

Assume first that $c \in \mu(A - B)$. Then $c \in A - B$ and hence $c \in A$, $c \notin B$ and a fortiori $c \notin \mu B$. Assume $c \notin \mu A$. Then there is a $d \in A$ such that $d \subset c$, however $d \notin A - B$ by the minimality of c, that is $d \in B$. But since $d \subset c$ it follows $c \in B$, which is a contradiction. Hence $c \in \mu A$ and therefore $c \in \mu A - \mu B$.

Conversely, assume $c \in \mu A - \mu B$. First we prove that $c \in A - B$. $c \in \mu A$ implies $c \in A$. Suppose that $c \in B$. Since $c \notin \mu B$, there is a $d \in \mu B$ such that $d \subset c$. However, as $B \subseteq A$, this implies $d \in A$ which contradicts $c \in \mu A$. Therefore $c \in A - B$. Now, suppose that c is not minimal in $A - B$. Then there exists $e \in A - B$ such that $e \subset c$, again contradicting $c \in \mu A$. Hence $c \in \mu(A - B)$. □

Lemma A5.3 *If c is a clause and P is a stable production field, then*

$$Newcarc(\Sigma, c, P) \subseteq Prod(\Sigma, c, P) \subseteq Th(\Sigma \cup \{c\}) \cap P. \qquad (A5.11)$$

Proof of Lemma A5.3

The proof is taken from Inoue (1991 b). Let $t \in Newcarc(\Sigma, c, P) = \mu(Th(\Sigma \cup \{c\}) \cap P) - \mu(Th(\Sigma) \cap P)$. Then, by lemma A5.2, $t \in Th(\Sigma \cup \{c\}) \cap P - Th(\Sigma) \cap P$. By theorem 5.20 (2) and (4) there is therefore a clause s in $Th(\Sigma \cup \{c\}) \cap P$ such that $s \subseteq t$. From the minimality of t with respect to

$Th(\Sigma \cup \{c\}) \cap P$ it follows that $t = s$ and therefore $t \in Prod(\Sigma, c, P)$ because there cannot exist $t' \subset t$ such that t' is produced from $\Sigma \cup \{c\}$ by theorem 5.20 (2) and the minimality of t with respect to $Th(\Sigma \cup \{c\}) \cap P$. This proves the first inclusion.

The second inclusion follows from theorem 5.20 (2). □

Proof of Theorem 5.21

The proof is adapted from Inoue, 1991 b. By lemmas A5.2 and A5.3,

$$Newcarc(\Sigma, c, P) \subseteq Prod(\Sigma, c, P) - Th(\Sigma) \cap P. \qquad (A5.12)$$

To prove the inverse inclusion, select $s \in Prod(\Sigma, c, P) - Th(\Sigma) \cap P$, and suppose $s \notin Newcarc(\Sigma, c, P)$. But $s \in Prod(\Sigma, c, P)$ implies $s \in Th(\Sigma \cup \{c\}) \cap P$ by lemma A5.3 and hence $s \in Th(\Sigma \cup \{c\}) \cap P - Th(\Sigma) \cap P$. Since $s \notin Newcarc(\Sigma, c, P)$, lemma A5.2 implies that s is not minimal in $Th(\Sigma \cup \{c\}) \cap P - Th(\Sigma) \cap P$, and there is a $s' \in Th(\Sigma \cup \{c\}) \cap P - Th(\Sigma) \cap P$ such that $s' \subset s$. By theorem 5.20 (4) however, there is a clause s'' produced from Σ, starting with c, such that $s'' \subseteq s' \subset s$. But this contradicts $s \in Prod(\Sigma, c, P)$. Therefore $s \in Newcarc(\Sigma, c, P)$ and this proves the theorem.

□

Lemma A5.4 *If P is a stable production field, then*

$$Carc(\Sigma \cup \{c\}, P) = \mu(Carc(\Sigma, P) \cup Newcarc(\Sigma, c, P)). \qquad (A5.13)$$

Proof of Lemma A5.4

$$\begin{aligned} Carc(\Sigma \cup \{c\}, P) &= \mu(Th(\Sigma \cup \{c\}) \cap P) \\ &= \mu((Th(\Sigma \cup \{c\}) \cap P) \cup (Th(\Sigma) \cap P)) \\ &= \mu(\mu(Th(\Sigma \cup \{c\}) \cap P) \cup \mu(Th(\Sigma) \cap P)) \\ &= \mu(Carc(\Sigma, P) \cup Carc(\Sigma \cup \{c\}, P)) \\ &= \mu(Carc(\Sigma, P) \cup (Carc(\Sigma \cup \{c\}, P) - Carc(\Sigma, P))) \\ &= \mu(Carc(\Sigma, P) \cup Newcarc(\Sigma, c, P)). \end{aligned}$$

□

Proof of Theorem 5.22

The proof is adapted from Inoue (1991, b). If Σ contains no clause, then this means that there is no restriction on the truth values of the propositions in N. Our knowledge is then represented by the tautology $\top$ and $N(\top) = B_{n+s}$. Thus $Carc(\emptyset, P) = Carc(\top, P) = PI(\top) \cap P$ by theorem 5.15. Now $PI(\top) = \top$ but $\top$ is not in P because P contains only clauses. However, if $p \vee \neg p \in P$, then $\top = p \vee \neg p$ is a prime implicate of $\top$ and hence $Carc(\emptyset, P) = \{p \vee \neg p : p \in N, p \vee \neg p \in P\}$.

Now, by lemma A5.4 and theorem 5.21, we have

$$\begin{aligned} Carc(\Sigma \cup \{c\}, P) &= \mu(Carc(\Sigma, P) \cup Newcarc(\Sigma, c, P)) \\ &= \mu(Carc(\Sigma, P) \cup (Prod(\Sigma, c, P) - Th(\Sigma) \cap P)) \\ &= \mu(Carc(\Sigma, P) \cup Prod(\Sigma, c, P)). \end{aligned} \tag{A5.14}$$

The last equality is proved as follows. To simplify notations, we write $Carc, Prod$ and Th for $Carc(\Sigma, P), Prod(\Sigma, c, P)$ and $Th(\Sigma) \cap P$ respectively. First, suppose that $x \in \mu(Carc \cup Prod)$. Then $x \in Carc \cup Prod$ and if $x \in Carc \cup (Prod - Th)$ then clearly $x \in \mu(Carc \cup (Prod - Th))$. Now we prove that it is impossible that $x \in \mu(Carc \cup Prod)$ and $x \notin Carc \cup (Prod - Th)$. In fact, this implies that $x \in Prod \cap Carc^c \cap Th$. But $x \in Th - Carc$ implies that there is $y \in Carc$ such that $y \subset x$ and hence $y \in Carc \cup Prod$. Then there is $z \in \mu(Carc \cup Prod)$ such that $z \subseteq y \subset x$, which is a contradiction to $x \in \mu(Carc \cup Prod)$.

Conversely, suppose that $x \in \mu(Carc \cup (Prod - Th))$. We must prove that there is no $y \in [Carc \cup Prod] - [Carc \cup (Prod - Th)]$ such that $y \subset x$. On the contrary, suppose that there is such an y. First, it can easily be verified that

$$[Carc \cup Prod] - [Carc \cup (Prod - Th)] = Carc^c \cap Prod \cap Th.$$

But $y \in Th - Carc$ implies that there is a $z \in Carc$ such that $z \subset y$. Hence there is a $z \in Carc \cup (Prod - Th)$ such that $z \subset y \subset x$, which is a contradiction to $x \in \mu(Carc \cup (Prod - Th))$. □

For the proof of theorem 5.23, we need the following lemma.

Lemma A5.5 *For a stable production field P,*

$$Newcarc(\Sigma, c, P) = Newcarc(PI(\Sigma), c, P). \tag{A5.15}$$

Proof of Lemma A5.5

First,

$$\begin{aligned} Carc(PI(\Sigma), P) &= \mu(Th(PI(\Sigma)) \cap P) \\ &= \mu(Th(\mu(Th(\Sigma))) \cap P) = \mu(Th(Th(\Sigma)) \cap P) \\ &= \mu(Th(\Sigma) \cap P) = Carc(\Sigma, P). \end{aligned} \tag{A5.16}$$

Furthermore,

$$\begin{aligned} Carc(PI(\Sigma) \cup \{c\}, P) &= \mu(Th(\mu Th(\Sigma) \cup \{c\}) \cap P) \\ &= \mu(Th(Th(\Sigma) \cup \{c\}) \cap P) \\ &= \mu(Th(\Sigma \cup \{c\}) \cap P) \\ &= Carc(\Sigma \cup \{c\}, P). \end{aligned} \tag{A5.17}$$

This proves the lemma by definition of *Newcarc*. □

Proof of Theorem 5.23

The proof is adapted from Inoue, 1991 b. The first equality is a direct consequence of the first equality in Eq. (5.32). The second equality follows from theorem 5.22 by using the production field P_N and applying lemmas A5.4 and A5.5 and theorem 5.21:

$$\begin{aligned} PI(\Sigma \cup \{c\}) &= \mu Th(\Sigma \cup \{c\}) \\ &= Carc(\Sigma \cup \{c\}, P_N) \\ &= \mu(Carc(\Sigma, P_N) \cup Newcarc(\Sigma, c, P_N)) \\ &= \mu(PI(\Sigma) \cup Newcarc(PI(\Sigma), c, P_N)) \\ &= \mu(PI(\Sigma) \cup (Prod(PI(\Sigma), c, P_N) - Th(\Sigma)) \\ &= \mu(PI(\Sigma) \cup Prod(PI(\Sigma), c, P_N)). \end{aligned}$$

The last equality follows from the application of the last equality in Eq. (A5.14) with Σ replaced by $PI(\Sigma)$ and P by P_N. □

Proof of Theorem 5.24 (see Kohlas, 1987)

The proof is made by induction on r. Equation (5.38) is valid for $r = 2$. Suppose that (5.38) is valid for $r - 1$. Then

$$\begin{aligned} P(\cup_{i=1}^{r} N(f_i)) &= P(\cup_{i=1}^{r-1} N(f_i) \cup N(f_r)) \\ &= P(\cup_{i=1}^{r-1} N(f_i)) + P(N(f_r)) - P(\cup_{i=1}^{r-1} N(f_i) \cap N(f_r)). \end{aligned} \tag{A5.15}$$

By the induction assumption, (5.38) can be applied to the first and third terms of the right-hand side of the last equality above. Let S'_k and S''_k denote the terms in the sum (5.38) when applied to the first and third terms above respectively. If these sums are introduced into the last equality above, we get

$$\begin{aligned} &\sum_{k=1}^{r-1} (-1)^{k+1} S'_k - \sum_{k=1}^{r-1} (-1)^{k+1} S''_k + P(N(f_r)) \\ &= S'_1 + P(N(f_r)) + \sum_{k=2}^{r-1} (-1)^{k+1} (S'_k + S''_{k-1}) + (-1)^{r+1} S''_{r-1}. \end{aligned} \tag{A5.18}$$

It is easy to verify that $S'_1 + P(N(f_r)) = S_1$, $S'_k + S''_{k-1} = S_k$ for $1 < k < r$ and $S''_{r-1} = S_r$. Thus (5.38) is also valid for r. □

Proof of Theorem 5.25

For $i = 1, \ldots, r$, let $A_i = N(f_i)$. Then

$$sp'(h) = P'(\cup_{i=1}^{r} A_i) \leq \sum_{i=1}^{r} P'(A_i) = S_1. \tag{A5.19}$$

If we define

$$A = S_1 - S_2 + S_3 - S_4 + \ldots - S_{2l}$$
$$B = S_1 - S_2 + \ldots \pm S_r$$
$$C = S_1 - S_2 + S_3 - \ldots + S_{2l+1},$$

then we must prove that $A \leq B \leq C$.

Now, let E be a point in $\cup_{i=1}^{r} A_i$ which is contained in exactly n subsets A_i among all the subsets $A_1, \ldots, A_r$. Then $P'(E)$ appears n times as a contribution to S_1, $\binom{n}{2}$ times as a contribution to S_2, ... , $\binom{n}{n}$ times as a contribution to S_n, and zero times as a contribution to S_k for $n < k \leq r$ (see Feller, 1968). Therefore, $P'(E)$ appears

$$\binom{n}{1} - \binom{n}{2} + \binom{n}{3} - \ldots \pm \binom{n}{n} \tag{A5.20}$$

times in the sum defining B. But the binomial expansion of $(1-1)^n$ shows that (A5.20) equals 1 and hence $P'(E)$ appears exactly once in the sum defining B.

Similarly, $P'(E)$ appears

$$\binom{n}{1} - \binom{n}{2} + \ldots \pm \binom{n}{min(2l, n)} \tag{A5.21}$$

times in the sum defining A and

$$\binom{n}{1} - \binom{n}{2} + \ldots \pm \binom{n}{min(2l+1, n)} \tag{A5.22}$$

times in the sum defining C.

Therefore, to prove that $A \leq B$, we show that (A5.21) is smaller or equal to 1 for all $n = 1, \ldots, r$. If $2l \geq n$, then this is trivially satisfied. If $2l < n$, then we must prove that

$$\binom{n}{1} - \binom{n}{2} + \ldots - \binom{n}{2l} \leq 1. \tag{A5.23}$$

To verify this inequality, we use the equality (18) page 57 or Knuth (1969):

$$\binom{n}{0} - \binom{n}{1} + \ldots + (-1)^j \binom{n}{j} = (-1)^j \binom{n-1}{j}. \tag{A5.24}$$

Since $\binom{n}{0} = 1$, (A5.23) follows from (A5.24) because $2l$ is pair and $\binom{n-1}{2l} \geq 0$.

In a similar way, to prove that $B \leq C$, we show that (A5.22) is greater or equal to 1 for all $n = 1, \ldots, r$. Again, if $2l + 1 \geq n$, then this is trivially satisfied. If $2l + 1 < n$, then we must prove that

$$\binom{n}{1} - \binom{n}{2} + \ldots + \binom{n}{2l+1} \geq 1. \tag{A5.25}$$

The validity of this inequality follows from (A5.24) and the theorem is proved. □

Proof of Theorem 5.26

The "if" part of the theorem is clear. To prove the "only if" part, suppose that no literal of c' appears negated in c''. Define the sets of literals

$$X' = c' - c''$$
$$X'' = c'' - c'$$
$$X = c' \cap c''.$$

Then $c' \wedge c'' = \wedge\{x' \in X'\} \wedge \{x'' \in X''\} \wedge \{x \in X\}$. Set all literals of $X' \cup X'' \cup X$ to true, then $c' \wedge c''$ is true which shows that c' and c'' are not disjoint. □

Proof of Theorem 5.27

(1) Define $X'' = c'' - c'$. Then $X' = \emptyset$ implies $c' \subseteq c''$ and then

$$c' \vee c'' = c' \vee (c' \wedge \{x'' \in X''\}) = c'. \tag{A5.26}$$

(2). To prove the equivalence, first note that for any interpretation for which c' is true, the right hand side of (5.48) is clearly true. Then consider an interpretation for which c' is false, but c'' is true, such that $c' \vee c''$ is true. For such an interpretation at least one literal x_j of X' must be false. Consider the first one, namely the one with the least index and let j be this index. Then $x_1 \dots x_{j-1}(\neg x_j)c''$ is true and the right hand side of (5.48) is true.

Conversely, if the right hand side of (5.48) is true, then either c' is true or c'' is true. In both cases $c' \vee c''$ is true.

It is evident that the terms on the right hand side of (5.48) are mutually disjoint according to theorem 5.26. □

Proof of Theorem 5.28

We prove by induction on i that P_{ij} has the following properties:

(i) $f_1 \vee \dots \vee f_i \vee f_j = f_1 \vee \dots \vee f_i \vee \{d_k \in P_{ij})$
(ii) $d_h \wedge d_k = \perp$ for all d_h, d_k in P_{ij}
(iii) $f_k \wedge d_h = \perp$ for all $k = 1, \dots, i$ and $d_h \in P_{ij}$.

This is clearly true for $P_{0j} = \{f_j\}$. Suppose P_{i-1j} has these properties. Then, using theorem 5.26, P_{ij} clearly has properties (ii) and (iii). To show (i), using the induction hypothesis, we must prove that

$$f_1 \vee \dots \vee f_i \vee \{d_k \in P_{i-1j}\} = f_1 \vee \dots \vee f_i \vee \{d_k \in P_{ij}\} \tag{A5.27}$$

First we show that if the left hand side is true then the right hand side is true. If $d_k \wedge f_i = \perp$ and d_k is true, then $d_k \in P_{ij}$. If X' is empty, then $f_i \vee d_k = f_i$ by theorem 5.27 (1). If X' is not empty, then theorem 5.27 (2) implies that

$$f_i \vee d_k = f_i + (\neg x_1)d_k + \dots + x_1 \dots x_{x-1}(\neg x_s)d_k \tag{A5.28}$$

and the terms of the right hand side of this equation (except the first one) are the terms which are added to P_{ij}. The converse is proved similarly. This proves (i) for P_{ij}.

It follows then with $i = j - 1$ that

$$f_1 \vee \ldots \vee f_j = f_1 \vee \ldots \vee f_{j-1} + \left(\sum \{d_k \in P_{j-1j}\}\right) \tag{A5.29}$$

for all $j \in \{1, \ldots, r\}$. Then, by induction on j, it follows that

$$f_1 \vee \ldots \vee f_r = \sum \{d_k \in P_{01}\} + \ldots + \sum \{d_k \in P_{r-1r}\}. \tag{A5.30}$$

□

Proof of Theorem 5.29

The procedure *find-disjoint* terminates because it will be proved below that in the recursive call $j(d, c') < j(m, c)$ and the basis of the recursion is reached when $j(m, c) = 0$ or the condition of the first "if" is satisfied: in both cases R can be computed directly.

But first we prove that the set R generated by *find-disjoint* is correct, namely that R is a collection of disjoint mix-products $\{q_1, \ldots, q_n\}$ satisfying equation (5.87). If there is $k_0 \in I_0 \cap I_c$ such that $e^m_{k_0} = -e^c_{k_0}$ then $m \wedge c = \perp$ and hence

$$m \wedge \neg c = (m \wedge \neg c) \vee (m \wedge c) = m \wedge (\neg c \vee c) = m \tag{A5.31}$$

and R is correct. If there is $s_0 \in \{1, \ldots, r\}$ such that $I_{s_0} \subseteq I_c$ and $e^m_k = e^c_k$ for all $k \in I_{s_0}$ then again $m \wedge c = \perp$ and hence $m \wedge \neg c = m$ and R is correct. If $j(m, c) = 0$, but the other two previous situations (referred to as case 1) do not occur, then $c = a'b'$ and $m = b'k'm'$ and hence

$$m \wedge \neg c = b'k'm' \wedge (\neg a' \vee \neg b') = b'k'm'\neg a' = m\neg a'. \tag{A5.32}$$

Now if $S'_1 \neq \emptyset$ then $m\neg a'$ is a mix-product and if $S'_1 = \emptyset$ then $m\neg a' = \perp$ and in both cases R is correct. If $j(m, c) > 0$ and case 1 does not occur, then arbitrarily select $s_0 \in \{1, \ldots, r\}$ such that $S = I_c \cap I_{s_0} \neq \emptyset$. It follows that $m = \neg(m_1 m_2 m_3)m''$, $c = m_1 m_2^* c'$ and hence

$$\begin{aligned} m\neg c &= \neg(m_1 m_2 m_3)m''\neg(m_1 m_2^* c') \\ &= [\neg(m_1 m_2 m_3)m''\neg(m_1 m_2^*)] \vee [\neg(m_1 m_2 m_3)m''\neg c'] \\ &= a \vee b = a + (b\neg a). \end{aligned} \tag{A5.33}$$

Now we show that R_1 is a decomposition of a and R_2 is a decomposition of $b \wedge \neg a$. First note that

$$\begin{aligned} &\neg(m_1 m_2 m_3)\neg(m_1 m_2^*) \\ &= (\neg m_1 \vee \neg(m_2 m_3))(\neg m_1 \vee \neg m_2^*) \\ &= \neg m_1 \vee (\neg m_1 \neg m_2^*) \vee (\neg m_1 \neg(m_2 m_3)) \vee \neg m_2^* \neg(m_2 m_3) \\ &= \neg m_1 \vee \neg m_2^* \neg(m_2 m_3) \\ &= \neg m_1 \vee \neg m_2^* \neg m_2 \vee \neg m_2^* \neg m_3 \\ &= \neg m_1 + (\neg m_2^* \neg m_2)\neg(\neg m_1)) + (\neg m_2^* \neg m_3 \neg(\neg m_1 \vee \neg m_2^* \neg m_2)) \end{aligned}$$

and hence $a = a_1 + a_2 + a_3$ where

$$\begin{aligned} a_1 &= \neg m_1 m'' \\ a_2 &= \neg m_2^* \neg m_2 \neg(\neg m_1) m'' \\ a_3 &= \neg m_2^* \neg m_3 \neg(\neg m_1 \vee \neg m_2^* \neg m_2) m''. \end{aligned}$$

Now we show that R_i' is a decomposition of a_i for all $i = 1, 2, 3$.

If $S_1 = \emptyset$ then $m_1 = \top$ and hence $a_1 = \neg m_1 m'' = \bot$ and R_1' is correct. If $S_1 \neq \emptyset$ then a_1 is itself a mix-product and hence R_1' is again correct. If $|S_2| = 0$, i.e. $S_2 = \emptyset$, then $m_2 = \top$ and hence $a_2 = \bot$ and R_2' is correct. If $|S_2| \geq 1$ then arbitrarily select $k_0 \in S_2$. Then

$$\begin{aligned} \neg m_2^* \neg m_2 &= \neg\left(\wedge_{k\in S_2} x_k^{-e_k^m}\right) \neg\left(\wedge_{k\in S_2} x_k^{e_k^m}\right) \\ &= \neg\left(x_{k_0}^{-e_{k_0}^m} \wedge_{k\in S_2-\{k_0\}} x_k^{-e_k^m}\right) \neg\left(x_{k_0}^{e_{k_0}^m} \wedge_{k\in S_2-\{k_0\}} x_k^{e_k^m}\right) \\ &= \left(x_{k_0}^{e_{k_0}^m} \vee \neg\left(\wedge_{k\in S_2-\{k_0\}} x_k^{-e_k^m}\right)\right) \wedge \left(x_{k_0}^{-e_{k_0}^m} \vee \neg\left(\wedge_{k\in S_2-\{k_0\}} x_k^{e_k^m}\right)\right) \\ &= d_1 \vee d_2 \vee d_3 \\ &= (d_1 \vee d_2) + (d_3 \wedge \neg(d_1 \vee d_2)) \\ &= d_1 \vee d_2 \end{aligned}$$

where the last equality holds because $d_3 \wedge \neg(d_1 \vee d_2) = \bot$:

$$\begin{aligned} d_3 \neg d_1 \neg d_2 &= \neg v^* \neg v \neg(z \neg v) \neg(\neg z \neg v^*) \\ &= (\neg v^* \neg v)(\neg z \vee v)(z \vee v^*) \\ &= (\neg v^* \neg v)(\neg z z \vee \neg z v^* \vee z v \vee v v^*) \\ &= \neg v^* \neg v \neg z v^* \vee \neg v^* \neg v z v \vee \neg v^* \neg v v v^* = \bot, \end{aligned}$$

which shows that $\neg m_2^* \neg m_2 = d_1 + d_2$. This implies that

$$a_2 = (d_1 + d_2) m_1 m'' = d_1 m_1 m'' + d_2 m_1 m''. \tag{A5.34}$$

Now if $|S_2| = 1$ then $v = v^* = \top$ and $\neg v = \neg v^* = \bot$ and $d_1 = d_2 = \bot$ and $a_2 = \bot$ which proves that R_2' is correct. If $|S_2| > 1$ then both $d_1 m_1 m''$ and $d_2 m_1 m''$ are mix-products and again R_2' is correct.

For the decomposition of a_3 first note that

$$\begin{aligned} a_3 &= \neg m_2^* \neg m_3 \neg(\neg m_1 \vee \neg m_2^* \neg m_2) m'' \\ &= \neg m_2^* \neg m_3 m_1 \neg(\neg m_2^* \neg m_2) m'' \\ &= \neg m_2^* \neg m_3 m_1 (m_2^* \vee m_2) m'' \\ &= m_1 m_2 \neg m_2^* \neg m_3 m''. \end{aligned} \tag{A5.35}$$

If $S_3 = \bot$ then $m_3 = \top$ and $a_3 = \bot$ and R_3' is correct. If $S_2 = \emptyset$ then $m_2^* = \top$ and $a_3 = \emptyset$ and R_3' is correct. If both $S_2 \neq \emptyset$ and $S_3 \neq \emptyset$ then we have

$$\begin{aligned} m_2 \neg m_2^* &= \left(\wedge_{k\in S_2} x_k^{e_k^m}\right) \neg \left(\wedge_{k\in S_2} x_k^{-e_k^m}\right) \\ &= \left(\wedge_{i\in S_2} x_i^{e_i^m}\right)\left(\vee_{k\in S_2} x_k^{e_k^m}\right) \\ &= \vee_{k\in S_2}\left(x_k^{e_k^m} \wedge_{i\in S_2} x_i^{e_i^m}\right) \\ &= \vee_{k\in S_2}\left(\wedge_{i\in S_2} x_i^{e_i^m}\right) \\ &= \wedge_{i\in S_2} x_i^{e_i^m} = m_2 \end{aligned} \tag{A5.36}$$

and hence $a_3 = m_1 m_2 \neg m_3 m''$, which has the form of a mix-product and hence R_3' is correct.

Now we show that R_2 is a decomposition of $b \wedge \neg a$. We have

$$\begin{aligned} b \wedge \neg a &= \neg(m_1m_2m_3)m''\neg c'\neg(\neg m_1m_2m_3m''\neg(m_1m_2^*)) \\ &= \neg(m_1m_2m_3)m''\neg c'(m_1m_2m_3 \vee \neg(m''\neg(m_1m_2^*)) \\ &= \neg(m_1m_2m_3)m''\neg c' m_1m_2m_3 \vee \neg(m_1m_2m_3)m''\neg c'\neg(m''\neg(m_1m_2^*)) \\ &= \neg(m_1m_2m_3)m''\neg c'(\neg m'' \vee m_1m_2^*) \\ &= \neg(m_1m_2m_3)m''\neg c'\neg m'' \vee \neg(m_1m_2m_3)m''\neg c' m_1m_2^* \\ &= (\neg m_1 \vee \neg(m_2m_3))m''\neg c' m_1m_2^* \\ &= \neg m_1 m''\neg c' m_1m_2^* \vee \neg(m_2m_3)m''\neg c' m_1m_2^* \\ &= (\neg m_2 \vee \neg m_3)m''\neg c' m_1m_2^* \\ &= \neg m_2m_2^*m''\neg c' m_1 \vee m_1m_2^*\neg m_3m''\neg c'. \end{aligned}$$

Now if $S_2 = \emptyset$ then $\neg m_2 = \perp$ and $m_2^* = \top$ and hence $b \wedge \neg a = m_1 \neg m_3 m'' \wedge \neg c'$. But $m_1 \neg m_3 m''$ is a mix-product and c' is a conjunction and hence R_2 is correct in the case $S_2 = \emptyset$. If $S_2 \neq \emptyset$ then

$$\begin{aligned} \neg m_2 m_2^* &= \neg\left(\wedge_{k\in S_2} x_k^{e_k^m}\right)\left(\wedge_{k\in S_2} x_k^{-e_k^m}\right) \\ &= \left(\vee_{k\in S_2} x_k^{-e_k^m}\right)\left(\wedge_{i\in S_2} x_i^{-e_i^m}\right) \\ &= \vee_{k\in S_2}\left(x_k^{-e_k^m} \wedge_{i\in S_2} x_i^{-e_i^m}\right) \\ &= \wedge_{i\in S_2} x_i^{-e_i^m} = m_2^* \end{aligned} \tag{A5.37}$$

and therefore

$$\begin{aligned} b \wedge \neg a &= m_2^* m''\neg c' m_1 \vee m_2^* m_1 \neg m_3 m''\neg c' \\ &= m_2^* m''\neg c' m_1 = m_1 m_2^* m'' \wedge \neg c'. \end{aligned} \tag{A5.38}$$

But $m_1 m_2^* m''$ is a mix-product and c' is a conjunction and hence R_2 is also correct if $S_2 \neq \emptyset$. This completes the proof that the set R generated by *find-disjoint* is correct.

Finally, it remains to prove that the recursive call is valid, i.e. $j(d, c') < j(m, c)$ when $d = m_1 \neg m_3 m''$ and $d = m_1 m_2^* m''$. We start with $d = m_1 \neg m_3 m''$. If we define $j'(m, c)$ to be the number or intervals I_s, $s \neq s_0$ such that $I_c \cap I_s \neq$

$\emptyset$ and $j'(d,c')$ the number of intervals $I_s, s \neq s_0$ such that $I_s \cap S_4 \neq \emptyset$ then $j(d,c') = j'(d,c')$ because $S_3 \cap S_4 = \emptyset$, and $j(m,c) = j'(m,c) + 1$ because $I_{s_0} \cap I_c \neq \emptyset$. Furthermore, $j'(d,c') \leq j'(m,c)$ because $S_4 \subseteq I_c$. But this implies that

$$j(d,c) = j'(d,c') \leq j'(m,c) = j(m,c) - 1 < j(m,c) \tag{A5.39}$$

and the assertion is proved when $d = m_1 \neg m_3 m''$. Now let's consider the case where $d = m_1 m_2^* m''$. First $j(d,c')$ is the number of intervals $I_s, s \neq s_0$, such that $I_s \cap S_4 \neq \emptyset$. Then $j(d,c') \leq j'(m,c)$ because $S_4 \subseteq I_c$. Furthermore, $j(m,c) = j'(m,c) + 1$ because $I_{s_0} \cap I_c \neq \emptyset$. But this implies that

$$j(d,c') \leq j'(m,c) = j(m,c) - 1 < j(m,c), \tag{A5.40}$$

which completes the proof of the correctness of the procedure *find-disjoint*. □

6. Rule-Based Systems With Unreliable Rules

6.1 Inference Networks and Monotone Systems

In this chapter assumption-based systems as introduced in Chap. 5 will be treated from a different point of view. Clauses containing assumptions can be considered as uncertain rules. For example, the clause $\neg b \vee \neg a_1 \vee a$ corresponds to the implication $b \rightarrow a$ if a_1 is true. If a_1 is false, then the rule is not valid. The probability of a_1 is the reliability of the rule. Thus in this chapter uncertain or unreliable rules are considered and the inference process with such rules is discussed. Rather than considering explicit assumptions as in the previous chapter and computing degrees of support for certain hypotheses, a more direct way of computing degrees of support using unreliable rules will be proposed in this chapter.

Consider a set $P = \{p_1, \ldots, p_n\}$ of atomic propositions which are related through a certain number of inference rules $R = \{r_1, \ldots, r_m\}$ of the form

$$\text{if } p_i \wedge \ldots \wedge p_j \text{ then } p_k. \tag{6.1}$$

One of the particularity of these rules is that no proposition appears negated and there is only one proposition as conclusion of the rule. The propositions in the if-part of the rule are called its conditions and the proposition in the then-part is called the conclusion. Of course, each rule r_k can be represented by the clause $\neg p_i \vee \ldots \vee \neg p_j \vee p_k$ in the language $\mathcal{L}_P$. Such clauses containing at most one non-negated literal are called Horn clauses and in this first section only Horn clauses are admitted.

For a rule r_k, let $Cd(r_k) \subseteq P$ denote the conditions of r_k and let $Cc(r_k) \in P$ denote its conclusion. Rules and propositions can be visualized in the bipartite graph $G = (P + R, U)$ where

$$U = \{(p_j, r_k) : p_j \in Cd(r_k), r_k \in R\} \cup \{(r_k, p_k) : p_k = Cc(r_k), r_k \in R\}. \tag{6.2}$$

This is the traditional structure of an *inference network*. Fig. 6.1 shows an example of such a network where squares represent rules and circles represent propositions. As a general assumption we suppose that the inference networks we are dealing with contain no circuit (oriented cycles). Using the inference network, if we learn that certain propositions $F \subseteq P$ are true, then we can deduce some other propositions. These are precisely the propositions which are logical consequences of $R \cup F$. For example, proposition 1 is a logical consequence of all rules and propositions 6 and 12 in Fig. 6.1.

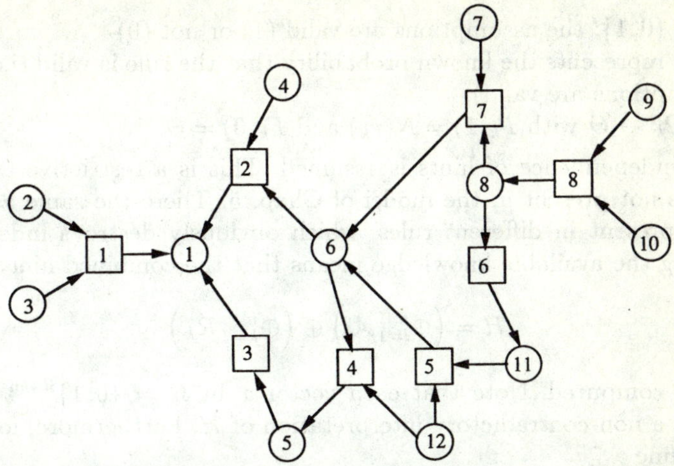

Fig. 6.1. An inference network

Instead of knowing that some proposition p_i is true with certainty, it is now assumed that we dispose of a not fully reliable assumption which would imply p_i. This could be represented by the implication $a_i \rightarrow p_i$. If a_i is true only with a probability $P(a_i)$, then p_i can be inferred with this probability, whereas with probability $1 - P(a_i)$ nothing can be inferred. Such an unreliable rule can alternatively be represented by a simple hint. Indeed, if 1 stands for "true" and 0 for "false", then to each proposition $p_i \in P$ a frame of discernment $\Theta_i = \{0,1\}$ corresponds. Then we have the simple hint

$$\mathcal{A}_i = (\Omega_i, P_i, \Gamma_i, \Theta_i), \tag{6.3}$$

where

(1) $\Omega_i = \{0,1\}$: the assumption is true (1) or false (0)
(2) $P_i(1)$ represents the known probability that the assumption is true and $P_i(0) = 1 - P_i(1)$ is the probability that the assumption is false
(3) $\Gamma_i : \Omega_i \rightarrow \Theta_i$ with $\Gamma_i(1) = \{1\}$ and $\Gamma_i(0) = \{0,1\}$.

Note that the hint $\mathcal{A}_i$ can also be viewed as a hint on the frame $\Theta = \prod_{i=1}^{n} \Theta_i = \{0,1\}^n$ corresponding to the Boolean cube induced by the propositions in P. Indeed, simply set $\Gamma_i(1) = N(p_i) \subseteq \Theta$ and $\Gamma_i(0) = \Theta$ (recall from Chap. 5 that $N(p_i)$ is the set of all vectors $\theta \in \Theta$ such that p_i is true under θ). This process is called the vacuous extension of a hint and it will be discussed in detail in the next chapter.

Furthermore, we also assume that the rules are not fully reliable. This means that a rule is valid only under some assumptions which cannot explicitly be identified. More precisely, the unreliable rule or clause r_k is represented by the simple hint

$$\mathcal{R}_k = (\Omega_k, P_k, \Gamma_k, \Theta), \tag{6.4}$$

where

(1) $\Omega_k = \{0,1\}$: the assumptions are valid (1) or not (0)

(2) $P_k(1)$ represents the known probability that the rule is valid (i.e. that the assumptions are valid).

(3) $\Gamma_k : \Omega_k \to \Theta$ with $\Gamma_k(1) = N(r_k)$ and $\Gamma_k(0) = \Theta$.

Here independence of hints is assumed. This is a restrictive assumption which was not present in the model of Chap. 5. There the same assumption could be present in different rules, which obviously destroys independence. Integrating the available knowledge means that the combined hint

$$\mathcal{H} = \left(\oplus_{i=1}^{n} \mathcal{A}_i\right) \oplus \left(\oplus_{k=1}^{m} \mathcal{R}_k\right) \tag{6.5}$$

has to be computed. Note that each vector x in $\Omega = \{0,1\}^{n+m}$ obviously represents a non-contradictory interpretation of $\mathcal{H}$. Furthermore, for a vector $x \in \Omega$, define

$$\Gamma(x) = \cap_{j=1}^{n+m} \Gamma_j(x_j) \tag{6.6}$$

and if $\mu_j = P_j(1)$ for all j, let

$$P(x) = \prod_{j=1}^{n+m} \mu_j^{x_j} (1-\mu_j)^{1-x_j}. \tag{6.7}$$

Then the combined hint defined in (6.5) is equal to $\mathcal{H} = (\Omega, P, \Gamma, \Theta)$.

Now suppose that we are more particularly interested in a proposition $p_j \in P$. The hypothesis that p_j is true is then represented by the subset $N(p_j) \subseteq \Theta$. This hypothesis can be judged in the light of the hint $\mathcal{H}$ by computing its degree of support $sp(p_j)$ and its degree of plausibility $pl(p_j)$. The second problem is immediately solved by the following theorem.

Theorem 6.1. *The degree of plausibility of every proposition $p_j \in P$ is equal to 1*

In order to compute $sp(p_j)$, we need the set of supporting interpretations

$$u(p_j) = \{x \in \Omega : \Gamma(x) \subseteq N(p_j)\}. \tag{6.8}$$

If $f : \Omega \to \{0,1\}$ is the Boolean function which assigns 1 to the elements of $u(p_j)$ and 0 to the other ones, then f is *monotone*, i.e. if $x' \leq x''$ then $f(x') \leq f(x'')$. Furthermore, $f(1,\ldots,1) = 1$ and $f(0,\ldots,0) = 0$. It is well known from reliability theory that such a monotone Boolean function generates a so-called *monotone system* $(M, \mathcal{S})$ where $M = \{1,\ldots,n+m\}$ is identified with the set of nodes of the inference network and $\mathcal{S}$ is a family of subsets of M defined as follows. A subset $S \subseteq M$ belongs to $\mathcal{S}$ if $f(x) = 1$ for the vector x defined by

$$x_j = \begin{cases} 1 & \text{if } j \in S \\ 0 & \text{if } j \notin S. \end{cases} \tag{6.9}$$

The system $(M, \mathcal{S})$ is called monotone because if $S \in \mathcal{S}$ and $S \subseteq S'$ then $S' \in \mathcal{S}$. Furthermore, note that $M \in \mathcal{S}$ and $\emptyset \notin \mathcal{S}$. Monotone systems are the fundamental structures of the theory of reliability (see Barlow, Proschan, 1975; Kohlas, 1987). This shows that we can consider the nodes j in M as possibly failing components of the technical system M. Node j is said to be functioning if $x_j = 1$ and μ_j represents the reliability of the component j. Then the system M is said to be functioning if its functioning nodes form a set which is in $\mathcal{S}$.

Conversely, given a monotone system $(M, \mathcal{S})$, we can define an associated monotone Boolean function as follows. Let Ω be the set of all Boolean vectors $x = (x_1, \ldots, x_t)$ of length equal to the number of elements of M. If $H(x) = \{i \in M : x_i = 1\}$, then the monotone Boolean function $f : \Omega \to \{0,1\}$ associated with the monotone system $(M, \mathcal{S})$ is given by

$$f(x) = \begin{cases} 1 & \text{if } H(x) \in \mathcal{S} \\ 0 & \text{if } H(x) \notin \mathcal{S}. \end{cases} \tag{6.10}$$

Therefore, monotone Boolean functions and monotone systems are essentially the same as far as reliability questions are concerned. This permits some useful abuse of language, like speaking of the functioning nodes of a vector x in Ω. The set $u(p_j)$ can easily be characterized in terms of logical consequences. In fact, a vector x is in $u(p_j)$ if and only if p_j is a logical consequence of the clauses corresponding to the functioning nodes of x, where the clause of a circular node is defined to be the proposition on that node and the clause of a rectangular node is the rule on that node. This is an immediate consequence of theorem 5.4 which allows us to identify the supporting interpretations of p_j in a graphical way. In fact, for a vector x in Ω, let $S(x)$ denote the nodes in the inference network corresponding to functioning propositions and $R(x)$ the nodes corresponding to functioning rules. Furthermore, if $Q(x)$ denotes the set of all nodes incident to a rule in $R(x)$, then define $G(x)$ as the subgraph of the inference network over the nodes $S(x) \cup R(x) \cup Q(x)$. Then it can easily be verified that a vector x is in $u(p_j)$ if and only if there is a bipartite subtree T of $G(x)$ such that

(1) p_j is anti-root and a leaf of T.
(2) For each non-leaf node p of T there is exactly one arc in T entering p and exactly one arc in T leaving p.
(3) Except p_j, all leaves of T are in $S(x)$.

The *reliability* of the system $(M, \mathcal{S})$ is classically defined as the probability that M is functioning (of course, this value depends on the reliabilities of the components $\mu = (\mu_1, \ldots, \mu_{n+m})$):

$$z_f(\mu) = P(f(x) = 1) = P(u(p_j)). \tag{6.11}$$

But since $P(u(p_j)) = sp(p_j)$, it follows immediately that the reliability of M coincides with the degree of support of p_j. This corresponds of course to

(5.35) in Chap. 5. But in the present case we have no contradictions such that $sp'(\circ) = 0$ and $sp(p_j) = sp'(p_j)$.

Computing the reliability $z_f(\mu)$ is a classical problem in reliability theory and several methods can be used (Barlow, Proschan, 1975; Kohlas, 1987). For example, theorem 5.8 implies that $f = f_1 \vee \ldots \vee f_r$ where $f_1, \ldots, f_r$ are the prime implicants of f (they all contain only positive literals according to theorem 5.7). Once the prime implicants of f have been found, the methods described in Sect. 5.7 can be used to compute $sp(p_j) = P(N(f))$ (e.g. the inclusion-exclusion method or some disjoint decomposition algorithm). It can easily be verified that the prime implicants of f correspond to the so-called *minimal links* of the monotone system $(M, \mathcal{S})$, i.e. the minimal elements in $\mathcal{S}$.

6.2 Computing Degrees of Support in Inference Trees and Networks

In this section, another method of computing $sp(p_j)$ will be presented. Given a monotone system $(M, \mathcal{S})$ corresponding to the monotone Boolean function f, a decomposition $J_1, \ldots, J_r$ of M is called a *modular decomposition* if there exist monotone Boolean functions $g_1, \ldots, g_r$ such that

$$f(x) = h(g_1(x_{J_1}), \ldots, g_r(x_{J_r})) \tag{6.12}$$

for some monotone Boolean function h (x_{J_i} denotes the projection of the vector x to J_i). Then the monotone system $B_i = (J_i, \mathcal{S}_i)$ derived from the monotone Boolean function g_i is called a *module*. For example, if

$$f(x) = g_1(x_{J_1}) \vee \ldots \vee g_r(x_{J_r}) \tag{6.13}$$

then $J_1, \ldots, J_r$ is called a *parallel* modular decomposition because at least one module must be functioning for the entire system to be functioning. Similarly, if

$$f(x) = g_1(x_{J_1}) \wedge \ldots \wedge g_r(x_{J_r}) \tag{6.14}$$

then $J_1, \ldots, J_r$ is called a *series* modular decomposition because all modules must be functioning for the entire system to be functioning. Formula (6.12) shows that it is sufficient to know the state of the modules B_i (functioning or not functioning) in order to determine the state of the entire system M. Let μ_{J_i} denote the vector of reliabilities μ_j for the nodes $j \in J_i$ and let z_i denote the reliability of the module B_i. If z_f and z_h are the reliabilities of the systems induced by the monotone Boolean functions f and h respectively, then it is well known (see Barlow, Proschan, 1975) that

$$z_f(\mu) = z_h(z_1(\mu_{J_1}), \ldots, z_r(\mu_{J_r})). \tag{6.15}$$

In order to compute the reliability of M, it is therefore possible to compute first the reliabilities of the modules, and then to use these values to compute the reliability of the system induced by h. In many cases this greatly facilitates the computations. For example, the reliability of the parallel system (6.13) is

$$z_f(\mu) = 1 - \prod_{i=1}^{r}(1 - z_i(\mu_{J_i}) \tag{6.16}$$

and the reliability of the series system (6.14) is

$$z_f(\mu) = \prod_{i=1}^{r} z_i(\mu_{J_i}). \tag{6.17}$$

Now consider an inference network $G = (P+R, E)$ and a particular proposition $t \in P$ for which we want to compute the degree of support $sp(t, G)$. Suppose that all nodes in $P + R$ are consecutively numbered from 1 to m, so that $M = \{1, \ldots, m\}$ and $G = (M, E)$. First, note that all arcs and nodes which do not lie on a path leading to t can be eliminated from G since they are clearly irrelevant for the computation of $sp(t, G)$. This generates a new network which will also be written $G = (M, E)$ in order to keep notations as simple as possible. As a general notation, let $f(t, G)$ denote the monotone Boolean function of the monotone system induced by the proposition t and the network G. If node t is removed from G, then the remaining graph decomposes into one or more connected subgraphs $G_1 = (M_1, E_1), \ldots, G_n = (M_n, E_n)$.

Theorem 6.2. *The subsets $M_1, \ldots, M_n, \{t\}$ of M form a parallel modular decomposition of M. In other words,*

$$f(t, G) = f(t, \{t\}) \vee f(t, G_1) \vee \ldots \vee f(t, G_n). \tag{6.18}$$

Using (6.16), this result implies that

$$sp(t, G) = 1 - (1 - \mu_t) \prod_{i=1}^{n}(1 - sp(t, G_i)). \tag{6.19}$$

Hence we only have to compute $sp(t, G_i)$ for all modules $i = 1, \ldots, n$. Two cases must be distinguished. In case (I), suppose that there is exactly one rule $r \in M_i$ which is incident to t in G and also that the subgraph of G_i defined by the nodes which are different from r is composed of one or more simply connected components H_j (i.e. there is a chain in H_j between every node x and y in H_j), each containing exactly one proposition q_j from $Cd(r)$ (see Fig. 6.2).

Theorem 6.3. *The subsets $H_1, \ldots, H_j, \ldots, \{r\}$ of M_i form a series modular decomposition of M_i. In other words,*

$$f(t, G_i) = f(r, \{r\}) \wedge f(q_1, H_1) \wedge \ldots \wedge f(q_j, H_j) \wedge \ldots \tag{6.20}$$

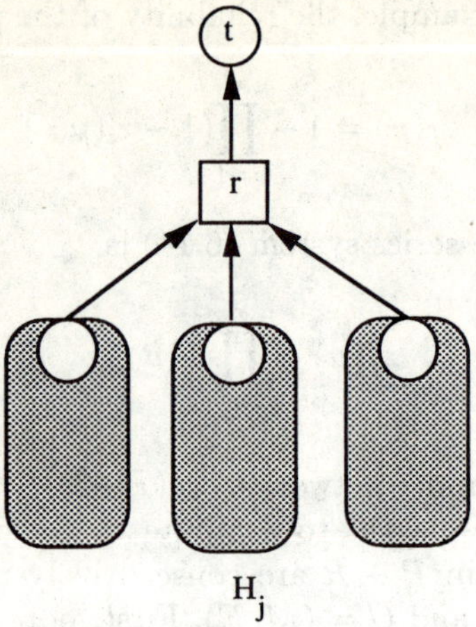

Fig. 6.2.

Using (6.17), this result implies that

$$sp(t, G_i) = \mu_r \prod \{sp(q_j, H_j) : q_j \in Cd(r)\}. \tag{6.21}$$

When G is a tree (in fact and AND-OR tree as it would be called in artificial intelligence), formulas (6.19) and (6.21) can be used to compute $sp(t, G)$ because in this case all H_j are also trees and computing $sp(t, G)$ is a similar task as computing $sp(q_j, H_j)$, the only difference is that H_j is a subtree of G. Hence (6.19) and (6.21) are recursive formulas whose basis is reached when trees are composed of one single node and then $sp(t, G) = \mu_t$ for such trees.

In case (II), there is either

(a) Exactly one rule $r \in M_i$ incident to t in G but there is at least one component H_j which contains two or more propositions in $Cd(r)$ (see Fig. 6.3 a)

(b) There are at least two rules in G_i which are incident to t in G (see Fig. 6.3 b).

A proposition $s \in G_i$ is called a *source of dependence* (abbreviated *sod*) with respect to t if there are two (edge) different paths from s to t in G. Note that there may exist several sources of dependence. For example, in Fig. 6.1 nodes 6 and 12 are both sources of dependence for $t = 1$.

Theorem 6.4. *In case (II), there is at least one source of dependence in G_i.*

A source of dependence is called *extremal* if there is no other source of dependence which leads to it.

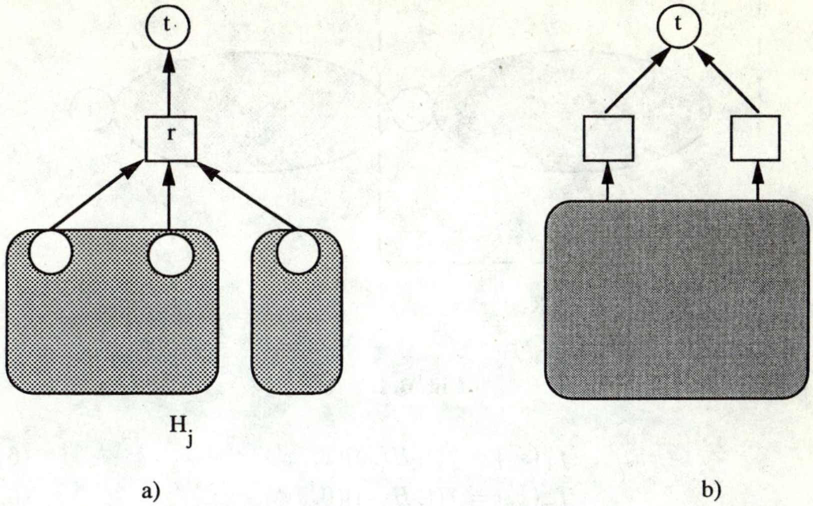

Fig. 6.3.

Theorem 6.5. *If there is a sod in G_i, then there is at least one extremal sod in G_i.*

For a node $s \in G_i$, let $E(s)$ denote the set of arcs in G_i which lie on a path leading to s . Furthermore, let $L(s)$ denote the set of nodes $l \in G_i$ which are incident to some arc in $E(s)$. This defines a connected graph $K(s) = (L(s), E(s))$. The important results about extremal sources of dependence are contained in the following two theorems. See Fig. 6.4 for a picture of the situation.

Theorem 6.6. *If s is an extremal source of dependence in G_i, then s is a cut-vertex of G_i. This means that every path from a node $q \in L(s)$ to t contains node s.*

Theorem 6.7. *If s is an extremal source of dependence, then $K(s)$ is a tree.*

Now suppose that s is an extremal source of dependence in G_i and let $H(s)$ denote the subgraph of G_i over the nodes $M_i - L(s) + \{s\}$. Furthermore, define

$$\Omega_s = \{0,1\}^{|M_i - L(s) + \{s\}|}, \ \Omega'_s = \{0,1\}^{|M_i - L(s)|}, \tag{6.22}$$

so that $\Omega_s = \{0,1\} \times \Omega'_s$. If the notation $x = (1_s, x')$ with x in Ω_s and x' in Ω'_s means that the s-component of x is 1 and the other components are those of x' and if $(0_s, x')$ is defined similarly, then we can define two monotone Boolean functions on Ω'_s:

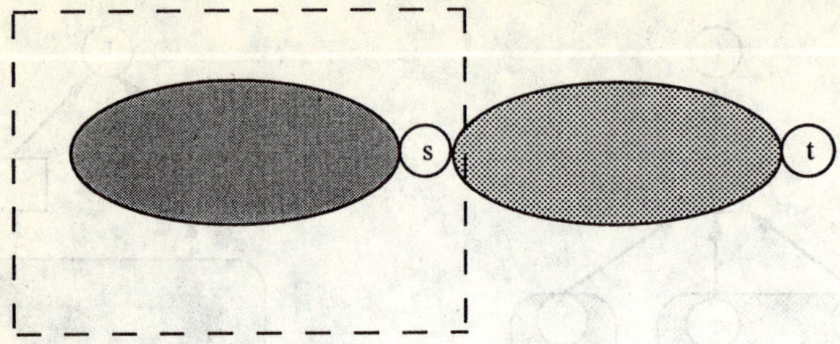

Fig. 6.4.

$$f_+(x') = f(t, H(s))(1_s, x') \tag{6.23}$$

$$f_-(x') = f(t, H(s))(0_s, x'). \tag{6.24}$$

If z, z^+ and z^- are the reliabilities of the monotone systems induced by $f(t, H(s)), f_+$ and f_- respectively, then it is well known in reliability theory that

$$z = \mu_s z^+ + (1 - \mu_s) z^-. \tag{6.24}$$

According to (6.23), when computing z^+ it is assumed that the proposition s is proved. But if s is proved, then $L(s)$ can be removed from the network G_i. This defines a new graph G_i^+ and the computation of z^+ can be performed using this reduced graph G_i^+. In a similar way, when computing z^-, it is assumed that s is not proved, which implies that the rules in $V(s) = \{r \in R : s \in Cd(r)\}$ cannot be applied to prove t. This shows that the computation of z^- can be performed using the reduced graph G_i^- obtained by eliminating $L(s)$ and $V(s)$ from G_i. Since by theorem 6.6 the proposition s is proved to the degree $sp(s, K(s))$, it follows from (6.24) that

$$sp(t, G_i) = sp(s, K(s)) sp(t, G_i^+) + (1 - sp(s, K(s)) sp(t, G_i^-). \tag{6.25}$$

If G_i is the subgraph containing rules 2 and 3 when node 1 is removed from the graph of Fig. 6.1, then Fig. 6.5 a) shows G_i^+ and Fig. 6.5 b) shows G_i^- for the extremal source of dependence 8 with respect to $t = 1$ (node 1 and its incident arcs are not in these networks, but they have nevertheless been drawn for clarity).

This method is called *factorization*. It is a well known computational technique in reliability theory. It has also been applied in probabilistic reasoning with Bayesian networks and influence diagrams (Pearl, 1988; Oliver, Smith, 1990), but is new in the context of evidential reasoning.

According to formula (6.19), in order to compute $sp(t, G)$ we have to compute the value of $sp(t, G_i)$ for all $i = 1, \ldots, n$. But if G_i satisfies the conditions of case (I), then (6.21) shows that computing $sp(t, G_i)$ requires computing

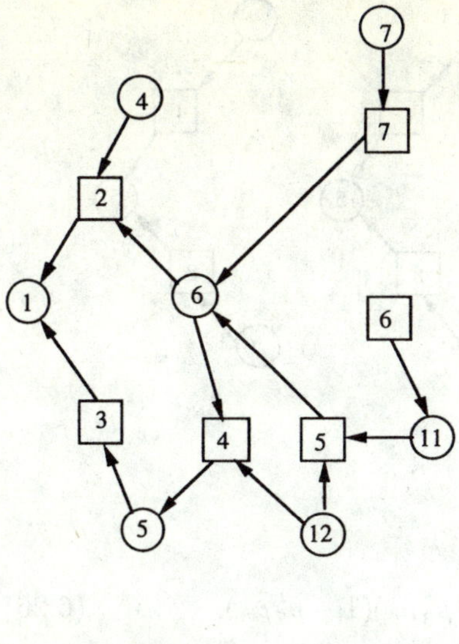

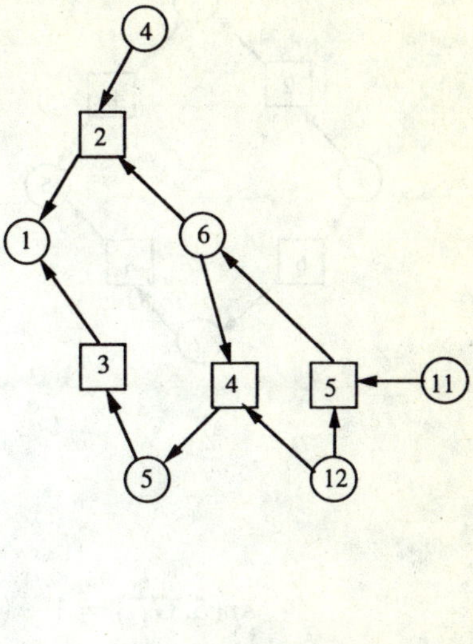

Fig. 6.5.

$sp(q_j, H_j)$, which is a similar problem as computing $sp(t, G)$ except that H_j is a subnetwork of G and is therefore easier to treat. Otherwise, according to theorems 6.4 and 6.5, there exists an extremal source of dependence s in G_i. Then formula (6.21) shows that it is sufficient to compute the degree of support of a proposition in three different inference networks, which are all proper subnetworks of G_i. Moreover, computing $sp(s, K(s))$ is particularly simple because $K(s)$ is a tree according to theorem 6.7. This demonstrates that $sp(t, G)$ can be computed recursively and the basis of the recursion is reached when the networks to be considered are composed of one single proposition. Thus we have proved the following theorem:

Theorem 6.8. *The degree of support $sp(t, G)$ can be computed recursively in a finite number of steps.*

As an example, consider the inference network pictured in Fig. 6.6 a) and suppose we want to compute the degree of support of proposition 5. We factorize on the extremal source of dependence 1. The graphs G_1^+ and G_1^- are pictured in Fig. 6.6 b) and c) respectively (node 5 is again drawn for clarity).

Note that rule 2 in G_i^+ induces a new simple hint on Θ_3 and the new reliability of node 3 equals $\mu_3' = 1 - (1 - \mu_2)(1 - \mu_3) = \mu_3 + \mu_2(1 - \mu_3)$. Similarly, the new reliability of node 7 equals $\mu_7' = \mu_7 + \mu_6(1 - \mu_7)$. Then we can easily compute

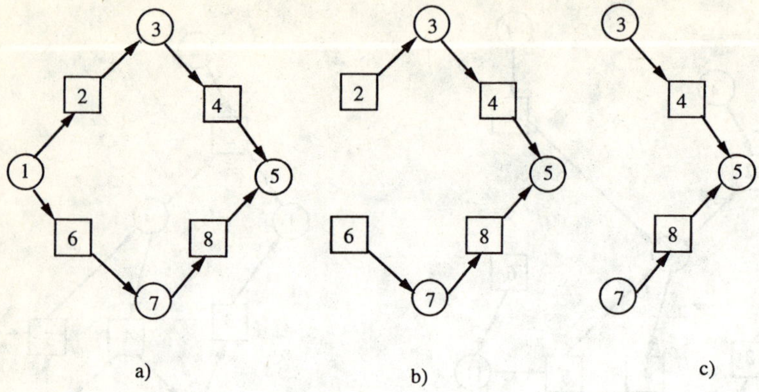

Fig. 6.6.

$$sp(5, G_1^+) = 1 - (1 - \mu_5)(1 - \mu_3'\mu_4)(1 - \mu_7'\mu_8) \tag{6.26}$$

and similarly

$$sp(5, G_i^-) = 1 - (1 - \mu_5)(1 - \mu_3\mu_4)(1 - \mu_7\mu_8). \tag{6.27}$$

Since $sp(1, K(s)) = \mu_1$, we obtain finally

$$sp(5, G) = \mu_1 sp(5, G_1^+) + (1 - \mu_1) sp(5, G_1^-). \tag{6.28}$$

6.3. Computing Bounds For The Degree of Support

The recursive computation of the degree of support of a proposition in a large network may take quite a long time. In order to keep computation times within reasonable limits, we consider an increasing sequence of subnetworks of the original nertwork. There are two values attached to each network of the sequence, one being a lower bound for the exact degree of support and the other an upper bound. Thus, starting from the network composed of the proposition t only, we compute a decreasing sequence of nested intervals containing the exact degree of support of t. If the length of the interval is judged to be sufficiently small for a particular network, then the process is terminated. Of course, the last network of the sequence may very well be the original network, in which case the interval is reduced to one point.

First, let $\hat{G} = (\hat{M}, \hat{U})$, $\hat{M} = \hat{P} + \hat{R}$ denote the original network and suppose that all irrelevant arcs have already been eliminated from $\hat{G}$. Given a subnetwork $G = (M, U)$ of $\hat{G}$, we must explain how to perform two operations:

(a) How is the next subnetwork G' used for computing the tighter bounds constructed (the initial network is the trivial network composed of the single proposition t).

(b) How are the lower and upper bounds for the exact degree of support computed.

The operation (a) is first considered. A proposition $q \in M$ is called *expandable* if there is a rule having q as its conclusion and let $E(G)$ denote the set of all expandable propositions of G. Then arbitrarily select an expandable proposition $q \in E(G)$: here different selection strategies can be applied, e.g. depth first search, breadth first search, most important first search etc. If

$$\begin{aligned} R(q) &= \{r \in \hat{R} : q = Cc(r)\} \\ P(q) &= \{p \in \hat{P} : \exists r \in R(q) \text{ such that } p \in Cd(r)\}. \end{aligned} \tag{6.29}$$

then define the new graph G' as the subnetwork of G over the nodes $M' = M + R(q) + P(q)$.

The operation (b) is now considered, i.e. how the bounds can be computed. In the discussion, we'll have to consider different reliability values on the nodes of certain networks. So, if $\hat{\mu}$ is the original vector of reliability values, let $G_{(l)} = (M, U, \mu_{(l)})$ and $G_{(u)} = (M, U, \mu_{(u)})$ be the network G with the reliabilities $\mu_{(l)}(i) = \hat{\mu}(i)$ for all $i \in M$ and

$$\mu_{(u)}(i) = \begin{cases} \hat{\mu}(i) & \text{if } i \notin E(G) \\ 1 & \text{if } i \in E(G). \end{cases} \tag{6.30}$$

Hence $G_{(l)}$ and $G_{(u)}$ differ only in the reliabilities of their nodes. The intuitive idea for computing upper and lower bounds is that the exact degree of support of an expandable proposition is necessarily larger than its original reliability and smaller than 1. So computing with the original reliabilities in G leads to a lower bound, whereas assuming the reliability 1 for the expandable propositions leads to an upper bound.

Theorem 6.9. *If $L = sp(t, G_{(l)})$ and $U = sp(t, G_{(u)})$ then*

$$L \le sp(t, \hat{G}) \le U. \tag{6.31}$$

So L and U can be used as lower and upper bounds for $sp(t, \hat{G})$.

The recursive procedure described in Sect. 6.2 can be used to compute these bounds. When passing from one network to another during the chaining process, it is not usually necessary to recompute the entire new network for updating the bounds because certain parts have already been computed in the old network. Also note that there is no expandable proposition in $E(G)$ if and only if $G = \hat{G}$. In this case, the two bounds coincide and are equal to $sp(t, \hat{G})$. When there is more than one expandable proposition, the selection of the best proposition to expand will not be discussed here. However, for a comparison of some selection strategies, see Cardona (1993).

6.4 The Factorization Method For Unreliable Rule-Based Systems

In contrast to the inference networks considered so far in this chapter, in this section we deal with more general rule-based systems where negations of propositions are permitted, i.e. we drop the assumption of Horn clauses. This provides more flexibility in the process of modeling knowledge, but it introduces the possibility of contradiction. This is the reason why monotone systems cannot be associated with such models. We generalize the factorization method of Sect. 6.3 for combining hints in this type of model. This approach can be seen as an alternative to the logic-based techniques which were presented in Chap. 5. Furthermore, it is also an alternative to the very general covering Markov tree approach to be presented in Chap. 8.

The method will be explained by means of an example and the corresponding general notions will be introduced when appropriate. The example is the alarm story introduced in Sect. 1.3 and revisited in Sect. 5.1. There is a set of propositions a, b, e, w, g, d, c and a set of assumptions $a_1, \ldots, a_6$. Each assumption a_i is true with a known probability q_i (and hence $1 - q_i$ is the probability that a_i is false). Furthermore, there is a set of clauses Σ describing the behaviour of the different actors in the story. These are the rules (or their equivalent clauses) R_1 to R_{14} given in Sect. 5.1:

$$\begin{array}{llll} R_1: & \neg b \vee \neg a_1 \vee a & R_8: & \neg a \vee a_4 \vee g \\ R_2: & \neg e \vee \neg a_1 \vee a & R_9: & \neg a_4 \vee \neg g \\ R_3: & \neg a_2 \vee a & R_{10}: & a \vee \neg g \\ R_4: & b \vee e \vee a_2 \vee \neg a & R_{11}: & \neg a \vee \neg a_5 \vee d \\ R_5: & \neg a \vee w & R_{12}: & a_5 \vee \neg d \\ R_6: & \neg a_3 \vee w & R_{13}: & \neg e \vee \neg a_6 \vee c \\ R_7: & a \vee a_3 \vee \neg w & R_{14}: & e \vee \neg c. \end{array}$$

To this set of rules describing the alarm story certain facts may be added. For example: there is a call from Mr. Watson (w), there is no call from Mrs. Gibbon ($\neg g$) and there is no call from Mr. Holmes' daughter ($\neg d$). In Sect. 5.1 the rules were assumed to be certain because the uncertainty was captured by the assumptions a_i involved in the rules and quantified by their probabilities. Alternatively, we could also leave the assumptions a_i implicit and consider the rules themselves as uncertain. For example, the rule "R_1 : if b then a" is valid with probability q_1 whereas with probability $1 - q_1$ the rule is void of information. However, the uncertain rules induced by the removal of assumptions are usually not stochastically independent: e.g. the uncertain rule "R_2 : if e then a" is not independent of R_1 because R_2 is valid if and only if R_1 is valid since the same assumption a_1 is present in both clauses. Independence is crucial for the correct computation of degrees of support by the factorization method. So it is an essential part of the modeling process to find a representation of the available knowledge by stochastically independent rules or clauses, possibly by introducing explicit assumptions to make clauses certain and hence independent. So it is assumed here that we dispose of a set of independent uncertain rules and there is a hint for each proposition or assumption in the model. These hints are again assumed to be independent and some may be vacuous, or precise like the assumptions a_i in the alarm story. Then we are interested in the degree of support that a certain proposition t is true or false. In the alarm story it is assumed that the clauses are certain, each assumption a_i is true with probability q_i and there are three certain facts: $w, \neg g, \neg d$. If we are interested in the presence of a burglary (proposition b), this model can first be simplified because $R_5 \wedge R_6 \wedge w = w$, $R_9 \wedge R_{10} \wedge \neg g = \neg g$ and $R_{12} \wedge \neg d = \neg d$. This leads to the simplified model

$$\begin{array}{llll} 1: & \neg b \vee \neg a_1 \vee a & 6: & \neg a \vee a_4 \vee g \\ 2: & \neg e \vee \neg a_1 \vee a & 7: & \neg a \vee \neg a_5 \vee d \\ 3: & \neg a_2 \vee a & 8: & \neg e \vee \neg a_6 \vee c \\ 4: & b \vee e \vee a_2 \vee \neg a & 9: & e \vee \neg c. \\ 5: & a \vee a_3 \vee \neg w & & \end{array}$$

with the facts $w, \neg g, \neg d$.

This model can be represented by a so-called variable and relation graph, also called rule graph. This is a bipartite graph which is constructed as follows. Its nodes are the assumptions a_j, the propositions p_i (both represented as circle nodes) and the rules (represented as rectangular nodes). There is an edge between a circular node c and a rectangular node R if c or $\neg c$ is a literal of R. Fig. 6.7 shows this graph for the alarm example. In contrast to an inference network like Fig. 6.1, this graph is undirected, reflecting the fact that it can be used to make inferences in all directions.

When the variable and relation graph is a tree (called a rule tree), then computations are easy because local computations are possible. This means that propositions (or more precisely hints on propositions) can be propagated through the uncertain rules until the proposition t is reached. Up to equivalence (see Chap. 3), hints can be represented by their basic probability assignments (bpa). Thus local computations, i.e. combination and propagation of hints, will be expressed in terms of bpas, or more precisely, in terms of unnormalized basic probability assignments (ubpa for short). A ubpa on an arbitrary frame X is a mapping $m_* : \mathcal{P}(X) \rightarrow [0,1]$ such that

$$\sum \{m_*(A) : A \subseteq X\} = 1. \tag{6.32}$$

In contrast to a traditional basic probability assignement, $m_*(\emptyset)$ need not necessarily be equal to zero. However, a ubpa m_* determines a bpa m in a canonical way by

$$m(A) = \begin{cases} \frac{m_*(A)}{1-m_*(\emptyset)} & \text{if } A \neq \emptyset \\ 0 & \text{if } A = \emptyset. \end{cases} \tag{6.33}$$

While working with a ubpa, the empty set is considered as an ordinary focal set, but of course degrees of support must be computed with respect to the corresponding bpa because the empty set does clearly not contribute to the

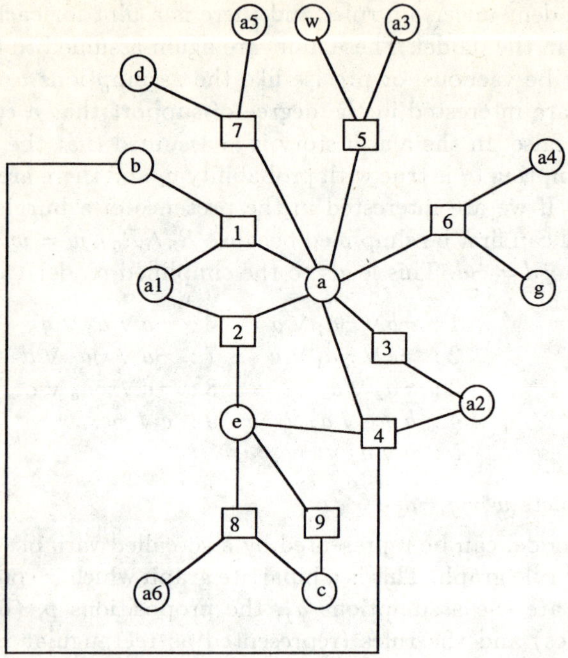

Fig. 6.7. A variable and relation graph

support of a hypothesis. Thus, once the ubpa relative to the proposition t has been determined, it must be normalized by (6.31) to get the exact bpa, but this normalization is done only at the end of the local computations.

Two local operations must be defined, namely propagation and combination. We start with the definition of propagation. First note that it is always possible to orient the edges of the rule tree in such a way that the query proposition t is an anti-root. This orientation permits us to consider each unreliable clause $p_1 \vee \ldots \vee p_k$ as an unreliable rule $R : \wedge_{i=2}^{k} \neg p_i \rightarrow p_1$ where the arc (ξ, p_1) belongs to the orientation. This is then a generalization of an inference tree and the method discussed here is in fact a generalization of the method discussed above for inference nets. Then, given the ubpas on the conditions $p_i, i = 2, \ldots, k$, the objective is to compute the ubpa relative to p_1 after propagation through the unreliable rule R. In general, consider the arbitrary rule

$$R : \wedge_{i \in I} l_i \rightarrow l_j \tag{6.34}$$

where l_i and l_j are literals. If m_{i*} is the ubpa relative to the proposition $|l_i|$ and m_* is the ubpa relative to the proposition $|l_j|$ corresponding to the propagation of the ubpas m_{i*} through the unreliable R having reliability r, then, in the spirit of Dempster's rule of combination, we obtain

$$\begin{aligned} m_*(1) &= r \prod_{i \in I} m_{i*}(1) \\ m_*(0) &= 0 \\ m_*(\emptyset) &= 1 - \prod_{i \in I}(1 - m_{i*}(\emptyset)) \\ m_*(\{0,1\}) &= 1 - (m_*(\emptyset) + m_*(0) + m_*(1)). \end{aligned} \tag{6.35}$$

These formulas define the first local computation procedure which is propagation through an uncertain rule.

Now consider the second local operation, namely the combination of two ubpas m_{1*} and m_{2*} on a frame $\Theta_j = \{0, 1\}$ (compare with theorem 4.1). If F equals $\{0\}$ or $\{1\}$, then by Dempster's rule we get

$$\begin{aligned} m_*(F) &= m_{1*}(F)m_{2*}(F) + m_{1*}(F)m_{2*}(\Theta_j) + m_{1*}(\Theta_j)m_{2*}(F) \\ m_*(\Theta_j) &= m_{1*}(\Theta_j)m_{2*}(\Theta_j). \end{aligned} \tag{6.36}$$

Starting from the leaves of the rule tree, propagations and combinations are applied sequentially until the query proposition t is reached. This gives us the ubpa relative to t. Then the corresponding bpa and hence finally the degree of support that t is true of false can be computed.

The local operations defined above can be proved to correspond exactly to the local computational prodedures used for propagating hints in Markov trees (see Chap. 8). This justifies these formulas because it is proved in Chap. 8 that if a variable and relation graph is a tree, then it is in fact a Markov tree and local computations in Markov trees give exact results. Such local computations are possible trees, but what about graphs which are not trees ? In this case, the *factorization* method may help. The intuitive idea is to assume hypothetically that a certain proposition has been proved to be true or on the contrary that it has been proved to be false. This permits the simplification of the model and makes the computations easier. Once all possible hypothetical cases have been considered, the corresponding results are weighted according to the probabilities of these cases to eventually get the exact result.

The goal of factorization is to create trees from the original variable and relation graph. Factorization is always performed with respect to a proposition or an assumption. Suppose that we factorize on the proposition or assumption t_i and let m_i denote the bpa of the hint originally pertaining to t_i. Now assume hypothetically that t_i is false. In general, this permits us to simplify the variable and relation graph G and a new graph $G(\neg t_i)$ can be drawn. If it is assumed that t_i is true, then a new graph $G(t_i)$ is obtained. For example, in the alarm story, if we assume that a_1 is false, then rules R_1 and R_2 can be eliminated because $R_1 \wedge R_2 \wedge a_1 = a_1$. The resulting graph $G(\neg a_1)$ is given in Fig. 6.8 a). Similarly, if a_1 is assumed to be true, then the rules R_1 and R_2 are transformed into $\neg b \vee a$ and $\neg e \vee a$ respectively: the corresponding graph $G(a_1)$ is given in Fig. 6.8 b).

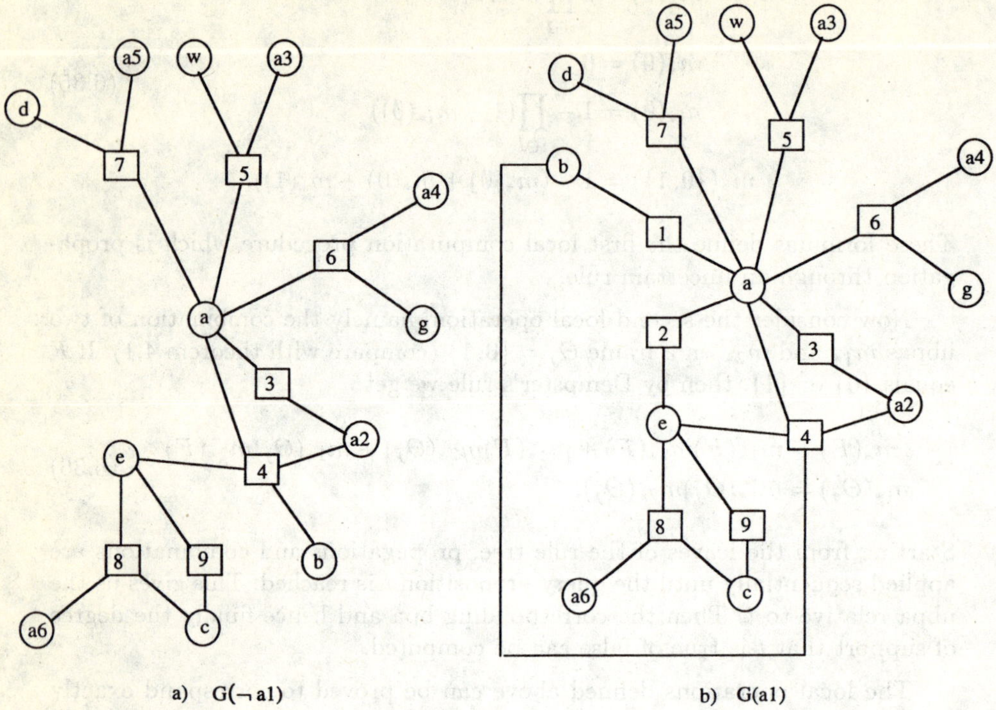

Fig. 6.8.

Normally, we should also consider the case where neither t_i nor $\neg t_i$ is proved to be true, i.e. nothing is known about t_i. This is a disturbing case because this time the graph G cannot be simplified. So we make an approximation and simply discard this case. Several examples have shown that the results obtained are usually still valid in spite of this simplification, correctness, however, is not guaranteed in general. For more information on this problem, refer to Cardona (1993). If exact results must absolutely be found, then covering Markov trees methods described in Chap. 8 can be used to cope with this third case.

Now, if m_* denotes the ubpa of the hint on t obtained by integrating all information in the model G (i.e. the original hints on the propositions and assumptions and the uncertain rules), then the formula of total probability implies that

$$m_*(A) = m_i(1)m_*^+(A) + m_i(0)m_*^-(A) \tag{6.37}$$

for all $A \subseteq \Theta_t = \{0,1\}$, possibly the emptyset, where m_*^+ and m_*^- denote the ubpa on Θ_t obtained by integrating the knowledge in the graphs $G(t_i)$ and $G(\neg t_i)$ respectively. The right-hand side of (6.37) is only an approximation of $m_*(A)$ because the third case is missing. As already stated, in most examples this has no consequence on the results.

Since the graphs $G(t_i)$ and $G(\neg t_i)$ are usually not trees yet, as in the alarm story after factorization on the assumption a_1, further factorizations must be performed until trees are obtained. Then local computations in these trees are performed and by repeated application of (6.37) the ubpa m_* is finally determined. From m_*, degrees of support for t can easily be computed.

The computational method described in this chapter has been implemented in the computer program CHRIS (Cardona, 1993). Among several other tasks, this program performs factorizations in a sequence which is automatically determined. Exact results are quickly obtained for most of the examples considered.

A6 Appendix: Proofs of Theorems in Chapter 6

Proof of Theorem 6.1

Let p_j be a proposition in P. We must prove that $pl(p_j) = 1 - sp(\neg p_j) = 1$, that is $sp(\neg p_j) = 0$. This holds because there is no vector $x \in \Omega$ such that $\Gamma(x) \subseteq N(\neg p_j)$. Indeed, if $x^* = (1, \ldots, 1)$, then $\Gamma(x^*) \not\subseteq N(\neg p_j)$ and $\Gamma(x^*) \subseteq \Gamma(x)$ for all $x \in \Omega$. □

Proof of Theorem 6.2

Let $C(G)$ denote the set of clauses associated with the nodes of G. Then t is a logical consequence of some set of clauses $C \subseteq C(G)$ if and only if there is a subgraph G_i such that $C \subseteq C(G_i)$ or $C = \{t\}$. This proves the theorem. □

Proof of Theorem 6.3

Let $C(G_i)$ denote the set of clauses associated with the nodes in G_i. The theorem follows from the fact that t is a logical consequence of a set of clauses $C \subseteq C(G_i)$ if and only if r is valid and each q_j is a logical consequence of $C \cap C(H_j)$, where $C(H_j)$ is the set of clauses associated with the nodes in H_j. □

Proof of Theorem 6.4

First suppose that we are in case (II) (a). Let q_1 and q_2 be two different propositions in $Cd(r) \cap H_j$. Since H_j is connected and q_1, q_2 are in H_j, there exists a chain (i.e. a non-oriented path) ω between q_1 and q_2 in H_j. Two cases must be distinguished: First, suppose that the arc $u \in \omega$ which is incident to q_1 is oriented from q_1 to another node in ω. Now, by the general hypothesis about arcs in G, u lies on some path from q_1 to t. But since $\{(q_1, r), (r, t)\}$ is another path from q_1 to t, this shows that q_1 is a source of dependence relative to t. Secondly, suppose that u is oriented from some node in ω to q_1. Let γ be the longest path on ω leading from a node q on ω to the node q_1, i.e. q is the most distant node from q_1 on ω. If $q = q_2$, then q_2 is clearly a source of dependence. Otherwise, if $q \neq q_2$, then there is an arc v on ω but not on

γ which has q as its starting node. But the path γ continued with the arcs $(q_1, r), (r, t)$ is a path from q to t and there is also a path from q to t starting with v. These are two different paths from q to t and thus q is a source of dependence.

Now suppose that we are in case (II) (b) and let $R_i(t)$ denote the rules in G_i which are incident to t. Since G_i is connected and there are two rules r_1 and r_2 in $R_i(t)$, there is a chain ω from r_1 to r_2 in G_i. Let γ be the longest path on ω leading from a node q on ω to the rule r_1. This path contains at least one arc and q is different from r_2. This implies that there is an arc u on ω but not on γ which has q as its starting node. But γ continued with the arc (r_1, t) is a path from q to t and there is also a path from q to t containing u. These are two different paths from q to t and thus q is a source of dependence. □

Proof of Theorem 6.5

If G_i contains exactly one sod, then it is of course extremal. Suppose now that there are $n \geq 2$ sods and that none of them is extremal. Arbitrarily select a sod s_1. Since it is not extremal there is another sod s_2 which leads to s_1. But since s_2 is not extremal, there must be another sod s_3, different from s_1, which leads to s_2. Sod s_3 is different from s_1 because otherwise there would be a circuit containing s_1 and s_2, which is excluded by our general assumption. This procedure continues in the same way until we arrive at the last sod s_n which is also not extremal by hypothesis. But this time there is no other sod which leads to it because all sods in $\{s_1, \ldots, s_{n-1}\}$ would generate a circuit. This is a contradiction and thus at least one sod must be extremal. □

Proof of Theorem 6.6

Suppose that there is a path γ from $q \in L(s)$ to t not passing by s. By definition there is a path $\varphi = (q \to s \to t)$ which goes from q to t and which contains s. Hence γ and φ are two different paths from q to t in G_i and so q is a source of dependence. But q leads to s and thus s is not extremal. This is a contradiction which shows that there is no such path γ. □

Proof of Theorem 6.7

Let q_1 and q_2 be two nodes in $K(s)$. Then by construction, there is a path from q_1 to s in $K(s)$ and a path from q_2 to s in $K(s)$, which implies that there is a chain between q_1 and q_2 in $K(s)$ and hence $K(s)$ is simply connected. It remains to prove that $K(s)$ contains no cycle. Suppose that this is not the case and let γ be a cycle in $K(s)$. Since γ is not a circuit, there must be a node q in γ which is the initial extremity of two arcs u and v on γ. But then by definition two paths exist from q to s in $K(s)$ containing u and v respectively. Hence q is a source of dependence which leads to s. But this is in contradiction with the fact that s is an extremal source of dependence. So such a cycle γ cannot exist and $K(s)$ is a tree. □

Proof of Theorem 6.9

First we prove that $L \leq sp(t, \hat{G})$. If $\hat{G}_{(l)} = (\hat{M}, \hat{U}, \hat{\mu}_{(l)})$ with

$$\hat{\mu}_{(l)}(i) = \begin{cases} 0 & \text{if } i \in \hat{M} - M \\ \hat{\mu}(i) & \text{if } i \in M, \end{cases} \tag{A6.1}$$

then $sp(t, G_{(l)}) = sp(t, \hat{G}_{(l)})$. Moreover, $\hat{\mu}_{(l)} \leq \hat{\mu}$ implies that $sp(t, \hat{G}_{(l)}) \leq sp(t, \hat{G})$ because the reliability of a monotone system is an increasing function of the reliabilities of its components (Barlow, Proschan, 1975; Kohlas, 1987). This shows that $sp(t, G_{(l)}) \leq sp(t, \hat{G})$ and so $sp(t, G_{(l)}) \leq sp(t, \hat{G})$.

Similarly, if $\hat{G}_{(u)} = (\hat{M}, \hat{U}, \hat{\mu}_{(u)})$ with

$$\hat{\mu}_{(u)}(i) = \begin{cases} 1 & \text{if } i \in \hat{M} - M + E(G) \\ \hat{\mu}(i) & \text{if } i \in M - E(G), \end{cases} \tag{A6.2}$$

then $sp(t, G_{(u)}) = sp(t, \hat{G}_{(u)})$. Moreover, $\hat{\mu}_{(u)} \geq \hat{\mu}$ implies that $sp(t, \hat{G}_{(u)}) \geq sp(t, \hat{G})$ and the theorem is proved. □

7. Compatible Frames of Discernment

7.1 Refinings and Coarsenings

There are plenty of examples which demonstrate that a reasoning process usually involves more than one single frame of discernment. Attention will often move from one frame to another as the discussion focusses on various aspects of a question. Also, varying degrees of precision of the answer need to be considered, depending on how crudely the question is formulated. Therefore, the available hints will often refer to distinct but linked and related frames. Moreover, hypotheses may be formulated with respect to even other frames. The goal of this chapter and the following one is to explain how to reason within a multi-frame environment. The following three introductory examples describe situations where it is natural and even necessary to introduce several frames.

Example 7.1. This is an adaptation of the "car that will not start" example described in Shafer et al. (1987). Suppose you are confronted with the problem of determining the cause of your car's failure to start. You begin by considering a fairly crude list of diagnoses like (faulty battery system, faulty fuel system, faulty starting system, something else). It is assumed that the diagnoses in the list are mutually exclusive and collectively exhaustive: exactly one is correct. But this list might not be fine enough to describe faithfully what you exactly know about this question. For example, it might be necessary to decompose the "faulty battery system" diagnosis into "weak or faulty battery" and "faulty battery connections", whereas the diagnosis "faulty starting system" is decomposed into "defective switch", "transmission not in park" and "some other fault in the starting system". At this stage the original list of diagnoses is replaced by a new list obtained by replacing the two diagnoses by their decomposition. We can even go one step further and replace "defective switch" by "defective ignition switch", "defective starter relay", "defect in some other switch". This second decomposition leads to a third list of possible diagnoses. If we regard the problem itself, namely "the car won't start", as a degenerate diagnose, then the decompositions can be organized in a so-called *tree of diagnoses* (see Fig. 7.1).

In fact, the three lists of diagnoses are the frames of discernment for the three increasingly refined questions: which element of the list is the cause of the problem ? We might also add the artificial frame $\{\emptyset\}$ corresponding to the

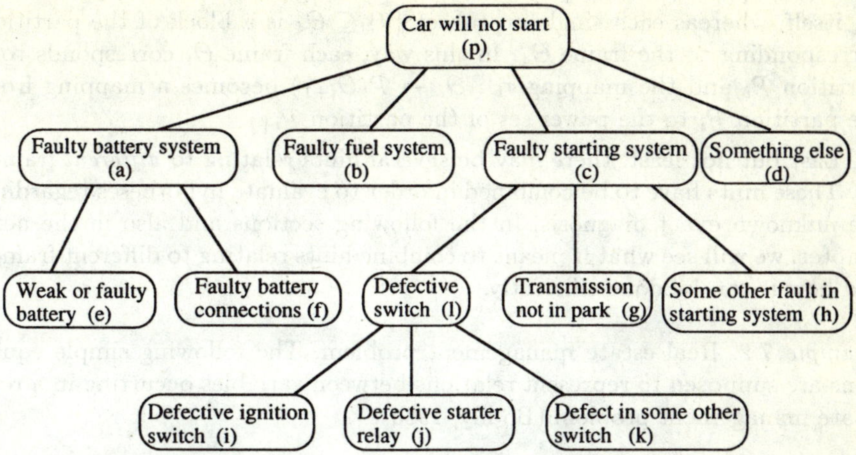

Fig. 7.1. A tree of diagnoses for the "car will not start" problem

question: what's the problem with the car ? More explicitely, if the diagnoses are abbreviated by the letters $\emptyset, a, \ldots, l$ (see Fig. 7.1), then the four frames are

$$\begin{aligned}
\Theta_1 &= \{\emptyset\}\\
\Theta_2 &= \{a, b, c, d\}\\
\Theta_3 &= \{e, f, b, l, g, h, d\}\\
\Theta_4 &= \{e, f, b, i, j, k, g, h, d\}.
\end{aligned}$$

In this example, when a frame Θ_i is replaced by another, finer frame Θ_{i+1}, the substitution of a diagnosis by a set of more precise diagnoses is described by a mapping

$$\tau_i : \Theta_i \to \mathcal{P}(\Theta_{i+1}). \tag{7.1}$$

For example, the passage from Θ_2 to Θ_3 is expressed by the mapping $\tau_2 : \Theta_2 \to \mathcal{P}(\Theta_3)$ given by

$$\tau_2(x) = \begin{cases} \{x\} & \text{if } x \in \{b, d\} \\ \{e, f\} & \text{if } x = a \\ \{l, g, h\} & \text{if } x = c. \end{cases} \tag{7.2}$$

In an alternative but equivalent point of view, this family of related frames can be seen as a collection of *partitions* of the frame Θ_4. Indeed, if the sets forming a partition are called *blocks*, then the blocks of the partition corresponding to Θ_2 are $\{e, f\}, \{b\}, \{i, j, k, g, h\}, \{d\}$ and the blocks of the partition corresponding to Θ_3 are

$$\{e\}, \{f\}, \{b\}, \{i, j, k\}, \{g\}, \{h\}, \{d\}.$$

The only block of the partition corresponding to the frame Θ_1 is the set Θ_4 itself, whereas each singleton subset $\{\theta\} \subseteq \Theta_4$ is a block of the partition corresponding to the frame Θ_4. In this way, each frame Θ_i corresponds to a partition $\mathcal{P}_i$ and the mapping $\tau_i : \Theta_i \rightarrow \mathcal{P}(\Theta_{i+1})$ becomes a mapping from the partition $\mathcal{P}_i$ to the power set of the partition $\mathcal{P}_{i+1}$.

Last but not least, there may be several hints relating to *different* frames Θ_i. These hints have to be combined in order to evaluate hypotheses regarding the unknown exact diagnosis. In the following sections and also in the next chapter, we will see what it means to combine hints relating to different frames and how it can be done efficiently.

Example 7.2. Real estate management problem. The following simple equations are supposed to represent relations between variables occurring in a real estate management problem (Bodily, 1985):

$$\begin{aligned} Revenue &= Occupancy \cdot Price \\ Variable\ Costs &= Occupancy \cdot Unit\ Costs \\ Profit &= Revenue - (Variable\ Costs + Fixed\ Costs). \end{aligned} \tag{7.3}$$

Here Price, Occupancy, Unit Costs and Fixed Costs are input variables. Profit is the result in which the manager is interested. Practical models are of course much larger and contain several dozen variables and equations. Input variables such as Occupancy and Unit Costs and even Price are seldom known exactly in advance. We only dispose of some hints about their values. Of course, these hints are relative to the possible values of each variable. These possible values are the elements of the frame of discernment for the question : what is the value of the variable ? Very often there is a certain degree of dependency between two or more input variables. An example is the relationship between Price and Occupancy. One would expect that the higher the Price, the smaller the Occupancy (note that this relationship is not explicitly written in (7.3)). Such a relationship might be given by an interval-valued function like the one pictured in Fig. 7.2.

The fact that this function is interval-valued means that, even if the Price is known exactly, no exact value for the Occupancy can be deduced: only an interval containing the exact value can be specified. If $\mathcal{R}_+$ is the frame for Price and Occupancy, then the dark region in Fig. 7.2 represents a relation $R \subseteq \mathcal{R}_+ \times \mathcal{R}_+$ between Price and Occupancy. This relation holds most of the time, but it is not impossible that sometimes Price does not at all influence Occupancy. This happens under special circumstances when the normal economic mechanism linking demand to price breaks down. If we are able to elicit probabilities for the occurrence of this special event, then we have a simple hint in favour of R for the question "What are the joint values of Price and Occupancy ? ". Similarly, the three functional relations (7.3) can be considered as deterministic hints relative to the values of the variables involved. Therefore, several hints relative to different frames have to be combined in

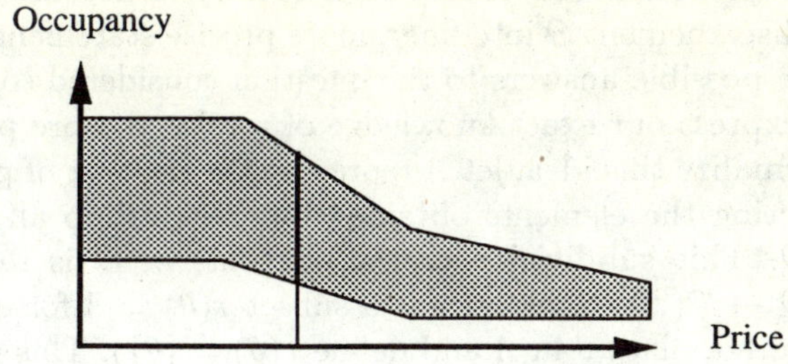

Fig. 7.2.

order to evaluate hypotheses about the value of Profit. The frames are also somehow related because they are indeed product spaces. For example, the frame $\mathcal{R}_+$ for the variable Price is extended to the product frame $\mathcal{R}_+ \times \mathcal{R}_+$ for the composed variable (Price, Occupancy). If we learn that Price = t, then the value of (Price, Occupancy) necessarily lies in $\{t\} \times \mathcal{R}_+$. This defines a mapping

$$\tau : \mathcal{R}_+ \rightarrow \mathcal{R}_+ \times \mathcal{R}_+ \tag{7.4}$$

such that the sets $\tau(t) = \{t\} \times \mathcal{R}_+$ form a partition of $\mathcal{R}_+ \times \mathcal{R}_+$ (compare 7.1).

Example 7.3. Propositional systems. Consider the set of clauses R_1 to R_{14} introduced in Sect. 5.1. For each proposition or assumption in the model, a natural frame of discernment is attached, namely $\{true, false\}$ because every proposition is either true or false. Then each clause R_i, considered as a statement, is a piece of knowledge relative to the propositions or assumptions involved in the clause. The frame of discernment for this knowledge is then the set of all possible truth values of the propositions or assumptions in the clause. This set is isomorphic to the Boolean cube $B_k = \{0,1\}^k$ if there are k atoms in the clause (see Chap. 5: 1 means true and 0 means false). Since the rules are not all composed of the same propositions and assumptions, we have to reason on a whole family of frames which are interrelated because a proposition may appear in several rules. As in the other examples, the notion of a partition is also present. As an illustration consider the rule R_5 involving propositions a and w. The frame for this rule is all four possible couples $c_1, \ldots, c_4$ in B_2. Since rule R_6 also involves the proposition w, then considering R_5 and R_6 simultaneously shows that we have to extend these vectors to the frame of discernment for a, w, a_3. Each vector c_i is then considered as the subset $c_i \times \{0,1\}$ of B_3. These four subsets form a partition of B_3. This example, like the preceding one, is an instance of a very important class of frames called multivariate families of compatible frames.

These examples show that it is often necessary to decompose the elements of a frame of discernement Θ into finer, more precise statements. This must be done when the possible answers to the question considered so far prove to be too crude to express our exact knowledge of a related more precise question. In order to formalize this idea, let Λ represent the new set of possible answers (frame), collecting the elements obtained by splitting up all or some of the elements of Θ. This subdivision of the elements of Θ is represented by a mapping $\tau : \Theta \to \mathcal{P}(\Lambda)$ which assigns a subset $\tau(\theta) \subseteq \Lambda$ for each $\theta \in \Theta$ (if θ is not decomposed, place θ in Λ and define $\tau(\theta) = \{\theta\}$). This mapping has to satisfy the following requirements in order to represent indeed a decomposition of the elements of Θ:

$$\begin{aligned}&1)\ \tau(\theta) \neq \emptyset \text{ for all } \theta \in \Theta\\ &2)\ \tau(\theta') \cap \tau(\theta'') = \emptyset \text{ whenever } \theta' \neq \theta''\\ &3)\ \cup \{\tau(\theta) : \theta \in \Theta\} = \Lambda.\end{aligned} \tag{7.5}$$

Remark that the family of sets $\tau(\theta), \theta \in \Theta$ is indeed a partition of Λ. Following Shafer (1976 a), we call the mapping $\tau : \Theta \to \mathcal{P}(\Lambda)$ a *refining*, whereas Λ is called a *refinement* of Θ and Θ is called a *coarsening* of Λ. Also, we say that Λ is *finer* than Θ and Θ is *coarser* than Λ. For convenience and by abuse of notation, a refining from Θ to $\mathcal{P}(\Lambda)$ will usually be written $\tau : \Theta \to \Lambda$ instead of $\tau : \Theta \to \mathcal{P}(\Lambda)$. Note that a refining τ can be extended from Θ to $\mathcal{P}(\Theta)$ by defining

$$\tau(A) = \cup \{\tau(\theta) : \theta \in A\} \tag{7.6}$$

for all non-empty subsets $A \subseteq \Theta$ and letting $\tau(\emptyset) = \emptyset$ by convention. The following theorem of Shafer (1976 a) collects some properties of refinings:

Theorem 7.1. *Let $\mathcal{F}$ be an arbitrary family of subsets of Θ and let A, B be two subsets of Θ. If $\tau : \Theta \to \mathcal{P}(\Lambda)$ is a refining, then*
(1) $\tau(\cup \{F : F \in \mathcal{F}\}) = \cup \{\tau(F) : F \in \mathcal{F}\}$
(2) $\tau(\cap \{F : F \in \mathcal{F}\}) = \cap \{\tau(F) : F \in \mathcal{F}\}$
(3) $\tau(F) = \emptyset$ if and only if $F = \emptyset$
(4) $\tau(A) \subseteq \tau(B)$ if and only if $A \subseteq B$
(5) $\tau(A^c) = \tau(A)^c$
(6) $\tau(\Theta) = \Lambda$.

In general, there are statements about the more precise question represented by Λ which cannot be expressed exactly with respect to the coarser frame Θ. These statements are precisely the subsets of Λ which are not images of subsets of Θ by the refining τ. Since a proposition $C \subseteq \Theta$ is clearly represented by the proposition $\tau(C)$ in Λ, it is natural to say that C implies a proposition $A \subseteq \Lambda$ if $\tau(C) \subseteq A$. Then it is easily proved that $C \subseteq \Theta$ implies $A \subseteq \Lambda$ if and only if C is a subset of

$$u(A) = \{\theta \in \Theta : \tau(\theta) \subseteq A\}. \tag{7.7}$$

This shows that $u(A)$ is the largest proposition discerned by Θ which implies the proposition A discerned by Λ. The set $u(A)$ is called the *inner reduction* of A (Shafer, 1976 a). Since the refining τ is in fact a multivalued mapping, then (7.6) has the same form as (3.1).

Similarly, a proposition $C \subseteq \Theta$ is implied by $A \subseteq \Lambda$ if $A \subseteq \tau(C)$. Then we can easily verify that C is implied by A if and only if C contains the set

$$v(A) = \{\theta \in \Theta : \tau(\theta) \cap A \neq \emptyset\}. \tag{7.8}$$

This shows that $v(A)$ is the smallest proposition discerned by Θ which is implied by the proposition A discerned by Λ. Therefore, if A is known to be true, then every proposition C in Θ which contains $v(A)$ is also true. It is thus very natural to choose $v(A)$ as the representative of A in Θ. Knowing that A is true, $v(A)$ is the smallest proposition in Θ which is certain to be true. This set $v(A)$ is called the *outer reduction* of A (Shafer, 1976 a). Again, note that (7.7) has the same form as (3.4).

The following theorem of Shafer (1976 a) collects a number of properties of the inner and outer reductions (compare with theorems 3.1 and 3.3).

Theorem 7.2. *Let $\mathcal{F}$ be an arbitrary family of subsets of Λ and let A, B be two subsets of Θ and C a subset of Λ. If $\tau : \Theta \to \mathcal{P}(\Lambda)$ is a refining having u and v as respective inner and outer reductions, then*

(1) $u(\cap \{F : F \in \mathcal{F}\} = \cap \{u(F) : F \in \mathcal{F}\}$
(2) $v(\cup \{F : F \in \mathcal{F}\} = \cup \{v(F) : F \in \mathcal{F}\}$
(3) $u(\cup \{F : F \in \mathcal{F}\} \supseteq \cup \{u(F) : F \in \mathcal{F}\}$
(4) $v(\cap \{F : F \in \mathcal{F}\} \subseteq \cap \{v(F) : F \in \mathcal{F}\}$
(5) If $A \subseteq B$, *then* $u(A) \subseteq u(B)$ *and* $v(A) \subseteq v(B)$
(6) $v(C) = u(C^c)^c$, $u(C) = v(C^c)^c$
(7) $u(\emptyset) = \emptyset$
(8) $u(\Lambda) = v(\Lambda) = \Theta$
(9) $u(C) \subseteq v(C)$
(10) $v(C) = \emptyset$ *if and only if* $C = \emptyset$
(11) $u(\tau(A)) = v(\tau(A)) = A$
(12) $\tau(u(C)) \subseteq C$
(13) $\tau(v(C)) \supseteq C$.

7.2 Families of Compatible Frames

A frame of discernment may have several different refinements and coarsenings. Also, once a frame Θ has been coarsened to some frame it can be refined to another frame Ω. In this way a whole network of related frames representing different but linked questions of a problem arises. The network of frames which is finally developed depends on the reasoning process, its organization and its goal. This represents an important task in the modeling process. However, not every family of frames and refinings is adapted for reasoning with hints because there must be a certain degree of compatibility between the frames and the refinings linking them. It becomes apparent that the structure of a *lattice* plays an important role in that respect. Using this concept we will be in a position to explain and to formalize what is to be understood by a family of compatible frames. The definition will be general enough to encompass all examples of hints relating to different frames we have encountered so far.

We begin with a review of the notion of a lattice. Let $(L, \leq)$ be a partly ordered set. By an *upper bound* to a subset $X \subseteq L$ we mean an element $a \in L$ such that $x \leq a$ for all $x \in X$. A *least upper bound*, also called a *supremum*, is an upper bound a such that $a \leq b$ for any other upper bound b. The notions of *lower bound* and *greatest lower bound*, also called *infimum*, are defined dually. It is clear that a subset X can have at most one infimum and supremum. A *lattice* is a partly ordered set L any two of whose elements x and y have an infimum (written $x \wedge y$) and a supremum (written $x \vee y$). The following theorem collects some properties of lattices:

Theorem 7.3. *Let x, y, z be elements of a lattice L.*

(1) $x \wedge x = x$, $x \vee x = x$ *(idempotence)*
(2) $x \wedge y = y \wedge x$, $x \vee y = y \vee x$ *(commutativity)*
(3) $x \wedge (y \wedge z) = (x \wedge y) \wedge z, x \vee (y \vee z) = (x \vee y) \vee z$ *(associativity)*
(4) $x \wedge (x \vee y) = x \vee (x \wedge y) = x$ *(absorption)*
(5) $x \leq y$ *is equivalent to any of the conditions (consistency):*
  a) $x \wedge y = x$
  b) $x \vee y = y$
(6) If $y \leq z$ *then* $x \wedge y \leq x \wedge z$ *and* $x \vee y \leq x \vee z$ *(isotony)*
(7)

$$x \wedge (y \vee z) \geq (x \wedge y) \vee (x \wedge z)$$
$$x \vee (y \wedge z) \leq (x \vee y) \wedge (x \vee z) \text{ (distributive inequalities)}$$

(8) If $x \leq z$ *then* $x \vee (y \wedge z) \leq (x \vee y) \wedge z$ *(modular inequality)*
(9) If $x \leq y$ *and* $x \leq z$ *then* $x \leq y \wedge z$
  If $x \geq y$ *and* $x \geq z$ *then* $x \geq y \vee z$

A lattice satisfying both conditions in (7) with equality instead of inequality is called *distributive*. We will work with distributive lattices in Chap. 8.

A lattice L is called $\wedge$-*complete* if $\wedge X$ exists for all subsets $X \subseteq L$. It is called $\vee$-complete if $\vee X$ exists for all $X \subseteq L$ and it is called complete if it is both $\wedge$-complete and $\vee$-complete. Moreover, we suppose that the lattices we consider in this book always have a *least element* denoted by 0 (i.e. $0 \leq x$ for all $x \in L$) and a *greatest element* denoted by 1 (i.e. $x \leq 1$ for all $x \in L$). Then we have $0 \wedge x = 0, 0 \vee x = x$ and $x \wedge 1 = x, x \vee 1 = 1$ for all $x \in L$. We will also need the concept of a *sublattice*. A subset K of L which is closed under finite applications of $\wedge$ and $\vee$ is itself a lattice. Since $K \subseteq L$, it is called a sublattice of L. If the lattice L is complete and if the sublattice $K \subseteq L$ is such that $\wedge X$ and $\vee X$ is in K for any $X \subseteq K$, then K is called a complete sublattice of L. Similar definitions can be given to define $\wedge$-complete and $\vee$-complete sublattices. One of the most important examples of a lattice is the family of all subsets of a set X partly ordered by set inclusion. The infimum of two sets is their intersection and the supremum is their union. Moreover, the least element is the empty set and the greatest element is the set X itself. This is of course a complete distributive lattice.

Finite lattices (and more generally finite partly ordered sets) can be graphically represented according to the following procedure. We say that *a covers* b if $a > b$ and $a > x > b$ is not satisfied by any x. An element a of a partly ordered set is an *atom* if a covers the least element. This leads to the following graphical representation. Small circles are drawn to represent the elements of L, so that a is higher than b whenever $a > b$. An arrow is then drawn from a to b whenever a covers b (Birkhoff, 1967). Then $a > b$ if and only if there is a path from a to b in the diagram. Figure 7.3 a) shows the diagram of the lattice composed of the subsets of a three-element set, and Fig. 7.3 b) shows the diagram of a lattice having 9 elements.

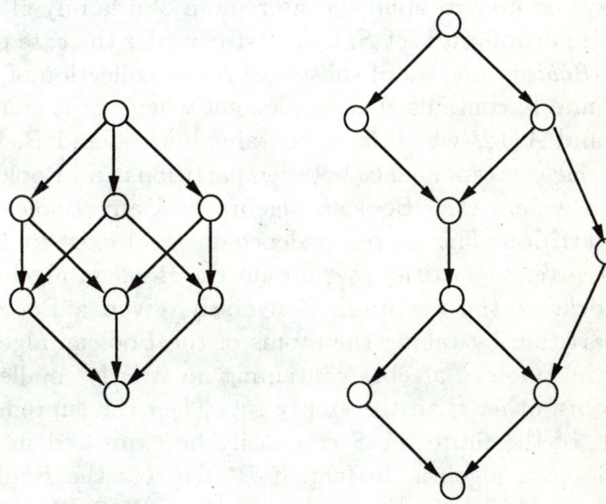

a) b)

Fig. 7.3.

In the introductory examples 7.1 to 7.3 we have seen that the notion of a partition is very strongly related to the notion of a refining. In fact, when we are using refinings we are indeed working with partitions since a coarsening Θ essentially corresponds to a partition of its refinement Λ. Moreover, if Λ

itself is a coarsening of another frame Ω, then both Θ and Λ can be seen as partitions of Ω. Therefore it is worthwile to analyze the structure of the family of partitions of a set before we give the definition of a family of compatible frames.

Let us begin by recalling that a partition of a set S is a set $\mathcal{P}$ of non-empty pairwise disjoint subsets of S whose union is S. As already mentioned, the members of $\mathcal{P}$ are called the *blocks* of $\mathcal{P}$ and a singleton as a block is called *trivial.* If two elements a and b in S belong to the same block, then we write $a \equiv b$ $(\mathcal{P})$. This is an equivalence relation on S and there is a one-to-one correspondence between the partitions of S and the equivalence relations on S. The set $Part(S)$ of all partitions of S is partly ordered by the relation $\mathcal{P}_1 \leq \mathcal{P}_2$ which holds if and only if for every $P_1 \in \mathcal{P}_1$ there exists $P_2 \in \mathcal{P}_2$ such that $P_1 \subseteq P_2$. In other words each block in $\mathcal{P}_2$ is a union of blocks in $\mathcal{P}_1$, or, that $x \equiv y(\mathcal{P}_1)$ implies $x \equiv y(\mathcal{P}_2)$. $Part(S)$ with this partial ordering forms a complete lattice called the *partition lattice* (Grätzer, 1978). The blocks of the infimum of an arbitrary collection $\mathcal{P}_i, i \in I$ of partitions are precisely the non-empty intersections of blocks in $\mathcal{P}_i$:

$$\wedge_{i \in I} \mathcal{P}_i = \{\cap_{i \in I} P_i : P_i \in \mathcal{P}_i, \cap_{i \in I} P_i \neq \emptyset\}. \tag{7.9}$$

Before we explain how to build the supremum of a family of partitions of an arbitrary (finite or infinite) set S, let's first consider the case of a *finite* set S. Recall that a *Boolean algebra* of subsets of S is a collection of subsets that contains both S and $\emptyset$, contains A's complement whenever it contains A, and contains $A \cap B$ and $A \cup B$ whenever it contains both A and B. If S is finite, there is a one-to-one correspondence between partitions and Boolean algebras, i.e. any partition generates a Boolean algebra and any Boolean algebra is generated by a partition. This correspondence does not exist for infinite S. In fact, for finite S, given a partition, we obtain the Boolean algebra by taking all unions of blocks of the partition. Conversely, given a Boolean algebra, we obtain the partition by taking the atoms of the Boolean algebra, namely the elements of the Boolean algebra containing no strictly smaller element of the Boolean algebra other than the empty set. Then the supremum $\mathcal{P}_1 \vee \mathcal{P}_2$ of two partitions of the finite set S can easily be expressed in terms of its corresponding Boolean algebra. In fact, if $\mathcal{P}^*$ denotes the Boolean algebra associated with a partition $\mathcal{P}$, then $(\mathcal{P}_1 \vee \mathcal{P}_2)^* = \mathcal{P}_1^* \cap \mathcal{P}_2^*$. To obtain the partition $\mathcal{P}_1 \vee \mathcal{P}_2$ we simply have to take the atoms of $\mathcal{P}_1^* \cap \mathcal{P}_2^*$. Moreover, note that $\mathcal{P}_1 \leq \mathcal{P}_2$ if and only if $\mathcal{P}_2^* \subseteq \mathcal{P}_1^*$.

Besides other properties, the following theorem (Grätzer, 1978) shows how to obtain the supremum of partitions in the case of an infinite set S.

Theorem 7.4. *Let $\mathcal{P}_i, i \in I$ be an arbitrary collection of partitions of a set S.*

(1) $x \equiv y(\vee_{i \in I} \mathcal{P}_i)$ if and only if there exists $n \geq 0$, $i_0, \ldots, i_n \in I$ and $x_0, \ldots, x_{n+1} \in S$ such that $x = x_0, y = x_{n+1}$ and $x_j \equiv x_{j+1}(\mathcal{P}_{i_j})$ for $0 \leq j \leq n$.

(2) *The least element 0 of $Part(S)$ has only trivial blocks and the greatest element 1 of $Part(S)$ has only one block, namely S itself.*
(3) *The atoms of $Part(S)$ are the partitions with exactly one non-trivial block and this block has exactly two elements.*
(4) *$\mathcal{P}_1$ covers $\mathcal{P}_2$ in $Part(S)$ if and only if $\mathcal{P}_1$ is the result of replacing two blocks of $\mathcal{P}_2$ by their union.*
(5) *$Part(S)$ is semi-modular, i.e. if b covers a then $b \vee c$ covers $a \vee c$ or $b \vee c = a \vee c$ for all $c \in Part(S)$.*

The proof of this theorem will not be given because it is essentially technical and it can be found in Grätzer, 1978. Finally, if $\mathcal{P}_1 \leq \mathcal{P}_2$, let $d : \mathcal{P}_2 \to \mathcal{P}(\mathcal{P}_1)$ denote the mapping which assigns to each block $P_2 \in \mathcal{P}_2$ the set of blocks in $\mathcal{P}_1$ whose union is P_2. This definition of d can be extended to $\mathcal{P}_2$ if we define $d(X) = \cup \{d(C) : C \in X\}$ for a set of blocks $X \subseteq \mathcal{P}_2$. The mapping d is called the *decomposition mapping.*

In view of the three introductory examples and in concordance with the properties of refinings, the following definition of a family of compatible frames can be given. A collection of frames of discernment $\mathcal{F}$ together with a collection $\mathcal{R}$ of refinings between them forms a *family of compatible frames* (f.c.f.) if the following four conditions hold:

(1) For each pair Θ, Λ of frames in $\mathcal{F}$, there is at most one refining $\tau : \Theta \to \Lambda$ in $\mathcal{R}$.
(2) There exists a set S and a mapping $\varphi : \mathcal{F} \to Part(S)$ such that $\varphi(\Theta_1) \neq \varphi(\Theta_2)$ whenever $\Theta_1 \neq \Theta_2$ (i.e. φ is injective) and such that $\varphi(\mathcal{F}) = \{\varphi(\Theta) : \Theta \in \mathcal{F}\}$ is a sublattice of $Part(S)$ containing the least partition 0 and the greatest partition 1.
(3) For each frame $\Theta \in \mathcal{F}$, there is a bijective mapping $b : \Theta \to \varphi(\Theta)$. By this mapping b, to each possible answer $\theta \in \Theta$ exactly one block $b(\theta) \in \varphi(\Theta)$ corresponds.
(4) There is a refining $\tau : \Theta \to \Lambda$ in $\mathcal{R}$ if and only if $\varphi(\Theta) \geq \varphi(\Lambda)$. In this case, $d(b(F)) = b(\tau(F))$ for all $F \subseteq \Theta$, i.e. the diagram pictured in Fig. 7.4 is commutative.

Remark. As we will see in the next chapter, the combination of an infinite number of hints requires $\varphi(\mathcal{F})$ to be a $\wedge$-complete sublattice of $Part(S)$. In this case, the second requirement for a f.c.f. is slightly modified to satisfy this stronger condition.

The first condition is rather intuitive and condition (2) means that the frames in $\mathcal{F}$ correspond to partitions of a set S. The mapping φ is an embedding of $\mathcal{F}$ into $Part(S)$. It will be used to make $\mathcal{F}$ a lattice (see below). Since the least partition of S is in $\varphi(\mathcal{F})$, there is exactly one frame Θ such that $\varphi(\Theta)$ is this partition. Then by the mapping b of condition (3) the elements of this frame Θ are in bijective correspondence with the blocks of the least partition of S, namely the singletons of S. So, from a mathematical point of view, the two sets S and Θ are isomorphic. According to condition (4), to each

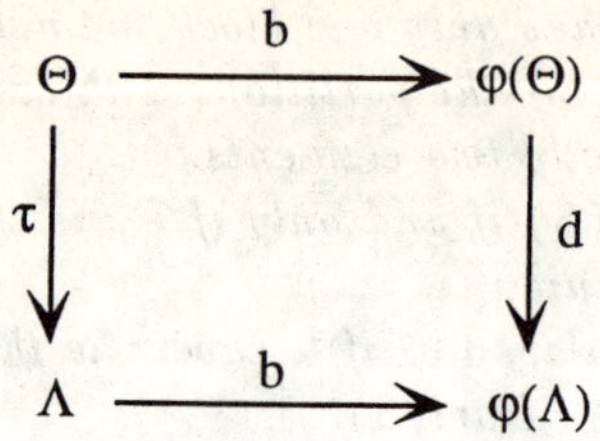

Fig. 7.4. The diagram is commutative

refining τ between the frames Θ_1 and Θ_2 a decomposition mapping between the partitions $\varphi(\Theta_1)$ and $\varphi(\Theta_2)$ corresponds.

The reason why frames and not simply partitions form the building stones of a family of compatible frames is that information originally pertains to frames and not to partitions (see examples 7.1 to 7.3). In the definition of a f.c.f., partitions are used as a technical means to make $(\mathcal{F}, \mathcal{R})$ a lattice, a structure which proves useful for combining hints relating to different frames as will be seen below. It would also have been possible to present a fully axiomatic theory of compatible frames without explicit reference to partitions. This is the point of view adopted in Shafer, 1991 and Shenoy, Shafer, 1990. However, the axiomatic approach makes it more difficult to keep the intuition behind the axioms. For example, the important and intuitive notion of conditional independence to be introduced below is not explicitly considered in these papers, see however Shenoy (1991 b). Conditional independence represents in some sense the semantics of the theory of local computations in Markov trees to be discussed in Chap. 8. On the other hand, the notion of a hypertree, which is an equivalent structure to that of a Markov tree, can be formulated without reference to qualitative independence: it is purely syntactic. The definition of a family of compatible frames given above is a trade-off between two extremas: partitions (the semantical point of view) and lattices (the syntactical point of view).

In a family of compatible frames we have the following result:

Theorem 7.5. *If $\tau : \Theta \rightarrow \Lambda$ and $\mu : \Lambda \rightarrow \Omega$ are two refinings in a family of compatible frames, then there is exactly one refining $\rho : \Theta \rightarrow \Omega$ in $\mathcal{R}$ and it satisfies $\rho = \mu \circ \tau$.*

This theorem asserts that the refinings present in a family of compatible frames are indeed defined in a compatible way. Furthermore, the frames in $\mathcal{F}$ can be partly ordered by

$$\Lambda \leq \Theta \text{ iff } \varphi(\Lambda) \leq \varphi(\Theta), \tag{7.10}$$

which means that $\Lambda \leq \Theta$ if and only if Λ is finer than Θ. Then the following theorem asserts that $(\mathcal{F}, \leq)$ is indeed a lattice.

Theorem 7.6. *In a f.c.f.* $(\mathcal{F}, \mathcal{R})$, *the partly ordered set* $(\mathcal{F}, \leq)$ *is a lattice. If* Θ_1 *and* Θ_2 *are two frames in* $\mathcal{F}$, *then*

$$\Theta_1 \wedge \Theta_2 = \varphi^{-1}(\varphi(\Theta_1) \wedge \varphi(\Theta_2))$$
$$\Theta_1 \vee \Theta_2 = \varphi^{-1}(\varphi(\Theta_1) \vee \varphi(\Theta_2)).$$

If $\varphi(\mathcal{F})$ is $\wedge$-complete (a stronger condition in requirement (2) of a f.c.f.), then $(\mathcal{F}, \mathcal{R})$ is a $\wedge$-complete lattice and $\wedge_{i \in I}\Theta_i = \varphi^{-1}(\wedge_{i \in I}\varphi(\Theta_i))$ for any collection of frames $\Theta_i, i \in I$ (the proof goes along the same lines as the one for theorem 7.6). Furthermore, remark that the greatest element in $\mathcal{F}$ is $\varphi^{-1}(1)$, i.e. $\varphi(1) = 1$.

Families of compatible frames are introduced in order to be able to combine hints relating to different frames. For $i = 1, \ldots, n$, let $\mathcal{H}_i = (\Omega_i, P_i, \Gamma_i, \Theta_i)$ be a hint relative to the frame Θ_i in the f.c.f. $(\mathcal{F}, \mathcal{R})$. These hints cannot be combined directly by Dempster's rule because this rule was defined in Chap. 4 only for hints relating to the same frame. As will be explained in detail in Chap. 8, the hints $\mathcal{H}_i$ must first be transported to a common frame Θ before they are combined. In principle, any common refinement Θ of $\{\Theta_1, \ldots, \Theta_n\}$ can be used, i.e. any frame $\Theta \in \mathcal{F}$ such that there is a refining $\tau_i : \Theta_i \rightarrow \Theta$ for all i and the refinings will be used to transport hints. According to requirement (4) of a f.c.f. and by equation (7.10), this means that any lower bound Θ of $\{\Theta_1, \ldots, \Theta_n\}$ can be used, but for computational efficiency issues the greatest lower bound (infimum) $\wedge_{i=1}^{n}\Theta_i$ turns out to be a better choice. Since lower bound is synonymous with common refinement and a frame Θ is greater than a frame Λ when Θ is coarser than Λ, the greatest lower bound $\Theta = \wedge_{i=1}^{n}\Theta_i$ is also called the *coarsest common refinement* of $\{\Theta_1, \ldots, \Theta_n\}$. Similarly, the least upper bound or *finest common coarsening* of $\{\Theta_1, \ldots, \Theta_n\}$ is defined to be the frame $\Theta = \vee_{i=1}^{n}\Theta_i$. As already mentioned, if $\varphi(\mathcal{F})$ is $\wedge$-complete, then $(\mathcal{F}, \mathcal{R})$ is a $\wedge$-complete lattice and $\wedge_{i \in I}\Theta_i$ exists in $\mathcal{F}$ for any collection of frames $\Theta_i : i \in I$ in $\mathcal{F}$ (I can be infinite). If I is infinite and $\mathcal{H}_i$ is a hint relative to Θ_i for all $i \in I$, then $\wedge_{i \in I}\Theta_i$ is used to combine the hints $\mathcal{H}_i : i \in I$ (see Chap. 8).

Example 7.4. Considering the four identity refinings (see the last requirement for a f.c.f.) and the three refinings obtained by composing the refinings τ_1, τ_2 and τ_3 described in example 7.1 together with the frames $\Theta_1, \ldots, \Theta_4$ defines a family of compatible frames $(\mathcal{F}, \mathcal{R})$. Indeed, condition (1) is clearly satisfied. For the second condition, let S be the set Θ_4 and define the mapping φ as follows:

$$\varphi(\Theta_1) = \{\Theta_4\}$$
$$\varphi(\Theta_2) = \{\{e, f\}, \{b\}, \{i, j, k, g, h\}, \{d\}\}$$
$$\varphi(\Theta_3) = \{\{e\}, \{f\}, \{b\}, \{i, j, k\}, \{g\}, \{h\}, \{d\}\}$$
$$\varphi(\Theta_4) = \{\{e\}, \{f\}, \{b\}, \{i\}, \{j\}, \{k\}, \{g\}, \{h\}, \{d\}\}.$$

This mappping is obviously injective. Since

$$\varphi(\Theta_1) \geq \varphi(\Theta_2) \geq \varphi(\Theta_3) \geq \varphi(\Theta_4), \tag{7.11}$$

it follows that

$$\begin{aligned} \varphi(\Theta_i) \wedge \varphi(\Theta_j) &= \varphi(\Theta_{max(i,j)}) \\ \varphi(\Theta_i) \vee \varphi(\Theta_j) &= \varphi(\Theta_{min(i,j)}) \end{aligned} \tag{7.12}$$

and this implies that $\varphi(\mathcal{F})$ is a complete sublattice of $Part(S)$. For the third condition, the bijective correspondence between Θ_i and $\varphi(\Theta_i)$ is given by associating elements at the same positions in these two sets, i.e. if $\theta_i \in \Theta_i$ is at the k-th position in Θ_i, then its corresponding block is the one which is located at the k-th position in $\varphi(\Theta_i)$. The first part of condition (4) is clearly satisfied and we will prove the second part for one refining only (proceed in a similar way for the other refinings). Consider the refining $\tau_2 : \Theta_2 \to \Theta_3$. Then, for example, we have

$$b(\tau_2(c)) = b(\{l, g, h\}) = \{\{i, j, k\}, \{g\}, \{h\}\} = d(\{i, j, k, g, h\}) = d(b(c)).$$

The equality $b(\tau_2(\theta_2)) = d(b(\theta_2))$ is proved similarly for the other elements $\theta_2 \in \Theta_2$. Finally, the diagram of this family of compatible frames is the chain pictured in Fig. 7.5. Note that the greatest element is Θ_1 and the least element is Θ_4.

Fig. 7.5.

Example 7.5 This example is a verification that the collection of all partitions of a set S can in fact be considered as a family of compatible frames. When the partition Θ is greater than the partition Λ (i.e. $\Theta \geq \Lambda$), the refining $\tau : \Theta \to \Lambda$ is simply the decomposition mapping d. The mapping φ is the identity on $Part(S)$ and $b : \Theta \to \varphi(\Theta)$ is the identity on Θ. If $\mathcal{R}$ denotes all refinings, then $(Part(S), \mathcal{R})$ is obviously a family of compatible frames.

Example 7.6. Multivariate models. Example 7.2 suggests that a set of variables always implicitly defines a family of compatible frames. Also, in example 7.3, if we look on each proposition and assumption as a variable with possible values true and false (or 0 and 1), then these variables also induce a f.c.f. as the following developments demonstrate. Consider a countable set of variables $\{X_i : i \in I\}$ and let Θ_i be the set of possible values of X_i representing the frame of discernment for the question: what is the value of X_i ? Then, for each non-empty subset $J \subseteq I$, consider the question: what is the value of the vector $(X_j, j \in J)$? The frame of discernment for this question is of course $\Theta_J = \prod_{j \in J} \Theta_j$. For convenience, we also define the frame $\Theta_\emptyset = \{\emptyset\}$. The frame Θ_J is therefore defined for all $J \subseteq I$ and we have a family of frames $\mathcal{F} = \{\Theta_J : J \subseteq I\}$. Furthermore, there is a refining $\tau : \Theta_J \to \Theta_K$ if and only if $J \subseteq K$. If $\emptyset \neq J \subset K$, the refining τ is defined by $\tau(t) = \{t\} \times \Theta_{K-J}$ for all $t \in \Theta_J$: the refining process corresponds to *cylindric extension* of a space. If $J = K$, then τ is the identity refining and if $J = \emptyset$, then $\tau(\emptyset) = \Theta_K$. Thus we dispose of a collection $\mathcal{R}$ of refinings linking the frames in $\mathcal{F}$. The following theorem asserts that this is in fact a family of compatible frames which is called a *multivariate family of compatible frames.*

Theorem 7.7. *The collection of frames and refinings $(\mathcal{F}, \mathcal{R})$ defined above is a family of compatible frames whose greatest and least elements are $\Theta_\emptyset = \{\emptyset\}$ and Θ_I respectively. If Θ_J and Θ_K are two frames in $\mathcal{F}$, then $\Theta_J \wedge \Theta_K = \Theta_{J \cup K}$ and $\Theta_J \vee \Theta_K = \Theta_{J \cap K}$. Furthermore, if $\mathcal{J}$ is an arbitrary collection of subsets of I and if $A = \cup\{J : J \in \mathcal{J}\}$, then $\wedge\{\Theta_J : J \in \mathcal{J}\} = \Theta_A$, which shows that $(\mathcal{F}, \mathcal{R})$ is $\wedge$-complete. However, $(\mathcal{F}, \mathcal{R})$ is not $\vee$-complete in general.*

Figure 7.3 a) shows the diagram of the corresponding lattice $(\mathcal{F}, \leq)$ for a set I composed of three elements. The highest node (greatest element) corresponds to the frame $\Theta_\emptyset$ and the lower node (least element) corresponds to the frame Θ_I.

The class of multivariate families of compatible frames is very frequently used and a lot of instances of this class will be considered in the rest of the book. Also, several examples of such families have already been considered implicitly in the six preceding chapters.

7.3 Independence of Frames

In a family of compatible frames $(\mathcal{F}, \mathcal{R})$, the frames in $\mathcal{F}$ usually represent not fully independent questions since the frames and therefore the questions are somehow related by the refinings. Consider a set of frames $\Theta_1, \cdots, \Theta_n$ in $\mathcal{F}$. Not all vectors $(\theta_1, \ldots, \theta_n)$ in $\prod_{i=1}^n \Theta_i$ can possibly represent the vector of the correct answers for the questions represented by the frames $\Theta_i, i = 1, \ldots, n$. In fact, let Λ be the coarsest common refinement of $\Theta_1, \ldots, \Theta_n$ and let $\tau_i : \Theta_i \to \Lambda$ denote the refining linking Θ_i and Λ. Since a possible answer $\theta_i \in \Theta_i$ is represented by $\tau_i(\theta_i)$ in the refinement Λ, if θ_i is correct, then the proposition $\tau_i(\theta_i)$ in Λ is also correct, i.e. $\tau_i(\theta_i)$ contains the answer to the question represented by Λ. Therefore, if

$$\cap_{i=1}^{n}\tau_i(\theta_i) = \emptyset, \tag{7.13}$$

then the vector $(\theta_1, \ldots, \theta_n)$ certainly does not represent the correct combined answer. A vector for which equation (7.13) does not hold is called *compatible* and the set of all compatible vectors

$$R(\Theta_1, \ldots, \Theta_n) = \{(\theta_1, \ldots, \theta_n) : \cap_{i=1}^{n}\tau_i(\theta_i) \neq \emptyset\} \tag{7.14}$$

is called the *compatibility relation* between the frames $\Theta_1, \ldots, \Theta_n$. If Λ' is another common refinement of $\Theta_1, \ldots, \Theta_n$, then the following theorem asserts that the compatibility relation can be computed with respect to Λ or Λ'.

Theorem 7.8. *Let Λ' be a common refinement of $\Theta_1, \ldots, \Theta_n$ and let $\tau'_i : \Theta_i \to \Lambda'$ denote the refining linking Θ_i and Λ'. If*

$$R'(\Theta_1, \ldots, \Theta_n) = \{(\theta_1, \ldots, \theta_n) : \cap_{i=1}^{n}\tau'_i(\theta_i) \neq \emptyset\} \tag{7.15}$$

then $R(\Theta_1, \ldots, \Theta_n) = R'(\Theta_1, \ldots, \Theta_n)$.

It follows from this result that the compatibility relation between a set of frames does not depend on which common refinement is used to compute it.

The compatibility relation R between a set of frames $\Theta_1, \ldots, \Theta_n$ formalizes the idea that there are answers in these frames which cannot be correct simultaneously. These are precisely the vectors outside R. In this case, knowledge of the answer in some frame may restrict possible answers in the other ones. Nevertheless, there are situations where all vectors can possibly be correct, i.e. $R = \prod_{i=1}^{n}\Theta_i$. In this case, we say that the frames $\Theta_1, \ldots, \Theta_n$ are *independent* (a justification of the term "independent" will be given below). This notion of independence is very important and will be widely used in the next chapter. When the frames $\Theta_1, \ldots, \Theta_n$ are independent, we write

$$\{\Theta_1, \ldots, \Theta_n\} \dashv . \tag{7.16}$$

Theorem 7.9. *Let $\mathcal{S} = \{\Theta_1, \ldots, \Theta_n\}$ be a collection of independent frames in a f.c.f.*

(1) Any subcollection $\mathcal{T} \subseteq \mathcal{S}$ is also independent.
(2) If $\tau_i : \Theta_i \to \Lambda$ is a refining for all $i = 1, \ldots, n$ and v_i is its corresponding outer reduction, then $v_j(\tau_i(F)) = \Theta_j$ for all $i \neq j$ and for all non-empty subsets $F \subseteq \Theta_i$.

The second assertion of this theorem shows that even if we know the answer in some $F \subseteq \Theta_i$, nothing can be inferred about the answer in other frames. This justifies the name "independent" for such a collection of frames.

Now we analyze, as an important particular case, the notion of independence in multivariate f.c.f. Let $\mathcal{F} = \{\Theta_J : J \subseteq I\}$ be a multivariate family of compatible frames where I is an arbitrary countable set. Recall that the refinings correspond to cylindrical extentions, see Sect. 7.2. For two frames Θ_J and Θ_K, if $J \cap K \neq \emptyset$, the elements in Θ_J and Θ_K cannot be selected indepently if they are required to be in $R(\Theta_J, \Theta_K)$: they must have identical components on $J \cap K$. But otherwise, if $J \cap K = \emptyset$, the frames are independent

Theorem 7.10. *Let $\mathcal{S} = \{\Theta_{J_1}, \ldots, \Theta_{J_n}\}, J_i \subseteq I$ for all $i = 1, \ldots, n$ be a collection of frames in a multivariate f.c.f. The frames in $\mathcal{S}$ are independent if and only if $J_i \cap J_k = \emptyset$ whenever i is different from k.*

This result provides a useful characterization of independence in multivariate f.c.f. Back to the general case, there is another way to characterize independence of frames.

Theorem 7.11. *Let $\mathcal{S} = \{\Theta_1, \ldots, \Theta_n\}$ be a collection of frames in a f.c.f. The frames in $\mathcal{S}$ are independent if and only if $\cap_{i=1}^n B_i \neq \emptyset$ for all $B_i \in \varphi(\Theta_i), i = 1, \ldots, n$.*

This result permits us to give an interesting consequence of the independence of a collection of frames.

Theorem 7.12. *If $\mathcal{S} = \{\Theta_1, \ldots, \Theta_n\}$ is an independent collection of frames in a f.c.f., then $\vee_{i=1}^n \Theta_i = 1$.*

However, as the following counterexample shows, the converse of this implication is false in general.

Example 7.7. We consider a family of five compatible frames: $\Theta_0 = \{t\}$, $\Theta_1 = \{a, \bar{a}\}$, $\Theta_2 = \{b, \bar{b}\}$, $\Theta_3 = \{c, \bar{c}\}$ and $\Theta_4 = \{a, b, c\}$ (take $S = \{a, b, c\}$ and $\varphi(\Theta_1) = \{\{a\}, \{b, c\}\}$ etc). The refinings are obvious from the diagram of the family pictured in Fig. 7.6. For example, $\tau_1(a) = \{a\}$, $\tau_1(\bar{a}) = \{b, c\}$.

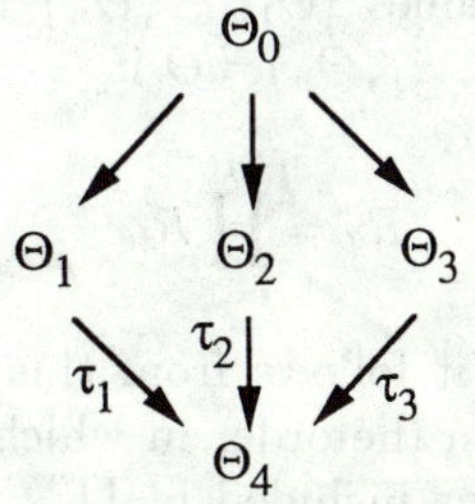

Fig. 7.6.

In this case we have $\Theta_1 \vee \Theta_2 \vee \Theta_3 = \Theta_0 = 1$ because using theorem 7.3 (1) it can easily be verified that the only block in $\varphi(\Theta_1) \vee \varphi(\Theta_2) \vee \varphi(\Theta_3)$ is

$S = \{a, b, c\}$. However, we have $\tau_1(a) \cap \tau_2(b) \cap \tau_3(c) = \emptyset$ and so the frames $\{\Theta_1, \Theta_2, \Theta_3\}$ are not independent. This example also shows that $\Theta_i \vee \Theta_k = 1$ for all different frames in a collection of frames is in general not sufficient to imply that this collection is independent. For multivariate f.c.f. however, since $\Theta_J \vee \Theta_K = \Theta_{J \cap K}$ and $\Theta_\emptyset = 1$, theorem 7.10 implies that these conditions are indeed necessary and sufficient for independence.

The following theorem examines the case where the family of compatible frames consists of all partitions $Part(S)$ of a set S (see example 7.5).

Theorem 7.13. *Let $\mathcal{S} = \{\mathcal{P}_1, \ldots, \mathcal{P}_n\}$ be a collection of frames in the f.c.f. composed of all partitions of a set. Then the partitions (frames) in $\mathcal{S}$ are independent if and only if $\cap_{i=1}^n P_i \neq \emptyset$ for all blocks $P_i \in \mathcal{P}_i, i = 1, \ldots, n$.*

A weaker notion of independence of frames called *conditional independence* proves fundamental for the design of computationally efficient algorithms for solving problems when reasoning with hints (see Chap. 8). Often, given the answer in some frame Θ, two other frames Θ_1 and Θ_2 are independent in the sense that knowing which is the correct answer in Θ_1 does not restrict the possible answers in Θ_2. Let's formalize this idea in terms of compatibility relations. Let $\mathcal{S} = \{\Theta_1, \ldots, \Theta_n\}$ be a collection of frames in a f.c.f. and let Θ be a frame not in $\mathcal{S}$. Furthermore, for $\theta \in \Theta$, denote by R_θ the relation between the frames in $\mathcal{S}$ obtained by imposing that the Θ-component of a vector in the compatibility relation $R(\mathcal{S} \cup \{\Theta\})$ equals θ:

$$R_\theta = \{(\theta_1, \ldots, \theta_n) : (\theta_1, \ldots, \theta_n, \theta) \in R(\mathcal{S} \cup \{\Theta\})\}. \tag{7.17}$$

Since $R_\theta \subseteq R(\mathcal{S})$, this set R_θ contains the tuples which remain compatible given θ, or, in other words, the vectors which are compatible with θ. Furthermore, let $R_{i\theta}$ denote the elements of Θ_i which are compatible with θ:

$$R_{i\theta} = \{\theta_i \in \Theta_i : (\theta_i, \theta) \in R(\Theta_i, \Theta)\}. \tag{7.18}$$

Then the collection of frames $\{\Theta_1, \ldots, \Theta_n\}$ is called *conditionally independent* given Θ, written $(\Theta_1, \ldots, \Theta_n) \dashv \Theta$ if

$$R_\theta = \prod_{i=1}^n R_{i\theta} \tag{7.19}$$

for all $\theta \in \Theta$. Remark that it follows from this definition that conditional independence does not rely on the order in which the frames $\Theta_1, \ldots, \Theta_n$ are considered. Since R_θ is always included in $\prod_{i=1}^n R_{i\theta}$, equation (7.19) really means that it is sufficient that each θ_i is compatible with θ (independently of the other θ_j) for the whole vector $(\theta_1, \ldots, \theta_n)$ to be compatible with θ.

Remark that independence is a special case of conditional independence because $\mathcal{S}$ is independent if and only if $\mathcal{S}$ is conditionally independent given 1. In fact, if θ denotes the unique element of 1, then $R_\theta = R(\mathcal{S})$ and $R_{i\theta} = \Theta_i$

because $\tau(\theta) = \Lambda$ for any common refinement Λ of $\Theta_1, \ldots, \Theta_n$ and 1 (see (7.15)).

We have the following results:

Theorem 7.14. *Let $\mathcal{S} = \{\Theta_1, \ldots, \Theta_n\}$ be a collection of frames in a f.c.f. If $\mathcal{S}$ is conditionally independent given Θ, then every subfamily $\mathcal{T} \subseteq \mathcal{S}$ is also conditionally independent given Θ.*

Theorem 7.15. *If $(\Theta_1, \ldots, \Theta_n) \dashv \Theta$, then $(\Theta_1 \wedge \Theta_2, \Theta_3, \ldots, \Theta_n) \dashv \Theta$.*

Theorem 7.16. *Suppose that $(\Theta_1, \ldots, \Theta_n) \dashv \Theta$. If $\Theta_1' \geq \Theta_1$ then*

$$(\Theta_1', \Theta_2, \ldots, \Theta_n) \dashv \Theta.$$

Theorem 7.17. *If $(\Theta_1, \Theta_2) \dashv \Theta$, then $(\Theta_1 \wedge \Theta, \Theta_2) \dashv \Theta$.*

Theorem 7.18. *If $\Theta_1 \geq \Theta \geq \Theta_2$ then*

$$(\Theta_1, \Theta_2) \dashv \Theta. \tag{7.20}$$

Now we analyze the notion of conditional independence in the context of *multivariate* families of compatible frames. Let $\mathcal{F} = \{\Theta_J : J \subseteq I\}$ be a multivariate f.c.f. There is a very simple characterization of conditional independence.

Theorem 7.19. *The frames in $\mathcal{S} = \{\Theta_{J_1}, \ldots, \Theta_{J_n}\}$, $J_i \subseteq I$ are conditionally independent given Θ_K if and only if $J_i \cap J_k \subseteq K$ whenever i is different from k.*

Since $J_i \cap J_k \subseteq K$ is equivalent to $(J_i - K) \cap (J_k - K) = \emptyset$, theorem 7.19 together with theorem 7.10 shows that the notion of conditional independence can be reduced to the notion of independence in multivariate families of compatible frames.

When we are working with multivariate f.c.f., it is sometimes convenient to write, by abuse of notation, only J for a frame $\Theta_J, J \subseteq I$. We also write sometimes $(J_1, \ldots, J_n) \dashv K$ instead of $\{\Theta_{J_1}, \ldots, \Theta_{J_n}\} \dashv \Theta_K$. If we adopt these conventions, then the next theorem given by Pearl (1988) collects some useful properties of conditional independence in multivariate f.c.f.

Theorem 7.20.

(1) Symmetry: $(U, W) \dashv V$ if and only if $(W, U) \dashv V$
(2) Decomposition: $(U, W \cup Y) \dashv V$ implies $(U, W) \dashv V$ and $(U, Y) \dashv V$
(3) Weak Union: $(U, W \cup Y) \dashv V$ implies $(U, W) \dashv (V \cup Y)$
(4) Contraction: $(U, W) \dashv V$ and $(U, Y) \dashv (V \cup W)$ implies $(U, W \cup Y) \dashv V$

(5) $(U, W) \dashv W$
(6) $(U, W) \dashv V$ *if and only if* $(U - V, W - V) \dashv$
(7) Strong Union: $(U, W) \dashv V$ *implies* $(U, W) \dashv (V \cup Y)$
(8) $(U, V) \dashv W$ *and* $(U, Y) \dashv W$ *imply* $(U, V \cup Y) \dashv W$.

Theorem 7.21. *Suppose that* $(V_1, \ldots, V_n) \dashv V$. *Then*

(1) Extended symmetry: $(V_{\sigma(1)}, \ldots, V_{\sigma(n)}) \dashv V$ *for all permutations* σ *of* $(1, \ldots, n)$.
(2) Extended decomposition: If $U_j \subseteq V_j$ *for all* $j = 1, \ldots, n$, *then*

$$(U_1, \ldots, U_n) \dashv V.$$

(3) Grouping: If $P_1, \ldots, P_s$ *is a partition of* $\{1, \ldots, n\}$ *and* $U_j = \cup\{V_i : i \in P_j\}$, *then* $(U_1, \ldots, U_s) \dashv V$.

An independence relation satisfying properties (1) to (5) in theorem 7.20 is called a *graphoid* by Pearl (1988). However, the property which he calls intersection, namely $(U, W) \dashv (V \cup Y)$ and $(U, Y) \dashv (V \cup Y)$ imply $(U, W \cup Y) \dashv V$, does not hold. As a consequence, there is in general no graphical representation of conditional independence (Pearl, 1988). The next theorem gives a general characterization of conditional independence of frames in terms of their associated partitions. It is the conditional independence version of theorem 7.11.

Theorem 7.22. *Let* $\mathcal{S} = \{\Theta_1, \ldots, \Theta_n\}$ *be a collection of frames in a f.c.f. The frames in* $\mathcal{S}$ *are conditionally independent given a frame* Θ *if and only if for all* $B \in \varphi(\Theta)$ *and all* $B_i \in \varphi(\Theta_i)$, $B \cap B_i \neq \emptyset$ *for all* i *implies that* $\cap_{i=1}^{n} B_i \cap B \neq \emptyset$.

Corollary 7.1. *If* $(\Theta_1, \ldots, \Theta_n) \dashv \Theta$ *then*

$$(\Theta, \Theta_1, \ldots, \Theta_n) \dashv \Theta. \tag{7.21}$$

Finally, the next theorem characterizes conditional independence in families of compatible frames consisting of all partitions of a set S.

Theorem 7.23. *Let* $\mathcal{S} = \{\mathcal{P}_1, \ldots, \mathcal{P}_n\}$ *be a collection of frames in the f.c.f.* $Part(S)$. *Then the partitions (frames) in* $\mathcal{S}$ *are conditionally independent given a partition* $\mathcal{P}$ *if and only if, for all* $P \in \mathcal{P}$ *and all* $P_i \in \mathcal{P}_i$, $P \cap P_i \neq \emptyset$ *for all* i *implies* $\cap_{i=1}^{n} P_i \cap P \neq \emptyset$.

To conclude this chapter, consider a partition $\mathcal{P}$ in $Part(S)$ and a non-empty subset $A \subseteq S$. Then we can define the partition $\mathcal{P}(A)$ of A whose blocks are precisely the non-empty intersections $A \cap P, P \in \mathcal{P}$. The next result then shows that for partitions conditional independence can be reduced to simple independence.

Theorem 7.24. *The partitions (frames) $\{\mathcal{P}_1, \ldots, \mathcal{P}_n\}$ are conditionally independent given $\mathcal{P}$ if and only if the partitions $\{\mathcal{P}_1(P), \ldots, \mathcal{P}_n(P)\}$ are independent for all $P \in \mathcal{P}$.*

A7 Appendix: Proofs of Theorems in Chapter 7

Proof of Theorem 7.1

(1). If $x \in \tau(\cup\{F : F \in \mathcal{F}\})$, then there exist $F \in \mathcal{F}$ and $\theta \in F$ such that $x \in \tau(\theta)$. Therefore $x \in \tau(F)$ and hence $x \in \cup\{\tau(F) : F \in \mathcal{F}\}$. Conversely, if $x \in \cup\{\tau(F) : F \in \mathcal{F}\}$, then there exists $F \in \mathcal{F}$ such that $x \in \tau(F)$. This implies that there exists $\theta \in F$ such that $x \in \tau(\theta)$. But this shows that $x \in \tau(\cup\{F : F \in \mathcal{F}\})$ and hence $\tau(\cup\{F : F \in \mathcal{F}\}) = \cup\{\tau(F) : F \in \mathcal{F}\}$.

(6). This equality follows directly from the third condition of a refining.

(5). According to the second condition of a refining, if the addition of sets means their disjoint union, then

$$\begin{aligned} \tau(A^c) &= \sum\{\tau(\theta) : \theta \in A^c\} \\ &= \sum\{\tau(\theta) : \theta \in \Theta\} - \sum\{\tau(\theta) : \theta \in A\}. \end{aligned} \tag{A7.1}$$

But (6) implies that $\sum\{\tau(\theta) : \theta \in \Theta\} = \Lambda$ and hence $\tau(A^c) = \tau(A)^c$.

(4). If $A \subseteq B$, then $\tau(A) \subseteq \tau(B)$ by definition. Conversely, if $\tau(A) \subseteq \tau(B)$, then

$$\sum_{\theta \in A} \tau(\theta) \subseteq \sum_{\theta \in B} \tau(\theta) \tag{A7.2}$$

and hence $A \subseteq B$.

(2). By de Morgan's laws and according to (1) and (5), we have

$$\begin{aligned} \tau(\cap\{F : F \in \mathcal{F}\}) &= \tau((\cup\{F^c : F \in \mathcal{F}\})^c) \\ &= \tau(\cup\{F^c : F \in \mathcal{F}\})^c \\ &= (\cup\{\tau(F^c) : F \in \mathcal{F}\})^c \\ &= \cap\{\tau(F^c)^c : F \in \mathcal{F}\} = \cap\{\tau(F) : F \in \mathcal{F}\}. \end{aligned} \tag{A7.3}$$

(3). If $F = \emptyset$ then $\tau(F) = \emptyset$ by convention. Conversely, if $F \neq \emptyset$ then $\tau(F) \neq \emptyset$ by the first condition of a refining. □

Proof of Theorem 7.2

First, remark that the refining $\tau : \Theta \to \Lambda$ can be seen as the multivalued mapping of a (Bayesian) hint $\mathcal{H} = (\Theta, P, \tau, \Lambda)$ where P remains undefined.

Then the definition of u given in (3.1) coincides with the one given in (7.7), and the definition of v given in (3.4) corresponds to the one given in (7.8). Therefore, assertions (1) to (4) in theorem 7.2 follow directly from theorem 3.3 (although Θ is not necessarily finite, as is assumed for the set of interpretations of a hint in the first part of the book. However, the proofs of theorems 3.1 and 3.3 trivially extend to the infinite case). Similarly, assertions (5) to (9) follow from theorem 3.1.

(10). If $C = \emptyset$, then $v(C) = \emptyset$ by the first assertion of theorem 3.1. Conversely, if $C \neq \emptyset$, then there exists $\theta \in \Theta$ such that $\tau(\theta) \cap C \neq \emptyset$ because $\{\tau(\theta) : \theta \in \Theta\}$ is a partition of Λ and hence $v(C)$ is not empty.

(11). By definition of $\tau(A)$ we have

$$u(\tau(A)) = \{\theta \in \Theta : \tau(\theta) \subseteq \tau(A)\} = A, \tag{A7.4}$$

and by theorem 7.2 (6) and theorem 7.1 (5) we can write

$$v(\tau(A)) = u(\tau(A)^c)^c = u(\tau(A^c))^c = (A^c)^c = A. \tag{A7.5}$$

(13). Let $x \in C$. There is a $\theta' \in \Theta$ such that $x \in \tau(\theta')$. But $x \in \tau(\theta') \cap C$ implies that $\theta' \in v(C)$, which shows that $x \in \tau(v(C))$.

(12). According to theorem 7.2 (6), theorem 7.1 (5) and theorem 7.2 (13) respectively, we have

$$\tau(u(C)) = \tau(v(C^c)^c) = \tau(v(C^c))^c \subseteq C. \tag{A7.6}$$

□

Proof of Theorem 7.3

The proof of this theorem can be found in almost any book on abstract algebra, e.g. MacLane, Birkhoff (1971) □

Proof of Theorem 7.4

The proof of this theorem can be found in Grätzer, 1978. □

Proof of Theorem 7.5

(a) Since $\tau : \Theta \to \Lambda$ and $\mu : \Lambda \to \Omega$ are refinings, then $\varphi(\Theta) \geq \varphi(\Lambda) \geq \varphi(\Omega)$ by requirement (4). This requirement then also implies that there is a refining $\rho : \Theta \to \Omega$ because $\varphi(\Theta) \geq \varphi(\Omega)$. The uniqueness of ρ follows from requirement (1).

(b) Consider the following diagram:

Let $X \subseteq \Theta$. Since $d_3 = d_2 \circ d_1$, the second part of requirement (4) implies that

$$\begin{aligned} b_3(\rho(X)) = d_3(b_1(X)) &= d_2(d_1(b_1(X))) \\ &= d_2(b_2(\tau(X))) = b_3(\mu(\tau(X))). \end{aligned} \tag{7.7}$$

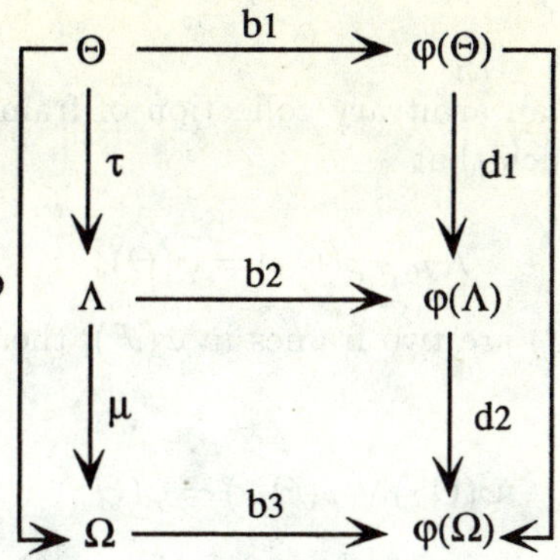

Fig. A7.1.

This implies that $\rho(X) = \mu(\tau(X))$ because b_3 is bijective. □

Proof of Theorem 7.6

To show that $(\mathcal{F}, \leq)$ is a lattice, we must prove the following two assertions

(1) There exists a frame $\Theta \in \mathcal{F}$ such that $\Theta \leq \Theta_i$ for $i \in \{1,2\}$, and if $\Theta' \leq \Theta_i$ for all $i \in \{1,2\}$, then $\Theta' \leq \Theta$.
(2) There exists a frame $\Theta \in \mathcal{F}$ such that $\Theta_i \leq \Theta$ for all $i \in \{1,2\}$, and if $\Theta_i \leq \Theta'$ for all $i \in \{1,2\}$, then $\Theta \leq \Theta'$.

The second requirement of a f.c.f. implies that there is exactly one frame Θ such that $\varphi(\Theta) = \varphi(\Theta_1) \wedge \varphi(\Theta_2)$. We show that $\Theta = \varphi^{-1}(\varphi(\Theta_1) \wedge \varphi(\Theta_2))$ satisfies (1) and hence $\Theta = \Theta_1 \wedge \Theta_2$. According to the definition of the order relation in $\mathcal{F}$, form $\varphi(\Theta_1) \wedge \varphi(\Theta_2) \leq \varphi(\Theta_i)$ it follows that

$$\Theta = \varphi^{-1}(\varphi(\Theta_1) \wedge \varphi(\Theta_2)) \leq \varphi^{-1}(\varphi(\Theta_i)) = \Theta_i. \tag{7.8}$$

This proves the first part of (1). Furthermore, $\Theta' \leq \Theta_i$ implies that $\varphi(\Theta') \leq \varphi(\Theta_i)$ and hence $\varphi(\Theta') \leq \varphi(\Theta_1) \wedge \varphi(\Theta_2)$. It follows that

$$\Theta' = \varphi^{-1}(\varphi(\Theta')) \leq \varphi^{-1}(\varphi(\Theta_1) \wedge \varphi(\Theta_2)) = \Theta. \tag{7.9}$$

The proof that $\varphi^{-1}(\varphi(\Theta_1) \vee \varphi(\Theta_2))$ satisfies (2) is completely similar and hence $\varphi^{-1}(\varphi(\Theta_1) \vee \varphi(\Theta_2)) = \Theta_1 \vee \Theta_2$. □

Proof of Theorem 7.7

The first condition for a f.c.f. is evident and to prove the second condition let S be the set $\Theta_I = \prod_{i \in I} \Theta_i$. Then, to the frame Θ_J with $\emptyset \neq J \subset I$ corresponds the partition $\varphi(\Theta_J)$ whose blocks are the sets $\{t\} \times \Theta_{I-J}, t \in \Theta_J$, whereas to $\Theta_\emptyset$ corresponds the partition whose unique block is $S = \Theta_I$ itself and to Θ_I corresponds the partition whose blocks are all trivial. The one-to-one correspondence between the elements of a frame and the blocks of its corresponding partition is obvious and φ is clearly injective.

Now we prove that $\varphi(\mathcal{F})$ is a $\wedge$-complete sublattice of $Part(S)$. We must prove two assertions:

(a) If $\varphi(\Theta_J), J \in \mathcal{J}$ is an arbitrary collection of frames in $\varphi(\mathcal{F})$, then there is a frame $\Theta \in \mathcal{F}$ such that

$$\wedge_{J\in\mathcal{J}}\varphi(\Theta_J) = \varphi(\Theta). \tag{A7.10}$$

(b) If $\varphi(\Theta_J)$ and $\varphi(\Theta_K)$ are two frames in $\varphi(\mathcal{F})$, then there is a frame Θ in $\mathcal{F}$ such that

$$\varphi(\Theta_J) \vee \varphi(\Theta_K) = \varphi(\Theta). \tag{A7.11}$$

As the following development shows, the frame Θ_A satisfies assertion (a). In fact, if $J = \emptyset$ for all $J \in \mathcal{J}$, then (7.9) implies (A7.10) since $\varphi(\Theta_\emptyset) = \{\Theta_I\}$. Otherwise, again according to (7.9), each block of $\wedge_{J\in\mathcal{J}}\varphi(\Theta_J)$ is of the form $\{\theta\} \times \Theta_{I-A}$ for some $t \in \Theta_A$, which shows that each block of $\wedge_{J\in\mathcal{J}}\varphi(\Theta_J)$ is also a block of $\varphi(\Theta_A)$. We prove the converse. For a vector $t \in \Theta_A$, let t_J denote the projection of t to Θ_J. Then each block $\{t\} \times \Theta_{I-A}$ in $\varphi(\Theta_A)$ can be written as the intersection $\cap\ \{\{t_J\} \times \Theta_{I-J} : J \in \mathcal{J}\}$. Then (7.9) implies that $\{t\} \times \Theta_{I-A}$ is in $\wedge_{J\in\mathcal{J}}\varphi(\Theta_J)$ and hence each block in $\varphi(\Theta_A)$ is also a block in $\wedge_{J\in\mathcal{J}}\varphi(\Theta_J)$. This shows (A7.10) and (a) is proved.

To prove (b), first remark that

$$x \equiv y\ (\varphi(\Theta_H)) \text{ if and only if } x_H = y_H \tag{A7.12}$$

where the subscript H in z_H indicates projection of the vector z to the frame Θ_H. If we select Θ as the frame $\Theta_{J\cap K}$, then (A7.11) is satisfied as the following development shows. First suppose that $J \cap K$ is not empty. To prove (A7.11), we must show that

$$x \equiv y\ (\varphi(\Theta_J) \vee \varphi(\Theta_K)) \text{ if and only if } x \equiv y\ (\varphi(\Theta_{J\cap K})) \tag{A7.13}$$

for all x and y in Θ_I. But according to theorem 7.4 (1), $x \equiv y\ (\varphi(\Theta_J)\vee\varphi(\Theta_K))$ implies that there exists a sequence of elements $x_1, \ldots, x_{n-1}$ in Θ_I such that

$$\begin{aligned} x \equiv x_1\ (\varphi(\Theta_{J_1})) \equiv x_2\ (\varphi(\Theta_{J_2})) \equiv \ldots \\ \equiv x_{n-1}\ (\varphi(\Theta_{J_{n-1}})) \equiv y\ (\varphi(\Theta_{J_n})) \end{aligned} \tag{A7.14}$$

where each J_i is either J or K. For a subscript $i \in I$, let $z(i)$ denote the i-th component of the vector $z \in \Theta_I$. If $i \in J \cap K$, then (A7.12) and (A7.14) imply that $x(i) = y(i)$ and hence $x_{J\cap K} = y_{J\cap K}$, which shows that $x \equiv y\ (\varphi(\Theta_{J\cap K}))$ according to (A7.12).

Conversely, if $x \equiv y\ (\varphi(\Theta_{J\cap K}))$, then $x_{J\cap K} = y_{J\cap K}$. This permits to define a vector $z \in \Theta_I$ as follows:

$$z(i) = \begin{cases} x(i) & \text{if } i \in J \\ y(i) & \text{if } i \in K \\ \textit{arbitrarily} & \text{otherwise.} \end{cases} \tag{A7.15}$$

Then we have

$$x \equiv z\ (\varphi(\Theta_J)) \equiv y\ (\varphi(\Theta_K)) \tag{A7.16}$$

and hence $x \equiv y\ (\varphi(\Theta_J) \vee \varphi(\Theta_K))$ according to theorem 7.4 (1). This shows (A7.13), and (b) is proved when $J \cap K \neq \emptyset$.

If $J \cap K = \emptyset$, then let $\Theta = \Theta_\emptyset = \{\emptyset\}$ and hence $\varphi(\Theta) = \{\Theta_I\}$. To prove (A7.11) we must show that $\varphi(\Theta_J) \vee \varphi(\Theta_K)$ has only one block, namely Θ_I. This means that we must show that

$$x \equiv y\ (\varphi(\Theta_J) \vee \varphi(\Theta_K)) \tag{A7.17}$$

for all vectors x and y in Θ_I. Since $J \cap K = \emptyset$, the vector z given in (A7.15) is well defined for any $x, y \in \Theta_I$ and (A7.16) holds, which implies (A7.17) according to theorem 7.4 (1). This completes the proof of b) and the second requirement for a f.c.f. is satisfied. Since the third and fourth requirements are clearly satisfied, this proves that $(\mathcal{F}, \mathcal{R})$ is indeed a family of compatible frames.

Moreover, $\varphi(\Theta_\emptyset) = \{\Theta_I\}$ implies that $\Theta_\emptyset = \{\emptyset\}$ is the greatest element in $(\mathcal{F}, \leq)$ because $\{\Theta_I\}$ is the greatest partition in $Part(S)$. Similarly, Θ_I is the least element in $(\mathcal{F}, \leq)$ because the elements of Θ_I are the blocks of the least partition in $Part(S)$.

Finally, the above developments together with theorem 7.6 imply the remaining assertions of the theorem, except that $(\mathcal{F}, \mathcal{R})$ is not $\vee$-complete in general. This last equation is proved by the exhibition of a counter-example. Let I be the set of all integer numbers let $\Theta_i = \{0, 1\}$ for all $i \in I$. So an element of Θ_I is an infinite sequence in $\{0, 1\}$. For $j \in I$, let $J_j = \{i \in I : i \geq j\}$. This defines an infinite number of frames $\{\Theta_{J_j} : j \in I\}$. Suppose that there is $R \subseteq I$ such that $\vee_{i=1}^{\infty} \Theta_{J_j} = \Theta_R$. Then $\Theta_{J_j} \leq \Theta_R$ for all j and hence $F \subseteq J_j$ for all j which implies that $R \subseteq \cap_{j=1}^{\infty} J_j = \emptyset$ and hence R is necessarily equal to $\emptyset$. So suppose that $\vee_{j=1}^{\infty} \Theta_{J_j} = \Theta_\emptyset$. Then

$$\vee_{j=1}^{\infty} \varphi(\Theta_{J_j}) = \{\Theta_I\} \tag{A7.18}$$

and we prove that this equality is impossible, which implies that there is no $R \subseteq I$ such that $\vee_{j=1}^{\infty} \Theta_{J_j} = \Theta_R$, i.e. $(\mathcal{F}, \mathcal{R})$ is not $\vee$-complete in general. Let s_0 be the sequence in Θ_I composed zeros: $s_0 = (0, 0, \ldots)$ and let s_1 be the sequence composed of ones: $s_1 = (1, 1, \ldots)$. If (A7.18) holds, then s_0 and s_1 belong to the same block of $\vee_{j=1}^{\infty} \varphi(\Theta_{J_j})$ and by theorem 7.4 (1) there is $t_0, \ldots, t_n$ in Θ_I and $\{S_1, \ldots, S_n\} \subseteq \{J_1, J_2, \ldots\}$ such that

$$s_0 = t_0 \equiv t_1(\Theta_{S_1}) \equiv \ldots \equiv t_{n-1}(\Theta_{S_{n-1}}) \equiv t_n(\Theta_{S_n}) = s_1. \tag{A7.19}$$

But $t_0 = s_0$ and $t_{k-1} \equiv t_k(S_k)$ implies that each $t_k, k = 0, \ldots, n$ has an infinite number of zeros, which is in contradiction to $T_n = s_1$ because s_1 has no zero. Hence (A7.18) cannot hold and the theorem is proved. □

Proof of Theorem 7.8

Let $\mu : \Lambda \to \Lambda'$ denote the refining linking Λ and Λ'. Then by theorem 7.5 and theorem 7.1 (2) we have

$$\cap_{i=1}^{n} \tau_i'(\theta_i) = \cap_{i=1}^{n} \mu(\tau_i(\theta_i)) = \mu(\cap_{i=1}^{n} \tau_i(\theta_i)). \tag{A7.20}$$

Therefore, using theorem 7.1 (3),

$$(\theta_1, \ldots, \theta_n) \in R'(\Theta_1, \ldots, \Theta_n) \text{ if and only if } (\theta_1, \ldots, \theta_n) \in R(\Theta_1, \ldots, \Theta_n).$$

□

Proof of Theorem 7.9

(1). This follows directly from the definition of independence.

(2). By definition of the outer reduction, we have

$$v_j(\tau_i(F)) = \{\theta_j \in \Theta_j : \tau_j(\theta_j) \cap \tau_i(F) \neq \emptyset\}. \tag{A7.21}$$

But by assertion (1) of theorem 7.9 $\tau_j(\theta_j) \cap \tau_i(\theta_i) \neq \emptyset$ for all $\theta_j \in \Theta_j$ and all $\theta_i \in \Theta_i$. Then, according to theorem 7.1 (4),

$$\emptyset \neq \tau_j(\theta_j) \cap \tau_i(\theta_i) \subseteq \tau_j(\theta_j) \cap \tau_i(F) \tag{A7.22}$$

for all $\theta_i \in F$ and all $\theta_j \in \Theta_j$. Since F is not empty, it follows that $\tau_j(\theta_j) \cap \tau_i(F) \neq \emptyset$ for all $\theta_j \in \Theta_j$ and hence $\Theta_j \subseteq v_j(\tau_i(F))$ and finally $\Theta_j = v_j(\tau_i(F))$. □

Proof of Theorem 7.10

Let $J = \cup_{i=1}^{n} J_i$ and let $\tau_i : \Theta_{J_i} \to \Theta_J$ denote the refining linking Θ_{J_i} to Θ_J. If $J_i \cap J_k \neq \emptyset$, then it is clearly possible to find $\theta_i \in \Theta_{J_i}$ and $\theta_k \in \Theta_{J_k}$ such that $\tau_i(\theta_i) \cap \tau_k(\theta_k) = \emptyset$: the only condition is that θ_i and θ_k do not coincide on the components in $J_i \cap J_k$. This implies that $R(\Theta_{J_i}, \Theta_{J_k})$ is different from $\Theta_{J_i} \times \Theta_{J_k}$ and thus the family $(\Theta_{J_i}, \Theta_{J_k})$ is not independent. By theorem 7.9 (1), it follows that $\mathcal{S}$ is not independent.

Conversely, suppose that $J_i \cap J_k = \emptyset$ whenever $i \neq k$ and let $\theta_i \in \Theta_{J_i}$ for all $i = 1, \ldots, n$. Then $\cap_{i=1}^{n} \tau_i(\theta_i)$ is the singleton in Θ_J consisting of the vector $(\theta_1, \ldots, \theta_n)$. This shows that $\mathcal{S}$ is independent. □

Proof of Theorem 7.11

Let $\Lambda = \wedge_{i=1}^{n} \Theta_i$ and let $\tau_i : \Theta_i \to \Lambda$ denote the refining linking Θ_i to Λ. Furthermore, let $b_i : \Theta_i \to \varphi(\Theta_i)$ and $b : \Lambda \to \varphi(\Lambda)$ denote the bijective mappings defined by requirement (3) of a f.c.f. and let $d_i : \varphi(\Theta_i) \to \varphi(\Lambda)$ denote the decomposition mapping. Since b is bijective, if $\theta_i \in \Theta_i$ and $B_i \in \varphi(\Theta_i)$ are such that $b_i(\theta_i) = B_i$ for all $i = 1, \ldots, n$, then

$$b(\cap\tau_i(\theta_i)) = \cap(b(\tau_i(\theta_i))) = \cap(d_i(b_i(\theta_i))) = \cap d_i(B_i) \tag{A7.23}$$

because $b \circ \tau_i = d_i \circ b_i$ according to requirement (4) of a f.c.f.

Suppose that $\mathcal{S}$ is an independent collection of frames and let B_i be a block in $\varphi(\Theta_i)$ for $i = 1, \ldots, n$. If $\theta_i = \varphi^{-1}(B_i)$ then $\cap\tau_i(\theta_i) \neq \emptyset$ and hence $\cap d_i(B_i) \neq \emptyset$ by (A7.23). But this implies that $\cap B_i \neq \emptyset$ because each block in $\cap d_i(B_i)$ is contained in $\cap B_i$.

Conversely, let θ_i be an element of Θ_i for $i = 1, \ldots, n$. If $B_i = b_i(\theta_i)$ then $\cap B_i \neq \emptyset$ by hypothesis. This shows that $\cap d_i(B_i) \neq \emptyset$ and hence $b(\cap\tau_i(\theta_i)) \neq \emptyset$ by (A7.23). But this implies that $\cap\tau_i(\theta_i)) \neq \emptyset$ and hence $R(\Theta_1, \ldots, \Theta_n) = \prod_{i=1}^n \Theta_i$ which proves that the frames in $\mathcal{S}$ are independent. □

Proof of Theorem 7.12

First remark that $\vee_{i=1}^n \Theta_i = 1$ if and only $\vee_{i=1}^n \varphi(\Theta_i) = \{S\}$ because $\varphi(1) = 1$. Furthermore, to prove that $\vee_{i=1}^n \varphi(\Theta_i) = \{S\}$, we have to show that the partition $\vee_{i=1}^n \varphi(\Theta_i)$ contains only one block, namely S. For this purpose, if x and y are two elements of S, then we show that they belong to the same block of $\vee_{i=1}^n \varphi(\Theta_i)$. Let B_1 denote the block of $\varphi(\Theta_1)$ which contains x and B_2 the block of $\varphi(\Theta_2)$ which contains y. Arbitrarily select blocks $B_i \in \varphi(\Theta_i)$ for all $3 \leq i \leq n$. Then theorem 7.11 implies that $\cap_{i=1}^n B_i \neq \emptyset$ and hence $B_1 \cap B_2 \neq \emptyset$. Arbitrarily select an element $z \in B_1 \cap B_2$. Then $x \equiv z\ (\varphi(\Theta_1))$, $z \equiv y\ (\varphi(\Theta_2))$ and theorem 7.4 (1) implies that $x \equiv y\ (\vee_{i=1}^n \varphi(\Theta_i))$. This shows that x and y belong to the same block of $\vee_{i=1}^n \varphi(\Theta_i)$ and the theorem is proved. □

Proof of Theorem 7.13

According to example 7.5, the mapping φ of this f.c.f. is the identity. This theorem is then a direct consequence of theorem 7.11. □

Proof of Theorem 7.14

Assume without loss of generality that $\mathcal{T} = \{\Theta_1, \ldots, \Theta_k\}$ for some $k \leq n$. If

$$R'_\theta = \{(\theta_1, \ldots, \theta_k) : (\theta_1, \ldots, \theta_k, \theta) \in R(\mathcal{T} \cup \{\Theta\})\}, \tag{A7.24}$$

then it is sufficient to show that $\prod_{i=1}^k R_{i\theta} \subseteq R'_\theta$. Let $(\theta_1, \ldots, \theta_k) \in \prod_{i=1}^k R_{i\theta}$. This vector can always be extended to a vector $(\theta_1, \ldots, \theta_n) \in \prod_{i=1}^n R_{i\theta}$. This vector is an element of R_θ since $\mathcal{S}$ is conditionally independent given Θ by hypothesis. But this implies that $(\theta_1, \ldots, \theta_k) \in R'_\theta$ and the theorem is proved. □

In order to prove theorems 7.15 to 7.17, we need the following lemma.

Lemma A7.1. *Let* $\Theta = \wedge_{i=1}^{n}\Theta_i$ *and let* $\tau_i : \Theta_i \to \Theta$ *for all* $i = 1,\ldots,n$. *If* $\theta \in \Theta$, *then for every* $i = 1,\ldots,n$ *there is exactly one* $\theta_i \in \Theta_i$ *such that* $\cap_{i=1}^{n}\tau_i(\theta_i) = \{\theta\}$.

Proof of Lemma A7.1

Let $b_i : \Theta_i \to \varphi(\Theta_i)$ and $b : \Theta \to \varphi(\Theta)$ denote the mappings defined by the second requirement of a f.c.f. and let $d_i : \varphi(\Theta_i) \to \varphi(\Theta)$ denote the decomposition mapping. If P is the block $b(\theta)$ in $\varphi(\Theta)$, then for every $i = 1,\ldots,n$ there is exactly one block P_i in $\varphi(\Theta_i)$ such that $P = \cap P_i$. Now let $\theta_i = b_i^{-1}(P_i)$. Since $d_i \circ b_i = b \circ \tau_i$ we have $d_i = b \circ \tau_i \circ b_i^{-1}$ and hence $b^{-1} \circ d_i = \tau_i \circ b_i^{-1}$. This implies that

$$\begin{aligned}\cap_{i=1}^{n}\tau_i(\theta_i) &= \cap\tau_i(b_i^{-1}(P_i)) = \cap(b^{-1}(d_i(P_i))) \\ &= b^{-1}(\cap d_i(P_i)) \supseteq b^{-1}(\{b(\theta)\}) = \{\theta\}\end{aligned} \tag{A7.25}$$

because $\{b(\theta)\} \subseteq d_i(P_i)$ since $b(\theta) = P$ is a block contained in P_i for all i. Now suppose that $\cap\tau_i(\theta_i)$ contains at least two elements θ and θ' and let $P' = b(\theta')$. Then we have

$$\{b(\theta), b(\theta')\} = b(\{\theta,\theta'\}) \subseteq b(\tau_i(\theta_i)) = d_i(b_i(\theta_i)) \tag{A7.26}$$

for all i and hence $P_i = P + P' + \ldots$ for all i, which implies that $P + P' \subseteq \cap P_i$. But this is impossible because $P = \cap P_i$ and hence $\cap\tau_i(\theta_i) = \{\theta\}$. Now suppose that there is a vector $(\theta'_1,\ldots,\theta'_n)$ different from $(\theta_1,\ldots,\theta_n)$ such that $\cap\tau_i(\theta'_i) = \{\theta\}$. Then

$$\begin{aligned}\{\theta\} &= (\cap\tau_i(\theta_i)) \cap (\cap\tau_i(\theta'_i)) = \cap(\tau_i(\theta_i) \cap \tau_i(\theta'_i)) \\ &= \cap(\tau_i(\{\theta_i\} \cap \{\theta'_i\})) = \emptyset,\end{aligned} \tag{A7.27}$$

which is a contradiction proving that $(\theta_1,\ldots,\theta_n)$ is the unique vector such that $\cap\tau_i(\theta_i) = \{\theta\}$. □

Proof of Theorem 7.15

Let $\Lambda = \wedge_{i=1}^{n}\Theta_i \wedge \Theta$ and consider the frames and refinings given in Fig. A7.2.

We must prove that $\lambda(x) \cap (\cap_{i=3}^{n}\tau_i(\theta_i)) \cap \mu(\theta) \neq \emptyset$ when

$$\lambda(x) \cap \mu(\theta) \neq \emptyset \text{ and } \tau_i(\theta_i) \cap \mu(\theta) \neq \emptyset \text{ for } i = 3,\ldots,n. \tag{A7.28}$$

By lemma A7.1, there are elements $\theta_1 \in \Theta_1$ and $\theta_2 \in \Theta_2$ such that $\tau_1(\theta_1) \cap \tau_2(\theta_2) = \{x\}$. But by theorem 7.1 (2), $\lambda(x) \cap \mu(\theta) = \lambda(\tau_1(\theta_1)) \cap \lambda(\tau_2(\theta_2)) \cap \mu(\theta) \neq \emptyset$ implies that $\lambda \circ \tau_1(\theta_1) \cap \mu(\theta) \neq \emptyset$ and $\lambda \circ \tau_2(\theta_2) \cap \mu(\theta) \neq \emptyset$. Then the second part of (A7.28) and $(\Theta_1,\ldots,\Theta_n) \dashv \Theta$ imply that

$$\lambda \circ \tau_1(\theta_1) \cap \lambda \circ \tau_2(\theta_2) \cap (\cap_{i=3}^{n}\tau_i(\theta_i)) \cap \mu(\theta) \neq \emptyset \tag{A7.29}$$

and hence the theorem is proved because $\lambda \circ \tau_1(\theta_1) \cap \lambda \circ \tau_2(\theta_2) = \lambda(x)$. □

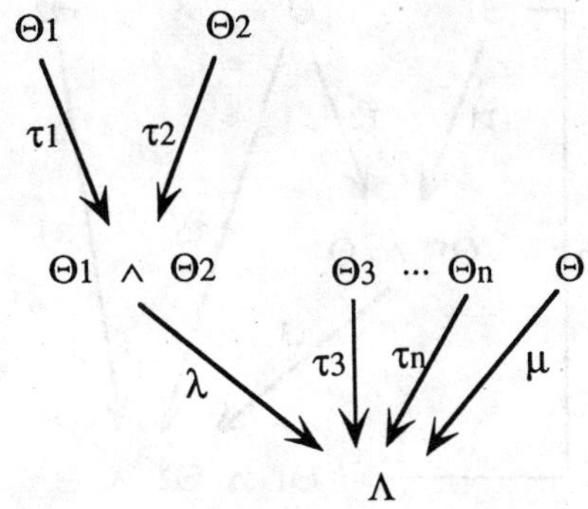

Fig. A7.2.

Proof of Theorem 7.16

Let $\Lambda = \wedge_{i=1}^{n}\Theta_i \wedge \Theta$ and $\tau_i : \Theta_i \to \Lambda$ for all i and $\lambda : \Theta_1' \to \Theta_1$, $\tau : \Theta \to \Lambda$. Let $x \in \Theta_1'$ and $\theta_i \in \Theta_i$ for all i and $\theta \in \Theta$. We must prove that

$$\tau_1 \circ \lambda(x) \cap (\cap_{i=2}^{n}\tau_i(\theta_i)) \cap \tau(\theta) \neq \emptyset \tag{A7.30}$$

when $\tau_1 \circ \lambda(x) \cap \tau(\theta) \neq \emptyset$ and $\tau_i(\theta_i) \cap \tau(\theta) \neq \emptyset$ for all $i = 2, \ldots, n$. If $\tau_1(\lambda(x)) \cap \tau(\theta) \neq \emptyset$ then there is $w \in \tau_1(\lambda(x)) \cap \tau(\theta)$, which implies that there is $t \in \lambda(x)$ such that $w \in \tau_1(t) \cap \tau(\theta)$ and hence $\tau_1(t) \cap \tau(\theta) \neq \emptyset$. Then $(\Theta_1, \ldots, \Theta_n) \dashv \Theta$ implies that

$$\tau_1(t) \cap (\cap_{i=2}^{n}\tau_i(\theta_i)) \cap \tau(\theta) \neq \emptyset \tag{A7.31}$$

and since $\tau_1(t) \subseteq \tau_1(\lambda(x))$ (A7.30) holds and the theorem is proved. □

Proof of Theorem 7.17

Consider the frames and refinings given in Fig. A7.3.

Let $x \in \Theta_1 \wedge \Theta$, $y \in \Theta_2$ and $\theta \in \Theta$. We must prove that $\tau_3(x) \cap \tau_4(y) \cap \tau_5(\theta) \neq \emptyset$. Then by lemma A7.1, there is $\theta_1 \in \Theta_1$ and $\theta' \in \Theta$ such that

$$\tau_1(\theta_1) \cap \tau_2(\theta') = \{x\}. \tag{A7.32}$$

Now assume that

$$\begin{aligned} \tau_3(x) \cap \tau_5(\theta) &\neq \emptyset \\ \tau_4(y) \cap \tau_5(\theta) &\neq \emptyset. \end{aligned} \tag{A7.33}$$

Then we have

$$\tau_3(x) = \tau_3(\tau_1(\theta_1) \cap \tau_2(\theta')) = \tau_3 \circ \tau_1(\theta_1) \cap \tau_3 \circ \tau_2(\theta') = \tau_6(\theta_1) \cap \tau_5(\theta') \tag{A7.34}$$

and hence

$$\emptyset \neq \tau_3(x) \cap \tau_5(\theta) = \tau_6(\theta_1) \cap \tau_5(\theta') \cap \tau_5(\theta). \tag{A7.35}$$

This implies that $\theta = \theta'$ and hence

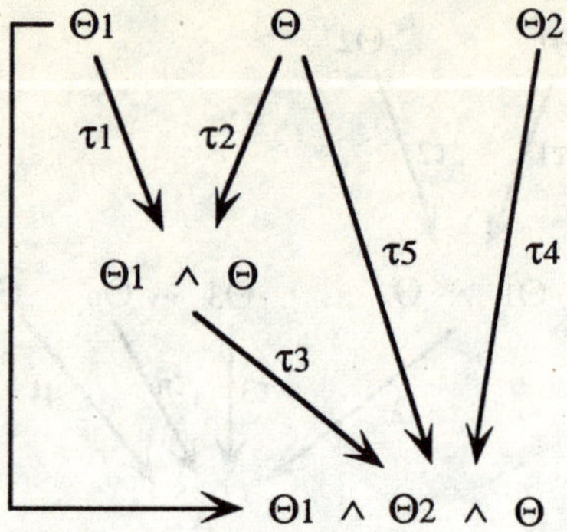

Fig. A7.3.

$$\tau_1(\theta_1) \cap \tau_2(\theta) = \{x\}. \tag{A7.36}$$

Equation (A7.35) implies that $\tau_6(\theta_1) \cap \tau_5(\theta) \neq \emptyset$, which together with (A7.33) and $(\Theta_1, \Theta_2) \dashv \Theta$ implies that

$$\tau_6(\theta_1) \cap \tau_4(y) \cap \tau_5(\theta) \neq \emptyset, \tag{A7.37}$$

or equivalently

$$\tau_3(\tau_1(\theta_1) \cap \tau_2(\theta)) \cap \tau_4(y) \cap \tau_5(\theta) \neq \emptyset. \tag{A7.38}$$

Then by (A7.36) we conclude that

$$\tau_3(x) \cap \tau_4(y) \cap \tau_5(\theta) \neq \emptyset \tag{A7.39}$$

and the theorem is proved. □

Proof of Theorem 7.18

Consider the frames and refinings given in Fig. A7.4.

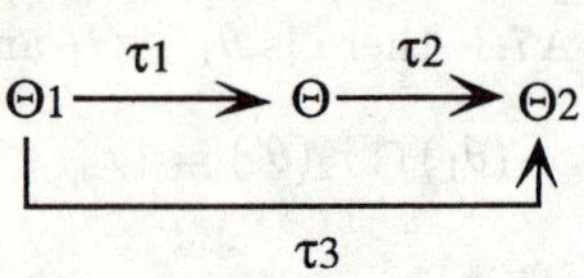

Fig. A7.4.

Let $\theta \in \Theta$. Since $\Theta_1 \wedge \Theta \wedge \Theta_2 = \Theta_2$ we have

$$\begin{aligned} R_\theta &= \{(\theta_1, \theta_2) \in \Theta_1 \times \Theta_2 : \tau_3(\theta_1) \cap \{\theta_2\} \cap \tau_2(\theta) \neq \emptyset\} \\ &= \{(\theta_1, \theta_2) \in \Theta_1 \times \Theta_2 : \theta_2 \in \tau_2(\tau_1(\theta_1) \cap \{\theta\})\} \end{aligned} \tag{A7.40}$$

because $\tau_3 = \tau_2 \circ \tau_1$ (theorem 7.5) and $\tau_2(A \cap B) = \tau_2(A) \cap \tau_2(B)$ (theorem 7.1 (2)). On the other hand, since $\Theta_1 \wedge \Theta = \Theta$, we have

$$R_{1\theta} = \{\theta_1 \in \Theta_1 : \tau_1(\theta_1) \cap \{\theta\} \neq \emptyset\} = \{\theta_1 \in \Theta_1 : \theta \in \tau_1(\theta_1)\} \tag{A7.41}$$

and since $\Theta \wedge \Theta_2 = \Theta_2$

$$R_{2\theta} = \{\theta_2 \in \Theta_2 : \tau_2(\theta) \cap \{\theta_2\} \neq \emptyset\} = \{\theta_2 \in \Theta_2 : \theta_2 \in \tau_2(\theta)\}. \tag{A7.42}$$

Now let $(\theta_1, \theta_2) \in R_{1\theta} \times R_{2\theta}$, i.e. $\theta \in \tau_1(\theta_1)$ and $\theta_2 \in \tau_2(\theta)$. Then $\theta \in \tau_1(\theta_1) \cap \{\theta\}$ and hence $\theta_2 \in \tau_2(\theta) \subseteq \tau_2(\tau_1(\theta_1) \cap \{\theta\})$ by theorem 7.1 (4). Hence $\theta_2 \in \tau_2(\tau_1(\theta_1) \cap \{\theta\})$, which proves that $(\theta_1, \theta_2) \in R_\theta$ and the theorem is proved. □

Proof of Theorem 7.19

Suppose that $J_i \cap J_k \subseteq K$ whenever $i \neq k$. Let $J = \cup_{i=1}^n J_i$ and let $\Lambda = \Theta_{J \cup K}$ be the coarsest common refinement of Θ_J and Θ_K. Furthermore, let $\tau_i : \Theta_{J_i} \to \Lambda$ and $\tau : \Theta_K \to \Lambda$ denote the corresponding refinings. If $\theta \in \Theta_K$ and $\theta_i \in \Theta_{J_i}$ for all i, then it must be proved that if $\tau_i(\theta_i) \cap \tau(\theta) \neq \emptyset$ for all i then

$$\cap_{i=1}^n \tau_i(\theta_i) \cap \tau(\theta) \neq \emptyset. \tag{A7.43}$$

But $\tau_i(\theta_i) \cap \tau(\theta) \neq \emptyset$ implies that θ_i and θ coincide on $K \cap J_i$ for every i. As a general notation, for a vector x in a frame Θ_L let $\pi_L(x)$ denote the Θ_l-component of x. Now we define the vector $t \in \Lambda = \Theta_{J \cup K}$ as follows

$$\pi_l(t) = \begin{cases} \pi_l(\theta) & \text{if } l \in K \\ \pi_l(\theta_i) & \text{if } l \in (J-K) \cap J_i. \end{cases} \tag{A7.44}$$

This vector t is well defined because if $l \in K \cap J_i$ then $\pi_l(\theta) = \pi_l(\theta_i)$ and if $\lambda \in (J-K) \cap J_i$ and $l \in (J-K) \cap J_k$ then $l \in J_i \cap J_k \subseteq K$ and again $\pi_l(\theta_i) = \pi_l(\theta) = \pi_l(\theta_k)$. Then clearly $\cap_{i=1}^n \tau_i(\theta_i) \cap \tau(\theta) = \{t\}$ and (A7.43) is satisfied.

Now we prove the converse. First note that $J_i \cap J_k \subseteq K$ if and only if $(J_i - K) \cap (J_k - K) = \emptyset$. Suppose that there exists an x in $(J_i - K) \cap (J_k - K)$ for some $i \neq k$. Remember that $\tau_i(\theta_i) \cap \tau_k(\theta_k) \neq \emptyset$ if and only if θ_i and θ_k coincide on $J_i \cap J_k$. Then, since $J_l = (J_l \cap K) + (J_l - K)$, it is possible to select vectors $\theta_l \in \Theta_{J_l}$, $l = 1, \ldots, n$ such that θ_l coincide with θ on $J_l \cap K$ for all $l = 1, \ldots, n$ and such that $\pi_x(\theta_i) \neq \pi_x(\theta_k)$. Then $\theta_l \in R_{l\theta}$ for all l but $(\theta_1, \ldots, \theta_l)$ is not in R_θ because $\tau_i(\theta_i) \cap \tau_k(\theta_k) = \emptyset$ since $x \in J_i \cap J_k$ and $\pi_x(\theta_i) \neq \pi_x(\theta_k)$. This proves that $(\Theta_{J_1}, \ldots, \Theta_{J_n}) \not\perp \Theta_K$ and the theorem is proved. □

Proof of Theorem 7.20

(1) This is an immediate consequence of the definition of conditional independence.

The other statements are proved by showing that the conditions of theorem 7.19 are satisfied:

(2) $U \cap (W \cup Y) = (U \cap W) \cup (U \cap Y) \subseteq V$ implies that $U \cap W \subseteq V$ and $U \cap Y \subseteq V$.

(3) $U \cap (W \cup Y) \subseteq V$ implies that $U \cap W \subseteq V$ and hence of course $U \cap W \subseteq V \cup Y$.
(4) $U \cap (W \cup Y) = (U \cap W) \cup (U \cap Y) \subseteq V$ because $U \cap W \subseteq V$ by hypothesis and $U \cap Y \subseteq V$ because if $x \in U \cap Y$, then by hypothesis $x \in V \cup W$ and if $x \in W$ then $x \in U \cap W \subseteq V$.
(5) This follows from $U \cap W \subseteq W$.
(6) $(U, W) \dashv V$ if and only if $U \cap W \subseteq V$ if and only if $U \cap W \cap V^c = \emptyset$ if and only if $(U - V) \cap (W - V) = \emptyset$. But the last equality means that $(U - V, W - V) \dashv$ by theorem 7.10.
(7) $U \cap W \subseteq V$ implies $U \cap W \subseteq V \cup Y$
(8) $U \cap (V \cup Y) = (U \cap V) \cup (U \cap Y) \subseteq W$. □

Proof of Theorem 7.21

The first assertion follows from the definition of conditional independence. Assertion (2) follows from theorem 7.16 and assertion (3) follows from theorem 7.15. □

Proof of Theorem 7.22

Let $\Lambda = \wedge_{i=1}^{n} \Theta_i \wedge \Theta$ and consider the frames and refinings given in Fig. A7.5.

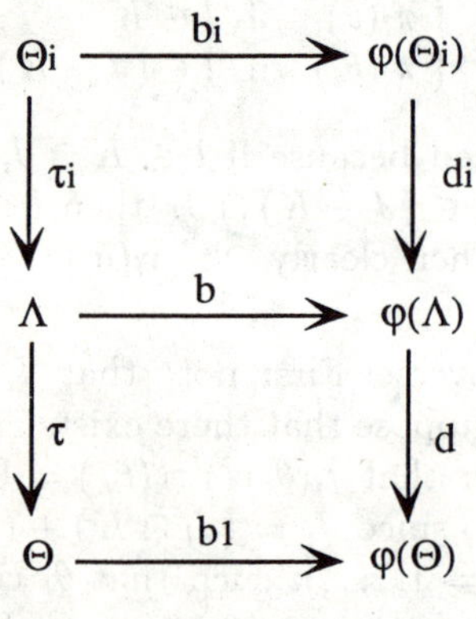

Fig. A7.5.

Let $\theta_i \in \Theta_i$ and $B_i \in \varphi(\Theta_i)$ such that $B_i = b_i(\theta_i)$ and let $\theta \in \Theta$ and $B \in \varphi(\Theta)$ such that $B = b_1(\theta)$. Then we have

$$\begin{aligned} \tau_i(\theta_i) \cap \tau(\theta) &= \tau_i(b_i^{-1}(B_i)) \cap \tau(b_1^{-1}(B)) \\ &= b^{-1}(d_i(B_i)) \cap b^{-1}(d(B)) \qquad (A7.45) \\ &= b^{-1}(d_i(B_i) \cap d(B)). \end{aligned}$$

Now, if $\tau_i(\theta_i) \cap \tau(\theta) \neq \emptyset$ then $d_i(B_i) \cap d(B) \neq \emptyset$ and hence $B \cap B_i \neq \emptyset$. Conversely, suppose that $B \cap B_i \neq \emptyset$ and let $x \in B \cap B_i$. Then the block in

$\varphi(A)$ containing x is a block in $d(B) \cap d_i(B_i)$. Hence $d(B) \cap d_i(B_i) \neq \emptyset$ and hence $\tau_i(\theta_i) \cap \tau(t) \neq \emptyset$ by (A7.45). Therefore we have

$$B_i \cap B \neq \emptyset \text{ if and only if } \tau_i(\theta_i) \cap \tau(\theta) \neq \emptyset \tag{A7.46}$$

and it can be proved similarly that

$$\cap_{i=1}^{n} B_i \cap B \neq \emptyset \text{ if and only if } \cap \tau_i(\theta_i) \cap \tau(\theta) \neq \emptyset. \tag{A7.47}$$

Then theorem 7.22 follows from these two last equivalences and from the definition of conditional independence. □

Proof of Corollary 7.1

This is a direct consequence of theorem 7.22 because $B \cap B' \neq \emptyset$ iff $B = B'$ aor any B, B' in $\varphi(\Theta)$. □

Proof of Theorem 7.23

According to example 7.5, the mapping φ of this f.c.f. is the identity. This theorem is then a direct consequence of theorem 7.22. □

Proof of Theorem 7.24

Suppose that $\{\mathcal{P}_1, \ldots, \mathcal{P}_n\} \dashv \mathcal{P}$. Given $P \in \mathcal{P}$, we use theorem 7.13 to show that $\{\mathcal{P}_1(P), \ldots, \mathcal{P}_n(P)\} \dashv$. Let $A_i \in \mathcal{P}_i(P)$ for all i, i.e. let $P_i \in \mathcal{P}_i$ such that $A_i = P_i \cap P$. We must prove that $\cap A_i \neq \emptyset$. Since A_i is not empty for all i, it follows from $\{\mathcal{P}_1, \ldots, \mathcal{P}_n\} \dashv \mathcal{P}$ and theorem 7.23 that $\emptyset \neq (\cap P_i) \cap P = \cap(P_i \cap P) = \cap A_i$ and hence $\cap A_i \neq \emptyset$.

Conversely, suppose that $\{\mathcal{P}_1(P), \ldots \mathcal{P}_n(P)\} \dashv$ for all $P \in \mathcal{P}$. We prove that $\{\mathcal{P}_1, \ldots, \mathcal{P}_n\} \dashv \mathcal{P}$ using theorem 7.23. Let $P \in \mathcal{P}$ and $P_i \in \mathcal{P}_i$ such that $P \cap P_i \neq \emptyset$ for all i. This implies that $P \cap P_i \in \mathcal{P}_i(P)$ and hence $\cap_{i=1}^{n}(P \cap P_i) \neq \emptyset$ by hypothesis and theorem 7.13. But this shows that $\cap_{i=1}^{n} P_i \cap P \neq \emptyset$ and hence $\{\mathcal{P}_1, \ldots \mathcal{P}_n\} \dashv \mathcal{P}$ by theorem 7.23. □

8. Reasoning on Compatible Frames

8.1 Reasoning With Constraints

In a family of compatible frames $(\mathcal{F}, \mathcal{R})$, suppose we learn that the answer to the question represented by the frame $\Theta \in \mathcal{F}$ lies in the subset $F \subseteq \Theta$. This set can be seen as the focal set of a deterministic hint representing a *constraint* for the answer. This constraint permits us to restrict the answers for some other frames in the family because of the refinings linking them. Indeed, if $\tau : \Theta \to \Lambda$ is in $\mathcal{R}$, then we deduce that the answer to the question represented by Λ lies in the set

$$\tau(F) = \cup \{\tau(\theta) : \theta \in F\} \tag{8.1}$$

which is called the *extension* of the constraint F from Θ to Λ and noted $F \downarrow \Lambda$. Obviously we cannot restrict the answer to a smaller subset if we want it to contain the answer with certainty. Similarly, if $\tau : \Lambda \to \Theta$ is in $\mathcal{R}$, then we have seen in Sect. 7.1 that the outer reduction $v(F) = F \uparrow \Lambda$ is the smallest proposition in Λ which is certain to be true. The process of computing the constraint $F \uparrow \Lambda$ is called *marginalization* because it corresponds to the projection to subspaces in multivariate families of compatible frames. The extension and marginalization operations can be combined into the notion of a *transport* of a constraint. In fact, if Λ is neither a refinement nor a coarsening of Θ, then we can use the refinings $\tau : \Theta \to \Theta \wedge \Lambda$ and $\mu : \Lambda \to \Theta \wedge \Lambda$ to compute $v(\tau(F))$, where v is the outer reduction associated with μ. This set is the constraint induced by F on the frame Λ. It is denoted by $F|\Lambda$ because

$$v(\tau(F)) = (F \downarrow \Theta \wedge \Lambda) \uparrow \Lambda. \tag{8.2}$$

$F|\Lambda$ is called the *transport* of F from Θ to Λ and it *represents* the constraint F within the frame Λ. The following theorem will be very useful on many occasions.

Theorem 8.1. *Consider the sequence of frames and refinings*

$$\Theta_1 \xrightarrow{\tau_1} \Theta_2 \xrightarrow{\tau_2} \Theta_3. \tag{8.3}$$

If v_1, v_2 and v are the outer reductions of τ_1, τ_2 and $\tau_2 \circ \tau_1$ respectively, then $v = v_1 \circ v_2$.

This result permits us to show that any common refinement of Θ and Λ can be used to compute the transport of $F \subseteq \Theta$ from Θ to Λ:

Theorem 8.2. *If $F \subseteq \Theta$, then any common refinement Θ' of Θ and Λ can be used to transport F from Θ to Λ:*

$$F|\Lambda = (F \downarrow \Theta') \uparrow \Lambda. \tag{8.4}$$

Note that the information contained in F is usually not the same as the information contained in $F|\Lambda$ because the transport of $F|\Lambda$ back to Θ is in general different from F. In fact, if v_1 and v_2 are the outer reductions of τ and μ respectively, then

$$(F|\Lambda)|\Theta = v_1(\mu(v_2(\tau(F)))) \supseteq v_1(\tau(F)) = F \tag{8.5}$$

by theorem 7.2 (13) and (11) and in general the inclusion is strict. When F and $F|\Lambda$ express the same information, i.e. $(F|\Lambda)|\Theta = F$, then we say that F *is carried* by Λ. A constraint can be carried by several frames, in particular $F \subseteq \Theta$ is carried by any frame Λ which is a refinement of Θ. It expresses exactly the same information relative to any frame by which it is carried.

Now, suppose that we have two frames Θ_1, Θ_2 and two constraints $F_1 \subseteq \Theta_1$ and $F_2 \subseteq \Theta_2$. If $\Theta = \Theta_1 \wedge \Theta_2$ then F_1 and F_2 are both carried by Θ and the constraint F_1 can be represented by $F_1|\Theta$ and F_2 can be represented by $F_2|\Theta$. Considering these two constraints simultaneously implies that the answer to the question represented by Θ is in $F = F_1|\Theta \cap F_2|\Theta$. By abuse of notation we usually simply write $F = F_1 \cap F_2$. So, the constraints F_1 and F_2 must first be expressed with respect to a common frame before they are combined. As the following theorem shows, any common refinement of Θ_1 and Θ_2 can be used to compute the combined constraint induced by F_1 and F_2.

Theorem 8.3. *Let Θ' be a common refinement of Θ_1 and Θ_2 and $\Theta = \Theta_1 \wedge \Theta_2$. If $F' = F_1|\Theta' \cap F_2|\Theta'$ and $F = F_1|\Theta \cap F_2|\Theta$, then*

$$F'|\Theta = F \quad \text{and} \quad F|\Theta' = F'. \tag{8.6}$$

The following theorem shows that the sequence of combination is irrelevant when three constraints are combined (associativity).

Theorem 8.4. *If $F_i \subseteq \Theta_i$ for $i = 1, 2, 3$, then*

$$(F_1 \cap F_2) \cap F_3 = F_1 \cap (F_2 \cap F_3). \tag{8.7}$$

Since the combination of two constraints is a commutative operation, this theorem allows one to speak of the combination of any number of constraints relating to different frames of a f.c.f. Before we go further, let us make an observation concerning the computational effort required to compute $F_1 \cap F_2$. As illustrated by multivariate families of compatible frames, the coarser is the frame Θ used to compute $F_1 \cap F_2$, the simpler is the computation. Therefore it is not only the number of constraints to combine which is crucial, but also how coarse is the frame used for computing the actual intersection. This point will be clarified when we discuss computations in multivariate families of compatible frames (Sect. 8.4). Now we give two theorems which serve as a preparation for a more efficient computation of $F_1 \cap \ldots F_n$ than the straight forward method using the frame $\wedge_{i=1}^n \Theta_i$. For this purpose assumptions involving conditional independence of frames will be needed and this notion plays an essential role in the improved method.

Theorem 8.5. *Let Θ_1, Θ_2 and Θ_3 be three frames such that $F_1 \subseteq \Theta_1$ and $F_2 \subseteq \Theta_2$. Then, in general,*

$$(F_1 \cap F_2)|\Theta_3 \subseteq F_1|\Theta_3 \cap F_2|\Theta_3. \tag{8.8}$$

But if $(\Theta_1, \Theta_2) \dashv \Theta_3$, then

$$(F_1 \cap F_2)|\Theta_3 = F_1|\Theta_3 \cap F_2|\Theta_3. \tag{8.9}$$

First (8.8) shows that in general it is not feasible to first transport F_1 and F_2 to Θ_3 and then to combine them there. This would give too large sets. However, if $(\Theta_1, \Theta_2) \dashv \Theta_3$, then the above procedure is allowed. And this greatly facilitates computations : Instead of combining F_1 and F_2 on $\Theta_1 \wedge \Theta_2$ and then transport the combined constraint to Θ_3 via $\Theta_1 \wedge \Theta_2 \wedge \Theta_3$ one is allowed to first transport F_1 and F_2 to Θ_3 via $\Theta_1 \wedge \Theta_3$ and $\Theta_2 \wedge \Theta_3$ and then combine these on Θ_3. In these latter operations much coarser frames are involved in general than in the first approach. The following result asserts that under some independence assumption a constraint in Θ_1 can transit by Θ_3 before it is transported to Θ_2.

Theorem 8.6. *Let Θ_1, Θ_2 and Θ_3 be three frames and let $F \subseteq \Theta_1$. Then, in general,*

$$F|\Theta_2 \subseteq (F|\Theta_3)|\Theta_2. \tag{8.10}$$

But if $(\Theta_1, \Theta_2) \dashv \Theta_3$, then

$$F|\Theta_2 = (F|\Theta_3)|\Theta_2. \tag{8.11}$$

Now we are in position to explain the fundamental notion of a *Markov tree*. As far as we know, the notion of a Markov tree has been first introduced and extensively studied by Shafer, Shenoy, Mellouli (1987). This idea also emerged independently in the theory of data bases, see Maier (1983), Beeri at al. (1983), Fagin (1983), and the concept was already implicit in the book of Bertele, Brioschi (1972). Most of the theorems given in Sect. 8.1 and 8.2 are adaptations of results taken from Shafer et al. (1987) and no explicit reference to this paper will be given each time such a theorem is formulated. As will be seen below, the Markov tree structure is crucial for an efficient computation of the intersection of several constraints. Let $N = \{\Theta_1, \Theta_2, \ldots\}$ be a collection of frames in a family of compatible frames $(\mathcal{F}, \mathcal{R})$. If we consider the frames in N as nodes, then adding edges permits the construction of spanning trees over N. One of them is pictured in Fig. 8.1.

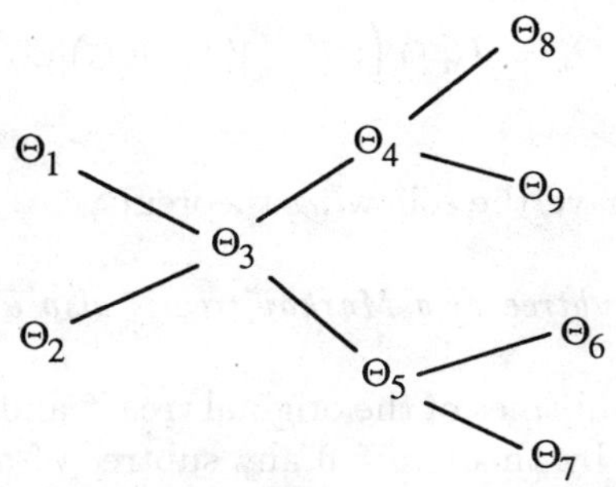

Fig. 8.1. A tree of frames

Consider one of these possible spanning trees $T = (N, E)$, where E denotes the set of edges of the tree. Let Θ_n be a node in T. If we remove Θ_n and its incident edges from T, then the remaining graph will consist of one, two or more isolated subtrees of T, one for each neighbour Θ_k of Θ_n. Let $T_{nk} = (N_{nk}, E_{nk})$ denote the subtree containing the neighbour frame Θ_k and let $V_n = \{k : \Theta_k \text{ is a neighbour of } \Theta_n\}$. Define

$$\Lambda_{nk} = \wedge \{\Theta_j : \Theta_j \in N_{nk}\}. \tag{8.12}$$

Then $T = (N, E)$ is called a *Markov tree* if

$$\{\Lambda_{nk} : k \in V_n\} \dashv \Theta_n \tag{8.13}$$

for all $\Theta_n \in N$.

Theorem 8.7. *If $T = (N, E)$ is a Markov tree, then*

$$(\Lambda_{nk}, \Theta_n) \dashv \Theta_k \tag{8.14}$$

for all neighbour frames Θ_k of Θ_n.

Now we are in position to formulate the fundamental theorem for combining constraints efficiently (Shafer et al., 1987).

Theorem 8.8. *Let*

$$F_1 \subseteq \Theta_1, F_2 \subseteq \Theta_2, \ldots, F_m \subseteq \Theta_m \tag{8.15}$$

be a collection of constraints in a family of compatible frames. Let $N = \{\Theta_1, \ldots, \Theta_m\}$ *and let* $T = (N, E)$ *be a spanning tree over the frames in* N. *If* $F = \cap_{i=1}^m F_i$ *and* $F_{nk} = \cap \{F_j : \Theta_j \in N_{nk}\}$, *then*

$$F|\Theta_n \subseteq F_n \cap \Big(\cap_{k \in V_n}[(F_{nk}|\Theta_k)|\Theta_n]\Big). \tag{8.16}$$

Furthermore, if $T = (N, E)$ *is a Markov tree, then*

$$F|\Theta_n = F_n \cap \Big(\cap_{k \in V_n}[(F_{nk}|\Theta_k)|\Theta_n]\Big). \tag{8.17}$$

Furthermore, we have the following theorem:

Theorem 8.9. *Any subtree of a Markov tree is also a Markov tree.*

The trees T_{nk} are subtrees of the original tree T and hence they are strictly smaller than T. Since by theorem 8.9 any subtree of a Markov tree is also a Markov tree, a formula similar to (8.17) can be used to compute $F_{nk}|\Theta_k$ with the tree T_{nk}. Formula (8.17) is therefore a *recursive procedure.* The basis of the recursion is attained when there is only one node Θ_k in the tree T_{nk} and in this case $F|\Theta_k = F_k$. Once $F_{nk}|\Theta_k$ is computed for all $k \in V_n$, according to formula (8.17), two operations remain to be done. The first one is to transport these constraints from Θ_k to Θ_n, a local operation involving only the frame $\Theta_k \wedge \Theta_n$. The second operation is to actually compute the intersection occurring in (8.17). The essential point is that this intersection is performed on the frame Θ_n and not on the finer frame $\Lambda = \wedge_{i=1}^m \Theta_i$ as would normally be necessary if the definition of $F|\Theta_n = (\cap_{i=1}^m (F_i|\Lambda))|\Theta_n$ is used. These facts explain why (8.17) is called a *local computation procedure.* This may reduce the computational effort considerably because computing intersections in a coarser frame is much easier than in finer frames (a coarser frame always has less elements than a finer frame).

As an illustration, suppose that the tree given in Fig. 8.1 is a Markov tree. Then computing $F|\Theta_3 = (\cap_{i=1}^9 F_i)|\Theta_3$ can be performed in the following three steps:

(1) $F_4' = F_4 \cap F_8|\Theta_4 \cap F_9|\Theta_4$
(2) $F_5' = F_5 \cap F_6|\Theta_5 \cap F_7|\Theta_5$
(3) $F|\Theta_3 = F_3 \cap F_1|\Theta_3 \cap F_2|\Theta_3 \cap F_4'|\Theta_3 \cap F_4'|\Theta_3$.

Theorem 8.10. *If the sequence of frames* $\Theta_1, \Theta_2, \Theta_3, \ldots, \Theta_m$ *is such that there is a refining* $\tau_i : \Theta_i \to \Theta_{i+1}$ *for all* $i = 1, \ldots, m-1$, *then the sequence is a Markov tree.*

According to this theorem, the four frames $\Theta_1, \ldots, \Theta_4$ given in example 7.1 can be arranged into the Markov tree $\Theta_1 - \Theta_2 - \Theta_3 - \Theta_4$. As another example, consider the multivariate family of compatible frames generated by four variables $X_1, \ldots, X_4$. If the frame $\Theta_J = \prod_{i \in J} \Theta_i$ is simply written J, then the four frames pictured in Fig. 8.2 are arranged into a Markov tree because if a leaf node is removed from the tree the condition is trivially satisfied since $\Theta \dashv \Theta$ for any frame Θ in a f.c.f., and if $K = \{1,2,3\}$ is removed, then $\{2,3\} \cap \{1,2\} \subseteq K$, $\{2,3\} \cap \{2,3,4\} \subseteq K$ and $\{1,2\} \cap \{2,3,4\} \subseteq K$, which proves that $(\{2,3\}, \{1,2\}, \{2,3,4\}) \dashv K$ by theorem 7.18.

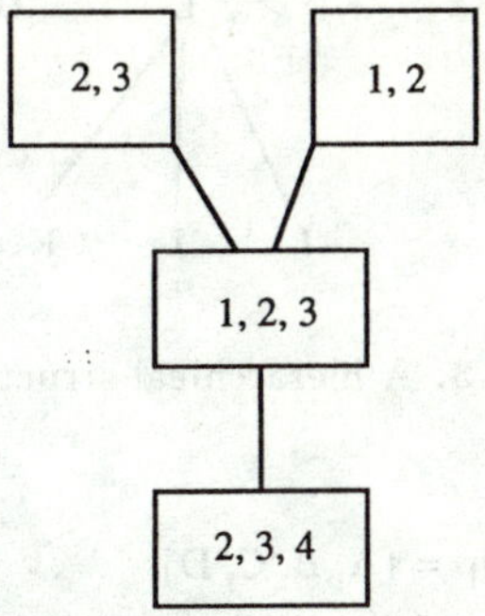

Fig. 8.2.

Example 8.1 Diagnostic Markov trees (this example is taken from Shafer et al., 1987). We continue example 7.1. Consider the f.c.f. consisting of all partitions of the frame $\Theta_4 = \{e, f, b, i, j, k, g, h, d\} = S$. From the tree of diagnoses given in Fig. 7.1, we construct a so-called *hierarchical structure* for S. This is a rooted tree in which the root is S, the other nodes are subsets of S, and the daughters of any non-leaf node form a partition of that node. To build a hierarchical structure from the tree of diagnoses given in Fig. 7.1 we simply replace each node with the set of leaves that lie below it. For example, node l is replaced by the set $L = \{i, j, k\}$ and the node c is replaced by $C = \{i, j, k, g, h\}$. This gives the hierarchical structure given in Fig. 8.3.

Then each non-leaf node M of the hierarchical structure generates a partition of S, namely $\mathcal{D}_M \cup \{M^c\}$ where $\mathcal{D}_M$ is the set of M's daughters. Each of these partitions is a frame of discernment in our f.c.f. $Part(S)$ and we suppose that there is a hint relative to each of these frames. For example, the hint on $\mathcal{D}_A \cup \{A^c\}$ may have three focal sets: one asserting that the fault is not in the battery system (focal set $\{e, f\}^c$) with probability 0.5, one asserting that the battery connections are faulty (focal set $\{f\}$) with probability 0.4 and one asserting that the battery itself is weak or faulty (focal set $\{e\}$) with probability 0.1. The problem is to find a Markov tree for this set of frames. From the hierarchical structure, we define the following tree called *tree of families*. First remove all leaf nodes from the hierarchical structure and then associate the partition $\mathcal{D}_M \cup \{M^c\}$ with any non-leaf node (the partition $\mathcal{D}_P$ is associated

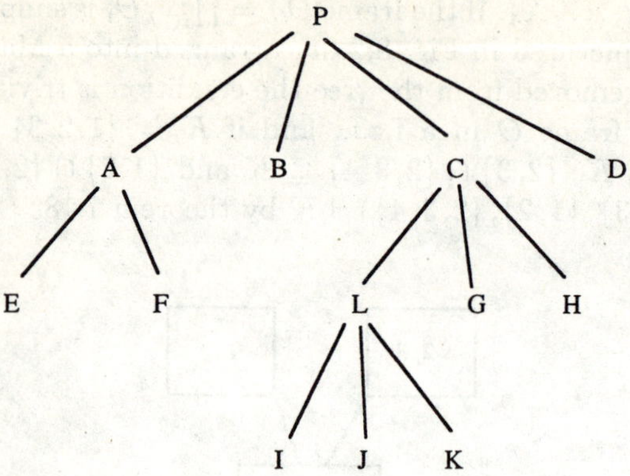

Fig. 8.3. A hierarchical structure

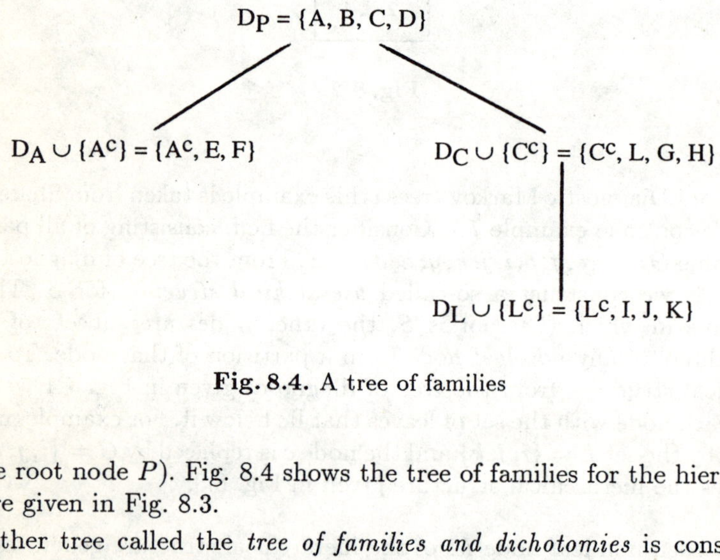

Fig. 8.4. A tree of families

with the root node P). Fig. 8.4 shows the tree of families for the hierarchical structure given in Fig. 8.3.

Another tree called the *tree of families and dichotomies* is constructed from the tree of families as follows. First we put a new node in the middle of each edge and we associate with it the partition $\{M, M^c\}$ where M is the node on the lower end of the edge (i.e. the one associated with the partition $\mathcal{D}_M \cup \{M^c\}$). Then we add back to the tree each leaf M that we removed from the hierarchical structure when we were constructing the tree of families, and we associate with it the partition $\{M, M^c\}$. Fig. 8.5 shows the tree of families and dichotomies constructed from Fig. 8.4. The next theorem completes example 1.

Theorem 8.11. *The tree of families and the tree of families and dichotomies constructed from a hierarchical structure are Markov trees.*

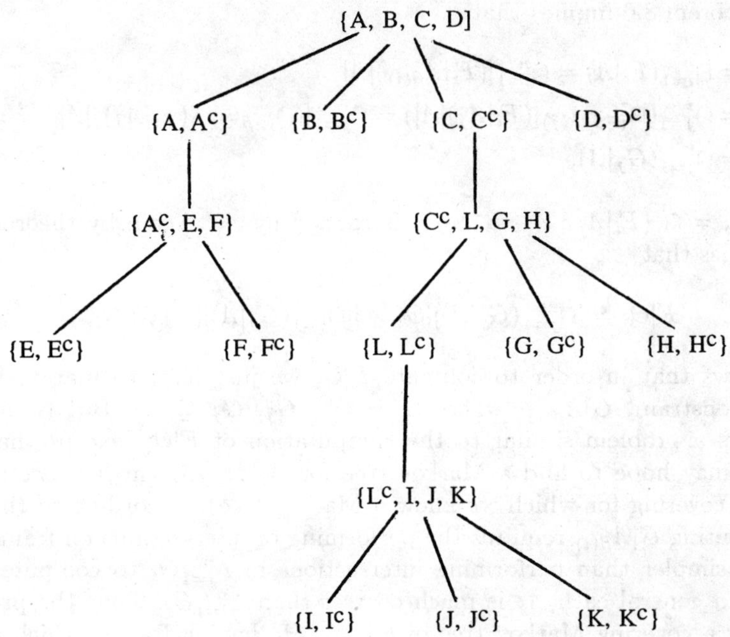

Fig. 8.5. A tree of families and dichotomies

It is perfectly possible that there is no Markov tree for a given collection of frames $\Theta_1, \ldots, \Theta_m$. For example, all three possible spanning trees given in Fig. 8.6 are clearly not Markov. Even in this case we can usually avoid computing the intersection $F = \cap_{i=1}^m F_i$ using the frame $\wedge_{i=1}^m \Theta_i$.

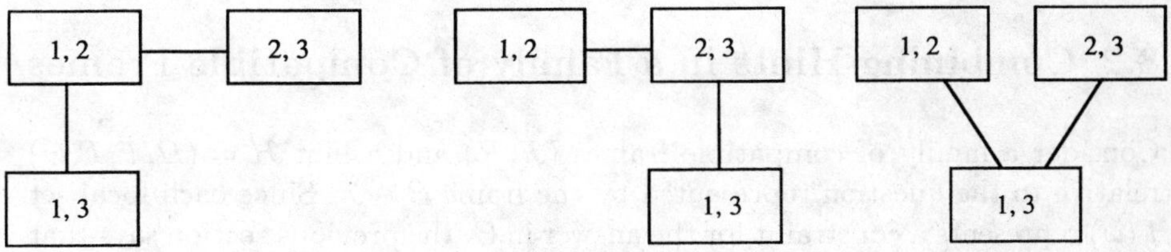

Fig. 8.6.

We proceed as follows.

A collection of frames $\mathcal{L} = \{\Lambda_1, \ldots, \Lambda_r\}$ is called a *covering* of the family $\mathcal{T} = \{\Theta_1, \ldots, \Theta_n\}$ if there is a surjective mapping $\beta : \{1, \ldots, n\} \rightarrow \{1, \ldots, r\}$ (i.e. $\beta^{-1}(j)$ exists for all $j \in \{1, \ldots, r\}$) such that $\Theta_i \geq \Lambda_{\beta(i)}$ for all $i = 1, \ldots, n$. The mapping is called an *embedding* of $\mathcal{T}$ into $\mathcal{L}$. If $\Lambda = \wedge_{j=1}^r \Lambda_j$, then theorem 8.3 says that we can use Λ to compute $F = \cap_{i=1}^n F_i$ because

$\wedge_{i=1}^{n}\Theta_i \geq \Lambda$. Since $\Theta_i \geq \Lambda_{\beta(i)} \geq \Lambda$, theorem 7.18 implies that $(\Theta_i, \Lambda) \dashv \Lambda_{\beta(i)}$. Then theorem 8.6 implies that

$$\begin{aligned} F &= \cap_{i=1}^{n}(F_i|\Lambda) = \cap_{i=1}^{n}[(F_i|\Lambda_{\beta(i)})|\Lambda] \\ &= \cap_{j=1}^{r}(\cap_{i\in\beta^{-1}(j)}[(F_i|\Lambda_j)|\Lambda]) = \cap_{j=1}^{r}[(\cap_{i\in\beta^{-1}(j)}(F_i|\Lambda_j))|\Lambda] \\ &= \cap_{j=1}^{r}(G_j|\Lambda), \end{aligned} \tag{8.18}$$

where $G_j = \cap \{F_i|\Lambda_j : i \in \beta^{-1}(j)\}$ is carried by Λ_j. Again by theorem 8.6, this implies that

$$F|\Theta_i = [\cap_{j=1}^{r}(G_j|\Lambda)]|\Theta_i = [[\cap_{j=1}^{r}(G_j|\Lambda)]|\Lambda_{\beta(i)}]|\Theta_i. \tag{8.19}$$

This shows that in order to compute $F|\Theta_i$ we just have to marginalize to Θ_i the constraint $G|\Lambda_{\beta(i)}$, where $G = \cap_{j=1}^{r}G_j$, $G_j \subseteq \Lambda_j$. But computing $G|\Lambda_{\beta(i)}$ is a problem similar to the computation of $F|\Theta_i$, except that this time we may hope to find a Markov tree for $\Lambda_1, \ldots, \Lambda_r$ (in fact, we always choose a covering for which we know a Markov tree). According to theorem 8.8 computing $G|\Lambda_{\beta(i)}$ requires the performing of intersections on frames Λ_j, which is simpler than performing intersections in $\wedge_{i=1}^{n}\Theta_i$ to compute $F|\Theta_i$ because in general each Λ_j is much coarser than $\wedge_{i=1}^{n}\Theta_i$. Thus the problem is finding a covering Markov tree of $\Theta_1, \ldots, \Theta_n$ having frames which are as coarse as possible. This is a difficult problem which will be discussed in Sect. 8.4 for the special case of multivariate families of compatible frames. Also, we frequently do not try to find a Markov tree for the original set of frames $\Theta_1, \ldots, \Theta_n$, but we directly use a covering family for which we know a Markov tree arrangement (see Sect. 8.4). Note that the single frame $\wedge_{i=1}^{n}\Theta_i$ is a trivial covering Markov tree of $\Theta_1, \ldots, \Theta_n$ and the procedure described above is reduced to the definition of $F|\Theta_i$.

8.2 Combining Hints in a Family of Compatible Frames

Consider a family of compatible frames $(\mathcal{F}, \mathcal{R})$ and a hint $\mathcal{H} = (\Omega, P, \Gamma, \Theta)$ relative to the question represented by the frame $\Theta \in \mathcal{F}$. Since each focal set $\Gamma(\omega)$ represents a constraint for the answer in Θ, the previous section says that $\mathcal{H}$ induces a hint relative to any other frame Λ in the family: simply transport each focal set $\Gamma(\omega)$ from Θ to Λ. The *transport* of $\mathcal{H}$ to Λ is thus defined as the hint $\mathcal{H}|\Lambda = (\Omega, P, \Gamma|\Lambda, \Lambda)$ where $\Gamma|\Lambda$ is defined by $(\Gamma|\Lambda)(\omega) = \Gamma(\omega)|\Lambda$ for all $\omega \in \Omega$. As a special case, if Λ is a refinement of Θ, then the transport of $\mathcal{H}$ to Λ is called the *extension* of $\mathcal{H}$ and it is written $\mathcal{H} \downarrow \Lambda$. Similarly, if Λ is a coarsening of Θ, then the transport is called the *marginal* of $\mathcal{H}$ and it is written $\mathcal{H} \uparrow \Lambda$. Furthermore, the hint $\mathcal{H}$ is said to be *carried* by Λ if $(\mathcal{H}|\Lambda)|\Theta = \mathcal{H}$. The information expressed by the hints $\mathcal{H}$ and $\mathcal{H}|\Lambda$ is the same when $\mathcal{H}$ is carried by Λ. Both theorems 8.5 and 8.6 carry over to hints (see below). The following is the generalization of theorem 8.6:

Theorem 8.12. *Let $\mathcal{H} = (\Omega, P, \Gamma, \Theta_1)$ be a hint relative to Θ_1. If $(\Theta_1, \Theta_2) \dashv \Theta_3$, then*

$$\mathcal{H}|\Theta_2 = (\mathcal{H}|\Theta_3)|\Theta_2. \tag{8.20}$$

Now suppose that we have two hints $\mathcal{H}_1 = (\Omega_1, P_1, \Gamma_1, \Theta_1)$ and $\mathcal{H}_2 = (\Omega_2, P_2, \Gamma_2, \Theta_2)$ relative to the two frames Θ_1 and Θ_2 respectively. These hints cannot be combined directly by Dempster's rule because they are not relative to the same frame of discernment. However, following the concept of Dempster's rule presented in Chap. 4, we consider pairs of interpretations $\omega = (\omega_1, \omega_2) \in \Omega_1 \times \Omega_2$. Such a pair generates two constraints $\Gamma_1(\omega_1) \subseteq \Theta_1$ and $\Gamma_2(\omega_2) \subseteq \Theta_2$ to be combined. Since these two constraints are carried by the frame $\Theta = \Theta_1 \wedge \Theta_2$, the information they express is the same as $\Gamma_1(\omega_1)|\Theta$ and $\Gamma_2(\omega_2)|\Theta$ respectively. This means that if $\omega = (\omega_1, \omega_2)$ is the correct combined interpretation in $\Omega_1 \times \Omega_2$, then the answer to the question represented by Θ is in $\Gamma_1(\omega_1)|\Theta$ and $\Gamma_2(\omega_2)|\Theta$, i.e. in $\Gamma(\omega) = \Gamma_1(\omega_1)|\Theta \cap \Gamma_2(\omega_2)|\Theta$. If $\mathcal{H}_1$ and $\mathcal{H}_2$ are independent, the probability that (ω_1, ω_2) is the correct combined interpretation is $P_1P_2(\omega) = P_1(\omega_1)P_2(\omega_2)$. But it is quite possible that $\Gamma(\omega) = \emptyset$ for some ω. Such an interpretation is called *contradictory*. If $\Gamma(\omega) = \emptyset$ then $\omega = (\omega_1, \omega_2)$ is certainly not the correct combined interpretation and hence we have to condition $(\Omega_1 \times \Omega_2, P_1P_2)$ on non-contradictory interpretations $\Omega = \Omega_1 \times \Omega_2 - K$ where

$$K = \{(\omega_1, \omega_2) : \Gamma_1(\omega_1)|\Theta \cap \Gamma_2(\omega_2)|\Theta = \emptyset\}. \tag{8.21}$$

This leads to the hint $\mathcal{H} = (\Omega, P, \Gamma, \Theta)$ where $P(\omega) = P_1P_2(\omega)/(1 - P_1P_2(K))$. Since $\mathcal{H}$ is obviously the combination of $\mathcal{H}_1|\Theta$ and $\mathcal{H}_2|\Theta$ by Dempster's rule, $\mathcal{H} = \mathcal{H}_1|\Theta \oplus \mathcal{H}_2|\Theta$ is called the combinationof $\mathcal{H}_1$ and $\mathcal{H}_2$ by Dempster's rule. Remark that this generalization of Dempster's rule is also commutative and there are no contradictory interpretations when Θ_1 and Θ_2 are independent, i.e. when $(\Theta_1, \Theta_2) \dashv$. As the following theorem shows, any common refinement of Θ_1 and Θ_2 can be used to compute $\mathcal{H}_1 \oplus \mathcal{H}_2$ (compare theorem 8.3).

Theorem 8.13. *Let Θ' be a common refinement of Θ_1 and Θ_2 and $\Theta = \Theta_1 \wedge \Theta_2$. If $\mathcal{H}' = \mathcal{H}_1|\Theta' \oplus \mathcal{H}_2|\Theta'$ and $\mathcal{H} = \mathcal{H}_1|\Theta \oplus \mathcal{H}_2|\Theta$, then $\mathcal{H}' = \mathcal{H}|\Theta'$ and $\mathcal{H} = \mathcal{H}'|\Theta$.*

We can easily prove by induction that this result still holds when more than two hints are combined. The following theorem shows that combining hints in a family of compatible frames is an associative operation such that the combination of $n > 2$ hints $\mathcal{H}_1, \ldots, \mathcal{H}_n$ can simply be denoted by $\mathcal{H}_1 \oplus \ldots \oplus \mathcal{H}_n$.

Theorem 8.14. *If $\mathcal{H}_i$ is a hint on Θ_i for $i = 1, 2, 3$, then*

$$(\mathcal{H}_1 \oplus \mathcal{H}_2) \oplus \mathcal{H}_3 = \mathcal{H}_1 \oplus (\mathcal{H}_2 \oplus \mathcal{H}_3). \tag{8.22}$$

The next theorem is a generalization of the second part of theorem 8.5.

Theorem 8.15. *Let $\mathcal{H}_1 = (\Omega_1, P_1, \Gamma_1, \Theta_1)$ and $\mathcal{H}_2 = (\Omega_2, P_2, \Gamma_2, \Theta_2)$ be two hints in a family of compatible frames. If Θ_3 is a frame such that $(\Theta_1, \Theta_2) \dashv \Theta_3$, then*

$$(\mathcal{H}_1 \oplus \mathcal{H}_2)|\Theta_3 = \mathcal{H}_1|\Theta_3 \oplus \mathcal{H}_2|\Theta_3. \tag{8.23}$$

This result is important because it shows that if $(\Theta_1, \Theta_2) \dashv \Theta_3$, then we don't need to consider $\Theta_1 \wedge \Theta_2 \wedge \Theta_3$ to compute $(\mathcal{H}_1 \oplus \mathcal{H}_2)|\Theta_3$ because we can use the coarser frames $\Theta_1 \wedge \Theta_3$ and $\Theta_2 \wedge \Theta_3$. Furthermore, the effective computation of the combination involving the intersections is actually performed locally on the coarser frame Θ_3. Of course, theorem 8.15 holds also when more than two hints are combined. Most important, theorem 8.8 concerning Markov trees carries over to hints:

Theorem 8.16. *Let $\mathcal{H}_i = (\Omega_i, P_i, \Gamma_i, \Theta_i), i = 1, \ldots, m$ be a collection of hints in a family of compatible frames. Let $T = (N, E)$ be a Markov tree over the frames $N = \{\Theta_1, \ldots, \Theta_m\}$. If $\mathcal{H} = \oplus_{i=1}^m \mathcal{H}_i$ and $\mathcal{H}_{nk} = \oplus \{\mathcal{H}_j : \Theta_j \in N_{nk}\}$, then*

$$\mathcal{H}|\Theta_n = \mathcal{H}_n \oplus \Big(\oplus_{k \in V_n} (\mathcal{H}_{nk}|\Theta_k)|\Theta_n\Big). \tag{8.24}$$

This is a fundamental result of Shafer et al. (1987) which permits the reduction of the computational effort required to compute the marginal of the combination of several hints. Just like formula (8.17), formula (8.24) is in fact a *recursive procedure* which allows *local computations* (combinations and transports) within the Markov tree structure (see the previous section). As an illustration, suppose that the tree given in Fig. 8.1 is in fact a Markov tree. Then the computation of $\mathcal{H}|\Theta_3 = (\oplus_{i=1}^9 \mathcal{H}_i)|\Theta_3$ can be done by the following three steps:

(1) $\mathcal{H}'_4 = \mathcal{H}_4 \oplus \mathcal{H}_8|\Theta_4 \oplus \mathcal{H}_9|\Theta_4$
(2) $\mathcal{H}'_5 = \mathcal{H}_5 \oplus \mathcal{H}_6|\Theta_8 \oplus \mathcal{H}_7|\Theta_5$
(3) $\mathcal{H}|\Theta_3 = \mathcal{H}_3 \oplus \mathcal{H}_1|\Theta_3 \oplus \mathcal{H}_3|\Theta_3 \oplus \mathcal{H}'_4|\Theta_3 \oplus \mathcal{H}'_5|\Theta_3$.

Furthermore, the theory of covering Markov trees developed at the end of the previous section can obviously be adapted to help combine hints instead of constraints. In fact, from a set of hints $\mathcal{H}_i = (\Omega_i, P_i, \Gamma_i, \Theta_i)$ to be combined, a covering Markov tree of $\Theta_1, \ldots, \Theta_n$ is first derived. Then this covering tree is used to compute the combination of the hints according to the local propagation procedure given in theorem 8.16. Methods for finding covering Markov trees in the important special case of multivariate f.c.f. will be presented in Sect. 8.4.

8.3 Markov Trees and Hypertrees

Under some additional conditions about the f.c.f. $(\mathcal{F}, \mathcal{R})$, it will be possible to give two characterizations of Markov trees. One of them will lead to another local propagation algorithm. The first additional condition, referred to as *Assumption (1)*, states that the lattice $(\mathcal{F}, \mathcal{R})$ is *distributive*, i.e.

Assumptions (1): For all $\Theta_1, \Theta_2, \Theta_3$ in $\mathcal{F}$,

$$\begin{aligned}(\Theta_1 \vee \Theta_2) \wedge \Theta_3 &= (\Theta_1 \wedge \Theta_3) \vee (\Theta_2 \wedge \Theta_3) \\ (\Theta_1 \wedge \Theta_2) \vee \Theta_3 &= (\Theta_1 \vee \Theta_3) \wedge (\Theta_2 \vee \Theta_3).\end{aligned} \tag{8.25}$$

We can verify that the two conditions in (8.25) are equivalent (Grätzer, 1978). By theorem 7.6, Assumption (1) means that $\varphi(\mathcal{F})$ is a distributive sublattice of $Part(S)$. Not every sublattice of $Part(S)$ is distributive: $Part(S)$ itself is not distributive (Grätzer, 1978). However, according to theorem 7.7, every multivariate f.c.f. satisfies Assumption (1) because $\cap$ is distributive with respect to $\cup$. Distributivity is thus a property which a particular sublattice of $Part(S)$ can or cannot have. Anyway, this property will be used in the proofs of theorems in the rest of this chapter. In order to be able to characterize Markov trees in terms of the lattice structure only and not in terms of conditional independence of frames, we need a further assumption. Consider the following theorem:

Theorem 8.17. *Let $\Theta_1, \ldots, \Theta_n$ be a collection of frames in a f.c.f. If $(\Theta_1, \ldots, \Theta_n) \dashv \Theta$ then $\Theta_i \vee \Theta_j \geq \Theta$ for all $i \neq j$.*

However, as the following counter-example shows, the converse is false in general. Consider the family of compatible frames $Part(S)$ and let $\mathcal{P}_1$ and $\mathcal{P}_2$ be the two partitions of S given in Fig. 8.7 a). (both partitions have three blocks and Fig. 8.7 b) shows the diagram of the corresponding f.c.f.). Using theorem 7.3 (1) we find that $\mathcal{P}_1 \vee \mathcal{P}_2 = 1$ but $(\mathcal{P}_1, \mathcal{P}_2) \not\dashv 1$ because not each block of $\mathcal{P}_1$ intersects with each block of $\mathcal{P}_2$. The second condition, referred to as *Assumption (2)*, states that the reverse implication holds in theorem 8.17, i.e.

Assumption (2):

$$(\Theta_1, \ldots, \Theta_n) \dashv \Theta \text{ iff } \Theta_i \vee \Theta_j \geq \Theta \text{ whenever } i \neq j. \tag{8.26}$$

Again, this assumption will be crucial in the proofs of theorems in the remaining sections of this chapter because it allows the conditional independence property to be expressed in terms of the lattice structure only. Theorems 7.7 and 7.19 imply that every multivariate f.c.f. satisfies Assumption (2), but $Part(S)$ does not satisfy it as we have seen above. We assume that both Assumptions (1) and (2) hold for the f.c.f. considered in the rest of this chapter, as is the case for every multivariate f.c.f. These assumptions will only

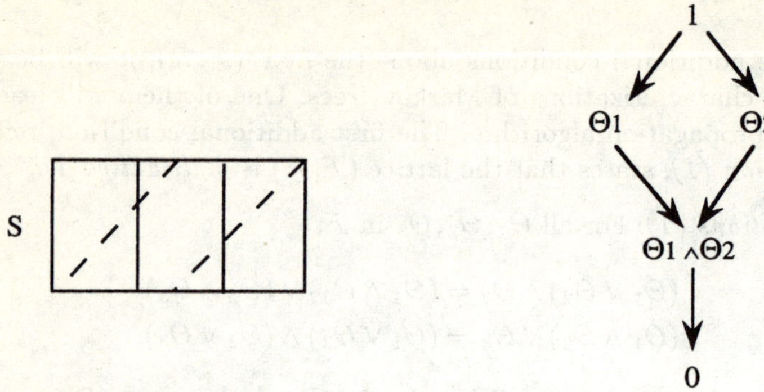

a) b)

Fig. 8.7.

be used in the proofs of theorems. Then we have the following characterization of Markov trees which does not make use of the concept of conditional independence:

Theorem 8.18. *A tree $T = (N, E)$ is a Markov tree if and only if $\Theta_1 \vee \Theta_2 \geq \Theta$ for all frames Θ_1 and Θ_2 in N and all frames Θ on the unique path between Θ_1 and Θ_2 in T.*

For multivariate f.c.f., this result is reduced to the following corollary (see also Shenoy, Shafer, 1990).

Corollary 8.1. *Let $(\mathcal{F}, \mathcal{R})$ be a multivariate f.c.f. A tree $T = (N, E)$ in $(\mathcal{F}, \mathcal{R})$ is a Markov tree if and only if $R \cap S \subseteq K$ for all frames Θ_R, Θ_S in N and all frames Θ_K on the unique path between Θ_R and Θ_S in T.*

Let $H = \{J_1, \ldots, J_n\}$ be a hypergraph in a multivariate f.c.f. and set $J = \cup_{i=1}^{n} J_i$. Then the *variable and relation graph* associated with H is defined as follows. Its node set is $J + H$ and there is an edge $\{x, y\} \in E$ if and only if there is $J_i \in H$ such that $y = J_i$ and $x \in J_i$.

Theorem 8.19. *Let $G = (J + H, E)$ be a variable and relation graph. If G is a tree, then it is a Markov tree.*

This theorem implies that if the variable and relation graph is a tree, then it is indeed a covering Markov tree of the hypergraph H.

Example 8.2. Suppose that the retail price R of a car is determined by the dealer's profit P and wholesale price W. The latter depends on the production cost U_1, the marketing cost U_2 and manufacturer's profit U_3 (Pearl, 1988). That is, we have the relations

$$\begin{aligned} R &= W + P \\ W &= U_1 + U_2 + U_3. \end{aligned} \tag{8.27}$$

Suppose that we have hints about what might be the production costs U_1, the marketing costs U_2, the industry profit U_3, the profits of two different dealers and that we know the retail prices for the car asked by the dealers. Then one might be interested in different questions or different variables: for example what can be said about the wholesale price, what about the industry profit or the dealer's profit ?

In any case, the available hints, together with the known values of the retail prices and the relations (8.27) must be combined. Note that the known values as well as the relations (8.27) can be considered as deterministic hints. The combined hint must then be coarsened to the variable of interest. This example induces a multivariate model with variables $U_1, U_2, U_3, W, P_1, P_2, R_1, R_2$. Note that there are two dealer's profits as well as two retail prices. Correspondingly, there are two relations of the type $R_i = W + P_i$. The available hints are relative to frames in the hypergraph

$$\begin{aligned} H =&\{\{U_1\},\{U_2\},\{U_3\},\{P_1\},\{P_2\},\{R_1\},\{R_2\},\{U_1,U_2,U_3,W\}, \\ &\{R_1,W,P_1\},\{R_2,W,P_2\}. \end{aligned} \tag{8.28}$$

Fig. 8.45 displays its variable and relation graph. It is a tree and by theorem 8.19 it is a covering Markov tree of H.

Let's consider a relation like $W = U_1 + U_2 + U_3$ and suppose that the hints $\mathcal{H}(U_1\}, \mathcal{H}(U_2), \mathcal{H}(U_3)$ relate to the variables U_1, U_2, U_3. Let $\mathcal{H}_4$ denote the deterministic hint representing the relation $W = U_1 + U_2 + U_3$. Suppose that the focal sets of the hints $\mathcal{H}(U_1\}, \mathcal{H}(U_2), \mathcal{H}(U_3)$ are closed real intervals. If for some interpretations of the three hints the focal intervals are I_1, I_2 and I_3 respectively, then it can easily be seen that propagation in the Markov tree leads to the focal interval

$$I_1 + I_2 + I_3 \tag{8.29}$$

for the combined hint

$$(\mathcal{H}(U_1) \oplus \mathcal{H}(U_2) \oplus \mathcal{H}(U_3) \oplus \mathcal{H}_4)|\{W\}. \tag{8.30}$$

The addition in (8.29) is the addition of intervals. Thus combining hints is reduced to interval arithmetics in this example. In this spirit we write $\mathcal{H}(U_1\} + \mathcal{H}(U_2) + \mathcal{H}(U_3)$ for the combined hint (8.30). This highlights the fact that Dempster's rule of combination of hints essentially corresponds to interval arithmetics in this case. For example, if the wholesale price W is the variable of interest and $\mathcal{H}$ denotes the combination of all available hints, then the Markov tree propagation algorithm implies that

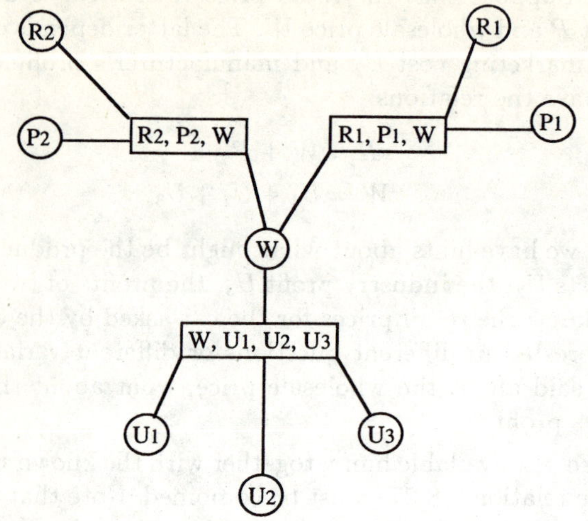

Fig. 8.8. A variable and relation graph

$$\mathcal{H}|\{W\} = (\mathcal{H}(U_1) + \mathcal{H}(U_2) + \mathcal{H}(U_3)) \oplus (\mathcal{H}(R_1) - \mathcal{H}(P_1)) \oplus (\mathcal{H}(R_2) - \mathcal{H}(P_2)) \tag{8.31}$$

if $\mathcal{H}(R_i)$ and $\mathcal{H}(P_i)$ denote the deterministic hint corresponding to the relations R_i and P_i respectively. If the industry profit U_3 is of interest, then

$$\mathcal{H}|\{U_3\} = ((\mathcal{H}(R_1) - \mathcal{H}(P_1)) \oplus (\mathcal{H}(R_2) - \mathcal{H}(P_2)) - (\mathcal{H}(U_1) + \mathcal{H}(U_2)). \tag{8.32}$$

It must be noticed that this arithmetic of hints crucially depends on the Markov tree property. It cannot be hoped in general that all numerical models of this type can be computed so easily if the exact solution is to be found, because generally the variable and relation graph is not a tree. If approximations are sufficient, then non-Markov covering trees and Monte Carlo sampling can be used.

As another example, the rule trees considered in Chap. 6 are also variable and relation graphs. Since they are trees, they are indeed Markov trees and the local propagation methods described there (essentially modus ponens genralized to hints) is nothing else than the usual local propagation algorithm in Markov trees. The only difference is that the empty set was considered as a usual focal set and normalization was only performed at the end of the propagation, but this makes no difference as can easily be verified. This justifies the computational procedures presented there.

There is another graph which can readily be obtained from a hypergraph $H = \{J_1, \ldots, J_n\}$, namely the so-called *relational graph* whose nodes are the hyperedges $J_i \in H$ and there is an edge between the nodes K and L if and only if $K \cap L \neq \emptyset$.

Theorem 8.20 *Let $G = (H, E)$ be a relational graph constructed from the hypergraph H. If G is a tree, then it is a Markov tree.*

So, if the relational graph is a tree, then it is a covering Markov tree of H.

In data base theory the notion of a *join tree* is sometimes considered. Its relation to Markov trees is the following. Maier (1983) defines an intersection graph of a hypergraph H as a graph $G = (H, E)$ where E is any subset of $H \times H$, and *join tree* of H as an intersection graph of H which is a Markov tree. Hence if the relation graph of H is a tree, then it is a join tree. Refer to Maier (1983) for more information on the relation between join trees, i.e. Markov trees having H as set of nodes, and data base theory. For example, assuming that there is one join tree for H, then a method for finding one is presented (the method relies on algorithms for finding so-called maximum spanning trees).

Now we introduce a certain number of standard graph-theoretical definitions which serve as a justification for the names given to some concepts used in the sequel. We follow the presentation given by Shenoy, Shafer (1990) for multivariate families of compatible frames. However, the results presented here concern arbitrary f.c.f. and several ideas and results are taken from Shafer (1991). Suppose that $v_1, v_2, \ldots, v_n$ is a sequence of the nodes of a graph $T = (V, E)$ and suppose that for $i = 2, \ldots, n$ the node v_i is a leaf (node of degree 1) in the subgraph of T containing the nodes $\{v_1, \ldots, v_i\}$. In this case we say that $v_1, \ldots, v_n$ is a *tree construction sequence in nodes* of T because it is easily verified that this is equivalent to T being a tree (i.e. a connected graph without cycle). We construct the tree by beginning with v_1, then adding v_2 and an edge between v_1 and v_2, then adding v_3 and an edge between v_3 and either v_1 or v_2 etc. (see Fig. 8.9). If l is a leaf in a graph, i.e. l is incident to exactly one edge of the graph, then the other node of this edge is called the *bud* of l. Given a tree construction sequence in nodes $v_1, \ldots, v_n$, for $i = 2, \ldots, n$, let $b(i) < i$ denote the integer such that $v_{b(i)}$ is the bud of v_i in the subgraph over the nodes $v_1, \ldots, v_i$. The function $b(i)$ is called the *budding* of the tree construction sequence in nodes.

Now suppose that t and b are two distinct edges in a graph $G = (V, E)$. Edges are considered as two-elements sets containing their extremal nodes. We say that t is a *twig* and b is a *branch* for t if one of t's elements is a leaf of G and the other is in b. Note that a twig may have many branches. For example, in Fig. 8.9, the edge $\{v_4, v_6\}$ is a twig with branches $\{v_2, v_4\}, \{v_7, v_4\}$ and $\{v_5, v_4\}$. The next theorem of Shafer (1991) gives a characterization of twigs and branches.

Theorem 8.21. *Let t and b be two distinct edges in the graph $G = (V, E)$. t is a twig and b is a branch for t if and only if $t \cap b \neq \emptyset$ and*

$$t \cap (\cup \{e : e \in E - \{t\}) = t \cap b. \tag{8.33}$$

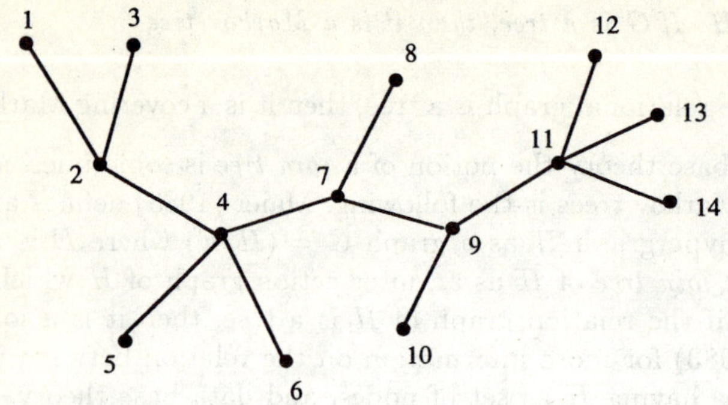

Fig. 8.9.

An ordering $e_1, e_2, \ldots, e_n$ of the edges of a graph $T = (V, E)$ is called a *tree construction sequence in edges* of T if for $i = 2, \ldots, n$, the edge e_i is a twig in the subtree T_i of T having edges $\{e_1, \ldots, e_i\}$ and nodes $e_1 \cup \ldots \cup e_i$. This name is justified because a tree is a graph for which a tree construction sequence in edges exists. The condition that $t \cap b \neq \emptyset$ says that a twig is connected with the rest of the graph. This implies that a graph obtained by adding a sequence of twigs is connected. On the other hand, condition (8.33) asserts that a twig is connected only through its branches, which implies that a graph build-up by adding a sequence of twigs is acyclic. This shows that if a tree construction sequence in edges exists, then the graph is a tree. Conversely, if the graph is a tree, then there is a tree construction sequence in nodes $v_1, \ldots, v_n$ with budding $b(i)$. Then $\{v_2, v_{b(2)}\}, \ldots, \{v_n, v_{b(n)}\}$ is a tree construction sequence in edges of the graph.

Since every twig has at least one branch, we can choose, for all $i = 1, \ldots, n$, an integer $b(i) < i$ such that $e_{b(i)}$ is a branch for e_i in the subtree $e_1 \cup \ldots \cup e_i$. Such a function is called a *branching* for E and $e_1, \ldots, e_n$.

In graph theory a *hypergraph* is a collection of subsets of a set. The elements of a hypergraph are called *hyperedges.* This generalizes the notion of a graph because edges are two-elements subsets of the set of nodes. Since a power set is a lattice, by analogy we also say that a subset of a lattice is a hypergraph. In particular, any collection of frames $H = \{\Theta_1, \ldots, \Theta_n\}$ in a f.c.f. $(\mathcal{F}, \mathcal{R})$ is a hypergraph because $(\mathcal{F}, \mathcal{R})$ is a lattice by theorem 7.6. In analogy to theorem 8.21, we say now in this context that a frame Θ_i in H is a *twig* if there is another frame Θ_j in H such that

$$\Theta_i \vee (\wedge \{\Theta_k : k \neq i\}) = \Theta_i \vee \Theta_j. \tag{8.34}$$

We call any frame Θ_j satisfying (8.34) a *branch* for the twig Θ_i.

A hypergraph $H \subseteq \mathcal{F}$ is called a *hypertree* if we can enumerate its frames in a sequence, say $\Theta_1, \ldots, \Theta_n$, so that Θ_i is a twig in $\{\Theta_1, \ldots, \Theta_i\}$ for all $i = 2, \ldots, n$. We call any such sequence a *hypertree construction sequence* for H

and Θ_1 is the *root* of the hypertree. Given such an ordering, for all $i = 2, \ldots, n$, there is an integer $b(i) < i$ such that $\Theta_{b(i)}$ is a branch for Θ_i in $\{\Theta_1, \ldots, \Theta_i\}$. Such a function $b(i)$ is called a *branching* for the sequence $\Theta_1, \ldots, \Theta_n$. A hypertree construction sequence may have more than one branching. Furthermore, for each frame of a hypertree there is at least one construction sequence beginning with this frame (see corollary 8.2 below). Note that in the definition of a twig we have not required that $\Theta_i \vee \Theta_j \neq 1$, a condition similar to $t \cap b \neq \emptyset$ (see theorem 8.21).

In multivariate f.c.f., if $\Theta_i = \Theta_{J_i}$ with $J_i \subseteq I$, then (8.34) is translated into

$$J_i \cap (\cup \{J_k : k \neq i\}) = J_i \cap J_j. \tag{8.35}$$

A hypergraph $J_1, \ldots, J_n$ satisfying condition (8.25) is called a hypergraph with the *running intersection property* (Lauritzen, Spiegelhalter, 1988), or also a *decomposable hypergraph* (Lauritzen et al., 1984; see also Leimer, 1985 and 1989).

Example 8.3. Consider the multivariate f.c.f. induced by the propositions and assumptions in the 14 clauses introduced in Sect. 5.1 to describe Pearl's alarm story (see also Chap. 6). Each of these clauses R_i defines a frame J_i, for example $J_1 = \{b, a_1, a\}$, $J_2 = \{e, a_1, a\}$ (more precisely the frame of discernment of R_i is the Boolean cube corresponding to the variables in J_i). So we have 14 frames of discernment and the following set of frames is a covering of this family:

$$\begin{aligned} K_1 = \{a, b, e, a_1, a_2\}, K_2 = \{a, a_5, d\}, K_3 = \{a, a_4, g\} \\ K_4 = \{a, a_3, w\}, K_5 = \{e, c, a_6\}. \end{aligned} \tag{8.36}$$

Then $(K_1, K_2, \ldots, K_5)$ is a hypertree construction sequence and $H = \{K_i : i = 1, \ldots, 5\}$ is a hypertree.

When a hypergraph $H = \{J_1, \ldots, J_n\}$ has not got too many hyperedges, the following algorithm of Graham is an easy test to know whether H is hypertree or not (see Graham (1979), Yu, Ozsoyoglu (1979), Maier (1983), Kruse et al. (1991)). The Graham reduction algorithm consists of a repeated application of two reduction rules to hypergraphs until neither can be applied any further. These two rules are

(1) If K and L are in H and $K \subseteq L$ then remove K from H.
(2) Let $J = \cup_{i=1}^{n} J_i$. If $j \in J$ appears in only one J_i then remove it from J_i (i.e. replace J_i by $J_i - \{j\}$ in the hypergraph H).

The Graham reduction is said to *succeed* on a hypergraph H if the result of applying the Graham reduction algorithm to H is the empty hypergraph. Before we are in a position to formulate a theorem about Graham algorithm and hypertrees, we first need to introduce the notion of a connected hypergraph. A *path* from J_r to J_s in H is a sequence of elements $E_1, \ldots, E_k$ in H such that $J_r = E_1, J_s = E_k$ and $E_i \cap E_j \neq \emptyset$ for all $1 \leq i < k$. The hypergraph H is called *connected* if there is a path between any two of its elements. The following theorem asserts that the Graham reduction algorithm can be used to test whether a hypergraph H is a hypertree.

Theorem 8.22 *Let H be a connected hypergraph. The following three conditions are equivalent:*

(1) H is a hypertree
(2) Graham's reduction algorithm succeeds on H
(3) H has a join tree.

The notion of a hypertree (and hence of a Markov tree) is also strongly related to the so-called acyclic hypergraphs or acyclic data base schemes (see Maier (1983), Fagin (1983)). It is also worth consulting the paper of Tarjan, Yannakakis (1984) for more information on this topic.

Theorem 8.23. *An ordering $\Theta_1, \ldots, \Theta_n$ of the frames in H is a hypertree construction sequence of H with branching b if and only if $(\Theta_i, \Theta_j) \dashv \Theta_{b(i)}$ for all $i = 2, \ldots, n$ and all $j < i$.*

This theorem characterizes hypertrees, which is an algebraic property (see (8.34)) in terms of conditional independence of frames. The next two theorems assert that hypertrees and Markov trees indeed represent the same structure.

Theorem 8.24. *Suppose that $T = (N, E)$ is a Markov tree. Then N is a hypertree and any leaf in T is a twig in N. If $\Theta_1, \ldots, \Theta_n$ is a tree construction sequence of T with budding b, then $\Theta_1, \ldots, \Theta_n$ is a hypertree construction sequence with branching b.*

Theorem 8.25. *Let $\Theta_1, \ldots, \Theta_n$ be a hypertree construction sequence with branching b for the hypertree $N \subseteq \mathcal{F}$. If $E = \{\{\Theta_2, \Theta_{b(2)}\}, \ldots, \{\Theta_n, \Theta_{b(n)}\}\}$, then $T = (N, E)$ is a Markov tree and $\Theta_1, \ldots, \Theta_n$ is a tree construction sequence for T with b as its budding.*

Corollary 8.2 *If H is a hypertree, then for every frame $\Theta \in H$ there is at least one hypertree construction sequence of H starting with Θ.*

If N is a hypertree and $T = (N, E)$ is a Markov tree, then T is called a Markov tree *representative* of the hypertree N. By theorem 8.25 every hypertree has at least one Markov tree representative, but in general there are more than one.

If we take the branching $b(K_i) = K_1$ for all $i = 2, \ldots, 5$ for the hypertree H given in example 8.3, then by theorem 8.25 the tree pictured in Fig. 8.10 is a Markov tree representative of H and also a covering Markov tree of the original set of frames induced by the 14 clauses in Pearl's alarm story. Then the Markov tree algorithm can be used to propagate the precise hints on the assumptions. This method is therefore an alternative to the logic-based procedure described in Chap. 5.

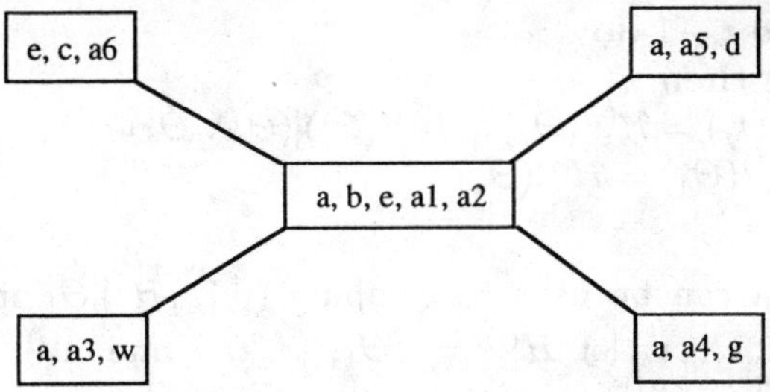

Fig. 8.10.

The following theorem is useful to prove the correctness of the algorithm based on the hypertree structure which can be used to compute the transport of the combination of several hints (see below).

Theorem 8.26.
(1) Suppose that $\Theta_1 \leq \Theta_2 \leq \Theta_3$ *and let* $\mathcal{H}$ *be a hint on* Θ_1. *Then*

$$\mathcal{H}|\Theta_3 = (\mathcal{H}|\Theta_2)|\Theta_3. \tag{8.37}$$

(2) Let $\mathcal{H}_1$ *be a hint on* Θ_1 *and* $\mathcal{H}_2$ *a hint on* Θ_2. *Then*

$$(\mathcal{H}_1 \oplus \mathcal{H}_2)|\Theta_1 = \mathcal{H}_1 \oplus [\mathcal{H}_2|(\Theta_1 \vee \Theta_2)]. \tag{8.38}$$

The second assertion of this theorem is important because it says that the computation of $(\mathcal{H}_1 \oplus \mathcal{H}_2)|\Theta_1$ can actually be performed on Θ_1 and not on the finer frame $\Theta_1 \wedge \Theta_2$.

We have the following theorem (Shafer, 1991).

Theorem 8.27. *For* $j = 1, \ldots, n$, *let* $\mathcal{H}_j$ *be a hint on* Θ_j *and let* $\mathcal{H} = \oplus_{j=1}^{n} \mathcal{H}_j$. *Let* Θ_t *be a twig in* $\{\Theta_1, \ldots, \Theta_n\}$ *with branch* Θ_b. *If* $\Lambda = \wedge_{j \neq t} \Theta_j$, *then*

$$\mathcal{H}|\Lambda = (\mathcal{H}_b \oplus [\mathcal{H}_t|\Theta_t \vee \Theta_b]) \oplus (\oplus \{\mathcal{H}_j : j \notin \{t, b\}\}). \tag{8.39}$$

This result is most valuable when applied to a hypertree. Suppose that H is a hypertree and that $\Theta_1, \ldots, \Theta_n$ is a hypertree construction sequence of H with branching b. For each $i = 1, \ldots, n$ we consider the sequence of hints $\mathcal{H}^{(i)}(\Theta_1), \ldots, \mathcal{H}^{(i)}(\Theta_i)$, where $\mathcal{H}^{(i)}(\Theta_j)$, $j \leq i$ is a hint on Θ_j to be defined by the algorithm given below. If $\mathcal{H}_j$ is our initial hint on Θ_j, then we compute these sequences of hints according to the following algorithm.

Algorithm hypertree;
begin
$\mathcal{H}^{(n)}(\Theta_j) := \mathcal{H}_j$ for all $j = 1, \ldots, n$;
For $i = n$ **downto** 2 **do**
For $j = 1$ **to** $i - 1$ **do**
if $j = b(i)$ **then**
$\mathcal{H}^{(i-1)}(\Theta_j) = \mathcal{H}^{(i)}(\Theta_j) \oplus \mathcal{H}^{(i)}(\Theta_i)|(\Theta_i \vee \Theta_j)$
else $\mathcal{H}^{(i-1)}(\Theta_j) = \mathcal{H}^{(i)}(\Theta_j)$
end.

This algorithm can be used to compute $(\oplus_{j=1}^{n}\mathcal{H}_j)|\Theta_1$ more efficiently. In fact, for $i = 1, \ldots, n$, let $H^{(i)} = \{\Theta_1, \ldots, \Theta_i\}$ and $\Lambda^{(i)} = \wedge_{j\leq i}\Theta_j$ and $\mathcal{H}^{(i)} = \oplus_{j\leq i}\mathcal{H}^{(i)}(\Theta_j)$ ($\mathcal{H}^{(i)}$ is a hint on $\Lambda^{(i)}$). Since Θ_i is a twig in $H^{(i)}$ with branch $\Theta_{b(i)}$, the algorithm and theorem 8.27 imply that

$$\mathcal{H}^{(i)}|\Lambda^{(i-1)} = \mathcal{H}^{(i-1)} \tag{8.40}$$

for all $i = n, \ldots, 2$. On the other hand, we have

$$\Lambda^{(n)} \leq \Lambda^{(n-1)} \leq \ldots \leq \Lambda^{(2)} \leq \Lambda^{(1)} = \Theta_1 \tag{8.41}$$

and

$$\mathcal{H} = \oplus_{i=1}^{n}\mathcal{H}_i = \oplus_{j=1}^{n}\mathcal{H}^{(n)}(\Theta_j) = \mathcal{H}^{(n)} \tag{8.42}$$

by the initialization of the algorithm. Then theorem 8.26 (1) implies that

$$\begin{aligned}\mathcal{H}|\Theta_1 &= (\mathcal{H}|\Lambda^{(n-1)})|\Theta_1 = (\mathcal{H}^{(n)}|\Lambda^{(n-1)})|\Theta_1 \\ &= \mathcal{H}^{(n-1)}|\Theta_1\end{aligned} \tag{8.43}$$

by (8.40). But again by theorem 8.26 (1), we have

$$\mathcal{H}^{(n-1)}|\Theta_1 = (\mathcal{H}^{(n-1)}|\Lambda^{(n-2)})|\Theta_1 = \mathcal{H}^{(n-2)}|\Theta_1 \tag{8.44}$$

by (8.40). Continuing in this way we obtain $\mathcal{H}^{(2)}|\Theta_1 = \mathcal{H}^{(1)}|\Theta_1$, and thus finally

$$\begin{aligned}\mathcal{H}|\Theta_1 &= \mathcal{H}^{(1)}|\Theta_1 = \mathcal{H}^{(1)}(\Theta_1)|\Theta_1 \\ &= \mathcal{H}^{(1)}(\Theta_1)\end{aligned} \tag{8.45}$$

because by definition $\mathcal{H}^{(1)} = \mathcal{H}^{(1)}(\Theta_1)$ and $\mathcal{H}^{(1)}(\Theta_1)$ is a hint on Θ_1. Thus we have the following theorem:

Theorem 8.28. *For $j = 1, \ldots, n$, let $\mathcal{H}_j$ be a hint on Θ_j and let $\mathcal{H} = \oplus_{j=1}^{n}\mathcal{H}_j$. Then*

$$\mathcal{H}|\Theta_1 = \mathcal{H}^{(1)}(\Theta_1). \tag{8.46}$$

When the hypergraph $H \subseteq \mathcal{F}$ induced by the hints we want to combine is a hypertree, then theorem 8.28 says that we can compute the transport of the combined hint to any frame in H by *local computations.* This results in a considerable gain in the computational effort. In the algorithm note that the hint $\mathcal{H}^{(i)}(\Theta_i)|(\Theta_i \vee \Theta_j)$ can be replaced by $\mathcal{H}^{(i)}(\Theta_i)|\Theta_j$ because of theorem 8.26 (2) and $\Theta_i - \Theta_j \wedge \Theta_i - \Theta_j$ is a Markov tree. Thus we just have to combine $\mathcal{H}^{(i)}(\Theta_j)$ with the transport of $\mathcal{H}^{(i)}(\Theta_i)$ to Θ_j. However, it is simpler to transit by the frame $\Theta_i \vee \Theta_j$ than the frame $\Theta_i \wedge \Theta_j$. In the algorithm each iteration i can be thought of as a deletion of the frame Θ_i from the hypertree $H^{(i)} = \{\Theta_1, \ldots, \Theta_i\}$, leaving a new collection of hints on the smaller hypertree $H^{(i-1)}$.

Example 8.4. Consider the multivariate f.c.f. induced by the variables $X_1, \ldots, X_9$. Within this f.c.f., we consider the frames $\Theta_1, \ldots, \Theta_9$ defined by the following subsets $J_1, \ldots, J_9$ of $I = \{1, \ldots, n\}$ respectively:

$$\begin{aligned} &\{1,2\}, \{3,4\}, \{2,3,4\}, \{2,4\}, \{2,5,6\} \\ &\{2,5\}, \{5,6,7\}, \{2\}, \{8,9\}. \end{aligned} \tag{8.47}$$

Then $H = \{\Theta_1, \ldots, \Theta_9\}$ is a hypertree and

$$(\Theta_3, \Theta_4, \Theta_5, \Theta_2, \Theta_7, \Theta_9, \Theta_8, \Theta_6, \Theta_1) \tag{8.48}$$

is a hypertree construction sequence with branching

$$(\Theta_3, \Theta_3, \Theta_3, \Theta_5, \Theta_4, \Theta_4, \Theta_5, \Theta_3), \tag{8.49}$$

i.e. the branch of Θ_4 is Θ_3, the branch of Θ_5 is Θ_3 etc. The corresponding Markov tree $T = (H, E)$ defined in theorem 8.25 is pictured in Fig. 8.11. Let $\mathcal{H}_i$ be a hint on Θ_i for $i = 1, \ldots, n$ and let $\mathcal{H} = \oplus_{i=1}^{9} \mathcal{H}_i$. Then this example clearly shows that the computation of $\mathcal{H}|\Theta_3$ is much simpler when using the hypertree algorithm than the straight forward method which needs the combination of hints on $\wedge_{i=1}^{9} \Theta_i = \{1, \ldots, 9\}$. For example, the deletion of Θ_9 requires the combination of $\mathcal{H}_3$ and $\mathcal{H}_1|\Theta_3$, an operation which can be performed on $\Theta_3 = \{2,3,4\}$ once $\mathcal{H}_1$ is first marginalized to $\Theta_1 \vee \Theta_3 = \{2\}$ and then extended to Θ_3. Compare also this procedure with the local propagation method in the Markov tree T given in Fig. 8.11 a) (see the description of the computations at the end of Sect. 8.2).

In the multivariate case, the *fusion algorithm* of Shenoy (1991 a) is a special case of the hypertree algorithm, the only difference is that the hypertree structure is implicit in the fusion algorithm. In fact, the hypertree is implicitly induced by an arbitrary ordering of the variables, each step of the algorithm corresponding to the elimination of a variable (see Sect. 8.4). This is the reason why the method is sometimes also called the *peeling algorithm* (Almond, Kong, 1991; Cannings et al., 1978). Furthermore, it can be verified that the algorithm proposed by Kong (1986) for combining hints with local computations in multivariate f.c.f. is a special case of the hypertree algorithm. Kong's algorithm also sequentially removes variables from a graph. The method proposed by Lauritzen, Spiegelhalter (1988) for computing marginal probabilities in conditional probabilistic networks is essentially the hypertree algorithm.

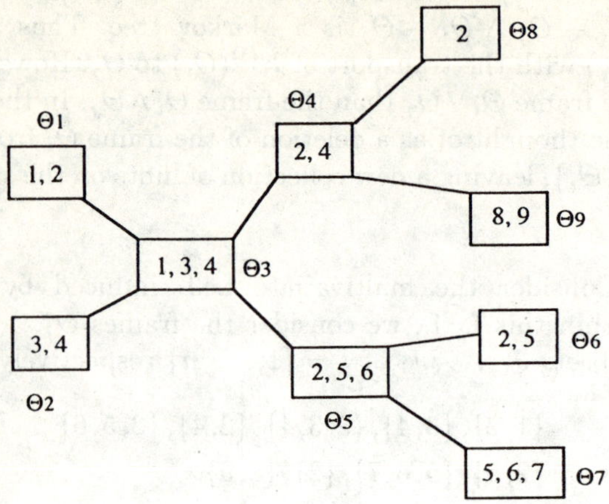

Fig. 8.11.

This method is used on the commercial software package HUGIN. The next section contains more information on the relation between elimination of variables and constructions of hypertrees or Markov trees.

8.4 Triangulated Graphs, Hypergraphs and Markov Trees

In this section two algorithms for finding covering Markov trees for multivariate f.c.f. will be presented. No optimality issues are considered here: refer to the next section for such questions. Both algorithms are based on graph theoretical notions, especially triangulated graphs. So we begin by introducing some graph theoretical terminology. Given a hypergraph $H = \{J_1, \ldots, J_n\}$, the graph $G(H) = (V, E)$ with $V = \cup_{i=1}^{n} J_i$ and

$$E = \{\{u, v\} : \exists J_i \in H \text{ such that } \{u, v\} \subseteq J_i\} \tag{8.50}$$

is called the *two-section graph* of H. For example, the two-section graph of the hypergraph

$$\{1,2\}, \{1,2,4\}, \{1,2,5\}, \{3,4,7\}, \{5,6,8\}, \{7,8,9\} \tag{8.51}$$

is given in Fig. 8.12.

A *complete subgraph* of a graph G is a set of nodes (vertices) all of which are pairwise adjacent, a node u being called *adjacent* to another node v if $\{u, v\}$ is an edge of G. Note that the termimology "complete subgraph" is abusive because a complete subgraph is defined to be a set of nodes and not a set of nodes and edges, i.e. a true subgraph of G. But this abuse of language is

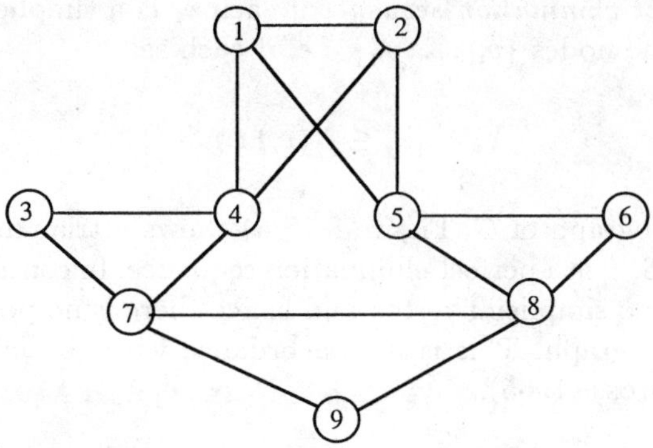

Fig. 8.12.

helpful and should not be disturbing. A *clique* is a maximal complete subgraph of G. For example, the cliques of the two-section graph given in Fig. 8.12 are the sets of vertices given in (8.51), except $\{1, 2\}$ which is a complete subgraph but not maximal. Mellouli (1988) proves the following theorem.

Theorem 8.29. *Let H be a hypergraph and H' the hypergraph formed by all cliques of the two-section graph $G(H)$. Then the class of all covering Markov trees of H coincides with the class of all covering Markov trees of H'.*

According to this result, finding covering Markov trees of the original hypergraph H is the same as finding covering Markov trees of the cliques of $G(H)$.

A graph $G = (V, E)$ is said to be *triangulated* if there does not exist a cycle of length four or above which does not have a *chord*, i.e. an edge connecting two non-adjacent vertices in a cycle. The graph in Fig. 8.12 is not triangulated because the cycle $2, 4, 7, 8, 5, 2$ has no chord. In the literature triangulated graphs have also been called *chordal, rigid-circuit, monotone transitive* and *perfect elimination* graphs (Golumbic, 1980). For a vertex $v \in V$, the nodes in

$$N(v) = \{w \in V : \{v, w\} \in E\} \tag{8.52}$$

are called the *neighbours* of v. Then the *closure* of v in G is defined as $Cl(v) = N(v) \cup \{v\}$. The closure of v is a clique of G. When we need to be more precise, we write $N(v, G)$ and $Cl(v, G)$ instead of $N(v)$ and $Cl(v)$. A vertex v is called *simplicial* if $N(v)$ is a complete subgraph of G. Dirac (1961) showed the following theorem.

Theorem 8.30 *Every triangulated graph $G = (V, E)$ has a simplicial vertex. Moreover, if G is not a complete graph, then it has two non-adjacent simplicial vertices.*

Let $\sigma = [v_1, \ldots, v_n]$ be an ordering of the vertices of a graph G. Then σ is called a *perfect elimination sequence* if each v_i is a simplicial vertex of the subgraph over the nodes $\{v_i, \ldots, v_n\}$, i.e. if each set

$$X_i = \{v_j \in N(v_i) : j > i\} \tag{8.53}$$

is a complete subgraph of G. Figure 8.13 a) shows a triangulated graph and $\sigma = [1, 2, 3, 4, 5, 6, 7]$ is a perfect elimination sequence. In contrast, the graph in Fig. 8.13 b) has no simplicial vertex and hence there is no perfect elimination sequence for this graph. This is in concordance with the following result of Fulkerson and Gross (1965).

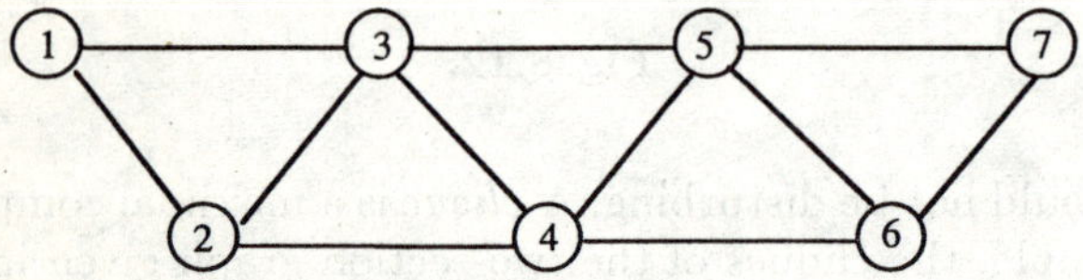

a)

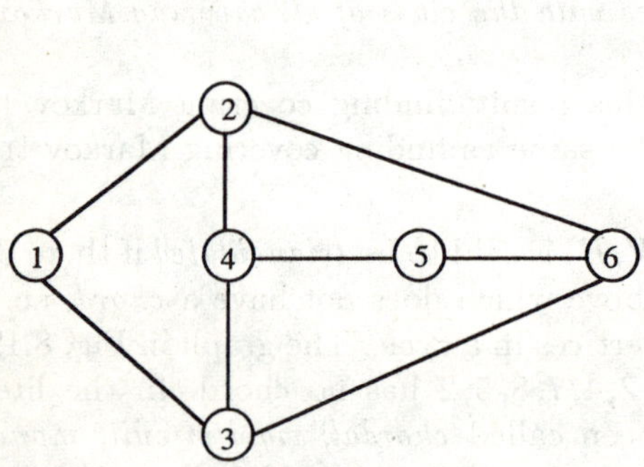

b)

Fig. 8.13.

Theorem 8.31. *A graph G is a triangulated graph if and only if it has a perfect elimination sequence. Moreover, any simplicial vertex can start a perfect elimination sequence.*

Triangulated graphs have been extensively studied in the literature and the following theorem is a typical result in this field (Walter, 1972; Gavril, 1974;

Buneman, 1974). For further results consult the excellent book of Golumbic (1980).

Theorem 8.32. *Let G be a graph. G is a triangulated graph if and only if there is a spanning tree $T = (\mathcal{C}, E)$ over the set of cliques $\mathcal{C}$ of G such that the subgraph of T over*

$$\mathcal{C}(v) = \{C \in \mathcal{C} : v \in C\} \tag{8.54}$$

is connected, and hence is a tree, for all vertices v of G.

The following theorem suggests that results about triangulated graphs may help to design procedures for finding covering Markov trees. For the rest of this section we follow the presentation given by Almond and Kong (1991). In particular, the two procedures given below to construct covering Markov trees are taken from this paper (see also Jensen, 1988).

Theorem 8.33. *Let G be a graph. Then the cliques of G can be arranged according to a Markov tree if and only if G is a triangulated graph.*

In view of theorem 8.31 the basic technique to triangulate a graph $G = (V, E)$ is to take an arbitrary elimination sequence σ of V and make it a perfect elimination sequence by adding the necessary edges to E (see the example 8.5 below). The sequence σ is then a perfect elimination sequence of the obtained triangulated graph G_σ. These added edges are called *fill-in edges* induced by the elimination sequence σ. The set of all fill-in edges is called the *triangulation* of G induced by σ. More generally, a triangulation of $G = (V, E)$ is a set of edges F such that $G'(V, E \cup F)$ is triangulated.

Example 8.5. Let's illustrate this triangulation technique on the non-triangulated graph $G = (V, E)$ given in Fig. 8.12 using the elimination sequence $\sigma = [9, 6, 3, 8, 7, 5, 4, 2, 1]$ (Almond, Kong, 1991). For $i = 1, \ldots, n$, let $G_i = (V_i, E_i)$ denote the graph when the vertices $[v_1, v_2, \ldots, v_{i-1}]$ have been sequentially eliminated from G and where E_i contains the necessary fill-in edges to make each vertex v_i a simplicial vertex in G_i. For each vertex v_i, Fig. 8.14 shows G_i, $N(v_i, G_i)$, and $Cl(v_i, G_i)$ (leave aside the last column for the moment). Fill-in edges added during the elimination process are drawn dotted in Fig. 8.14. The set of filled-in edges $F = \{\{7, 5\}, \{4, 5\}\}$ is then a triangulation and $G_\sigma = (V, E \cup F)$ is a triangulated graph.

Given an arbitrary elimination sequence σ, let G_σ denote the induced triangulated graph and let $G_{i\sigma} = (V_i, E_{i\sigma})$ denote the remaining graph when nodes $[v_1, \ldots, v_{i-1}]$ have been sequentially removed from G_σ. Then $V_i = \{v_i, \ldots, v_n\}$ and by construction $Cl(v_i, G_{i\sigma}) = Cl(v_i, G_i)$ (this observation will often be used in the sequel).

Theorem 8.34 *The cliques of the triangulated graph G_σ induced by the elimination σ are the maximal sets (with respect to set inclusion) among the closures $\{Cl(v_i, G_{i\sigma}) : i = 1, \ldots, n\}$.*

For example, the cliques of the graph G_σ defined in example 8.5 are

$$\begin{array}{c}\{1,2,4,5\}, \{4,5,7\}, \{5,7,8\} \\ \{3,4,7\}, \{5,6,8\}, \{7,8,9\},\end{array} \tag{8.55}$$

they are easily derived from the fifth column in Fig. 8.14 because $Cl(v_i, G_{i\sigma}) = Cl(v_i, G_i)$.

Now, given a graph G and an arbitrary elimination sequence

$$\sigma = [v_1, \ldots, v_n] \tag{8.56}$$

we define a procedure for building a spanning tree over the cliques of the triangulated graph G_σ. This procedure will be used to find a covering Markov tree for any hypergraph H. It will be referred to as the *Tree-Alg* procedure.

To construct a tree over the cliques of G_σ, the procedure Tree-Alg builds a sequence of trees $T_n, \ldots, T_1$ corresponding to the elimination sequence σ in the reverse order: first v_n, then v_{n-1}, ..., then v_1. The first tree T_n is the single node $\{v_n\}$. Then, given a tree T_{i+1}, the tree T_i is constructed as follows. According to theorem 8.35 below, there is always at least one node $\mathcal{N}$ in T_{i+1} such that $N(v_i, G_{i\sigma}) \subseteq \mathcal{N}$. Arbitrarily select such a node $\mathcal{N}$. If $N(v_i, G_{i\sigma}) = \mathcal{N}$, then T_i is obtained by replacing $\mathcal{N}$ by $Cl(v_i, G_{i\sigma})$ in T_{i+1}. Otherwise, if $N(v_i, G_{i\sigma}) \subset \mathcal{N}$, then T_i is obtained from T_{i+1} by creating a new node $Cl(v_i, G_{i\sigma})$ and by linking it to $\mathcal{N}$. The result of the Tree-Alg procedure is then defined to be the tree T_1.

Since $Cl(v_i, G_{i\sigma}) = Cl(v_i, G_i)$, the tree produced by Tree-Alg for the graph G defined in example 8.5 can easily be constructed from the fifth column in Fig. 8.14 : the sequence of trees $T_n, \ldots, T_1$ is given in the sixth column of Fig. 8.14 (remember that the sequence starts with the tree T_n which is at the bottom of the column). The construction of the tree T_1 produced by the Tree-Alg procedure is always possible as the following theorem shows:

Theorem 8.35. *For all $i \leq n-1$, there is a node $\mathcal{N}$ in T_{i+1} such that $N(v_i, G_{i\sigma}) \subseteq \mathcal{N}$.*

As already mentioned, the result of the Tree-Alg procedure is a spanning tree over the cliques of G_σ:

Theorem 8.36. *The nodes of the tree T_1 produced by the Tree-Alg procedure are exactly the cliques of the triangulated graph G_σ.*

Furthermore, we have the following important theorem:

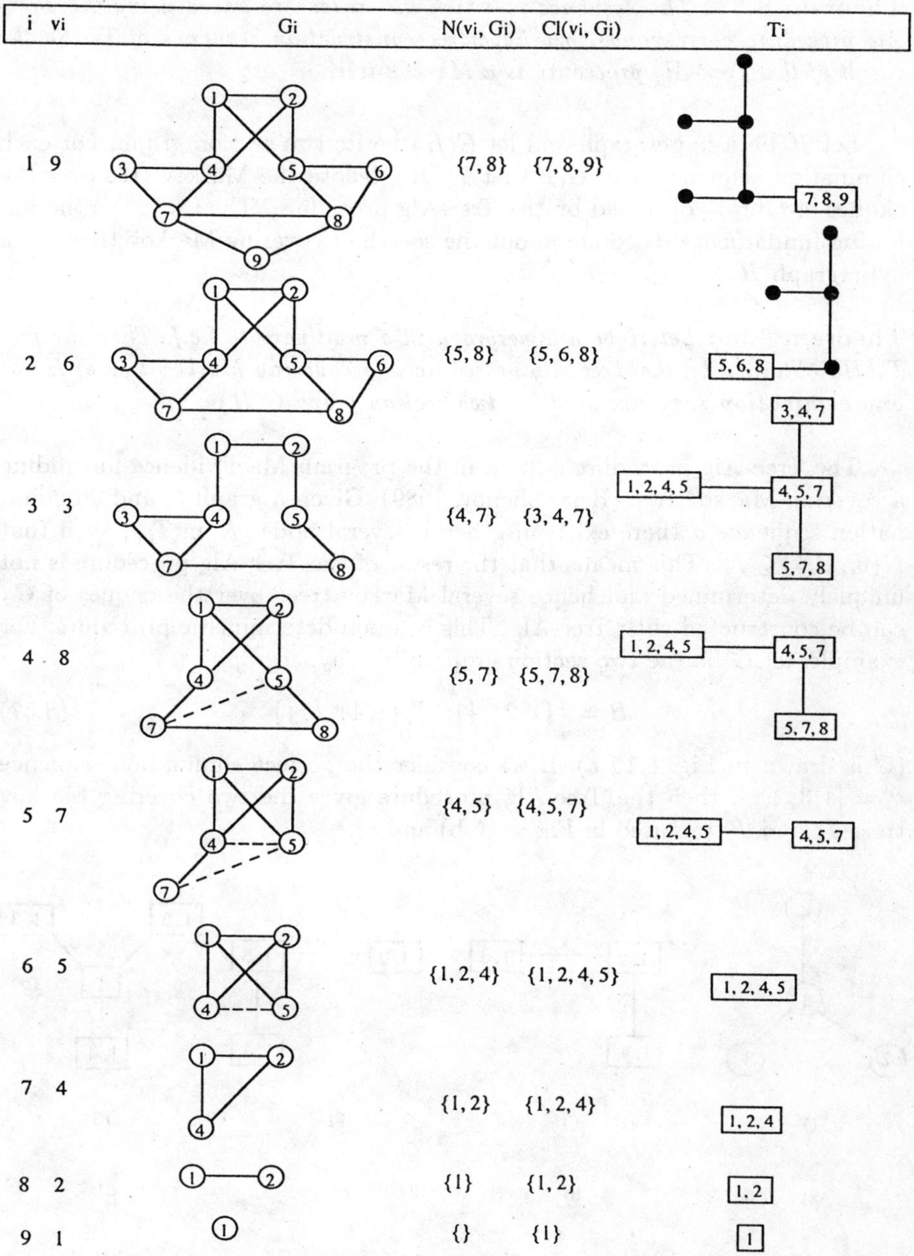
i vi
Gi
N(vi, Gi)
Cl(vi, Gi)
Ti
1 9
{7, 8}
{7, 8, 9}
7, 8, 9
2 6
{5, 8}
{5, 6, 8}
5, 6, 8
3 3
{4, 7}
{3, 4, 7}
3, 4, 7
1, 2, 4, 5
4, 5, 7
5, 7, 8
4 8
{5, 7}
{5, 7, 8}
5 7
{4, 5}
{4, 5, 7}
6 5
{1, 2, 4}
{1, 2, 4, 5}
7 4
{1, 2}
{1, 2, 4}
1, 2, 4
8 2
{1}
{1, 2}
1, 2
9 1
{}
{1}
1

Fig. 8.14.

Theorem 8.37. *The sequence in which the nodes are added using the Tree-Alg procedure corresponds to a hypertree construction sequence of T_1. So the result of the Tree-Alg procedure is a Markov tree.*

Let H be a hypergraph and let $G(H)$ be its two-section graph. For each elimination sequence σ of $G(H)$ let $T_\sigma(H)$ denote the Markov tree over the cliques of $G(H)_\sigma$ obtained by the Tree-Alg procedure. Then we have the following fundamental theorem about the search of covering Markov trees for a hypergraph H.

Theorem 8.38. *Let H be a hypergraph in a multivariate f.c.f. Then the tree $T_\sigma(H)$ obtained by the Tree-Alg procedure is a covering Markov tree of H for any elimination sequence σ of the two-section graph $G(H)$.*

The Tree-Alg procedure is used in the program MacEvidence for finding a covering Markov tree (Hsia, Shenoy, 1989). Given a graph G and an elimination sequence σ there exist, in general, several nodes $\mathcal{N}$ in T_{i+1} such that $N(v_i, G_{i\sigma}) \subseteq \mathcal{N}$. This means that the result of the Tree-Alg procedure is not uniquely determined and hence several Markov trees over the cliques of G_σ can be constructed with Tree-Alg. This is a non-deterministic procedure. For example, let G be the two-section graph of the hypergraph

$$H = \{\{1,2\},\{1,3\},\{1,4\},\{1\}\} \tag{8.57}$$

(G is drawn in Fig. 8.15 a). If we consider the perfect elimination sequence $\sigma = [4,3,2,1]$, then the Tree-Alg procedure gives the two covering Markov trees T_1 and T_2 pictured in Fig. 8.15 b) and c).

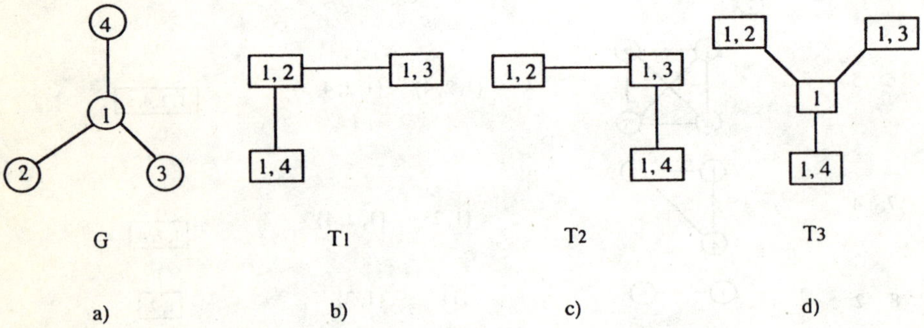

Fig. 8.15.

However, if $\mathcal{H}_1, \ldots, \mathcal{H}_4$ are four hints relative to the four hyperedges (8.57), then it is clear that for computing $(\oplus_{i=1}^4 \mathcal{H}_i)|\{1\}$ it is better to use the covering Markov tree T_3 given in Fig. 8.15 d) because Dempster's rule is actually performed on the smaller frame $\{1\}$. Since $\{1\}$ is not a clique of G, T_3 can never be found by the Tree-Alg procedure. Starting from the observation that $\{1\}$ is the intersection of the three cliques of G and $\{1\}$ is the frame which is added to T_1 and T_2 to get T_3, Jensen (1988) proposes the following method of constructing Markov trees. Starting from a graph G, find the cliques of G and for any pair of cliques C_1, C_2 having a non-empty intersection C, link C_1 to C and C_2 to C. This gives the so-called *junction graph* of G. Fig. 8.16 a) shows the junction graph of the graph G given in Fig. 8.15 a) (the intersection nodes are marked with squared boxes). Jensen then defines a *junction tree* as a Markov spanning tree of the junction graph . Fig. 8.16 b) shows a junction tree obtained from the junction graph given in Fig. 8.16 a). This shows that a frame may appear more than once in a junction tree.

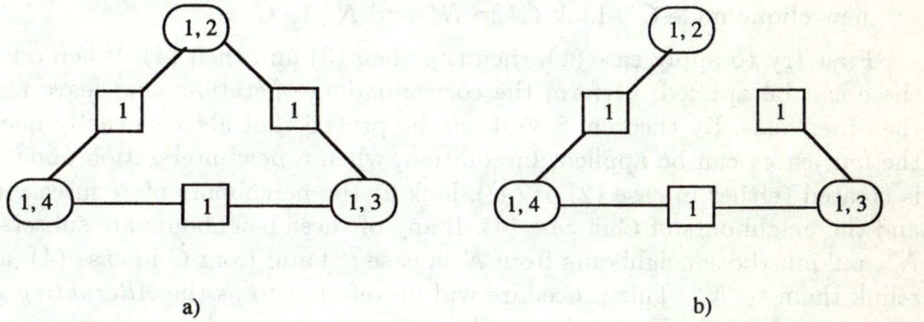

Fig. 8.16.

Remark that if the junction graph is not connected, then no junction tree can be formed from it. Mellouli (1988) calls a spanning tree over the hyperedges of a hypergraph *alternating* if for every pair of neighbouring nodes, one is a subset of the other. Then of course every junction tree is an alternating Markov tree. The tree T_3 in Fig. 15 d) is an alternating Markov tree which is not a junction tree derived from the graph G in Fig. 8.15 a). This shows that the class of alternating covering Markov trees is strictly larger than the class of junction trees. According to Almond and Kong (1991), it is often the case that the use of a well chosen alternating covering Markov tree requires less computational effort than a covering Markov tree found by the Tree-Alg procedure. They propose the following modification of the Tree-Alg procedure to find good covering alternating Markov trees.

Building this tree requires the introduction of an artificial distinction between *clique nodes* (shown in figures with round boxes) and *intersection nodes* (shown with square boxes). Once again, the procedure starts with an elimination sequence $\sigma = [v_1, \ldots, v_n]$ of the nodes of a graph G, which in many cases is the two-section graph of a hypergraph. Just as in the Tree-Alg procedure, once every vertex has been eliminated, one moves backwards to build a sequence of trees $T_n, \ldots, T_1$. The first tree T_n is the single clique node $\{v_n\}$. Then, given a tree T_{i+1}, we construct the tree T_i as follows. Let $N^* = N(v_i, G_i)$ and

$C^* = Cl(v_i, G_i)$ with G_i defined as in example 8.5. There are four distinct cases:

(1) If N^* is an existing intersection node in T_{i+1}, then create the new clique node C^* and link it to N^*.
(2) If N^* is a strict subset of an intersection node in T_{i+1}, then let N be the smallest among such intersection nodes (break ties arbitrarily). Create a new intersection node N^* and a new clique node C^*. Link C^* to N^* and N^* to N.
(3) If N^* is an existing clique node in T_{i+1}, then replace it by the clique node C^*.
(4) If N^* is a proper subset of an existing clique node C in T_{i+1} (but not of an existing intersection node), create a new intersection node N^* and a new clique node C^*. Link C^* to N^* and N^* to C.

First try to apply case (1), then (2), then (3) and then (4). When one of these can be applied, perform the corresponding operations and leave aside the other ones. By theorem 8.35 it can be proved that always exactly one of the four cases can be applied. In addition, when a new intersection node N^* is created (either in case (2) or (4)), look at the neighbours of N in case (2) and the neighbours of C in case (4). If any of these neighbours are subsets of N^*, unkink those neighbours from N in case (2) and from C in case (4), and relink them to N^*. This procedure will be referred to as the *Alternating-Alg* procedure. Like the Tree-Alg procedure the Alternating-Alg procedure is not completely determined (e.g. the node C in case (4) may not be unique). So in general several trees can be constructed by Alternating-Alg. Figure 8.17 shows the alternating covering Markov tree derived from the top left triangulated graph using the perfect elimination sequence $\sigma = [1,2,3,4,5,6]$. Note that when passing from T_3 to T_2 the link joining $\{6\}$ to $\{3,5,6\}$ is redirected to $\{5,6\}$.

Theorem 8.39. *Let T be a tree produced by the Alternating-Alg procedure using the elimination sequence σ on the graph G. Then*

(1) T is an alternating Markov tree whose clique nodes are the cliques of the induced triangulated graph G_σ and each intersection node is the intersection of some pair of cliques of G_σ.
(2) The frames in T are all different and the neighbours of a clique node are all intersection nodes.
(3) If G is the two-section graph of a hypergraph H, then T is a covering Markov tree of H.

Almond and Kond (1991) give other properties of Markov trees produced by the Alternating-Alg procedure.

To conclude this section, let us mention that Shenoy, Shafer (1990) explain how to expand the Markov tree structure into an architecture for parallel computation. Each vertex of the tree is thought of as a processor and the propagation (transport) of hints is considered as message passing from one

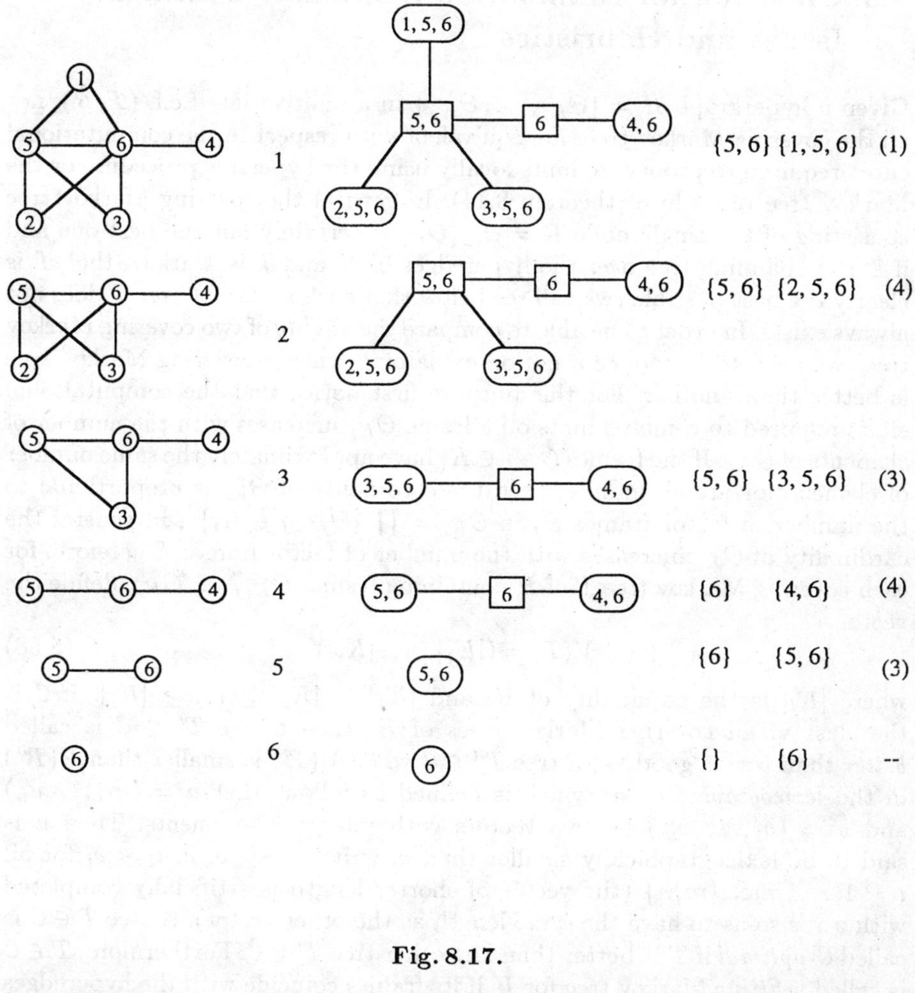

Fig. 8.17.

processor to another. So far, the problem was to compute the marginalization of the combined hint to a single frame of the tree. When the marginalization has to be computed for all frames of the tree, Shenoy, Shafer (1990) show how this can be done efficiently with only slight changes in the architecture they propose for a single frame. The marginalization is computed for all frames simultaneously. The paper of Shenoy and Shafer discusses only the multivariate case. For a more general presentation with partitions refer to Shafer et al. (1987) and for an axiomatic study of these questions on lattices the paper of Shafer (1991) is an excellent reference.

8.5 Choosing an Elimination Sequence: Optimality Issues and Heuristics

Given a hypergraph $H = \{\Theta_{K_1}, \ldots, \Theta_{K_n}\}$ in a multivariate f.c.f. $(\mathcal{F}, \mathcal{R})$, not all the covering Markov trees are equivalent with respect to the computational effort required to propagate hints locally using the hypertree procedure or the Markov tree procedure (theorem 8.14). In general the covering Markov tree consisting of the single node $K = \cup_{i=1}^{n} \Theta_{K_i}$ is certainly not the best one and if T is a spanning tree over the hyperedges of H and T is Markov, then T is clearly the best possible (we will see below that such a Markov tree T does not always exist). In order to be able to compare the *quality* of two covering Markov trees we need to introduce a notion explaining when a covering Markov tree is better than another. For this purpose first notice that the computational effort required to combine hints on a frame Θ_{K_i} increases with the number of elements of Θ_{K_i}. If the frames $\Theta_j, j \in K_i$ have approximately the same number of elements or are all infinite, then the cardinality of Θ_{K_i} is proportional to the number of factor frames Θ_j in $\Theta_{K_i} = \prod \{\Theta_j : j \in K_i\}j$; in general the cardinality of Θ_{K_i} increases with the number of factor frames. Therefore, for each covering Markov tree T of H containing frames $\Theta_{K_1}, \ldots, \Theta_{K_r}$, define the vector

$$V(T) = (|K_1|, \ldots, |K_r|) \tag{8.58}$$

where $|K_i|$ is the cardinality of K_i and $|K_1| \geq |K_2| \geq \ldots \geq |K_1|$. If $\mathcal{C}$ is the class of all covering Markov trees of H, then a tree $T' \in \mathcal{C}$ is called *better* than (or as good as) a tree $T'' \in \mathcal{C}$ when $V(T')$ is smaller than $V(T'')$ in the *lexicographic* order which is defined as follows. Let $u = (u_1, \ldots, u_n)$ and $v = (v_1, \ldots, v_m)$ be two vectors with integer components. Then u is said to be lexicographically smaller than v, written $u \leq v$, if $u_i \leq v_i$ for all $i = 1, \ldots, \max \{m, n\}$ (the vector of shorter length is artificially completed with zeros so as to have the same length as the other vector). A tree $T \in \mathcal{C}$ is called *$\mathcal{C}$-optimal* if T is better than any other tree $T' \in \mathcal{C}$. Furthermore, $T \in \mathcal{C}$ is called a *fitting* Markov tree for H if its frames coincide with the hyperedges in H. Of course, any fitting tree is $\mathcal{C}$-optimal. However, the converse is not true because there is no fitting tree for $H = \{\{1,2\}, \{1,3\}, \{2,3\}\}$ whereas the trivial covering Markov tree $K = \{1,2,3\}$ is clearly $\mathcal{C}$-optimal.

A weaker notion of optimality is sometimes also considered. Indeed, it has been noticed that it is essentially the size of the largest frame, namely $|K_1|$, which is crucial for the computational effort when propagating in the tree. A tree $T \in \mathcal{C}$ for which the size of the largest frame is as small as possible is called *$\mathcal{C}$-first order optimal*, i.e. $|K_1| \leq |K'_1|$ for all $T' \in \mathcal{C}$ $(V(T') = (|K'_1|, \ldots, |K'_s|)$. Of course, if T is $\mathcal{C}$-optimal then it is $\mathcal{C}$-first order optimal.

Instead of trying to find optimal or first order optimal trees in the class $\mathcal{C}$ of all covering Markov trees, let's be modest and try to find optimal or first order optimal trees within the restricted class $\mathcal{C}_1$ of all covering Markov trees which can be produced by the procedure Tree-Alg. A tree $T \in \mathcal{C}_1$ which is optimal or first order optimal within $\mathcal{C}_1$ is called $\mathcal{C}_1$-optimal or $\mathcal{C}_1$-first order optimal.

The notions of $\mathcal{C}_2$-optimality and $\mathcal{C}_2$-first order optimality are defined similarly for the class $\mathcal{C}_2$ of all trees which can be produced by the Alternating-Alg procedure. All trees that can be produced by the Tree-Alg procedure using a fixed elimination sequence σ are equivalent in the lexicographic order because their nodes are the cliques of the triangulated graph $G(H)_\sigma$ induced by σ on the two-section graph $G(H)$ (theorem 8.31). However, different elimination sequences will in general induce non-equivalent trees because the triangulated graphs will be different. In consequence, certain elimination sequences are better than others. Arnborg et al. (1987) have shown that finding a sequence for which the size of the largest clique of $G(H)_\sigma$ is as small as possible is a NP-complete problem. This proves that finding a $\mathcal{C}_1$-first order optimal tree is a NP-complete problem. This is a rather pessimistic result which shows that finding optimal covering Markov trees is a difficult task. Moreover, theorem 8.39 (1) shows that if we apply the Alternating-Alg procedure to $G(H)$ using the elimination sequence σ we obtain a tree which is not better (in the lexicographic sense) than the tree obtained by the Tree-Alg procedure using σ (it may, however, be better for transporting and combining hints, see Fig. 8.15). Therefore, instead of trying to find optimal or first order optimal trees, we consider another criterion by looking for elimination sequences σ for which the corresponding set of filled-in edges of $G(H)$, denoted as $\Delta(\sigma)$, is as small as possible. It is hoped that if $\Delta(\sigma)$ has few edges, then the cliques of $G(H)_\sigma$ will be small and hence the tree produced by Tree-Alg using the elimination sequence σ will be good. We have the following theorem:

Theorem 8.40. *Let H be a hypergraph. If $G(H)$ is triangulated and σ is a perfect elimination sequence, then the tree $T_\sigma(H)$ generated by the Tree-Alg procedure is $\mathcal{C}$-optimal.*

Example 8.6. Consider the multivariate f.c.f. induced by the seven variables Revenue (R), Occupancy (O), Price (PRI), Variable Costs (VC), Unit Costs (UC), Fixed Costs (FC), Profit (P) introduced in the real estate management problem described in example 7.2. The three equations in (7.3) can be represented by three deterministic hints on the frames

$$\{R, O, PRI\}, \{VC, O, UC\}, \{FC, UC, R, PRO\}$$

and the influence of Price on Occupancy can be modelled by a simple hint on $\{PRI, O\}$ (see example 7.2). These four frames form a hypergraph H whose two-section graph is pictured in Fig. 8.18.

The graph $G(H)$ is triangulated and the elimination sequence

$$\sigma = (PRI, UC, O, VC, R, FC, PRO) \tag{8.59}$$

is perfect. Then by theorem 8.40 the tree $T_\sigma(H)$ generated by the Tree-Alg procedure is $\mathcal{C}$-optimal. This tree is pictured in Fig. 8.19.

Unfortunately, the problem of finding an elimination sequence such that $\Delta(\sigma)$ is as small as possible is also a difficult problem: Yannakakis (1981) has

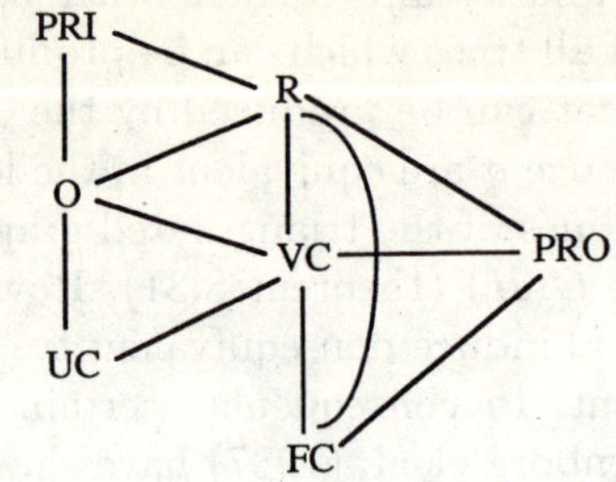

Fig. 8.18.

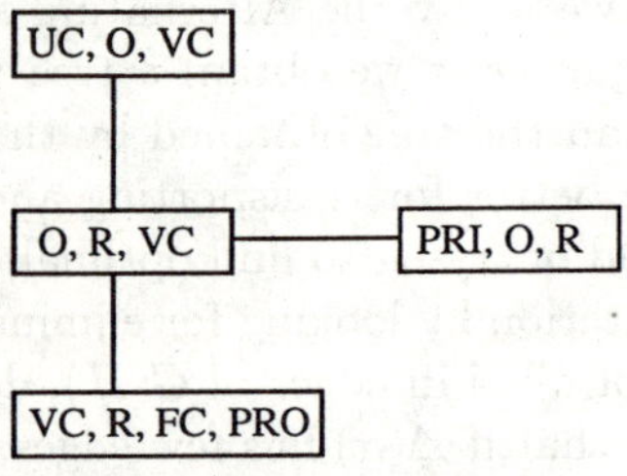

Fig. 8.19.

shown that finding a minimum fill-in is a NP-complete problem (a fill-in $\Delta(\sigma)$ is called minimum if $|\Delta(\sigma)| \leq |\Delta(\sigma')|$ for all elimination sequences σ'). If $G(H)$ is triangulated, then σ is a perfect elimination sequence of $G(H)$ if and only if $\Delta(\sigma)$ is a minimum fill-in (see theorem 8.31). Sometimes the notion of minimal fill-in is also considered: a fill-in $\Delta(\sigma)$ is called minimal if there is no sequence σ' such that $\Delta(\sigma') \subset \Delta(\sigma)$. Of course, every minimum fill-in is minimal but the converse is false in general. This dicussion shows that we need heuristics to determine elimination sequences to be used in the Tree-Alg procedure or the Alternating-Alg procedure.

The first heuristics we want to discuss is the Maximum Cardinality Search (MCS) introduced by Tarjan and Yannakakis (1984). The procedure works as follows. First give the number 1 to an arbitrary node. Then number the nodes consecutively, choosing as the next to number a node with a maximum number of previously numbered neighbours. Break ties arbitrarily. Figure 8.20 a) shows such a numbering (this graph is taken from Lauritzen, Spiegelhalter, 1988).

Its cliques are

$$\{\lambda, \beta, \sigma\}, \; \{\epsilon, \beta, \delta\}, \; \{\lambda, \epsilon, \beta\}, \; \{\tau, \lambda, \epsilon\}, \; \{\epsilon, \xi\}, \; \{\alpha, \tau\}. \tag{8.60}$$

Tarjan and Yannakakis (1984) have shown that if G is triangulated and if the cliques of G are ranked in ascending order according to the highest number in each clique, then we can arrange these cliques in a tree such that this

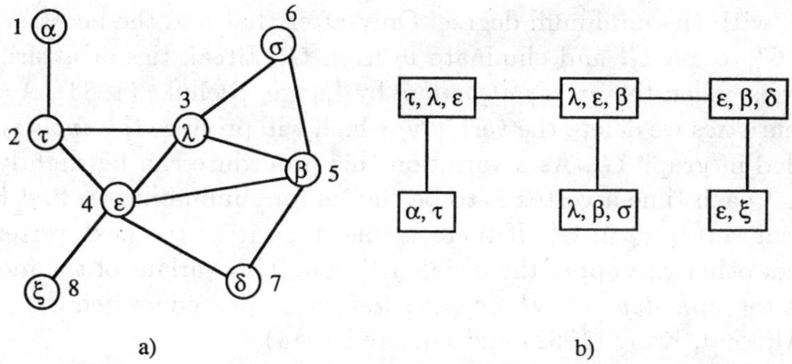

Fig. 8.20.

ranking is a hypertree construction sequence. In the example of Fig. 8.20 a), the highest number in the clique $\{\lambda, \beta, \sigma\}$ is 6, in $\{\epsilon, \beta, \delta\}$ it is 7, etc..., so that finally the cliques of (8.60) must be ranked in ascending order according to the vector (6, 7, 5, 4, 8, 2). This gives the following sequence of cliques which is a hypertree construction sequence:

$$\{\alpha, \tau\},\ \{\tau, \lambda, \epsilon\},\ \{\lambda, \epsilon, \beta\},\ \{\lambda, \beta, \sigma\},\ \{\epsilon, \beta, \delta\},\ \{\epsilon, \xi\}.$$

Then it is easy to construct a tree corresponding to this hypertree construction sequence. In general several trees can be constructed, one of them is given in Fig. 8.20 b) (note that the last clique $\{\epsilon, \xi\}$ could also have been linked to the node $\{\tau, \lambda, \epsilon\}$ or to $\{\lambda, \epsilon, \beta\}$, giving two other different trees. Now the question is whether these trees can be computed by our Tree-Alg procedure. Indeed this is possible. If u_1 is the node having number 1 according to the MCS procedure described above (for example $u_1 = \alpha$ in the graph of Fig. 8.20 a)), u_2 is the node with number 2, etc..., then the procedure Tree-Alg applied with the elimination sequence $\sigma = [u_n, u_{n-1}, \ldots, u_1]$ precisely gives the trees found with the hypertree construction sequence of the cliques. For example, applying Tree-Alg with the elimination sequence $\sigma = [\xi, \delta, \sigma, \beta, \epsilon, \lambda, \tau, \alpha]$ on the graph of Fig. 8.20 a) gives the tree of Fig. 8.20 b). Tarjan and Yannakakis (1984) also prove that if the graph is triangulated, then the sequence $\sigma = [u_n, u_{n-1}, \ldots, u_1]$ is a perfect elimination sequence and the resulting covering Markov tree is optimal according to theorem 8.40. The fill-in $\Delta(\sigma)$ corresponding to an elimination sequence found by MCS is in general neither minimal nor minimum. Finally, they have shown that MCS runs in O(n+e) time, where n is the number of nodes e is the number of edges of G.

Another heuristics called Minimum Degree works as follows (Bertele, Brioschi, 1972). For $i = 1, \ldots, n$, let G'_i denote the graph obtained from G when nodes $v_1, \ldots, v_{i-1}$ have been eliminated, but the necessary fill-in edges to make v_i a simplicial vertex in G_i have *not* yet been added. First eliminate the vertex with the minimum number of neighbours in G. Choose as the next vertex v_i to be eliminated the one having a minimum number of neighbours in G'_i, i.e. with the minimum degree. Only after that add the necessary fill-in edges to G'_i to get G_i and eliminate v_i from G_i. Break ties arbitrarily. This procedure is called the *greedy algorithm* by Lange, Blehnke (1983). According to this heuristics we delete the vertex v_i which will produce the *smallest clique* in the filled-in graph G_i. As a variation this procedure can be slightly modified in that each time a vertex is to be chosen for elimination we first look for a simplicial vertex v_i in G'_i. If there is one, take it as the next vertex to be eliminated, otherwise apply the usual criterion. This variant of the method is known as the *one-step look ahead, smallest clique* procedure defined by Kong (1986), Almond, Kong (1991) and Olmsted 1983).

A third heuristics suggested by Bertele, Brioschi (1972) called *Minimum Deficiency* also uses the graphs G'_i. Choose as the next vertex v_i to be eliminated the one which requires the minimum number of fill-in edges to make v_i a simplicial vertex in G_i. Almond and Kong (1991) call this heuristics *one-step look ahead, fewest fill-in.* Notice that if G is triangulated, then this procedure yields a perfect elimination sequence and the corresponding covering Markov tree is optimal by theorem 8.40.

There is another heuristics called *Lexicographic Search* proposed by Rose et al. (1976), see also Tarjan, Yannakakis (1984) and Wen (1991). It gives a mimimal fill-in but runs in $O(ne)$ time.

Although none of these heuristics are guaranteed to produce a $\mathcal{C}$-optimal covering Markov tree (see the counter-example in Almond, Kong (1991) borrowed from Bertele, Brioschi, 1972), they work well in most cases. The Minimum Deficiency heuristics gives particularly good covering Markov trees, but it is slower than the MCS heuristics. For more information on vertex elimination refer to Rose (1970, 1972, 1974) and Rose, Tarjan (1975, 1978).

On the other hand, notice that we only need to know the partial sequence $[v_1, \ldots, v_i]$ in order to compute $N(v_i, G_i)$ and $Cl(v_i, G_i)$, which are both used in the Tree-Alg and Alternating-Alg procedures. Therefore, as the sequence is determined stepwise, the necessary information for Tree-Alg and Alternating-Alg can be computed and recorded.

In our search for criteria to compare covering Markov trees we introduced the lexicographic order on the cliques of $G(H)_\sigma$, which in turn is defined via the *size* of a clique, namely the number of vertices in the clique. But we can envisage another way to compare covering Markov trees. Indeed, if $C = \{v_1, \ldots, v_k\} \subseteq I$ and each v_i corresponds to a frame $\Theta_i, i \in I$ having η_i elements, then we define the *weight* of C as

$$w(C) = ln_2 \,(\prod_{i=1}^{k} \eta_i). \tag{8.61}$$

In general, if $C_1, \ldots, C_r$ are the cliques of $G(H)_\sigma$, then its weight is

$$w(G(H)_\sigma) = \sum_{i=1}^{r} w(C_i). \tag{8.62}$$

Kjaerulff (1990) considers this criterion to compare elimination sequences and presents experimental results based on several heuristics, including those described above. Moreover, he gives a method to convert any non-minimal triangulation into a minimal one. He also presents a probabilistic algorithm based on simulated annealing to find a sequence giving the minimal weight, see also Thomas (1985). A similar criterion to $w(G(H)_\sigma)$ is considered in Wen (1991).

A8 Appendix: Proofs of Theorems in Chapter 8

Proof of Theorem 8.1

If $A \subseteq \Theta_3$, we must prove that $v(A) = v_1(v_2(A))$. Let X denote the set

$$\begin{aligned} v_1(v_2(A)) &= \{\theta_1 \in \Theta_1 : \tau_1(\theta_1) \cap v_2(A) \neq \emptyset\} \\ &= \{\theta_1 \in \Theta_1 : \exists\, \theta_2 \in \Theta_2 \text{ such that } \theta_2 \in \tau_1(\theta_1) \text{ and } \tau_2(\theta_2) \cap A \neq \emptyset\}, \end{aligned} \tag{A8.1}$$

and let Y denote the set

$$v(A) = \{\theta_1 \in \Theta_1 : \tau_2(\tau_1(\theta_1)) \cap A \neq \emptyset\}. \tag{A8.2}$$

If $\theta_1 \in X$, then there is $\theta_2 \in \Theta_2$ such that $\theta_2 \in \tau_1(\theta_1)$ and $\tau_2(\theta_2) \cap A \neq \emptyset$. Hence $\tau_2(\theta_2) \subseteq \tau_2(\tau_1(\theta_1))$ and $\tau_2(\tau_1(\theta_1)) \cap A \neq \emptyset$ which shows that $\theta_1 \in Y$.

Conversely, if $\theta_1 \in Y$, then select $x \in A$ such that $x \in \tau_2(\tau_1(\theta_1)) = \sum \{\tau_2(\theta_2) : \theta_2 \in \tau_1(\theta_1)\}$. It follows that there is exactly one $\theta_2 \in \tau_1(\theta_1)$ such that $x \in \tau_2(\theta_2)$ and therefore $\tau_2(\theta_2) \cap A \neq \emptyset$, which proves that $Y \subseteq X$ and hence $X = Y$. □

Proof of Theorem 8.2

Consider the frames and refinings given in Fig. A8.1.

According to theorems 7.5 and 8.1, we must prove that

$$v(v'(\lambda(\tau(F)))) = v(\tau(F)), \tag{A8.3}$$

which is true because $v' \circ \lambda = id$ by theorem 7.2 (11). □

Proof of Theorem 8.3

Let τ and μ denote the refinings used to define $F = \tau(F_1) \cap \mu(F_2)$ and let v be the outer reduction of the refining $\lambda : \Theta \to \Theta'$ (see Fig. A8.2).

If $\alpha : \Theta_1 \to \Theta'$ and $\beta : \Theta_2 \to \Theta'$ are the refinings used to compute $F' = \alpha(F_1) \cap \beta(F_2)$, then theorem 7.4 implies that $\lambda \circ \tau = \alpha$ and $\lambda \circ \mu = \beta$.

First, we prove that $F'|\Theta = F$. Using theorem 7.2 (4) and (11) we conclude that

$$\begin{aligned} F'|\Theta &= v(\lambda(\tau(F_1)) \cap \lambda(\mu(F_2))) \\ &\subseteq v(\lambda(\tau(F_1))) \cap v(\lambda(\mu(F_2))) \\ &= \tau(F_1) \cap \mu(F_2) = F. \end{aligned} \tag{A8.4}$$

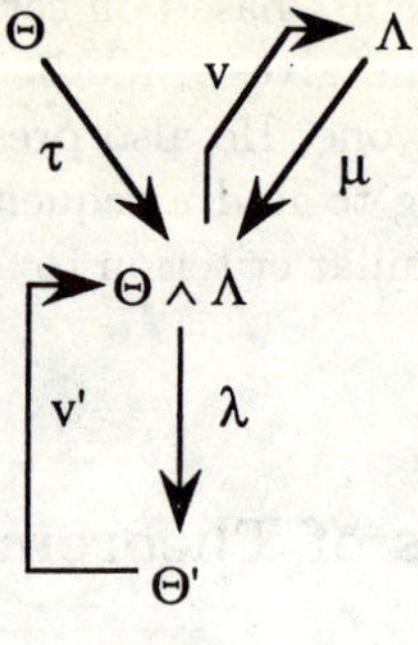

Fig. A8.1.

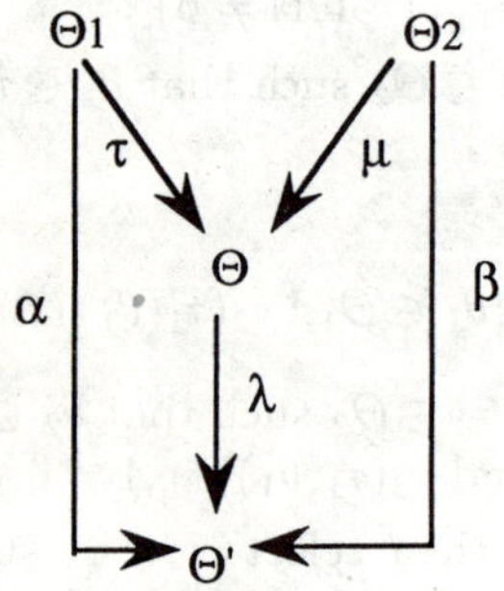

Fig. A8.2.

Conversely, if $x \in F$ we must prove that $\lambda(x) \cap \lambda(\tau(F_1)) \cap \lambda(\mu(F_2)) \neq \emptyset$. But this intersection equals $\lambda(x)$ because $x \in \tau(F_1) \cap \mu(F_2)$ and hence it is not empty because $\lambda(x) \neq \emptyset$. Hence in fact $F'|\Theta = F$ and the first part of the theorem is proved.

Second, using theorem 7.1 (2), we have

$$F|\Lambda' = \lambda(\tau(F_1)\cap\mu(F_2)) = \lambda(\tau(F_1))\cap\lambda(\mu(F_2)) = \alpha(F_1)\cap\beta(F_2) = F', \quad (A8.5)$$

which proves the second part of the theorem. □

Proof of Theorem 8.4

Consider the frames and refinings given in Fig. A8.3.

Using theorems 7.1 and 7.5 we conclude that

$$\begin{aligned}\tau_3(\tau_1(F_1) \cap \tau_2(F_2)) \cap \tau_4(F_3) &= \tau_3 \circ \tau_1(F_1) \cap \tau_3 \circ \tau_2(F_2) \cap \tau_4(F_3) \\ &= \tau_8(F_1) \cap \tau_7 \circ \tau_5(F_2) \cap \tau_7 \circ \tau_6(F_3) \qquad (A8.6) \\ &= \tau_8(F_1) \cap \tau_7(\tau_5(F_2) \cap \tau_6(F_3)),\end{aligned}$$

which proves the theorem. □

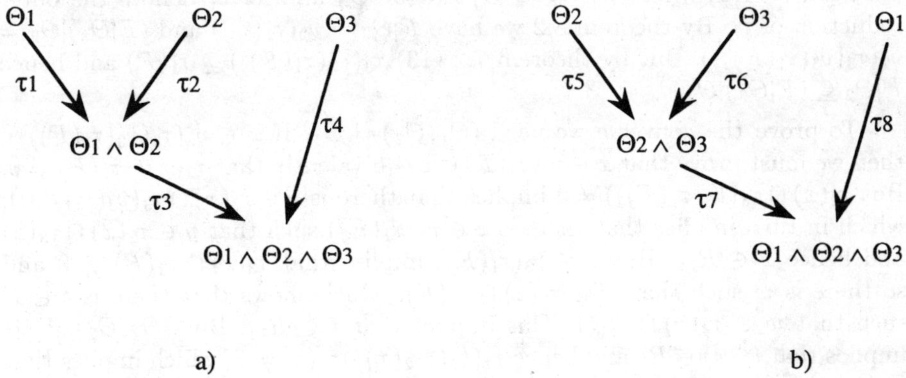

Fig. A8.3.

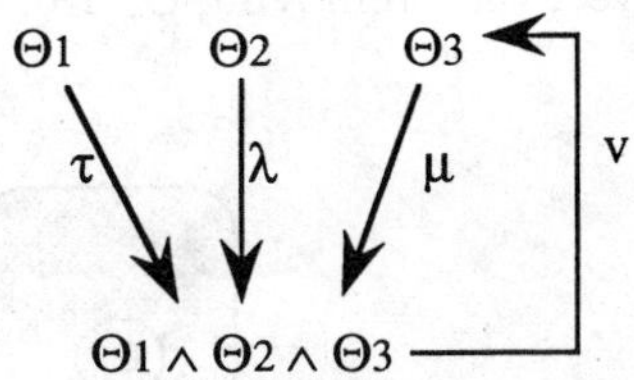

Fig. A8.4.

Proof of Theorem 8.5

Consider the frames and refinings given in Fig. A8.4.

According to theorem 8.3, we have $(F_1 \cap F_2)|\Theta_1 \wedge \Theta_2 \wedge \Theta_3 = \tau(F_1) \cap \lambda(F_2)$ and hence by theorem 7.2 (4)

$$\begin{aligned}(F_1 \cap F_2)|\Theta_3 &= v(\tau(F_1) \cap \lambda(F_2)) \\ &\subseteq v(\tau(F_1)) \cap v(\lambda(F_2)) \\ &= F_1|\Theta_3 \cap F_2|\Theta_3,\end{aligned} \tag{A8.7}$$

where the last equation holds in virtue of theorem 8.2.

Now suppose that $(\Theta_1, \Theta_2) \dashv \Theta_3$. If $x \in v(\tau(F_1)) \cap v(\lambda(F_2))$, then we have to prove that $x \in v(\tau(F_1) \cap \lambda(F_2))$, that is $\mu(x) \cap \tau(F_1) \cap \lambda(F_2) \neq \emptyset$. But $x \in v(\tau(F_1))$ implies that $\mu(x) \cap \tau(F_1) \neq \emptyset$. Select $y \in \mu(x) \cap \tau(F_1)$. Then there is $\theta_1 \in F_1$ such that $y \in \tau(\theta_1)$. Since $\tau(\theta_1) \cap \mu(x) \neq \emptyset$, we have $\theta_1 \in R_{1x}$ (see (7.18)).

Similarly, we can prove that there is $\theta_2 \in F_2$ such that $\theta_2 \in R_{2x}$. Since $(\Theta_1, \Theta_2) \dashv \Theta_3$, it follows by (7.19) that $(\theta_1, \theta_2) \in R_x$, which means that $(\theta_1, \theta_2, x) \in R(\Theta_1, \Theta_2, \Theta_3)$. This means that $\tau(\theta_1) \cap \lambda(\theta_2) \cap \mu(x) \neq \emptyset$ and therefore $\tau(F_1) \cap \lambda(F_2) \cap \mu(x) \neq \emptyset$ because $\theta_1 \in F_1$ and $\theta_2 \in F_2$. □

Proof of Theorem 8.6

For $i = 1, 2, 3$, let $\tau_i : \Theta_i \to \Theta_1 \wedge \Theta_2 \wedge \Theta_3$ and let v_i denote the outer reduction of τ_i. By theorem 8.2 we have $F|\Theta_2 = v_2(\tau_1(F))$ and $(F|\Theta_3)|\Theta_2 = v_2(\tau_3(v_3(\tau_1(F))))$. But by theorem 7.2 (13) $\tau_3(v_3(\tau_1(F))) \supseteq \tau_1(F)$ and hence $F|\Theta_2 \subseteq (F|\Theta_3)|\Theta_2$.

To prove the converse we need $(\Theta_1, \Theta_2) \dashv \Theta_3$. If $x \in v_2(\tau_3(v_3(\tau_1(F))))$, then we must prove that $x \in v_2(\tau_1(F))$ or equivalently that $\tau_2(x) \cap \tau_1(F) \neq \emptyset$. But $\tau_2(x) \cap \tau_3(v_3(\tau_1(F))) \neq \emptyset$ implies that there is $y \in \tau_2(x) \cap \tau_3(v_3(\tau_1(F)))$, which in turn implies that there is $z \in v_3(\tau_1(F))$ such that $y \in \tau_2(x) \cap \tau_3(z)$ and hence $x \in R_{2z}$. But $z \in v_3(\tau_1(F))$ implies that $\tau_3(z) \cap \tau_1(F) \neq \emptyset$ and so there is w such that $w \in \tau_3(z) \cap \tau_1(F)$, which shows that there is $t \in F$ such that $w \in \tau_3(z) \cap \tau_1(t)$. This implies that $t \in R_{1z}$. But $(\Theta_1, \Theta_2) \dashv \Theta_3$ implies that $(t, x) \in R_z$ and hence $\tau_1(t) \cap \tau_2(x) \cap \tau_3(z) \neq \emptyset$, which implies that $\emptyset \neq \tau_1(t) \cap \tau_2(x) \subseteq \tau_2(x) \cap \tau_1(F)$. □

Proof of Theorem 8.7

Since T is a Markov tree, then removing Θ_k from T implies that $(\Lambda_{kj}, j \in V_k) \dashv \Theta_k$ (see Fig. A8.5).

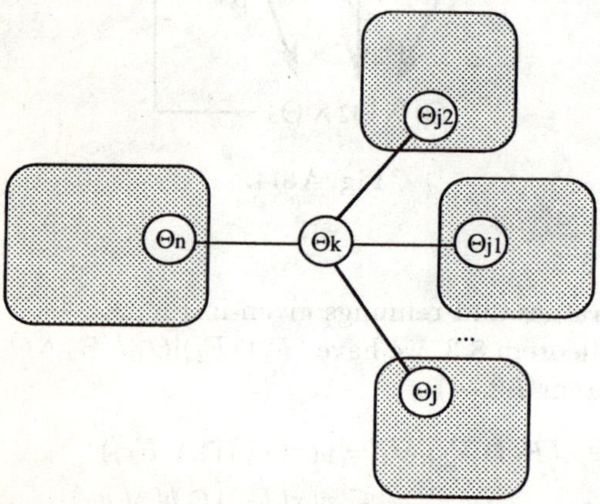

Fig. A8.5.

Then, by theorem 7.16, we have

$$(\wedge \{\Lambda_{kj} : j \in V_k - \{n\}\}, \Lambda_{kn}) \dashv \Theta_k. \tag{A8.8}$$

Since

$$\wedge \{\Lambda_{kj} : j \in V_k - \{n\}\} \wedge \Theta_k = \Lambda_{nk}, \tag{A8.9}$$

theorem 7.18 implies that $(\Lambda_{nk}, \Lambda_{kn}) \dashv \Theta_k$. Using theorem 7.17, this implies that

$$(\Lambda_{nk}, \Theta_n) \dashv \Theta_k \tag{A8.10}$$

because $\Theta_n \geq \Lambda_{kn}$. □

Proof of Theorem 8.8

If $\Lambda = \wedge \{\Theta_i : \Theta_i \in N\}$, then by definition

$$F = F_n|\Lambda \cap \left(\cap_{k \in V_n}(\cap \{F_j|\Lambda : \Theta_j \in N_{nk}\})\right). \tag{A8.11}$$

and since $(\Theta_j, \Lambda) \dashv \Lambda_{nk}$ for all $\Theta_j \in N_{nk}$, we can replace $F_j|\Lambda$ by $(F_j|\Lambda_{nk})|\Lambda$ in (A8.11). Since $\Theta_j \geq \Lambda_{nk} \geq \Lambda$ for all $\Theta_j \in N_{nk}$, theorem 7.18 implies that $(\Theta_j, \Lambda) \dashv \Lambda_{nk}$. Then by theorem 8.6 we can replace $F_j|\Lambda$ by $(F_j|\Lambda_{nk})|\Lambda$ in (A8.11). Since Λ is a common refinement of all Λ_{nk}, it follows from theorem 8.3 that

$$\cap \{(F_j|\Lambda_{nk})|\Lambda : \Theta_j \in N_{nk}\} = (\cap \{F_j|\Lambda_{nk} : \Theta_j \in N_{nk}\})|\Lambda = F_{nk}|\Lambda \tag{A8.12}$$

and hence

$$F = F_n|\Lambda \cap (\cap_{k \in V_n}(F_{nk}|\Lambda)) = F_n \cap (\cap_{k \in V_n} F_{nk}). \tag{A8.13}$$

Then, by the first part of theorem 8.5, we have

$$F|\Theta_n \subseteq F_n \cap (\cap_{k \in V_n}(F_{nk}|\Theta_n)) \tag{A8.14}$$

and by the first part of theorem 8.6, it follows that

$$F|\Theta_n \subseteq F_n \cap \left(\cap_{k \in V_n}(F_{nk}|\Theta_k)|\Theta_n\right). \tag{A8.15}$$

However, if $T = (N, E)$ is a Markov tree, then $(\Lambda_{nk} : k \in V_n) \dashv \Theta_n$ and hence $(\Theta_n, \Lambda_{nk} : k \in V_n) \dashv \Theta_n$ by corollary 7.1. Then equality holds in (A8.14) by the second part of theorem 8.5. Then theorems 8.6 and 8.7 imply that equality holds in (A8.15) and the theorem is proved. □

Proof of Theorem 8.9

Let $T = (N, E)$ be a Markov tree and let $T' = (N', E')$ be a subtree of T. Let $\Theta_n \in N'$ and let V'_n, N'_{nk} and Λ'_{nk} denote the sets V_n, N_{nk} and Λ_{nk} with respect to the tree T' instead of T. Then $V'_n \subseteq V_n$ and $N'_{nk} \subseteq N_{nk}$ for all $k \in V'_n$. Since T is a Markov tree we have $(\Lambda_{nk} : k \in V_n) \dashv \Theta_n$ and hence $(\Lambda_{nk} : k \in V'_n) \dashv \Theta_n$ by theorem 7.15. But $N'_{nk} \subseteq N_{nk}$ for all $k \in V'_n$ implies that $\Lambda'_{nk} \geq \Lambda_{nk}$ for all $k \in V'_n$ and hence $(\Lambda'_{nk} : k \in V'_n) \dashv \Theta_n$ by theorem 7.17. This shows that T' is a Markov tree and the theorem is proved. □

Proof of Theorem 8.10

If we remove Θ_1 or Θ_n, then the condition is trivially satisfied because there is only one neighbour frame. If the frame Θ_i with $1 < i < m$ is removed, then it must be proved that

$$(\wedge_{j<i}\Theta_j, \wedge_{j>i}\Theta_j) \dashv \Theta_i, \tag{A8.16}$$

which is equivalent to $(\Theta_{i-1}, \Theta_m) \dashv \Theta_i$ because $\wedge_{j<i}\Theta_j = \Theta_{j-1}$ and $\wedge_{j>i}\Theta_j = \Theta_m$. But theorem 7.18 implies that this condition is satisfied because $\Theta_{i-1} \geq \Theta_i \geq \Theta_m$. □

To prove theorem 8.11 we need the following Lemma of Shafer et al. (1987):

Lemma A8.1. *Let $\mathcal{P}$ be a partition of S. If we define*

$$\mathcal{P}_A^{\downarrow} = \{A^c\} \cup \{\{\theta\} : \theta \in A\} \tag{A8.17}$$

for any $A \in \mathcal{P}$, then $(\mathcal{P}_A^{\downarrow} : A \in \mathcal{P}) \dashv \mathcal{P}$.

Proof of Lemma A8.1

Let $P \in \mathcal{P}$ and let $P_A \in \mathcal{P}_A^{\downarrow}$ such that $P \cap P_A \neq \emptyset$ for all $A \in \mathcal{P}$. Then $P_P \subseteq P$ and $P_A = A^c$ for all $A \neq P, A \in \mathcal{P}$. Then

$$P \cap (\cap\{P_A : A \in \mathcal{P}\}) = P_P \neq \emptyset \tag{A8.18}$$

and the lemma is proved by theorem 7.22. □

Proof of Theorem 8.11

First we prove that the tree of families and dichotomies is a Markov tree, i.e.

$$\{\Lambda_{nk} : k \in V_n\} \dashv \mathcal{P}_n \tag{A8.19}$$

for all partition $\mathcal{P}_n$ of the tree with $\Lambda_{nk} = \wedge\{\mathcal{P}_j : \mathcal{P}_j \in N_{nk}\}$. We distinguish three cases: (1) $\mathcal{P}_n$ is the root node, (2) $\mathcal{P}_n$ is a family node with associated partition $\mathcal{D}_M \cup \{M^c\}$, (3) $\mathcal{P}_n$ is a dichotomy node $\{M, M^c\}$. In case (1), if $\mathcal{P} = \mathcal{D}_S$, then (A8.19) is reduced to

$$(\mathcal{P}_X^{\downarrow} : X \in \mathcal{P}) \dashv \mathcal{P}. \tag{A8.20}$$

In case (2), if $\mathcal{P} = \mathcal{D}_M \cup \{M^c\}$, then (A8.19) is reduced to

$$(\mathcal{P}_X^{\downarrow} : X \in \mathcal{P}) \dashv \mathcal{P}. \tag{A8.21}$$

In case (3), if $\mathcal{P} = \{M, M^c\}$, then (A8.19) is reduced to

$$(\mathcal{P}_M^{\downarrow}, \mathcal{P}_{M^c}^{\downarrow}) \dashv \{M, M^c\}. \tag{A8.22}$$

All three conditions (A8.20) to (A8.22) are special cases of Lemma A8.1, which proves that the tree of families and dichotomies is a Markov tree. The proof that the tree of families is also a Markov tree can be found in Shafer et al. (1987). □

Proof of Theorem 8.12

This is a direct consequence of the definition and of theorem 8.6. □

Proof of Theorem 8.13

Let $\mathcal{H}_i = (\Omega_i, P_i, \Gamma_i, \Theta_i)$ for $i = 1, 2$. According to theorem 8.3, for all $(\omega_1, \omega_2) \in \Omega_1 \times \Omega_2$ we have

$$\begin{aligned} \Gamma_1(\omega_1)|\Theta' \cap \Gamma_2(\omega_2)|\Theta' &= (\Gamma_1(\omega_1)|\Theta \cap \Gamma_2(\omega_2)|\Theta)|\Theta' \\ \Gamma_1(\omega_1)|\Theta \cap \Gamma_2(\omega_2)|\Theta &= (\Gamma_1(\omega_1)|\Theta' \cap \Gamma_2(\omega_2)|\Theta')|\Theta \end{aligned} \tag{A8.23}$$

and hence

$$\begin{aligned} \mathcal{H}_1|\Theta' \oplus \mathcal{H}_2|\Theta' &= (\mathcal{H}_1|\Theta \oplus \mathcal{H}_2|\Theta)|\Theta' \\ \mathcal{H}_1|\Theta \oplus \mathcal{H}_2|\Theta &= (\mathcal{H}_1|\Theta' \oplus \mathcal{H}_2|\Theta')|\Theta, \end{aligned} \tag{A8.24}$$

or $\mathcal{H}' = \mathcal{H}|\Theta'$ and $\mathcal{H} = \mathcal{H}'|\Theta$. □

Proof of Theorem 8.14

Let $\Theta = \Theta_1 \wedge \Theta_2 \wedge \Theta_3$. Then theorem 8.13 implies that

$$\begin{aligned} (\mathcal{H}_1 \oplus \mathcal{H}_2) \oplus \mathcal{H}_3 &= (\mathcal{H}_1|\Theta_1 \wedge \Theta_2) \oplus (\mathcal{H}_2|\Theta_1 \wedge \Theta_2))|\Theta \oplus \mathcal{H}_3|\Theta \\ &= (\mathcal{H}_1|\Theta \oplus \mathcal{H}_2|\Theta) \oplus \mathcal{H}_3|\Theta \\ &= \mathcal{H}_1|\Theta \oplus (\mathcal{H}_2|\Theta \oplus \mathcal{H}_3|\Theta) \\ &= \mathcal{H}_1|\Theta \oplus (\mathcal{H}_2|\Theta_2 \wedge \Theta_3 \oplus \mathcal{H}_3|\Theta_2 \wedge \Theta_3)|\Theta \\ &= \mathcal{H}_1 \oplus (\mathcal{H}_2 \oplus \mathcal{H}_3). \end{aligned} \tag{A8.25}$$

□

Proof of Theorem 8.15

Consider the frames and refinings given in Fig. A8.6.

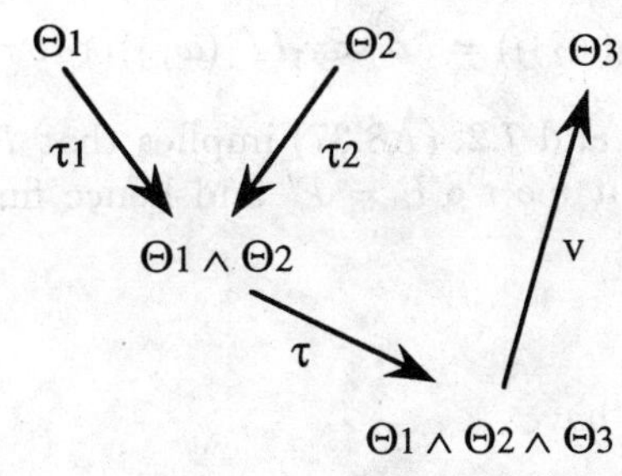

Fig. A8.6.

By definition, we have $\mathcal{H}_1 \oplus \mathcal{H}_2 = (\Omega, P, \Gamma, \Theta_1 \wedge \Theta_2)$, where $\Omega = \Omega_1 \times \Omega_2 - K$ with

$$K = \{(\omega_1, \omega_2) : \tau_1(\Gamma_1(\omega_1)) \cap \tau_2(\Gamma_2(\omega_2)) = \emptyset\}, \tag{A8.26}$$

and

$$P(\omega_1, \omega_2) = P_1(\omega_1)P_2(\omega_2)/(1 - P_1P_2(K)) \tag{A8.27}$$

and

$$\Gamma(\omega_1,\omega_2) = \tau_1(\Gamma_1(\omega_1)) \cap \tau_2(\Gamma_2(\omega_2)). \tag{A8.28}$$

Hence

$$(\mathcal{H}_1 \oplus \mathcal{H}_2)|\Theta_3 = (\Omega, P, v \circ \tau \circ \Gamma, \Theta_3). \tag{A8.29}$$

On the other hand, by theorem 8.2, we have

$$\mathcal{H}_1|\Theta_3 = (\Omega_1, P_1, v \circ \tau \circ \tau_1 \circ \Gamma_1, \Theta_3) \tag{A8.30}$$

and

$$\mathcal{H}_2|\Theta_3 = (\Omega_2, P_2, v \circ \tau \circ \tau_2 \circ \Gamma_2, \Theta_3). \tag{A8.31}$$

This implies that

$$\mathcal{H}_1|\Theta_3 \oplus \mathcal{H}_2|\Theta_3 = (\Omega', P', \Gamma', \Theta_3) \tag{A8.32}$$

where $\Omega' = \Omega_1 \times \Omega_2 - K'$ with

$$K' = \{(\omega_1,\omega_2) : v \circ \tau \circ \tau_1 \circ \Gamma_1(\omega_1) \cap v \circ \tau \circ \tau_2 \circ \Gamma_2(\omega_2) = \emptyset\}, \tag{A8.33}$$

and

$$P'(\omega_1,\omega_2) = P_1(\omega_1)P_2(\omega_2)/(1 - P_1P_2(K')) \tag{A8.34}$$

and

$$\Gamma'(\omega_1,\omega_2) = v \circ \tau \circ \tau_1 \circ \Gamma_1(\omega_1) \cap v \circ \tau \circ \tau_2 \circ \Gamma_2(\omega_2). \tag{A8.35}$$

If we show that $K = K'$ then $\Omega = \Omega'$ and $P = P'$. By theorem 8.5, for all (ω_1,ω_2) in $\Omega_1 \times \Omega_2$, we have

$$(\Gamma_1(\omega_1) \cap \Gamma_2(\omega_2))|\Theta_3 = \Gamma_1(\omega_1)|\Theta_3 \cap \Gamma_2(\omega_2)|\Theta_3, \tag{A8.36}$$

which can be written

$$v \circ \tau(\tau_1(\Gamma_1(\omega_1)) \cap \tau_2(\Gamma_2(\omega_2))) = v \circ \tau \circ \tau_1(\Gamma_1(\omega_1)) \cap v \circ \tau \circ \tau_2(\Gamma_2(\omega_2)). \tag{A8.37}$$

But using theorems 7.1 and 7.2, (A8.37) implies that $K = K'$. Furthermore, (A8.37) also implies that $v \circ \tau \circ \Gamma = \Gamma'$ and hence finally $(\mathcal{H}_1 \oplus \mathcal{H}_2)|\Theta_3 = (\mathcal{H}_1|\Theta_3) \oplus (\mathcal{H}_2|\Theta_3)$. □

Proof of Theorem 8.16

By theorem 8.14 we have

$$\mathcal{H} = \mathcal{H}_n \oplus (\oplus_{k \in V_n} \mathcal{H}_{nk}). \tag{A8.38}$$

Since T is a Markov tree, theorem 8.15 implies that

$$\mathcal{H}|\Theta_n = \mathcal{H}_n \oplus (\oplus_{k \in V_n} \mathcal{H}_{nk}|\Theta_n). \tag{A8.39}$$

Then, by theorem 8.7 and 8.12, we have

$$\mathcal{H}|\Theta_n = \mathcal{H}_n \oplus (\oplus_{k \in V_n} (\mathcal{H}_{nk}|\Theta_k)|\Theta_n). \tag{A8.40}$$

□

Proof of Theorem 8.17

By the definition of $\geq$ in $\mathcal{F}$, we have

$$\Theta_i \vee \Theta_j \geq \Theta \text{ iff } \varphi(\Theta_i) \vee \varphi(\Theta_j) \geq \varphi(\Theta). \tag{A8.41}$$

In order to prove that $\varphi(\Theta_i) \vee \varphi(\Theta_j) \geq \varphi(\Theta)$, we must prove that if x, y are two points in S which belong to the same block P in $\varphi(\Theta)$, then they belong to a same block of $\varphi(\Theta_i) \vee \varphi(\Theta_j)$, or, in other words, if $x \equiv y(\varphi(\Theta))$ then $x \equiv y(\varphi(\Theta_i) \vee \varphi(\Theta_j))$. Let P_i in $\varphi(\Theta_i)$ such that $x \in P_i$ and P_j in $\varphi(\Theta_j)$ such that $y \in P_j$. Since $(\Theta_1, \ldots, \Theta_n) \dashv \Theta$ it follows that $(\Theta_i, \Theta_j) \dashv \Theta$ by theorem 7.15. Let $P_i \in \varphi(\Theta_i)$ such that $x \in P_i$ and $P_j \in \varphi(\Theta_j)$ such that $y \in P_j$. Since $x \in P \cap P_i$ and $y \in P \cap P_j$ it follows that $P \cap P_i \cap P_j \neq \emptyset$ according to theorem 7.20. Let $z \in P \cap P_i \cap P_j$. Then by theorem 7.4 (1) we have

$$x \equiv z(\varphi(\Theta_i)) \equiv y(\varphi(\Theta_j)) \tag{A8.42}$$

and hence $x \equiv y(\varphi(\Theta_i) \vee \varphi(\Theta_j))$. □

Proof of Theorem 8.18

If Θ is on the path between Θ_1 and Θ_2, then removing Θ from T implies that Θ_1 and Θ_2 are in two different subtrees T_1 and T_2. If $\Lambda_1 = \wedge \{\Theta_i \in T_1\}$ and $\Lambda_2 = \wedge \{\Theta_i \in T_2\}$, then $(\Lambda_1, \Lambda_2) \dashv \Theta$ because T is a Markov tree and by theorem 7.15. Hence $(\Theta_1, \Theta_2) \dashv \Theta$ by theorem 7.17 and $\Theta_1 \vee \Theta_2 \geq \Theta$ (by theorem 8.17).

Conversely, remove a frame Θ from T. This generates several subtrees T_1, $\ldots, T_n$ and let $\Lambda_i = \wedge \{\Theta_j \in T\}$ for all $i = 1, \ldots, n$. Consider two frames Λ_i and Λ_j. Then the distributivity of the lattice (Assumption (2)) implies that

$$\Lambda_i \vee \Lambda_j = \wedge \{\Theta_1 \vee \Theta_2 : \Theta_1 \in T_i, \Theta_2 \in T_j\}. \tag{A8.43}$$

But by hypothesis $\Theta_1 \vee \Theta_2 \geq \Theta$ for all $\Theta_1 \in T_i$ and $\Theta_2 \in T_j$ and hence $\Lambda_i \vee \Lambda_j \geq \Theta$ by (A8.43). Then by Assumption (2) we have $(\Lambda_1, \ldots, \Lambda_n) \dashv \Theta$ and T is a Markov tree. □

Proof of Corollary 8.1

This is a direct consequence of theorem 8.18 and theorem 7.7. □

Proof of Theorem 8.19

We use corollary 8.1. Let v_1 and v_2 be two nodes in G and suppose that $\{v_1, v_2\}$ is not in E because in this case the condition of corollary 8.1 is trivially satisfied. Then the only situation where $v_1 \cap v_2$ is not empty is when both v_1 and v_2 are in H and there is exactly one intermediate node $v \in J$ between them. In this case $v_1 \cap v_2 = v$ and the theorem is proved. □

Proof of Theorem 8.20

We use the definition of a Markov tree (see 8.13). Suppose that we remove a node J_n from G and let T_{nk} be the tree containing node J_n. Then by theorem 7.7

$$\Lambda_{nk} = \cup\{J_l : J_l \in N_{nk}\}. \tag{A8.44}$$

Since there is no edge between different nodes belonging to different trees T_{nk}, it follows that $\Lambda_{nk} \cap \Lambda_{nk'} = \emptyset$ for $k \neq k'$ in $N(n)$ and hence

$$(\Lambda_{nk} : k \in N(n)) \dashv \tag{A8.45}$$

by theorem 7.10. Then necessarily

$$(\Lambda_{nk} : k \in N(n)) \dashv J_n \tag{A8.46}$$

and G is a Markov tree. □

Proof of Theorem 8.21

If $t \neq b$ and $t \cap b \neq \emptyset$, then there exist three different nodes v_1, v_2, v_3 in V such that $t = \{v_1, v_2\}$ and $b = \{v_1, v_3\}$. We prove that v_2 is a leaf. Suppose that this is not the case, i.e. there is $e \in E - \{t\}$ such that $v_2 \in e$. Then $v_2 \in t \cap (\cup \{e \in E - \{t\}\})$ and hence $v_2 \in b$ by (8.33), which is a contradiction.

Conversely, suppose that $t = \{v_1, v_2\}$ where v_2 is a leaf and $v_1 \in b$. Since $t \neq b$ we have $b = \{v_1, v_3\}$. Then $t \cap b = \{v_1\} \neq \emptyset$ and v_1 is in the right-hand side of (8.33). If $v \in \cup \{t \cap e : e \in E - \{t\}\}$, then there is $e \neq t$ such that $v \in e \cap t$. Since $v_2 \notin e$ for all $e \in E - \{t\}$ because v_2 is a leaf, v equals v_1 and equation (8.33) is valid. □

Proof of Theorem 8.22

The proof of this theorem can be found in Maier, 1983. □

Proof of Theorem 8.23

First recall that $(\mathcal{F}, \leq)$ is a lattice by theorem 7.5. If $\Theta_1, \ldots, \Theta_n$ is a hypertree construction sequence and $j < i$, then by theorem 7.3 (6) we have

$$\Theta_i \vee \Theta_j \geq \Theta_i \vee (\wedge_{k<i}\Theta_k) = \Theta_i \vee \Theta_{b(i)} \geq \Theta_{b(i)} \tag{A8.47}$$

and hence $(\Theta_i, \Theta_j) \dashv \Theta_{b(i)}$ by Assumption (2).

Conversely, we must prove that

$$\Theta_i \vee (\wedge_{k<i}\Theta_k) = \Theta_i \vee \Theta_{b(i)} \tag{A8.48}$$

for all $i = 2, \ldots, n$. But by the distributivity (Assumption (1)),

$$\Theta_i \vee (\wedge_{k<i}\Theta_k) = \wedge_{k<i}(\Theta_i \vee \Theta_k) \leq \Theta_i \vee \Theta_{b(i)}. \tag{A8.49}$$

On the other hand, $\Theta_i \vee \Theta_k \geq \Theta_{b(i)}$ for all $k < i$ (Assumption (2)) implies that

$$\Theta_i \vee (\wedge_{k<i}\Theta_k) = \wedge_{k<i}(\Theta_i \vee \Theta_k) \geq \Theta_{b(i)} \tag{A8.50}$$

(by definition of the infimum). Since $\Theta_i \vee (\wedge_{k<i}\Theta_k) \geq \Theta_i$ we have

$$\Theta_i \vee (\wedge_{k<i}\Theta_k) \geq \Theta_i \vee \Theta_{b(i)}. \tag{A8.51}$$

by theorem 7.3 (6). But inequalities (A8.49) and (A8.51) imply (A8.48). □

Proof of Theorem 8.24

Let $N = \{\Theta_k : k \in K\}$. First we prove that any leaf of T is a twig in N. Let Θ_i be a leaf of T and let $\Theta_{b(i)}$ denote the bud of Θ_i. We show that

$$\Theta_i \vee (\wedge \{\Theta_k : k \in K - \{i\}\}) = \Theta_i \vee \Theta_{b(i)}. \tag{A8.52}$$

Then by theorem 8.7 and Assumption (2), we have

$$(\wedge \{\Theta_k : k \in K - \{i\}\}) \vee \Theta_i \geq \Theta_{b(i)} \tag{A8.53}$$

and hence

$$(\wedge \{\Theta_k : k \in K - \{i\}\}) \vee \Theta_i \geq \Theta_{b(i)} \vee \Theta_i \tag{A8.54}$$

by theorem 7.3 (6) and (1). Since the other inequality is trivial, (A8.49) is proved.

Now suppose that $\Theta_1, \ldots, \Theta_n$ is a tree construction sequence of T with budding b. Since Θ_i is a leaf in the subtree over the nodes $\Theta_1, \ldots, \Theta_i$ and this subtree is also a Markov tree (theorem 8.9), we have just proved above that Θ_i is a twig in $\{\Theta_1, \ldots, \Theta_i\}$ with branch $\Theta_{b(i)}$. This shows that $\Theta_1, \ldots, \Theta_n$ is a hypertree construction sequence with branching b. This proves also that N is a hypertree. □

Proof of Theorem 8.25

Since $b(i) < i$ for all i, the definition of E clearly implies that $\Theta_1, \ldots, \Theta_n$ is a tree construction sequence of $T = (N, E)$ with b as its budding. It remains to prove that T is a Markov tree. We prove by induction on k that the tree $T_k = (N_k, E_k)$ with

$$N_k = \{\Theta_1, \ldots, \Theta_k\},\ E_1 = \emptyset,\ E_k = \{\{\Theta_2, \Theta_{b(2)}\}, \ldots, \{\Theta_k, \Theta_{b(k)}\}\} \tag{A8.55}$$

is a Markov tree.

T_1 is clearly a Markov tree and suppose that T_k is a Markov tree. We remove a frame Θ from T_{k+1} and distinguish two cases.

(1). Suppose that $\Theta = \Theta_{b(k+1)}$. Let $S_1, \ldots, S_r$ denote the subtrees induced by the removal of Θ and suppose that $S_1 = \{\Theta_{k+1}\}$. If $\Lambda_i = \wedge \{\Theta_j \in S_i\}$ for all $i = 1, \ldots, r$, then we must prove that

$$(\Lambda_1, \ldots, \Lambda_r) \dashv \Theta, \tag{A8.56}$$

which is equivalent to

$$\Lambda_i \vee \Lambda_j \geq \Theta \tag{A8.57}$$

whenever $i \neq j$ by Assumption (2). Condition (A8.57) is satisfied when i and j are different from 1 because T_k is a Markov tree. If $\Lambda_i = \Lambda_1$ and $\Lambda = \wedge_{j<k+1}\Theta_j$, then we have

$$\Theta_{k+1} \vee \Lambda_j \geq \Theta_{k+1} \vee \Lambda = \Theta_{k+1} \vee \Theta_{b(k+1)} \geq \Theta_{b(k+1)} = \Theta \tag{A8.58}$$

because Θ_{k+1} is a twig and (A8.57) holds also for i or $j = 1$.

(2). Suppose that $\Theta \neq \Theta_{b(k+1)}$ and let $S_1, \ldots, S_r$ again be the subtrees induced by removing Θ. Again we assume that S_1 contains the frames Θ_{k+1} and $\Theta_{b(k+1)}$ (see Fig. A8.7).

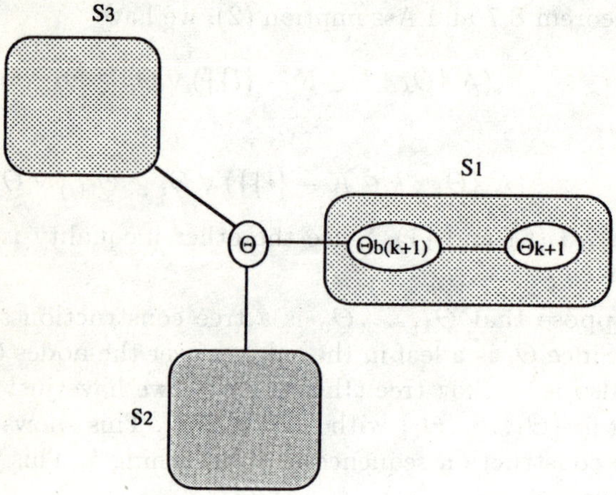

Fig. A8.7.

Let $\Lambda_1, \ldots, \Lambda_r$ and Λ be defined as above. We must prove that $\Lambda_i \vee \Lambda_l \geq \Theta$ whenever $i \neq l$. Again, this condition is satisfied if both i and l are different from 1 because T_k is a Markov tree. So, assume that $i = 1$ and $\Lambda_1 = \Theta_{k+1} \wedge (\wedge_{i \in I}\Theta_i)$ for some $I \subseteq \{1, \ldots, k\}$ and let $\Lambda_l = \wedge_{j \in J}\Theta_j$ for some $J \subseteq \{1, \ldots, k\}$. Then by the distributivity (Assumption (1)) we have

$$\begin{aligned} \Lambda_1 \vee \Lambda_l &= (\Theta_{k+1} \wedge (\wedge_{i \in I}\Theta_i)) \vee \Lambda_l \\ &= (\Lambda_l \vee \Theta_{k+1}) \wedge (\Lambda_l \vee (\wedge_{i \in I}\Theta_i)) \\ &= (\Lambda_l \vee \Theta_{k+1}) \wedge (\wedge_{i \in I}(\Lambda_l \vee \Theta_i)). \end{aligned} \tag{A8.59}$$

But $\Lambda_l \vee \Theta_i = \wedge_{j \in J}(\Theta_j \vee \Theta_i)$ and hence, by definition of the infimum, $\Lambda_l \vee \Theta_i \geq \Theta$ because $\Theta_j \vee \Theta_i \geq \Theta$ for all $j \in J$ according to theorem 8.18 (every path from a node of S_1 to one of S_j contains Θ). This implies that $\Lambda_l \vee \Theta_i \geq \Theta$ for all $i \in I$ and therefore

$$\wedge_{i \in I}(\Lambda_l \vee \Theta_i) \geq \Theta. \tag{A8.60}$$

On the other hand, the fact that Θ_{k+1} is a twig implies that

$$\Lambda_l \vee \Theta_{k+1} \geq \Lambda \vee \Theta_{k+1} = \Theta_{b(k+1)} \vee \Theta_{k+1} \geq \Theta_{b(k+1)}. \tag{A8.61}$$

Since $\Lambda_l \vee \Theta_{k+1} \geq \Lambda_l$ we conclude that

$$\Lambda_l \vee \Theta_{k+1} \geq \Lambda_l \vee \Theta_{b(k+1)} = \wedge_{j\in J}(\Theta_j \vee \Theta_{b(k+1)}) \geq \Theta \tag{A8.62}$$

by theorem 8.18 (again every path from a node $\Theta_j, j \in J$ to $\Theta_{b(k+1)}$ contains Θ). But (A8.59) with (A8.60) and (A8.62) imply that $\Lambda_1 \vee \Lambda_l \geq \Theta$. □

Proof of Corollary 8.2

Let $T = (H, E)$ be the Markov tree defined in theorem 8.25. There is obviously at least one tree construction sequence of T starting with Θ. Then by theorem 8.24 this is a hypertree construction sequence and it starts with Θ. □

Proof of Theorem 8.26

(1). Since $\Theta_1 \vee \Theta_3 \geq \Theta_3 \geq \Theta_2$, we have $(\Theta_1, \Theta_3) \dashv \Theta_2$ (by Assumption (2)). Then (8.37) follows from theorem 8.12.

(2). $\Theta_1 - \Theta_1 \vee \Theta_2 - \Theta_2$ is a Markov tree because $\Theta_1 \vee \Theta_2 \geq \Theta_1 \vee \Theta_2$. Therefore, the local propagation algorithm in Markov trees (theorem 8.16) implies that

$$(\mathcal{H}_1 \oplus \mathcal{H}_2)|\Theta_1 = (\mathcal{H}_2|\Theta_1 \vee \Theta_2)|\Theta_1 \oplus \mathcal{H}_1. \tag{A8.63}$$

But since $\Theta_1 \wedge (\Theta_1 \vee \Theta_2) = \Theta_1$, the definition of $\oplus$ implies that

$$(\mathcal{H}_2|\Theta_1 \vee \Theta_2)|\Theta_1 \oplus \mathcal{H}_1 \tag{A8.64}$$

can be written $\mathcal{H}_2|\Theta_1 \vee \Theta_2 \oplus \mathcal{H}_1$ and the theorem is proved. □

Proof of Theorem 8.27

The hint $\mathcal{H}' = \oplus_{i\neq t}\mathcal{H}_i$ is a hint on Λ and $\mathcal{H} = \mathcal{H}' \oplus \mathcal{H}_t$. Then by theorem 8.26 (2) we have

$$\mathcal{H}|\Lambda = \mathcal{H}' \oplus (\mathcal{H}_t|\Lambda \vee \Theta_t). \tag{A8.65}$$

But since Θ_t is a twig with branch b in H, we have $\Lambda \vee \Theta_t = \Theta_b \vee \Theta_t$ and (8.39) follows from the associativity of the combination operation. □

Proof of Theorem 8.29 (adapted from Mellouli, 1988)

Since we are in a multivariate family of compatible frames, we simplify notations by writing J instead of Θ_J for a frame Θ_J in the family. Let T be a covering Markov tree of H' and let $X \in H$. We must prove that there is a frame $X' \in T$ such that $X \subseteq X'$ (by abuse of notation $X' \in T$ means that X' is a node of the tree T). But $X \in H$ implies that X is a complete subgraph of $G(H)$ and therefore there is a clique C in H' such that $X \subseteq C$. Since T is a covering Markov tree of H', there is a frame X' in T such that $X \subseteq C \subseteq X'$.

Conversely, suppose that T is a covering Markov tree of H. We prove that if $C = \{i_1, \ldots, i_n\}$ is a complete subgraph of $G(H)$ (in particular a clique $C \in H'$), then there is a frame X in T such that $C \subseteq X$

We prove this by induction on n. Start with $n = 2$. If $C = \{i_1, i_2\}$ is an edge of $G(H)$ there is a hyperedge K in H such that $C \subseteq K$. Since T is a covering Markov tree of H, there is a frame $X \in T$ such that $K \subseteq X$ and hence $C \subseteq X$. Now suppose that the assertion is proved for all cliques having $n-1$ elements with $n - 1 \geq 2$ (in particular it is true for all cliques having two elements). Then let $C_j = C - \{i_j\}$ for $j = 1, 2, 3$. Since each C_j has $n - 1$ elements, then by the induction hypothesis there exist X_1, X_2, X_3 in T such that $C_j \subseteq X_j$ for $j = 1, 2, 3$. If two frames among X_1, X_2, X_3 are identical, say $X_1 = X_2$, then $C_1 = C - \{i_1\} \subseteq X_1$ and $i_1 \in C_2 \subseteq X_2 = X_1$ and hence $C \subseteq X_1$. If not, let us first suppose that one of these frames, say X_1, lies on the path between X_2 and X_3 in T. Since T is a Markov tree, corollary 8.1 implies that

$$C - \{i_2, i_3\} = C_2 \cap C_3 \subseteq X_2 \cap X_3 \subseteq X_1 \qquad (A8.66)$$

and since $\{i_2, i_3\} \subseteq C_1 \subseteq X_1$ we have $C \subseteq X_1$.

If none of these nodes X_1, X_2, X_3 lies on the path between the other two, then there is always a node X different from X_1, X_2, X_3 in T that lies on the three paths between any two nodes out of $\{X_1, X_2, X_3\}$. Since T is a Markov tree, corollary 8.1 implies that $C - \{i_1, i_2\} = C_1 \cap C_2 \subseteq X_1 \cap X_2 \subseteq X$. But $\{i_1, i_2\} \subseteq C_3 \subseteq X_3$ together with $\{i_1, i_3\} \subseteq C_2 \subseteq X_2$ implies that $i_1 \in X_3 \cap X_2 \subseteq X$. Similarly, $\{i_1, i_2\} \subseteq C_3 \subseteq X_3$ together with $\{i_2, i_3\} \subseteq C_1 \subseteq X_1$ implies that $i_2 \in X_3 \cap X_1 \subseteq X$. Thus $\{i_1, i_2\} \subseteq X$ and therefore $C \subseteq X$. □

Proof of Theorem 8.30

The proof of this theorem can be found in Golumbic (1980). □

Proof of Theorem 8.31

The proof of this theorem can be found in Golumbic (1980). □

Proof of Theorem 8.32

The proof of this theorem can be found in Golumbic (1980). □

Proof of Theorem 8.33

Suppose that G is a triangulated graph. We show that the tree T of theorem 8.32 is a Markov tree using corollary 8.1. Let C_1 and C_2 be two nodes in T. We have to show that $C_1 \cap C_2 \subseteq C$ for any C on the unique path γ from C_1 to C_2 in T. Suppose that $v \in C_1 \cap C_2$. Then C_1 and C_2 are in $\mathcal{C}(v)$ and γ is the unique path between C_1 and C_2 in $\mathcal{C}(v)$ by theorem 8.32. Hence $v \in C$ for all $C \in \gamma$, which proves that T is a Markov tree.

Conversely, suppose that $T = (\mathcal{C}, E)$ is a Markov tree over the cliques of G. Then we must show that the subgraph over $\mathcal{C}(v)$ is connected in order to apply theorem 8.32. Let C_1 and C_2 be two cliques in $\mathcal{C}(v)$. Then $v \in C_1 \cap C_2$ and since T is a Markov tree, $v \in C$ for all C on the path γ between C_1 and C_2 by corollary 8.1, which shows that γ is also a path between C_1 and C_2 in the subgraph over $\mathcal{C}(v)$. This proves that $\mathcal{C}(v)$ is connected. □

Proof of Theorem 8.34

Suppose that $Cl(v_i, G_{i\sigma})$ is a maximal closure. Then by construction $Cl(v_i, G_{i\sigma})$ is a complete subgraph of G_σ and hence there is a clique C of G_σ such that

$$Cl(v_i, G_{i\sigma}) \subseteq C. \tag{A8.67}$$

Now let v_l be the first vertex of C to be eliminated in the sequence σ and let x be a vertex in $C - \{v_l\}$. Since C is a complete subgraph of G_σ, there is an edge e in G_σ linking x and v_l. Since $x \in C$ and v_l is the first vertex in C to be eliminated, it follows that $x \in G_{l\sigma}$ and e is an edge of $G_{l\sigma}$. This implies that $C - \{v_l\} \subseteq N(v_l, G_{l\sigma})$ and hence $C \subseteq Cl(v_l, G_{l\sigma})$. Now suppose that the inclusion is strict in (A8.67). Then $Cl(v_i, G_{i\sigma}) \subset C \subseteq Cl(v_l, G_{l\sigma})$ which is a contradiction to the maximality of $Cl(v_i, G_{i\sigma})$. This shows that equality holds in (A8.67) and hence $Cl(v_i, G_{i\sigma})$ is a clique of G_σ.

On the other hand, suppose that C is a clique of G_σ and let v_l be the first vertex of C to be eliminated. We have just seen above that $C \subseteq Cl(v_l, G_{l\sigma})$. We prove the converse, i.e. $Cl(v_l, G_{l\sigma}) \subseteq C$.

Suppose that $x \in Cl(v_l, G_{l\sigma})$. We show that x is linked (in G_σ) to any vertex $y \in C$ which is different from x. This proves that $x \in C$ because assuming the contrary implies a contradiction with the maximality of C. Let $G_\sigma = (V_\sigma, E_\sigma)$ and $G_{l\sigma} = (V_{l\sigma}, E_{l\sigma})$. Since $x \in Cl(v_l, G_{l\sigma})$ we have $\{v_l, x\} \in E_{l\sigma}$ (if $x = v_l$ then $x \in C$ because $v_l \in C$). Since $v_l \in C$ and C is a clique of G_σ we have $\{v_l, y\} \in E_\sigma$. But v_l being the first vertex of C to be eliminated implies that $y \in V_{l\sigma}$ and $\{v_l, y\} \in E_{l\sigma}$. This shows that both x and y are in $N(v_l, G_{l\sigma})$ and since v_l is a simplicial vertex of $G_{l\sigma}$, it follows that $\{x, y\} \in E_{l\sigma} \subseteq E_\sigma$, which shows that x and y are linked in G_σ. Therefore, the clique C equals $Cl(v_l, G_{l\sigma})$.

It remains to prove that $Cl(v_l, G_{l\sigma})$ is a maximal closure. Suppose that $Cl(v_l, G_{l\sigma})$ is not maximal. Then there is a maximal closure $Cl(v_t, G_{t\sigma})$ such that

$$Cl(v_l, G_{l\sigma}) \subset Cl(v_t, G_{t\sigma}). \tag{A8.68}$$

By the first part of the theorem, there is a clique K of G_σ such that

$$Cl(v_l, G_{l\sigma}) \subset Cl(v_t, G_{t\sigma}) = K. \tag{A8.69}$$

But $Cl(v_l, G_{l\sigma}) = C$ and hence $C \subset K$ which contradicts the maximality of the clique C. This shows that $Cl(v_l, G_{l\sigma})$ is a maximal closure and the theorem is proved. □

Proof of Theorem 8.35

Let v_l be the first vertex in $N(v_i, G_{i\sigma})$ to be eliminated in the sequence σ. Since $v_i \in G_{i\sigma}$ and $v_l \in N(v_i, G_{i\sigma})$, there is an edge $e' \in E_{i\sigma}$ linking v_i and v_l. If we prove that $Cl(v_l, G_{l\sigma}) \supseteq N(v_i, G_{i\sigma})$ and that v_l is eliminated after v_i in the sequence σ (i.e. $\sigma = [\ldots, v_i, \ldots, v_l, \ldots]$), then the theorem is proved because the procedure Tree-Alg implies that $Cl(v_l, G_{l\sigma})$ is certainly contained in a node $\mathcal{N}$ of T_{i+1}. The fact that v_l is eliminated after v_i is clear because they are both vertices of $G_{i\sigma}$ and v_i is the next vertex to be eliminated. Now let's prove that $N(v_i, G_{i\sigma}) \subseteq Cl(v_l, G_{l\sigma})$. If $x \in N(v_i, G_{i\sigma})$, then there is an edge $e'' \in E_{i\sigma}$ linking x and v_i. Since v_i is a simplicial vertex in $G_{i\sigma}$, there is an edge $e \in E_{i\sigma}$ linking x and v_l ($v_l \in N(v_i, G_{i\sigma})$). But e is in $E_{l\sigma}$ because v_l is the first vertex in $N(v_i, G_{i\sigma})$ to be eliminated and both x and v_l are in $N(v_i, G_{i\sigma})$. This shows that $x \in N(v_l, G_{l\sigma})$ and hence $x \in Cl(v_l, G_{l\sigma})$ and the proof is complete. □

Proof of Theorem 8.36

Let $\mathcal{M}$ be the set of nodes of T_1 and let $\mathcal{C} = \{Cl(v_i, G_{i\sigma}) : i = 1, \ldots, n\}$ and $Max\ \mathcal{C}$ the maximal sets of $\mathcal{C}$. According to the procedure Tree-Alg we have clearly $\mathcal{M} \subseteq \mathcal{C}$ and $Max\ \mathcal{C} \subseteq \mathcal{M}$. If $\mathcal{K}$ denotes the cliques of G_σ, we must prove that $\mathcal{K} = \mathcal{M}$. If $K \in \mathcal{K}$, then $K \in\ Max\ \mathcal{C}$ by theorem 8.34 and hence $K \in \mathcal{M}$.

It will be useful to show that no element of $\mathcal{M}$ is a proper subset of another element of $\mathcal{M}$. To show this, we prove by induction that this property holds for all trees $T_n, T_{n-1}, \ldots, T_1$. The property clearly holds for T_n and suppose that it holds for T_{i+1}. We prove that it holds also for T_i. In a first case, assume that $\mathcal{N} = N(v_i, G_{i\sigma})$. Then the induction hypothesis together with $v_i \notin H$ for all $H \in T_{i+1}$ implies that T_i, obtained by replacing $\mathcal{N}$ by $Cl(v_i, G_{i\sigma})$, also satisfies the property (use the fact that the sets (nodes) of T_{i+1} are all different). Secondly, assume that $\mathcal{N} \supset N(v_i, G_{i\sigma})$: the tree T_i is obtained by linking $\mathcal{N}$ with $Cl(v_i, G_{i\sigma})$. The fact that $v_i \notin H$ for all $H \in T_{i+1}$ implies that $Cl(v_i, G_{i\sigma})$ is not contained in any such H. Furthermore, assuming that there is $H \in T_{i+1}$ such that $H \subset Cl(v_i, G_{i\sigma})$ implies that $H \subseteq N(v_i, G_{i\sigma})$ because $v_i \notin H$. Hence $H \subseteq N(v_i, G_{i\sigma}) \subset \mathcal{N}$ which is impossible by induction hypothesis. Therefore, the condition is always satisfied for T_{i+1} and hence also for T_1.

Now we prove that $\mathcal{M} \subseteq \mathcal{K}$, i.e. $\mathcal{M} \subseteq Max\ \mathcal{C}$ by theorem 8.34. Let $H \in \mathcal{M}$. Since we know that $\mathcal{M} \subseteq \mathcal{C}$, there is v_i such that $H = Cl(v_i, G_{i\sigma})$. It remains to prove that $Cl(v_i, G_{i\sigma})$ is maximal. Suppose that $Cl(v_i, G_{i\sigma})$ is not maximal, i.e. there is $v_j \neq v_i$ such that $Cl(v_i, G_{i\sigma}) \subset Cl(v_j, G_{j\sigma})$ and $Cl(v_j, G_{j\sigma}) \in Max\ \mathcal{C}$. Since $Max\ \mathcal{C} \subseteq \mathcal{M}$, there is $H' \in \mathcal{M}$ such that

$$H = Cl(v_i, G_{i\sigma}) \subset Cl(v_j, G_{j\sigma}) = H' \qquad (A8.70)$$

and hence H is a proper subset of H', which is impossible by the assertion proved above. Therefore, $H = Cl(v_i, G_{i\sigma})$ is in $Max\ \mathcal{C}$. □

Proof of Theorem 8.37

First, we show that if $H_1, \ldots, H_t, \ldots, H_r$ is a hypertree construction sequence of T_{i+1}, and if T_i is obtained by replacing the node $\mathcal{N} = H_t$ by $Cl(v_i, G_{i\sigma})$ (i.e. when $N(v_i, G_{i\sigma}) = H_t$), then $H_1, \ldots, Cl(v_i, G_{i\sigma}), \ldots, H_r$ is a hypertree construction sequence of T_i. The condition $H_k \cap (\cup_{j<k} H_j) = H_k \cap H_s$

for some $s < k$ is clearly satisfied for $k < t$ because $H_1, \ldots, H_{t-1}$ is a hypertree construction sequence. Now we prove that

$$Cl(v_i, G_{i\sigma}) \cap (\cup_{j<t} H_j) = Cl(v_i, G_{i\sigma}) \cap H_s \tag{A8.71}$$

for some $s < t$. By hypothesis, there is a $s < t$ such that $H_t \cap (\cup_{j<t} H_j) = H_t \cap H_s$. Since $Cl(v_i, G_{i\sigma}) = H_t + \{v_i\}$ (because $N(v_i, G_{i\sigma}) = H_t$ and v_i is not in H_t), we have

$$\begin{aligned} Cl(v_i, G_{i\sigma}) \cap (\cup_{j<t} H_j) &= (H_t \cap (\cup_{j<t} H_j)) \cup (\{v_i\} \cap (\cup_{j<t} H_j)) \\ &= H_t \cap H_s \subseteq Cl(v_i, G_{i\sigma}) \cap H_s \end{aligned} \tag{A8.72}$$

because $\{v_i\} \cap H_j = \emptyset$ for all $j \leq r$, in particular for all $j < t$. Since the other inclusion is trivial, equation (A8.71) is proved.

If we abbreviate $Cl(v_i, G_{i\sigma})$ by Cl, it remains to prove that, for all $t+1 \leq k \leq r$,

$$H_k \cap (Cl \cup (\cup_{j<k, j\neq t} H_j)) \tag{A8.73}$$

equals $H_k \cap H_s$ for some $s < k, s \neq t$ or equals $H_k \cap Cl$. By hypothesis, there is $s < k$ such that $H_k \cap (\cup_{j<k} H_j) = H_k \cap H_s$. Since $\{v_i\} \cap H_k = \emptyset$ for $1 \leq k \leq r$ and $Cl = H_t + \{v_i\}$, we have

$$\begin{aligned} H_k \cap (Cl \cup (\cup_{j<k, j\neq t} H_j)) &= (H_k \cap (\cup_{j<k} H_j)) \cup (H_k \cap \{v_i\}) \\ &= H_k \cap (\cup_{j<k} H_j) = H_k \cap H_s. \end{aligned} \tag{A8.74}$$

If $s \neq t$ then the condition is satisfied. Otherwise,

$$H_k \cap (Cl \cup (\cup_{j<k, j\neq t} H_j)) = H_k \cap H_t \subseteq H_k \cap Cl \tag{A8.75}$$

because $Cl = H_t + \{v_i\}$. Since the other inclusion is trivial, the condition is also satisfied. This shows that $H_1, \ldots, Cl, \ldots, H_r$ is a hypertree construction sequence of T_i.

Now we prove that if $H_1, \ldots, H_t, \ldots, H_r$ is a hypertree construction sequence of T_{i+1}, and if T_i is obtained by linking the new node Cl to the node $\mathcal{N} = H_t$ (i.e. when $N = N(v_i, G_{i\sigma}) \subset H_t$), then $H_1, \ldots, H_r, Cl$ is a hypertree construction sequence of T_i. By hypothesis, we only have to prove that there is $s \leq r$ such that

$$Cl \cap (\cup_{j\leq r} H_j) = Cl \cap H_s. \tag{A8.76}$$

Since $\{v_i\} \cap H_j = \emptyset$ for $1 \leq j \leq r$ and $Cl = N + \{v_i\}$ we have

$$\begin{aligned} Cl \cap (\cup_{j\leq r} H_j) &= N \cap (\cup_{j\leq r} H_j) \subseteq N \\ &= N \cap H_t \subseteq Cl \cap H_t \end{aligned} \tag{A8.77}$$

because $N \subset H_t$. Since the other inclusion is trivial, (A8.76) is proved.

Finally, since each time a new node is added in the Tree-Alg procedure there is a corresponding new step in the hypertree construction sequence, it follows that this procedure builds T_1 according to a hypertree construction sequence. Theorem 8.25 implies that T_1 is a Markov tree. □

Proof of Theorem 8.38

Let $K \in H$. Then K is a complete subgraph of $G(H)$ and hence there is a clique C of $G(H)$ such that $K \subseteq C$. But each clique of $G(H)$ is contained in a clique of $G(H)_\sigma$ for any sequence σ. Since each clique of $G(H)_\sigma$ is a node of $T_\sigma(H)$ according to theorem 8.36, the assertion is proved. □

Proof of Theorem 8.39

The proof of the first two assertions can be found in Almond and Kong (1991). Using theorems 8.38 and 8.36, the first assertion implies that T is a covering Markov tree of H. □

Proof of Theorem 8.40

By theorem 8.31 $T_\sigma(H)$ is a spanning tree over the cliques of $G(H)_\sigma$. Since $G(H)$ is triangulated we have $G(H)_\sigma = G(H)$ and hence $T_\sigma(H)$ is a spanning tree over the cliques of $G(H)$. If H' denotes this set of cliques (a hypergraph), then this shows that $T_\sigma(H)$ is fitting for H' and hence $T_\sigma(H)$ is $\mathcal{C}$-optimal for the hypergraph H'. Then theorem 8.29 implies that $T_\sigma(H)$ is also $\mathcal{C}$-optimal for H. □

Part II

Applications

9. Statistical Inference

9.1 Introduction: The Basic Model

Statistical methods nowadays are basically decision oriented. In parameter estimation a specific value or possibly a range of values are to be selected as probable or plausible values of an unknown parameter. In hypothesis testing, a decision whether to accept or reject a null hypothesis has to be made. Thus, statistical theory is basically about choices and their qualities and risks. In contrast to this point of view, an approach to reasoning about hypotheses rather than to deciding on hypotheses will be put forward here. It is argued that the true job of a statistician is not to make decisions. It is rather to provide sound methods for decision support, to look for arguments in favour and against a hypothesis and to judge its credibility and plausibility. The statistician must also face the fact that statistical data are rarely the only information available about a given question. Hence, decisions based on statistical data only may be biased. Therefore, methods for judging hypotheses based on statistical data are more appropriate than methods for deciding them because these judgments may be complemented with further information which is not of purely statistical nature.

In order to judge hypotheses, arguments in favour and against the hypotheses must be elaborated and weighted. Such arguments may be provided by statistical data. This presupposes however the existence and knowledge of a model which links the observations to the relevant question to be considered. In the guise of an intuitive introduction into the subject, such a model will be proposed and the basic approach of analysis will be illustrated in this section. The model and its analysis will be compared to other, but related methods. In the next section, the model and the underlying theory will be developed more formally and put into the framework of the theory of hints. Sections 9.3 and 9.4 contain some illustrations and applications of the approach using classical problems of statistical analysis. This permits the comparison with well known classical results and methods.

Let Π be the set of admissible values of an unknown parameter π. Inference about this parameter is attempted by means of an observation process which yields an observation x in some observation space X. It is assumed that the actual observation x depends only on the unknown parameter π and a further unknown element ω. This is expressed by a known function $x = f(\omega, \pi)$ where

ω is assumed to be a chance element belonging to some set Ω on which a known probability distribution P is defined. In order to simplify the discussion, the sets Π, X and Ω are all assumed to be finite in this section and the probability measure P on Ω is defined by the discrete probabilities $p(\omega)$.

Suppose now that in the actual observation process the observation x is made. Using the observation x together with the process model $x = f(\omega, \pi)$, what inference can be made about π ? First, note that by the observation of x some chance elements ω must be excluded as impossible. In fact, if there is no parameter value $\pi \in \Pi$ such that $x = f(\omega, \pi)$, then ω is clearly excluded as a possible chance element generating x. Denote by $v_x(\Pi)$ the set of all ω which remain possible given the observation x:

$$v_x(\Pi) = \{\omega \in \Omega : \text{ there is a } \pi \in \Pi \text{ such that } x = f(\omega, \pi)\}. \tag{9.1}$$

Thus the observation x represents information which makes it possible to condition the probability measure P on Ω to $v_x(\Pi)$. This leads to the revised probabilities $p'(\omega) = p(\omega)/k$ where

$$k = P(v_x(\Pi)) = \sum\{p(\omega) : \omega \in v_x(\Pi)\}. \tag{9.2}$$

Within $v_x(\Pi)$ it remains unknown which element ω caused the observation x, only their respective probabilities $p'(\omega)$ are known. However, suppose for the moment that one knew the element ω which caused the observation x. Then the set of possible parameter values π could be restricted to the set

$$T_x(\omega) = \{\pi \in \Pi : x = f(\omega, \pi)\}. \tag{9.3}$$

In some cases $T_x(\omega)$ will be a one-element set, but in general it will contain more than one element, such that even the knowledge of ω does not permit to identify the unknown parameter π. Surely, the reader will recognize in this and the subsequent discussion the ingredients of hints. The link to the theory of hints will be formally established in the next section.

Consider a hypothesis about the unknown parameter π. Such a hypothesis is represented by a subset $H \subseteq \Pi$. H expresses the hypothesis that the unknown parameter value π is an element of H. This hypothesis may be true or false and in general this question cannot be decided unambiguously on the basis of the observation x. However, one may look for the elements ω which would make H true. The hypothesis H is necessarily true for elements ω satisfying $T_x(\omega) \subseteq H$. Thus, define

$$u_x(H) = \{\omega \in v_x(\Pi) : T_x(\omega) \subseteq H\}. \tag{9.4}$$

If one knew that the ω which caused the observation x were in the set $u_x(H)$, then one could confirm H and the hypothesis would be proved. Each ω in $u_x(H)$ is an argument in favour of H. As already stated, it is usually not known whether indeed an argument in favour of H is valid, but at least the *probability* of ω being in $u_x(H)$ is known. This means the probability that the hypothesis H could be proved on the basis of the observation x and the observation process is known. This probability is in fact the reliability with which the hypothesis can be logically derived or deduced from the data and the model. The larger this probability, the more the hypothesis is *supported* by the available information, the more *credible* is the hypothesis in the light of the observation and the model. This probability is the *degree of support* or the *degree of credibility* of the hypothesis. Let's denote this degree of support as usual by

$$sp_x(H) = P'(u_x(H)), \tag{9.5}$$

where the probability measure P' is defined by the conditional probabilities $p'(\omega)$ on $v_x(\Pi)$.

Any $\omega \in u_x(H^c)$ is an argument in favour of the negation of H, thus an argument against H. The more probable the event $u_x(H^c)$, the more doubt is shed on the hypothesis H. Therefore $P'(u_x(H^c))$ is called the *degree of doubt* of H. The less doubt is cast on H, the more *plausible* is the hypothesis H. Thus one can also define the *degree of plausibility* of H by

$$pl_x(H) = 1 - P'(u_x(H^c)) = 1 - sp_x(H^c). \tag{9.6}$$

This definition is even more strongly justified if one considers all the elements $\omega \in v_x(\Pi)$ which do not exclude H (i.e. which do not prove H^c), namely

$$v_x(H) = \{\omega \in v_x(\Pi) : T_x(\omega) \cap H \neq \emptyset\}. \tag{9.7}$$

This is the set of all arguments ω for which H is *possible*, albeit not necessarily proved. It is obvious that $v_x(H) = u_x(H^c)^c$ and thus $pl_x(H) = P'(v_x(H))$. Clearly $u_x(H) \subseteq v_x(H)$ such that always $sp_x(H) \leq pl_x(H)$. A hypothesis is always supported to at most its degree of plausibility.

Let's illustrate this analysis by a first, very simple example. Consider a fair coin with faces designated by 1 and 2, which is tossed n times. Suppose there are only two possible simple hypotheses: either face 1 carries heads and face 2 carries tails (hypothesis π_0) or both faces carry heads (hypothesis π_1). Thus $\Pi = \{\pi_0, \pi_1\}$. The observed value x is the number of heads turning up during the n throws. The chance element ω can be defined as the number of times face 1 turns up, $\omega \in \Omega = \{0, 1, \ldots, n\}$; it has a binomial distribution $bi(1/2, n)$. The observation process is then defined by

$$f(\omega, \pi_0) = \omega; f(\omega, \pi_1) = n \tag{9.8}$$

for all $\omega \in \Omega$. Now suppose at first that the observed value x is strictly smaller than n. Then clearly the parameter value π_1 is excluded and the only possible value for ω is $\omega = x$. It follows $p'(x) = 1$ and $T_x(x) = \{\pi_0\}$ if $x < n$. Hence $u_x(\{\pi_0\}) = v_x(\{\pi_0\}) = \{x\}$ and both the degree of support and the degree of plausibility of the hypothesis $\{\pi_0\}$ equals one,

$$sp_x(\{\pi_0\}) = pl_x(\{\pi_0\}) = 1. \tag{9.9}$$

In this case the observation allows for an unambiguous determination of the unknown parameter. The situation is different, if n times heads is observed,

$x = n$. In this case all values of $\omega \in \{0, 1, 2, \ldots, n\}$ are possible ($v_n(\Pi) = \Omega$) and the observation does not change the binomial distribution of ω. Now, the elements $\omega < n$ imply π_1 and therefore $u_n(\{\pi_1\}) = \{\omega : \omega < n\}$. On the other hand, there is no ω which implies π_0, hence $u_n(\{\pi_0\}) = \emptyset$. Thus, there is no support for π_0, $sp_n(\{\pi_0\}) = 0$, whereas $sp_n(\{\pi_1\}) = 1 - (1/2)^n$. The support for π_1 is nearly one, if n is large. This implies also that the degree of plausibility of π_0 equals $(1/2)^n$, which is nearly zero, if n is large. Thus, although in this case the observation does not unambiguously determine the unknown parameter, there is nevertheless, for large n, overwhelming evidence that π_1 must be the correct value. However, note that this analysis does not force one to make a decision, to accept or to reject a hypothesis. It only points out the degree of reliability of inferring a hypothesis from the observation and the model.

9.2 Statistical Inference in the Framework of Hints

In the model introduced in the previous section there are at least two distinct bodies of information or knowledge which were used in the analysis. One piece of knowledge is the process model of observation, described by the function $f(\omega, \pi)$ and the probability P on the random elements ω in Ω. The other piece is the observation x. Sometimes there might even be other elements; for example some information about the unknown parameter π, known independently and possibly (but not necessarily) prior to the observation x. In this section it will be shown that the concept of hints serves to describe uniformly those bodies of information. In particular and most importantly, it will be shown how such bodies of information must be combined by Dempster's rule in order to get a synthesis of the distinct pieces of information. And furthermore it will be argued that the support and plausibility functions can be used to judge hypotheses about the unknown value. This generalizes and formalizes the intuitive analysis introduced in the previous section.

Let's start with the process model of observation. One way to look at this model is to note that each $\omega \in \Omega$ determines, by the equation $x = f(\omega, \pi)$, a *relation* between x and π, i.e. a subset of the product space $X \times \Pi$. Thus, this model provides some information about the possible values of x and π. This information is, of course, uncertain because the value of ω is not known and it is also imprecise because, even if ω were known, this would in general not limit the possible values of (x, π) to a unique value. Define

$$\Gamma(\omega) = \{(x, \pi) \in X \times \Pi : x = f(\omega, \pi)\}. \tag{9.10}$$

In this way it becomes clear that the observation model defines a hint $\mathcal{H}_{pr} = (\Omega, P, \Gamma, X \times \Pi)$ relative to $X \times \Pi$. Here the "interpretations" are the chance elements which determine the outcome of the experiment. The observation x is simply a deterministic and precise hint $\mathcal{H}_x$ on X in this framework.

The model underlying our statistical analysis is represented so far by two distinct hints, one for the process model referring to the frame $X \times \Pi$ and one for the observation referring to the frame X. There could possibly be a third hint relative to Π describing information about the unknown parameter value independently of the observation model. This would correspond to the Bayesian a priori information (it is however not clear why such an information should be necessarily prior to the observation). In any case we are faced with the problem of combining different hints, possibly referring to different frames, to synthesize the information. Dempster's rule proposes itself for this task. It will be demonstrated that this is indeed the appropriate way of combining hints in the present context of statistical inference.

Dempster's rule in this context applies in the form

$$((\mathcal{H}_x \uparrow X \times \Pi) \oplus \mathcal{H}_{pr}) \downarrow \Pi. \tag{9.11}$$

According to Chap. 7 this is to be interpreted as follows. The hint relative to X describing an actual observation x is vacuously extended to $X \times \Pi$. This gives simply the deterministic hint with focal set $\{x\} \times \Pi$. The extended hint can then be combined using Dempster's rule with the hint relative to $X \times \Pi$ describing the process model. For each $\omega \in \Omega$ the intersection of $\Gamma(\omega) = \{(y, \pi) \in X \times \Pi : y = f(\omega, \pi)\}$ with the cylinder set $\{x\} \times \Pi$ has to be formed. This gives simply the new focal sets $\Gamma'(\omega) = \{(x, \pi) : \pi \in \Pi, x = f(\omega, \pi)\}$ for all ω for which the equation $x = f(\omega, \pi)$ has a solution for π. All ω for which this equation has no solution are in contradiction with the observation x and have to be excluded. The probability on Ω has to be conditioned on the event that ω is non-contradictory. This defines then the combined hint of the process model and the observation.

The combined hint represents information relative to the frame $X \times \Pi$. However, we are interested in information about the unknown parameter π alone. So the combined hint is marginalized to the frame Π. The marginalization leads to a hint with focal sets $\Gamma''(\omega) = \{\pi \in \Pi : x = f(\omega, \pi)\}$. This corresponds exactly to $T_x(\omega)$ in the notation of the previous section. This construction in terms of hints and Dempster's rule corresponds therefore exactly to the analysis introduced in the previous section. The fact that the result of the analysis is a hint shows that it makes sense to judge hypotheses $H \subseteq \Pi$ about the unknown parameter π by their degrees of support and plausibility relative to the hint (9.11). And it is clear that these degrees of support and plausibility correspond to those defined in (9.5) and (9.6).

Now we return to our statistical model. Let $x = f(\omega, \pi), \pi \in \Pi, x \in X, \omega \in \Omega, P$ define a process model of observation. It describes the result obtained by a single observation. But it is usual in statistics to assume that such an observation (or experiment) can be repeated independently several times. In each iteration $i = 1, 2, \ldots, n$ another chance element ω_i is selected and another corresponding observation $x_i = f(\omega_i, \pi)$ arises. Of course, the value of the unknown parameter π is the same during all experiments. Define $\mathbf{x} = (x_1, x_2, \ldots, x_n) \in X^n, \omega = (\omega_1, \omega_2, \ldots, \omega_n) \in \Omega^n$ and let

$f(\omega, \pi) = (f(\omega_1, \pi), f(\omega_2, \pi), \ldots, f(\omega_n, \pi))$. Furthermore, define the product probability space (Ω^n, P^n) where P^n denotes the product probability. The equation $\mathbf{x} = f(\omega, \pi)$ defines a new process model of observation describing the n-fold repetition of the original observation process.

This way of composing models is easily described in the language of hints. First we consider the multivariate family of compatible frames induced by the frames $X_1 = X, X_2 = X, \ldots, X_n = X, \Pi$. Then let $\mathcal{H}_i$ denote the hint relative to $X_i \times \Pi$ that describes the process model for the i-th observation and let $\mathcal{O}_i$ denote the hint on X_i describing the observation x_i. Then we have to combine all these hints and coarsen the result to the parameter space Π. So let

$$\mathcal{H} = \Big((\oplus_{i=1}^n \mathcal{H}_i) \oplus (\oplus_{i=1}^n \mathcal{O}_i)\Big) | \Pi). \tag{9.12}$$

The tree given in Fig. 9.1 a) is clearly a covering Markov tree of the hypergraph induced by the hints.

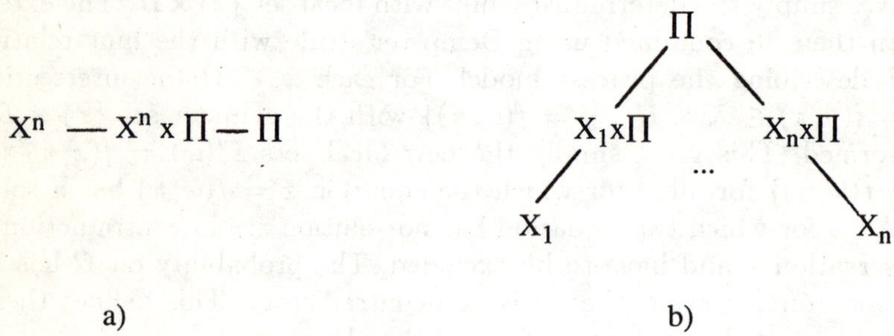

Fig. 9.1.

Therefore, if we define

$$\mathcal{H}' = (\mathcal{H}_1 \uparrow X^n \times \Pi) \oplus (\mathcal{H}_2 \uparrow X^n \times \Pi) \oplus \cdots \oplus (\mathcal{H}_n \uparrow X^n \times \Pi$$

and

$$\mathcal{O}' = (\mathcal{O}_1 \uparrow X^n) \oplus (\mathcal{O}_2 \uparrow X^n) \oplus \cdots \oplus (\mathcal{O}_n \uparrow X^n) \tag{9.13}$$

then theorem 8.16 implies that

$$\mathcal{H} = ((\mathcal{O}' \uparrow X^n \times \Pi) \oplus \mathcal{H}') \downarrow \Pi. \tag{9.14}$$

In view of equation (9.11), this way of computing $\mathcal{H}$ shows that the vector x can be considered as a simple compound observation (represented by the hint $\mathcal{O}'$ to be combined with the single compound process model $x = f(\omega, \pi)$ (represented by the hint $\mathcal{H}'$). But the tree given in Fig. 9.1 b) is another covering Markov tree which leads to another way of computing $\mathcal{H}$. In fact, if

$$\mathcal{P}_i = ((\mathcal{O}_i \uparrow X \times \Pi) \oplus \mathcal{H}_i) \downarrow \Pi \tag{9.15}$$

is the hint obtained by analyzing each observation individually, then again by theorem 8.16 we have

$$\mathcal{H} = \mathcal{P}_1 \oplus \mathcal{P}_2 \oplus \cdots \oplus \mathcal{P}_n. \tag{9.16}$$

This may be important for actual computations. Sometimes (9.14) is better suited, sometimes (9.16) is simpler. Applications of this result will be found in Sect. 9.3. Let us conclude this section with some remarks about how our approach to statistical inference relates to the more classical one. If $x = f(\omega, \pi), x \in X, \pi \in \Pi, \omega \in \Omega, P$ is a process model of observation, then it determines, loosely speaking, for any $\pi \in \Pi$ a conditional probability distribution P_π on X. In fact, for subsets B of X and any $\pi \in \Pi$ we have $P_\pi(B) = P(\{\omega \in \Omega : f(\omega, \pi) \in B\}$. Statistical models are usually specified in terms of a family of probability distributions P_π on X. It is well known that different process models may lead to the same family of distributions P_π (Bunke, 1975; Fraser 1968). A process model contains therefore more information than the usual statistical specifications P_π.

Consider the case of a finite set X. Then P_π is defined by the family of elementary probabilities

$$p_\pi(x) = P(\{\omega \in \Omega : f(\omega, \pi) = x\}. \tag{9.17}$$

For a fixed observation x, $p_\pi(x)$, considered as a function of π, is classically called the *likelihood function* of the statistical specification $p_\pi(x)$. On the other hand, also for a fixed observation x and a fixed π, the set $\{\omega \in \Omega : \pi \in T_x(\omega)\}$ is the set of all ω for which the parameter value π is possible. This set can also be written as $\{\omega \in \Omega : f(\omega, \pi) = x\}$. Then the degree of plausibility of the hypothesis $\{\pi\}$ is proportional to $P(\{\omega \in \Omega : f(\omega, \pi) = x\})$ (it is in general not equal, because some ω may have to be excluded as contradictory to the observation x). Let's write these plausibilities of singletons as $pl_x(\pi)$ (as an abbreviation of $pl_x(\{\pi\})$). Thus we always have

$$pl_x(\pi) = kp_\pi(x) = kP(\{\omega \in \Omega : f(\omega, \pi) = x\} \tag{9.18}$$

for some constant k independent of π.

This means that the likelihood function indicates in fact the relative plausibilities of single parameter values π. This gives the likelihood function a clear *probabilistic interpretation*, at least for discrete sample spaces: up to a normalization constant, the likelihood of a parameter value π equals the probability that, given the process model and the observation x, the value is possible (not excluded). Likelihoods are proportional to *degrees of plausibility* or *possibility*. These relative plausibilities of singleton values depend thus only on the family of distributions p_π and not on the full process model.

In the context of our general process model of observation and the related analysis, it is also without difficulty conceivable to assume additional, extra-statistical information about the unknown parameter π. This information can be described by a hint $\mathcal{H}_0$ relative to the frame Π. If $\mathcal{H}$ denotes the hint about π obtained from the analysis of the process model of observation, then the global information about π can be described by the combined hint $\mathcal{H}_0 \oplus \mathcal{H}$, provided of course, that the information in $\mathcal{H}_0$ does not stochastically depend on the random element of the process model of observation. Note that $\mathcal{H}_0$ needs not be precise.

In classical Bayesian statistics, $\mathcal{H}_0$ is assumed to be a precise hint (using our terminology), described by some probability distribution on Π. It is called the a priori distribution, which emphasizes the idea that it represents the knowledge about the unknown parameter π before the statistical observation is available. Bayesian analysis is in fact only possible if the a priori distribution is known at the time the statistical data of the observation is processed. But there is no reason why the extra-statistical information should always be prior to the information obtained from the statistical observation. There are surely situations, where extra-statistical information becomes available only afterwards. There is no reason why there should be a particular time sequence relating two sources of information. Anyway, in our context, the information contained in the statistical observations can be analyzed, if necessary, independently of any other information, and only afterwards combined at any time with extra-statistical information, or information from further statistical experiments or observations. The combination sequence is irrelevant because the Dempster's rule of combination is commutative, which – let's repeat it – is different in Bayes rule.

In order to place our analysis into the perspective of Bayesian analysis, suppose that $\mathcal{H}_0$ is indeed a precise hint. For the sake of simplicity, we assume Π to be a finite set, such that the information in $\mathcal{H}_0$ can be subsumed into a discrete density $p_0(\pi)$. According to theorem 4.8 the combined hint $\mathcal{H}_0 \oplus \mathcal{H}$ is also precise and it has the discrete density $p(\pi) = kp_0(\pi)pl_x(\pi)$ where $pl_x(\pi)$ is the plausibility of the singleton sets according to the hint $\mathcal{H}$ representing the process model of observation and the observation x. According to (9.18), this plausibility is proportional to $p_\pi(x)$, the conditional probability of the observation x given the parameter π. It follows that

$$p(\pi) = kp_0(\pi)p_\pi(x), \tag{9.19}$$

which is of course nothing else than Bayes' formula. Thus our approach based on Dempster's rule generalizes the Bayesian approach to statistics. This is not a big surprise since Dempster (1967) developed his rule as a generalization of Bayes' formula (see also Shafer, 1976 b). Note that in the particular case of an extra-statistical information $\mathcal{H}_0$ represented by a precise hint, it is not necessary to know all the details of the process model of observation; it is sufficient to know $p_\pi(x)$.

9.3 Sampling Models

As a first class of applications of our approach to statistical analysis we discuss a few sampling problems.

9.3.1 Simple Alternatives: A General Model

First the simple case of only two possible parameter values (or hypotheses) π_0 (null hypothesis) and π_1 (alternative) will be considered in some detail. This is both conceptually and practically an important problem. To start with, a quite general process model will be introduced and analyzed. In the next section, a particular model of this type will then be discussed.

As in the illustrative model sketched in Sect. 9.1, we define the parameter space $\Pi = \{\pi_0, \pi_1\}$ containing the two possible alternatives. It is unknown which one of these two possible alternatives is actually valid. In order to judge the likelihood of each of these two alternatives, some observation x in an observation space X will be made. The set X is assumed to be a finite set. The observation x depends not only on which alternative is valid, but also on a random element ω out of some known probability space $(\Omega, \mathcal{A}, P)$. The observation x is then determined by a known function

$$x = f(\omega, \pi), \; x \in X, \omega \in \Omega, \pi \in \Pi. \tag{9.20}$$

This is the process model of observation.

The model can be analyzed according to the method sketched in Sect. 9.1. Given an observation x, the set of possible random elements ω is $v_x(\Pi) = \{\omega : f(\omega, \pi) = x \text{ for } \pi = \pi_0 \text{ or } \pi = \pi_1\}$. This is an event that is known to have occurred, and the probability P must be conditioned on this event. Let $k = P(v_x(\Pi))$ (or more generally the exterior probability $k = P^*(v_x(\Pi))$, if $v_x(\Pi)$ is not $\mathcal{A}$-measurable). The new probability space $(v_x(\Pi), \mathcal{A} \cap v_x(\Pi), P' = P/k)$ will then be used in the sequel of the analysis.

Any random element ω out of $v_x(\Pi)$ determines then a non-empty subset of possible parameter values: $T_x(\omega) = \{\pi : f(\omega, \pi) = x\}$. Define then the sets of random elements ω which necessarily imply π_0 or π_1:

$$\begin{aligned} u_x(\pi_0) &= \{\omega : T_x(\omega) = \{\pi_0\}\} = \{\omega : f(\omega, \pi_0) = x, f(\omega, \pi_1) \neq x\} \\ u_x(\pi_1) &= \{\omega : T_x(\omega) = \{\pi_1\}\} = \{\omega : f(\omega, \pi_1) = x, f(\omega, \pi_0) \neq x\}. \end{aligned} \tag{9.21}$$

Similarly, define the sets of random elements ω for which π_0 or π_1 are possible:

$$\begin{aligned} v_x(\pi_0) &= \{\omega : \pi_0 \in T_x(\omega)\} = \{\omega : f(\omega, \pi_0) = x\} \\ v_x(\pi_1) &= \{\omega : \pi_1 \in T_x(\omega)\} = \{\omega : f(\omega, \pi_1) = x\}. \end{aligned} \tag{9.22}$$

It is then possible to define the degrees of support and plausibility of the two alternatives:

$$\begin{aligned} sp_x(\pi_0) &= P'(u_x(\pi_0)),\ sp_x(\pi_1) = P'(u_x(\pi_1)) \\ pl_x(\pi_0) &= P'(v_x(\pi_0)),\ pl_x(\pi_1) = P'(v_x(\pi_1)). \end{aligned} \tag{9.23}$$

One would tend to accept an alternative when its degree of support is large (nearly one) and reject the hypothesis when its degree of plausibility is small (nearly zero). Note that

$$\begin{aligned} &sp_x(\pi_0) + sp_x(\pi_1) \leq 1,\ pl_x(\pi_0) + pl_x(\pi_1) \geq 1 \\ &pl_x(\pi_0) - sp_x(\pi_0) = pl_x(\pi_1) - sp_x(\pi_1) \end{aligned} \tag{9.24}$$

such that a high support for one alternative entails a small plausibility for the other one and vice versa. It is however very well possible that both alternatives have a small degree of support and a high degree of plausibility, in which case the available information does not make it possible to decide between the alternatives.

As explained in section 9.2, the plausibilities $pl_x(\pi)$ are proportional to the likelihoods $p_\pi(x) = P(\omega : f(\omega, \pi) = x)$. Sometimes it will not be possible to specify the whole process model as defined above, but only the likelihoods $p_\pi(x)$. Then it will not be possible to determine unambiguously the degrees of support sp_x and only the plausibility ratio $pl_x(\pi_0)/pl_x(\pi_1) = p_{\pi_0}(x)/p_{\pi_1}(x)$ can be computed. This is the situation usually considered in statistics, where this ratio is called likelihood ratio. Now, if the likelihood ratio is small (large), then $pl_x(\pi_0)$ must be nearly zero (nearly one) and hence $pl_x(\pi_1)$ nearly one (nearly zero). According to (9.24) in such a situation $sp_x(\pi_1)$ equals nearly $pl_x(\pi_1)$, $sp_x(\pi_0)$ equals nearly $pl_x(\pi_0)$ and hypothesis π_1 (π_0) is strongly supported. This explains the usual likelihood ratio test from the point of view of evidential reasoning.

9.3.2 Simple Alternatives: Urn Models

As an application of the general model introduced above, consider an urn containing m balls, of which it is known that either m_0 or m_1 are white ones, but it is unknown which one of the two alternatives is the true one. In order to clarify this question balls are drawn at random from the urn and their colour is observed. To begin with, assume that only one single ball is drawn.

In selecting a particular process model great care must be exercised in correctly describing the actual physical process of sampling. For example the following process model can be proposed. Number the balls from 1 to m and identify the random element ω of the sampling with the numbers on the balls, $\Omega = \{1, \ldots, m\}$. The probability on Ω is given by the discrete uniform density $p(\omega) = 1/m$. Suppose that the white balls have the numbers from 1 to m_0 or m_1, depending on the alternative which is valid, and assume $m_0 < m_1$, $\Pi = \{m_0, m_1\}$. The observation x is the colour of the ball drawn at random from the urn, $X = \{white, black\}$. The process model of observation is then defined by

$$x = f(\omega, m_i) = \begin{cases} white & \text{if } \omega \leq m_i \\ black & \text{otherwise} \end{cases} \tag{9.25}$$

for $i = 1, 2$. What does this model actually describe ? In fact it states that there is a subpopulation of cardinality m_0 of white balls and these can always be numbered from 1 to m_0. There may be a further subpopulation of $m_1 - m_0$ other white balls (such that in total there are m_1 white balls). And the elements of this second subpopulation can be numbered from $m_0 + 1$ to m_1. Finally the remaining balls are black and can be numbered from $m_1 + 1$ to m. This is exactly the information which is encoded in the process model. We will contrast this below to a seemingly identical situation, which however is different and leads to a different analysis. This model can be analyzed according to the method discussed in the previous subsection. Suppose a white ball is drawn, $x = white$. Then the possible random elements are $v_{white}(\Pi) = \{\omega : \omega \leq m_1\}$. This event has the probability m_1/m and the new, conditional probabilities become $p'(\omega) = 1/m_1$ for $\omega \in v_{white}(\Pi)$. Furthermore, the following elements are obtained:

$$T_{white}(\omega) = \begin{cases} \{m_0, m_1\} & \text{if } \omega \leq m_0 \\ \{m_1\} & \text{if } m_0 < \omega \leq m_1. \end{cases}$$

$$\begin{aligned} u_{white}(m_0) &= \emptyset, \; u_{white}(m_1) = \{\omega : m_0 < \omega \leq m_1\} \\ v_{white}(m_0) &= \{\omega : \omega \leq m_0\}, \; v_{white}(m_1) = \{\omega : \omega \leq m_1\}. \end{aligned} \tag{9.26}$$

This implies

$$\begin{aligned} sp_{white}(m_0) &= 0, \; sp_{white}(m_1) = (m_1 - m_0)/m_1 \\ pl_{white}(m_0) &= m_0/m_1, \; pl_{white}(m_1) = 1. \end{aligned} \tag{9.27}$$

In other words, the sampling of a white ball defines a simple hint pointing to the alternative m_1 with a probability of $p_1 = (m_1 - m_0)/m_1$.

By a similar analysis it is found that the sampling of a black ball, $x = black$, defines a simple hint pointing to the alternative m_0 with a probability of $p_0 = (m_1 - m_0)/(m - m_0)$.

Of course, with a single sample from the urn not very much can be concluded about the two possible alternatives. More interesting is the case of a n times repeated sampling. The sampling is assumed to be done with replacement. Denote the n observations by $x_1, x_2, \ldots, x_n$, where each x_i has either the value white or black. Each observation x_i defines a simple hint $\mathcal{H}_i$ on the parameter space $\Pi = \{m_0, m_1\}$. According to section 9.2, the analysis of the combined process model of n independent, repeated observations leads to the combined hint $\mathcal{H}_1 \oplus \mathcal{H}_2 \oplus \cdots \oplus \mathcal{H}_n$ for the unknown alternative. The commutativity of the operation $\oplus$ suggests that the number x of white balls drawn in the n samples is a sufficient statistics for the judgment of the two possible alternatives m_0 and m_1.

For the computation of a n-fold combination of hints by Dempster's rule note that each simple hint represents a probable, but not infallible argument for either m_1 (if the corresponding ball is white) or m_0 (if the ball is black).

Each argument has a known reliability, p_1 for m_1 and p_0 for m_0. A sample of x white balls represents x probable arguments for m_1, $n-x$ black balls are $n-x$ probable arguments for m_0. However, it is impossible that an argument for m_1 and one for m_0 are valid simultaneously. This would represent a contradiction and the analysis has to be conditioned on the event that there is no contradiction. The probability that out of x arguments for m_1 and out of $n-x$ arguments for m_0 at least one of both types of arguments is valid, equals $(1-q_1^x)(1-q_0^{n-x})$ with $q_0 = 1-p_0$, $q_1 = 1-p_1$. Thus, let k be the probability of no contradiction

$$k = 1-(1-q_1^x)(1-q_0^{n-x}). \tag{9.28}$$

Now, the alternative m_1 is proved, if at least one of the x arguments in favour of m_1 is valid and no argument in favour of m_0 holds (in order to avoid a contradiction). The probability for this event must be conditioned on the event of no contradiction, which finally gives the degree of support

$$sp_x(m_1) = (1-q_1^x)q_0^{n-x}/k. \tag{9.29}$$

By a similar analysis, the degree of support for m_0 can also be obtained:

$$sp_x(m_0) = q_1^x(1-q_0)^{n-x}/k. \tag{9.30}$$

The plausibilities of the two alternatives are $pl_x(m_0) = 1-sp_x(m_1)$, $pl_x(m_1) = 1-sp_x(m_0)$. These results allow one to judge the credibility and plausibility of the two alternatives in the light of the combined hint.

A simplified analysis based on the relative plausibilities, i.e. the plausibility ratio of the two alternatives is also possible. According to theorem 4.9, the relative plausibilities or likelihoods for a combined hint can be obtained by multiplying those of the individual hints. Thus, for the n-fold sample with replacement which produces x white balls and $n-x$ black ones, we obtain

$$\begin{aligned} pl_x(m_0) &= c(pl_{white}(m_0))^x(pl_{black}(m_0))^{n-x} = cq_1^x \\ pl_x(m_1) &= c(pl_{white}(m_1))^x(pl_{black}(m_1))^{n-x} = cq_0^{n-x}. \end{aligned} \tag{9.31}$$

This leads to a plausibility or likelihood ratio of q_1^x/q_0^{n-x} to decide about the acceptance or rejection of the alternatives.

The same analysis can of course also be carried out for sampling from an infinite population. In this case Ω is defined to be the unit interval $[0,1]$ with the uniform distribution on it. The alternatives are then π_0 and π_1, $0 < \pi_0 < \pi_1 < 1$. For this case, exactly the same results as above hold, if $p_0 = (\pi_1-\pi_0)/(1-\pi_0)$ and $p_1 = (\pi_1-\pi_0)/\pi_1$.

9.3.3 Alternative Urn Models

It has been said that the process model must be selected with great care because the analysis and the results depend strongly on it. This issue will be discussed in this subsection. Consider the case of a population of m balls containing balls of two colours, white and black say, and it is known there are m_0 balls of one colour and $m - m_0$ of the other colour. The question is whether there are m_0 white or black balls. To get information relative to this question a ball is drawn at random from the urn and its colour is observed.

What is the correct model of this sampling ? One might be tempted to apply the model of the former subsection, namely there are either m_0 or $m_1 = m - m_0$ white balls. Suppose $m_0 < m_1$. Then (9.25) could again be taken as the process model. But the interpretation of this model given in subsection 9.3.2 does not sound correct for the new problem at hand. Consider the following alternative model. There are two possible urns, one with m_0 white and one with m_0 black balls. The ball (or the balls if sampling is repeated) is drawn from one of these urns, but it is unknown from which one. Let then π_0 denote the case that the ball is drawn from the first urn (m_0 white balls) and π_1 that the ball is drawn from the second urn (m_0 black balls). In both urns the balls are assumed to be numbered from 1 to m such that the white balls have numbers 1 to m_0 in π_0 and the black balls are numbered from 1 to m_0 in π_1. The random element ω is the number of the ball drawn and we have $p(\omega) = 1/m$ for all $\omega \in \Omega = \{1, 2, \ldots, m\}$. The process model can then be formulated as follows:

$$x = f(\omega, \pi) = \begin{cases} white & \text{if } \pi = \pi_0 \text{ and } \omega \leq m_0 \text{ or } \pi = \pi_1 \text{ and } \omega > m_0 \\ black & \text{if } \pi = \pi_0 \text{ and } \omega > m_0 \text{ or } \pi = \pi_1 \text{ and } \omega \leq m_0 \end{cases}$$

A simple analysis shows that this is different from the first model (compare with (9.25)). But note that both models have the same likelihood function $p_{\pi_0}(white) = m_0/m$, $p_{\pi_1}(white) = (m - m_0)/m$ or $p_{\pi_0}(black) = (m - m_0)/m$, $p_{\pi_1}(black) = m_0/m$.

The real reason for the uneasiness one feels in the question of how to select the process model is that the usual description of statistical problems in textbooks is too sloppy for our purposes. In order to compute a solution, it is sufficient to have $p_\pi(x)$, the conditional probabilities of the observation given the parameter, but this is not sufficient to determine the process model. Much work remains to be done in analyzing real sampling problems and in developing appropriate process models for them. For this purpose the physical process of sampling must be determined much more precisely. As a reward the analysis exposes more clearly the real information contained in the sample.

9.3.4 Composite Hypotheses, Finite Populations

In this section the study of the urn model of subsection 9.3.2 will be continued. Now, however, it will be assumed that any number of white balls from zero to m are possible; the assumption of a simple alternative will be abandoned. As before let $\Omega = \{1, \ldots, m\}$ be the space of random elements representing the m balls in the urn and $p(\omega) = 1/m$. The unknown parameter π is the number of white balls in the urn, $\pi \in \Pi = \{0, 1, \ldots, m\}$. The π white balls (if any) are again assumed to be numbered from 1 to π. This defines then the following process model of observation for a single sample:

$$x = f(\omega, \pi) = \begin{cases} white & \text{if } \omega \leq \pi \\ black & \text{otherwise,} \end{cases} \tag{9.32}$$

$$\omega \in \Omega, \pi \in \Pi, \ X = \{white, black\}.$$

We begin again by the analysis of a single sample. Suppose that a white ball is drawn, $x = white$. Then all ω are possible random elements, $v_{white}(\Pi) = \Omega$ because $\pi = m$ is not excluded. No conditioning is necessary. The focal sets are

$$T_{white}(\omega) = \{\pi : \pi \geq \omega\} \text{ for } \omega = 1, 2, \ldots, m. \tag{9.33}$$

These focal sets are nested and the hint is consonant (see Sect. 2.2). For a consonant hint it is sufficient to know the plausibility of singletons (see theorem 3.7),

$$pl_{white}(\pi) = P(\omega \leq \pi) = \pi/m \text{ for } \pi = 0, 1, \ldots, m \tag{9.34}$$

and we have in this case $pl_{white}(H) = \max\{pl_{white}(\pi) : \pi \in H\}$ for any subset $H \subseteq \Pi$ and $sp_{white}(H) = 1 - pl_{white}(H^c)$.

If a black ball is drawn, then a similar analysis shows that this leads to a consonant hint with focal sets

$$T_{black}(\omega) = \{\pi : \pi < \omega\} \text{ for } \omega = 1, 2, \ldots, m \tag{9.35}$$

and its plausibility function on singletons is

$$pl_{black}(\pi) = P(\omega > \pi) = (m - \pi)/m \text{ for } \pi = 0, 1, \ldots, m. \tag{9.36}$$

The interesting case is again the case of a n-fold repeated sampling with replacement. Let the n observations be $x_1, x_2, \ldots, x_n$, where each x_i has either the value *white* or *black*. Each observation x_i defines a consonant hint $\mathcal{H}_i$ on the parameter space Π. According to (9.16) the analysis of the combined process model of n independent, repeated observations leads to the combined hint $\mathcal{H} = \mathcal{H}_1 \oplus \mathcal{H}_2 \oplus \cdots \oplus \mathcal{H}_n$ for the unknown parameter. The commutativity of the operation $\oplus$ implies again that the number x of white balls drawn in the n samples is a sufficient statistics.

To simplify the analysis we may therefore assume without loss of generality that the first x hints represent white balls and the last $n - x$ hints black balls. Let's then first look at the partial combination $\mathcal{H}_{white} = \mathcal{H}_1 \oplus \mathcal{H}_2 \oplus \cdots \oplus \mathcal{H}_x$ of the "white" hints. Clearly, no combined interpretation $(\omega_1, \ldots, \omega_x)$ of this combined hint is contradictory and their focal sets are

$$T_{white}(\omega_1, \ldots, \omega_x) = \cap_{i=1}^{x} \{\pi \geq \omega_i\} = \{\pi \geq \max\{\omega_i : i = 1, \ldots, x\}\}. \tag{9.37}$$

Note that different combined interpretations $(\omega_1, \ldots, \omega_x)$ lead to the same focal set (9.37). In order to compute the probability of these focal sets, consider the random variable $y = \max\{\omega_i : i = 1, \ldots, x\}$ which has the probability distribution $P(y \leq k) = (k/m)^x$, such that the focal set $\{\pi \geq s\}$ gets the probability $p_{white}(s) = (s^x - (s-1)^x)/m^x$ for $s = 1, \ldots, m$. This is the basic probability assignment (bpa) for the hint $\mathcal{H}_{white}$ (see Sect. 3.3).

Similarly, let's look at the combination of the $n-x$ "black" hints, $\mathcal{H}_{black} = \mathcal{H}_{x+1} \oplus \cdots \oplus \mathcal{H}_n$. As before, there are no contradictory combined interpretations $(\omega_{x+1}, \ldots \omega_n)$ and the focal sets are

$$\begin{aligned} T_{black}(\omega_{x+1}, \ldots \omega_n) &= \cap_{i=x+1}^n \{\pi < \omega_i\} \\ &= \{\pi < min\ \{\omega_i : i = x+1, \ldots, n\}\}. \end{aligned} \tag{9.38}$$

Defining $z = min\ \{\omega_i : i = x+1, \ldots, n\}$ one finds that the focal sets $\{\pi < t\}$, $t = 1, \ldots, m$ have the probabilities $p_{black}(t) = ((m - t + 1)^{n-x} - (m - t)^{n-x})/m^{n-x}$, which defines the bpa for $\mathcal{H}_{black}$.

Finally, the overall combined hint $\mathcal{H} = \mathcal{H}_{white} \oplus \mathcal{H}_{black}$ can be formed. There is a contradiction between the interpretations of these two hints when $y \geq z$. The probability of no contradiction is

$$k = P(y < z) = \sum_{s<t} p_{white}(s) p_{black}(t). \tag{9.39}$$

Furthermore, the focal sets of $\mathcal{H}$ are obtained as intersections of those of $\mathcal{H}_{white}$ and $\mathcal{H}_{black}$:

$$\{s \leq \pi\} \cap \{\pi < t\} = \{s \leq \pi < t\} \tag{9.40}$$

for s and t in $\{1, \ldots, m\}$ and $s < t$. These focal sets have the probabilities

$$\frac{p_{white}(s) p_{black}(t)}{k} = \frac{(s^x - (s-1)^x)((m-t+1)^{n-x} - (m-t)^{n-x})}{k m^n}. \tag{9.41}$$

Note that this hint is no longer consonant. These results allow for the computation of the degrees of support and plausibility for any hypothesis about the unknown parameter π in $\Pi = \{0, 1, \ldots, m\}$.

9.3.5 Composite Hypotheses, Infinite Population

It is quite natural to pass from the finite urn model of the last section to a model of an infinite population, where we choose Ω to be the unit interval $[0, 1]$ with the uniform distribution on it. There is an unknown parameter $\pi \in \Pi = [0, 1]$ such that

$$x = f(\omega, \pi) = \begin{cases} white & \text{if } \omega \leq \pi \\ black & \text{otherwise.} \end{cases} \tag{9.42}$$

This defines a new process model. Note that this model no longer fits into the framework of hints considered so far because Ω is now *infinite*. However it is clear that the analysis carries over to this case without any problem. The full theory of general hints with arbitrary Ω will be developed in Part III of the book.

We pass immediately to the model corresponding to n repeated, independent trials according to this model:

$$x_i = f(\omega_i, \pi), \ i = 1, \ldots, n. \tag{9.43}$$

Let $x_1, x_2, \ldots, x_n$ represent the actual observations in these n trials. Suppose that there are x "white" and $n - x$ "black" samples. Just as before, we may without loss of generality assume that the "white" results correspond to the first x trials and the "black" ones to the last $n - x$ samples. The x "white" samples imply that $\{\pi \geq \max\{\omega_i : i = 1, \ldots, x\}\}$ and the $n - x$ "black" samples that $\{\pi < \min\{\omega_i : i = x+1, \ldots, n\}\}$. We define

$$y = \max\{\omega_i : i = 1, \ldots, x\}, \ z = \min\{\omega_i : i = x+1, \ldots, n\}\}. \tag{9.44}$$

The distribution functions of these two random variables are

$$P(y \leq t) = t^x, t \in [0,1], \ P(z > s) = (1-s)^{n-x}, s \in [0,1] \tag{9.45}$$

and their respective density functions are xt^{x-1} and $(n-x)(1-s)^{n-x-1}$. The condition of no contradiction is that $y \leq z$, which restricts the possible random elements ω_i. Its probability is

$$k = P(y \leq z) = \int_0^1 xt^{x-1}(1-t)^{n-x}dt = \frac{x\Gamma(x)\Gamma(n-x+1)}{\Gamma(n+1)}. \tag{9.46}$$

Note that k^{-1} equals the binomial coefficient $\binom{n}{x}$. The focal sets are then the intervals $\{y \leq \pi < z\}$. This permits the computation of the degrees of support for intervals $[a,b]$, $sp_x[a,b] = P(a \leq y < z \leq b | y \leq z)$:

$$\begin{aligned} sp_x[a,b] &= k^{-1} \int_a^b xt^{x-1}((1-t)^{n-x} - (1-b)^{n-x})dt \\ &= \int_a^b \frac{\Gamma(n+1)}{\Gamma(x)\Gamma(n-x+1)} t^{x-1}(1-t)^{n-x}dt - \\ & \binom{n}{x}(b^x - a^x)(1-b)^{n-x}. \end{aligned}$$

The first term is the beta distribution with parameters x and $n-x+1$, which we will denote as $be_{x,n-x+1}$, such that finally

$$sp_x[a,b] = be_{x,n-x+1}(b) - be_{x,n-x+1}(a) - \binom{n}{x}(b^x - a^x)(1-b)^{n-x}. \tag{9.47}$$

Alternatively, we may also write

$$sp_x[a,b] = k^{-1}\int_a^b (s^x - a^x)(n-x)(1-s)^{n-x-1}ds$$

$$= \int_a^b \frac{\Gamma(n+1)}{\Gamma(x+1)\Gamma(n-x)} s^x(1-s)^{n-x-1}ds + \binom{n}{x} a^x((1-b)^{n-x} - (1-a)^{n-x})$$

$$= be_{x+1,n-x}(b) - be_{x+1,n-x}(a) + \binom{n}{x} a^x((1-b)^{n-x} - (1-a)^{n-x}). \tag{9.48}$$

In particular, we obtain from these results

$$sp_x[0,b] = be_{x,n-x+1}(b) - \binom{n}{x} b^x(1-b)^{n-x}$$
$$sp_x[a,1] = 1 - be_{x+1,n-x}(a) - \binom{n}{x} a^x(1-a)^{n-x}. \tag{9.49}$$

Note, that for large n the terms $\binom{n}{x}a^x((1-b)^{n-x} - (1-a)^{n-x})$ and $\binom{n}{x}(b^x - a^x)(1-b)^{n-x}$ become very small and can be neglected. In this case $sp_x[a,b]$ equals approximatively $pl_x[a,b]$ and sp_x becomes nearly a probability measure. Also, for large n (and not too small x or $n-x$) $be_{x,n-x+1}$ and $be_{x+1,n-x}$ approach the normal distribution. Similar analyses can be found in Dempster (1965, 1968) and Shafer (1982).

For an illustration of Sect. 9.2 let's also look at the plausibility of singletons, $pl_x(\pi) = P(y \leq \pi \leq z | y \leq z) = \binom{n}{x}\pi^x(1-\pi)^{n-x}$. This is (up to the normalization constant $\binom{n}{x}$) the usual likelihood function, which in this way obtains a clear probabilistic interpretation.

9.4 Measurement Models

9.4.1 Location Problems

In this section process models of observation, arising from modeling diverse measurement devices which measure physical or other values with some error, are defined and analyzed. Some of the models considered here are examples of structural or functional models as introduced by Fraser (see for example Fraser, 1968; Bunke, 1975; Dawid, Stone, 1982). The analysis of these models uses methods similar to those described by Fraser and the results obtained correspond formally to those of Fraser (and others). The difference lies in their interpretation, which in our approach enables us to embed the structural models in a more general setting as special cases of a general process model of observation. The models will again lead to hints with infinite Ω. But once more the analysis can be made just as in the finite case. Refer, however, to Part III for a general theory of hints with arbitrary Ω.

Consider first an instrument to measure a real value $\mu \in \mathcal{R}$ whose error pattern has been investigated and found to be describable as independent random variables ω with a density function $g(\omega)$ on the real line. This type of model has already been considered in Sect. 1.5. For a single measurement x, the process model of observation or measurement in this case is defined by

$$x = \mu + \omega; \; x, \mu, \omega \in \mathcal{R}; \; \omega \sim g(\omega). \tag{9.50}$$

The analysis of this model starts with the remark that a given measurement x excludes no error value ω as impossible; no conditioning of $g(\omega)$ is necessary after measurement. Furthermore, for all ω, $T_x(\omega)$ contains the single element $\mu = x - \omega$. This model leads therefore to a precise hint for μ. Its support function sp_x is a *probability measure* (it is intuitively clear that theorem 3.8 carries over to this case) and in particular

$$\begin{aligned} sp_x\{\mu \leq t\} &= P\{T_x(\omega) \subseteq (-\infty, t]\} = P\{x - \omega \leq t\} \\ &= P\{\omega \geq x - t\} = \int_{x-t}^{+\infty} g(\omega) d\omega. \end{aligned} \tag{9.51}$$

This shows that the support function for μ has the density function $f(\mu) = g(x - \mu)$.

Let's emphasize once again the interpretation of $sp_x(H)$. It is the degree of support of the hypothesis $\mu \in H$. This means that, given the model (9.50) and the observation x, the set of the unobservable errors ω, which would imply the hypothesis, has probability $sp_x(H)$. In other words, this hypothesis can be proved or justified with a probability of $sp_x(H)$; the hypothesis H can be derived from the observation x with a reliability of $sp_x(H)$. The fact that $sp_x(H)$ is a probability measure does not signify that μ is random. Because the hint for μ is precise, it can also be said that the statement $\mu \in H^c$ is proved with probability $1 - sp_x(H)$. There is a degree of doubt of $1 - sp_x(H)$ in the hypothesis $\mu \in H$ and the degree of plausibility of this hypothesis equals also $sp_x(H)$.

Consider next n independent measurements of the unknown value μ:

$$x_i = \mu + \omega_i; \; x_i, \mu, \omega_i \in \mathcal{R}; \; \omega_i \sim g(\omega); \; i = 1, \ldots, n. \tag{9.52}$$

Let $\mathbf{x}$ and ω be the vectors with components x_i and ω_i respectively and $\mathbf{1}$ the vector with components 1. Then (9.52) can be written as

$$\mathbf{x} = \mu \mathbf{1} + \omega, \; \omega \sim \prod_{i=1}^{n} g(\omega_i). \tag{9.53}$$

Thus, given the observation $\mathbf{x}$, the error vector ω becomes restricted to the one-dimensional linear manifold $\omega = \mathbf{x} - \mu\mathbf{1}, \mu \in \mathcal{R}$. This determines the set $v_{\mathbf{x}}(\mathcal{R})$ of possible errors, given the observation $\mathbf{x}$. The observation $\mathbf{x}$ fixes the relative positions $\omega_i - \omega_1 = x_i - x_1, i = 2, \ldots, n$ of the errors.

The distribution of the vector ω has therefore to be conditioned on this information. In order to compute the corresponding conditional density let's introduce the transformation $\xi_1 = \omega_1, \xi_i = \omega_i - \omega_1, i = 2, \ldots, n$ which has the inverse transformation $\omega_1 = \xi_1, \omega_i = \xi_i + \xi_1, i = 2, \ldots, n$. Its Jacobian equals one and the density of the vector ξ becomes

$$h(\xi) = g(\xi_1)g(\xi_2 + \xi_1) \cdots g(\xi_n + \xi_1). \tag{9.54}$$

Given the observation $\mathbf{x}$, the variables ξ_i become fixed to $x_i - x_1$ for $i = 2, \ldots, n$. The corresponding conditional density is then

$$h(\xi_1|\mathbf{x}) = kg(\xi_1)g((x_2 - x_1) + \xi_1) \cdots g((x_n - x_1) + \xi_1), \tag{9.55}$$

where k is the normalization constant which normalizes the product of the densities g to a density function.

Now, $T_{\mathbf{x}}(\omega)$ is clearly still one-valued and its value is $\mu = x_1 - \omega_1 = x_1 - \xi_1$. The model leads therefore again to a precise hint and sp_x is a probability measure with a density $f(\mu)$ which can be computed from (9.55) using the transformation $\mu = x_1 - \xi_1$. It follows that

$$f(\mu) = kg(x_1 - \mu) \cdots g(x_n - \mu). \tag{9.56}$$

Let's illustrate this result by two examples.

In the case where $g(\omega)$ is a normal density with mean zero and (known) standard deviation σ, one obtains

$$\begin{aligned} f(\mu) &= k \prod_{i=1}^{n} exp(-\frac{(\mu - x_i)^2}{2\sigma^2}) = k\ exp\ \{-1/(2\sigma^2) \sum_{i=1}^{n} (\mu - x_i)^2\} \\ &= k\ exp\ \{(-1/(2\sigma^2)) \sum_{i=1}^{n} (\mu^2 - 2\mu x_i + x_i^2)\} \\ &= k'\ exp\ \{(-n/(2\sigma^2))(\mu^2 - 2\mu\bar{x})\} \\ &= k''\ exp\ \{(-n/(2\sigma^2))(\mu^2 - 2\mu\bar{x} + \bar{x}^2)\} \\ &= k''\ exp\ \{(-n/(2\sigma^2))(\mu - \bar{x})^2\}. \end{aligned} \tag{9.57}$$

$f(\mu)$ is thus a normal density with mean $\bar{x}$, the arithmetic mean of the n measurement values, and variance σ^2/n. The normalization constant k'' equals therefore $(2\pi)^{-1/2}(\sigma^2/n)^{-1/2}$. Thus, if one forms the hypothesis $\mu \in H = [\bar{x} - 1.96\sigma/\sqrt{n}, \bar{x} + 1.96\sigma/\sqrt{n}]$, then the normal table tells us that the degree of support of this hypothesis is $sp_{\mathbf{x}}(H) = 0.95$. This corresponds to the more familiar confidence interval. Note however the difference of the interpretation between the two points of view. In the usual confidence interval statement, $\bar{x}$ and therefore also the interval $[\bar{x} - 1.96\sigma/\sqrt{n}, \bar{x} + 1.96\sigma/\sqrt{n}]$ are considered as a sample of a random variable or a random interval and it is assured that this random interval contains or covers μ with a probability of 0.95. From our point of view, we state that the set of measurement errors ω

which, given the observation $\mathbf{x}$, imply that $\mu \in [\bar{x} - 1.96\sigma/\sqrt{n}, \bar{x} + 1.96\sigma/\sqrt{n}]$ has probability 0.95. In other words, this hypothesis can be proved or derived from the observation $\mathbf{x}$ with a reliability of 0.95. So, we reason conditionally on a given observation $\mathbf{x}$, and not on an a priori probability that a sampled interval does or does not cover the unknown parameter. Our approach belongs to the philosophy of reasoning: arguments for a given hypothesis are collected and then weighted. The second approach is one of decision: we adopt the hypothesis that the unknown parameter belongs to the interval and then we evaluate the risk of this decision.

In a second example assume that $g(\omega)$ is a uniform density between the limits -1 and $+1$. Then (9.56) becomes clearly a uniform density between the limits $\max x_i - 1$ (lower limit) and $\min x_i + 1$ (upper limit). Note that this interval cannot be empty, because this would indicate observations which are in contradiction with the process model. In such a case, the model would have to be rejected.

The model (9.52) as well as the models considered in subsections 9.4.2 and 9.4.3 below are, as already mentioned at the beginning of this section, particular cases of functional models or more precisely, structural models which were introduced and studied in detail by Fraser (see Fraser 1968 for a general introduction and Fraser, 1988 for further references). The mathematical analysis of these models can be borrowed from Fraser. However, the last step, the computation of the density functions for the degrees of support for hypotheses about the unknown parameters and their interpretation as reliabilities of proofs for these hypotheses, has not been made by Fraser and other authors who study functional models. Moreover, their methods depend crucially on the assumption that for any observation $\mathbf{x}$ and error ω the functional model has at most one solution $T_{\mathbf{x}}(\omega)$; in other words, the resulting hint is precise. This assumption is not necessary in our approach, as illustrated in Sect. 9.2 and 9.3.

9.4.2 Scale Problems

Consider now the process model

$$x = \omega\sigma; \; x, \omega \in \mathcal{R}, \sigma \in \mathcal{R}^+; \omega \sim g(\omega), \tag{9.58}$$

for the unknown scale parameter σ. The analysis of this single observation or measurement model starts by noting that an observation x eliminates the errors ω of opposite sign to x as impossible because σ is known to be positive. The density $g(\omega)$ has thus to be conditioned on the event $\omega x > 0$. Let $g'(\omega)$ be this conditional density. $T_x(\omega)$ contains for all ω of the same sign as x the single element $\sigma = x/\omega$. The resulting hint for σ is again *precise*. By a similar analysis as in the previous subsection, it follows that the support function of σ has a density function $f(\sigma)$ which can be obtained by applying the transformation $\sigma = x/\omega$ to the density $g'(\omega)$:

$$f(\sigma) = (|x|/\sigma^2)g'(x/\sigma). \tag{9.59}$$

Let's look at different examples:

(1) *Normal density.* Suppose that $g(\omega)$ is a normal density with mean zero and standard deviation one, $\omega \sim N(0,1)$. This density is symmetric about 0 and a conditioning on the positive or negative half axis amounts only to the division of the normal density by 1/2 and then we have to restrict it to the corresponding half axis. One obtains thus

$$f(\sigma) = (2/\pi)^{1/2}(|x|/\sigma^2)exp(-x^2/(2\sigma^2)),\ \sigma > 0. \tag{9.60}$$

(2) *Exponential density.* Suppose next that $g(\omega) = exp(-\omega)$ for $\omega \geq 0$. In this model a negative value x would be in complete contradiction to the model. No error ω would be compatible with such an observation. It can thus be assumed that $x > 0$. Then $g'(\omega) = g(\omega)$. It follows

$$f(\sigma) = (x/\sigma^2)exp(-x/\sigma),\ \sigma > 0. \tag{9.61}$$

(3) *Uniform density.* Here it is assumed that $g(\omega)$ is the uniform density between the limits minus one and one. In this case, $g'(\omega)$ simply becomes the uniform density between 0 and 1 or -1 and 0, depending on whether x is positive or negative and

$$f(\sigma) = \begin{cases} |x|/\sigma^2 & \text{if } \sigma \geq |x| \\ 0 & \text{otherwise.} \end{cases} \tag{9.62}$$

Next consider n repeated, independent measurements according to the process model (9.58):

$$x_i = \omega_i\sigma;\ x_i, \omega_i \in \mathcal{R},\ \sigma \in \mathcal{R}^+;\ \omega_i \sim g(\omega). \tag{9.63}$$

In vector notation this reads $\mathbf{x} = \sigma\omega$. Thus, an observation x restricts the possible error vectors ω to the half line $\omega = \lambda\mathbf{x}, \lambda > 0$. In other words, the errors ω_i are known in relative magnitude $\omega_i/\omega_1 = x_i/x_1, i = 2, \ldots, n$. This determines $v_{\mathbf{x}}(\mathcal{R}^+)$ and the distribution of the errors ω has to be conditioned on this information.

In order to compute the corresponding conditional density the transformation $\xi_1 = \omega_1, \xi_i = \omega_i/\omega_1, i = 2, \ldots, n$, or, inversely, $\omega_1 = \xi_1, \omega_i = \xi_i\xi_1, i = 2, \ldots, n$, is introduced. Its Jacobian equals ξ_1^{n-1} and it follows that ξ has the density

$$h(\xi) = |\xi_1|^{n-1}g'(\xi_1)g'(\xi_2\xi_1)\cdots g'(\xi_n\xi_1) \tag{9.64}$$

where g' is the conditional density of g, given $x\omega > 0$. The observation $\mathbf{x}$ binds ξ_i to the value x_i/x_1 for $i = 2, \ldots, n$. The corresponding conditional density is then

$$h(\xi_1|\mathbf{x}) = k|\xi_1|^{n-1}g'(\xi_1)g'(\xi_1x_2/x_1)\cdots g'(\xi_1x_n/x_1). \tag{9.65}$$

For all possible ω in $v_{\mathbf{x}}(\mathcal{R}^+)$ there is only one possible value in $T_{\mathbf{x}}(\omega)$, namely $\sigma = x_1/\omega_1 = x_1/\xi_1$. The model leads again to a precise hint for the unknown parameter σ and $sp_{\mathbf{x}}$ is a probability measure with a density function $f(\sigma)$ which can be computed from (9.65) using the transformation $\sigma = x_1/\xi_1$:

$$f(\sigma) = k'\sigma^{-(n+1)} g'(x_1/\sigma) \cdots g'(x_n/\sigma). \tag{9.66}$$

Let's apply this result to the examples considered above:

(1) *Normal density.* Here (9.66) gives

$$\begin{aligned} f(\sigma) &= k'\sigma^{-(n+1)} \prod_{i=1}^{n} exp(-x_i^2/(2\sigma^2)) \\ &= k''\sigma^{-(n+1)} exp(-s^2/(2\sigma^2)) \end{aligned} \tag{9.67}$$

where $s^2 = x_1^2 + x_2^2 + \ldots + x_n^2$ is the sum of the squares of the observations. The normalization gives

$$f(\sigma) = \frac{s^n}{2^{(n/2)-1}\Gamma(n/2)\sigma^{n+1}} exp(-s^2/(2\sigma^2)). \tag{9.68}$$

As an application let's compute $sp_{\mathbf{x}}(\sigma \geq s/\sqrt{z})$ by integrating (9.68) over the corresponding range. It follows, using the transformation $y = s^2/\sigma^2$

$$sp_{\mathbf{x}}(\sigma \geq s/\sqrt{z}) = \frac{1}{2^{n/2}\Gamma(n/2)} \int_0^z y^{(n/2)-1} e^{-y/2} dy. \tag{9.69}$$

$sp_{\mathbf{x}}(\sigma \geq s/\sqrt{z})$ is nothing but the Chisquare distribution with n degrees of freedom and this establishes again the connection with the usual confidence interval. But note once more the difference of interpretation between the confidence interval as a sample from a random interval and the degree of support as a reliability of an inference (see the discussion above with respect to the mean of a normal distribution).

This result can also be explained more directly. From the model $\mathbf{x} = \sigma\omega$ it follows immediately that $s^2 = \sigma^2 s_\omega^2$. Here s_ω^2 is the square sum of the errors. In the case of normal distributions it is well known that s_ω^2 is independent of all ω_i and has the Chisquare distribution. Hence the information that $\omega_i/\omega_1 = x_i/x_1, i = 2, \ldots, n$ does not change the distribution of s_ω^2 and it follows immediately that the set of all possible ω which imply that $\sigma \geq s/\sqrt{z}$ has indeed the probability given by (9.69).

(2) *Exponential density.* In this example (9.65) yields

$$\begin{aligned} f(\sigma) &= k'\sigma^{-(n+1)} \prod_{i=1}^{n} exp(-x_i/\sigma) \\ &= k'\sigma^{-(n+1)} exp(-\bar{x}n/\sigma) \end{aligned} \tag{9.70}$$

with $\bar{x} = (x_1 + x_2 + \ldots + x_n)/n$.

If the exponential distribution represents for example the life-span of a failing equipment, then the mean rate of failure $\lambda = 1/\sigma$ is more interesting than σ. Note that the hint for σ can easily be transformed into a hint for λ; the latter in particular is still precise. The density function (9.70) of the support function for σ can in fact be transformed into the density function $f(\lambda)$ of the support function for λ by the usual transformation for density functions:

$$f(\lambda) = k' \lambda^{n-1} e^{-n\bar{x}\lambda}. \tag{9.71}$$

The normalization gives finally

$$f(\lambda) = \frac{(n\bar{x})^n}{(n-1)!} \lambda^{n-1} e^{-n\bar{x}\lambda}. \tag{9.72}$$

(3) *Uniform density.* In this example (9.65) leads to

$$f(\sigma) = \begin{cases} k'/\sigma^{n+1} & \text{if } \sigma > \max |x_i| \\ 0 & \text{otherwise.} \end{cases} \tag{9.73}$$

The normalization gives

$$f(\sigma) = \begin{cases} n(max\ |x_i|)^n/\sigma^{n+1} & \text{if } \sigma > max\ |x_i| \\ 0 & \text{otherwise.} \end{cases} \tag{9.74}$$

Of course, the larger n, the more strongly we expect the unknown σ to be near the value $max\ |x_i|$.

9.4.3 Location and Scale

For a single measurement the model considered is now

$$x = \mu + \sigma\omega; \; x, \mu, \omega \in \mathcal{R}; \; \sigma \in \mathcal{R}^+; \; \omega \sim g(\omega). \tag{9.75}$$

Here μ and σ are the unknown parameters. A single observation x is a hint for (μ, σ) with $T_x(\omega) = \{(\mu, \omega) : x = \mu + \sigma\omega\}$. This hint is no longer precise, for any ω there is a linear manifold of possible values for (μ, σ).

More interesting is of course the case of n independent measurements, each according to the model (9.75):

$$x_i = \mu + \sigma\omega_i; \; x_i, \mu, \omega_i \in \mathcal{R}; \; \sigma \in \mathcal{R}^+; \; \omega_i \sim g(\omega). \tag{9.76}$$

Let $\mathbf{x}, \omega$ as above be the vectors with components $x_i, \omega_i, i = 1, 2, \ldots, n$ and $\mathbf{1}$ the n-tuple $(1, 1, \ldots, 1)$. Then (9.76) reads

$$\mathbf{x} = \mu\mathbf{1} + \sigma\omega, \omega \sim g(\omega) = \prod_{i=1}^{n} g(\omega_i) \tag{9.77}$$

or

$$\omega = (1/\sigma)(\mathbf{x} - \mu\mathbf{1}). \tag{9.78}$$

This result shows that the observation $\mathbf{x}$ restricts the random vector ω to the half hyperplane defined by $\lambda\mathbf{x} + \tau\mathbf{1}, \lambda > 0$. Or, in other words, the relative position of the errors ω_i is known by

$$\frac{\omega_i - \omega_1}{\omega_i - \omega_2} = \frac{x_i - x_1}{x_i - x_2}, i = 3, \ldots, n. \tag{9.79}$$

In order to compute the conditional density of the restricted vector ω, the transformation $\xi_1 = \omega_1, \xi_2 = \omega_2, \xi_i = (\omega_i - \omega_1)/(\omega_i - \omega_2), i = 3, \ldots, n$ or inversely, $\omega_1 = \xi_1, \omega_2 = \xi_2, \omega_i = (\xi_2 \xi_i - \xi_1)/(\xi_i - 1), i = 3, \ldots, n$, is used. Its Jacobian equals $(\xi_2 - \xi_1)^{n-2} / \prod_{i=3}^n (\xi_i - 1)^2$. This gives the following density function for the vector ξ:

$$h(\xi) = \frac{|\xi_2 - \xi_1|^{n-2}}{\prod_{i=3}^n (\xi_i - 1)^2} g(\xi_1) g(\xi_2) g(\frac{\xi_2 \xi_3 - \xi_1}{\xi_3 - 1}) \cdots g(\frac{\xi_2 \xi_n - \xi_1}{\xi_n - 1}). \tag{9.80}$$

The observation $\mathbf{x}$ restricts ξ_i to $(x_i - x_1)/(x_i - x_2), i = 3, \ldots, n$ and the corresponding conditional density function is therefore

$$h(\xi_1, \xi_2 | \mathbf{x}) = k |\xi_2 - \xi_1|^{n-2} g(\xi_1) g(\xi_2) \cdot \\ g(\frac{(x_3 - x_1)\xi_2 - (x_3 - x_2)\xi_1}{x_2 - x_1}) \cdots g(\frac{(x_n - x_1)\xi_2 - (x_n - x_2)\xi_1}{x_2 - x_1}). \tag{9.81}$$

Next, note that $\xi_1 = \omega_1 = (x_1 - \mu)/\sigma$ and $\xi_2 = \omega_2 = (x_2 - \mu)/\sigma$. This transformation has the Jacobian $(x_2 - x_1)/\sigma^3$. This permits the conclusion that this model leads once more to a precise hint and hence $sp_{\mathbf{x}}$ is a probability measure possessing a density function which can be obtained from (9.81) using this transformation:

$$f(\mu, \sigma) = k' \sigma^{-(n+1)} \prod_{i=1}^n g(x_i - \mu)/\sigma. \tag{9.82}$$

As an application of this result let's consider the case of normal densities $g(\omega)$ with mean zero and standard deviation one. In this case (9.82) becomes

$$f(\mu, \sigma) = k'' \sigma^{-(n+1)} exp(-(1/2\sigma^2) \sum_{i=1}^n (x_i - \mu)^2). \tag{9.83}$$

This density makes it possible, once the normalization constant k'' is determined, to compute degrees of support of combined hypotheses about the parameters μ and σ simultaneously. However, in practice, most of the time hypotheses about one parameter alone are considered. For this purpose, it is sufficient to know the marginal distributions of each parameter alone. This is of course a case of coarsening of a frame of discernment (see Chap. 7)

For the mean μ the marginal density is obtained by integrating (9.83) over σ from 0 to ∞. Using the transformation $y = s_\mu^2/(2\sigma^2)$, with $s_\mu^2 = \sum_{i=1}^n (x_i - \mu)^2$ one obtains

$$f(\mu) = k\Gamma(n/2)/(s_\mu^2)^{n/2} = k\Gamma(n/2)/(s^2 + n(\bar{x} - \mu)^2)^{n/2}, \tag{9.84}$$

where $s^2 = \sum_{i=1}^n (x_i - \bar{x})^2$ and $\bar{x} = (x_1 + x_2 + \ldots + x_n)/n$. The normalization constant is obtained by integrating (9.84) over μ. Using the transformation $y = \sqrt{n}(\bar{x} - \mu)$ it is found that this integral equals

$$(2k\Gamma(n/2)/(s^n\sqrt{n}))\int_0^\infty (1+y^2/s^2)^{-n/2}dy. \tag{9.85}$$

This integral in turn can be computed using the transformation $z = (1 + y^2/s^2)^{-1}$ and noting that

$$\int_0^1 (1-x)^{a-1}x^{b-1}dx = \frac{\Gamma(a)\Gamma(b)}{\Gamma(a+b)} \tag{9.86}$$

It follows finally

$$f(\mu) = \frac{\sqrt{n}\Gamma(n/2)}{\Gamma(1/2)\Gamma((n-1)/2)}\frac{1}{s(1+(\bar{x}-\mu)^2/(s^2/n))^{n/2}}. \tag{9.87}$$

For example, using this result, the degree of support of the hypothesis that $\mu \leq \bar{x} + ts/\sqrt{n(n-1)}$ can be computed by integrating (9.87). Using in sequence the transformations $y = (\bar{x}-\mu)/(s/\sqrt{n})$ and $z = y\sqrt{n-1}$ one obtains

$$\begin{aligned} sp_{\mathbf{x}}(\mu \leq \bar{x} + ts/\sqrt{n(n-1)}) &= \frac{\Gamma(n/2)}{\sqrt{n-1}\Gamma(1/2)\Gamma((n-1)/2)}\cdot \\ &\int_{-t}^\infty \frac{1}{(1+z^2/(n-1))^{n/2}}dz. \end{aligned} \tag{9.88}$$

But this is nothing but the Student distribution with $n-1$ degrees of freedom. This result establishes the link to the classical theory of confidence intervals (see the case of normal distributions with respect to the location and the scale models above, subsections 9.4.1 and 9.4.2).

In a similar vein, the marginal density with respect to the parameter σ can be obtained from (9.83) by integrating over μ. It follows that

$$f(\sigma) = k'\sigma^{-n}exp(-s^2/(2\sigma^2)). \tag{9.89}$$

This result corresponds to (9.67) with n instead of $n+1$. The normalization constant can therefore be deduced from (9.68):

$$f(\sigma) = \frac{s^{n-1}}{2^{((n-1)/2)-1}\Gamma((n-1)/2)\sigma^n}exp(-s^2/(2\sigma^2)). \tag{9.90}$$

Similarly, (9.69) can be used to compute

$$sp_{\mathbf{x}}(\sigma \geq s/\sqrt{z}) = \frac{1}{2^{(n-1)/2}\Gamma((n-1)/2)}\int_0^z y^{((n-1)/2)-1}e^{-y/2}dy \tag{9.91}$$

which is this time a Chisquare distribution with $n-1$ degrees of freedom in concordance with the usual confidence interval theory for this case.

10. Describing Uncertainty in Dynamical Systems

10.1 Describing Dynamical Systems by Hints

It is possible to describe uncertain time-discrete dynamical systems by uncertain restrictions. This leads to system models characterized by uncertain inclusions. These models contain both dynamical systems with stochastic disturbances as well as dynamical systems with unknown but bounded disturbances as special cases. Solutions to filtering, smoothing and prediction problems with respect to the general model can be developed. These solutions provide for a unified and generalized approach to filtering, prediction and smoothing problems of stochastic dynamical systems and systems with bounded disturbances because the general solution scheme applies to both cases. The possibility to look at the Kalman filter from this point of view has already been mentioned by Dempster (1988). This chapter is based on Kohlas (1991 b).

In this section the general model of uncertain dynamical systems described by uncertain restrictions is introduced and the solution to the related filtering, smoothing and prediction problems is presented. In Sect. 10.2 and 10.3 the specialization of the general solution scheme to stochastic systems and systems with bounded disturbances (in fact for a slightly more general system) is given.

Consider a sequence of variables $X_i, i = 0, 1, 2, \ldots$ describing subsequent states of a system in discrete time. Let Θ_0 be the *state space* such that $\Theta_i = \Theta_0$ for $i = 1, 2, \ldots$. For an uncertain, dynamical system it will be assumed that there are *uncertain transitions restrictions* described by hints $\mathcal{T}_i = (\Omega_i, P_i, \Gamma_i, \Theta_{i-1} \times \Theta_i)$ between state at time $i - 1$ and $i, i = 1, 2, \ldots$ If $\omega_i \in \Omega_i$ is the correct interpretation, then the restriction $(X_{i-1}, X_i) \in \Gamma_i(\omega_i) \subseteq \Theta_{i-1} \times \Theta_i$ is assumed to hold. Let $\Gamma_i(\omega_i|X) = \{\theta \in \Theta_i : (X, \theta) \in \Gamma_i(\omega_i)\}$ for any $X \in \Theta_{i-1}$. Then $(X_{i-1}, X_i) \in \Gamma_i(\omega_i) \subseteq \Theta_{i-1} \times \Theta_i$ is equivalent to $X_i \in \Gamma_i(\omega_i|X_{i-1})$. Given the old system state, an uncertain transition restriction thus corresponds to an uncertain inclusion of the new system state.

Consider furthermore a sequence of variables $Y_i, i = 0, 1, 2, \ldots$ taking values in some *observation space* $\Psi_i = \Psi_0$. *Uncertain observations relations* described by hints $\mathcal{O}_i = (\Omega'_i, Q_i, \Delta_i, \Theta_i \times \Psi_i), i = 0, 1, 2, \ldots$ link observations Y_i to states X_i, $(X_i, Y_i) \in \Delta_i(\omega'_i)$. As above, define $\Delta_i(\omega'_i|X) = \{\psi \in \Psi_i : (X, \psi) \in \Delta_i(\omega'_i)\}$. Given the system state, the uncertain observation relation then leads to the *uncertain inclusion* $Y_i \in \Delta_i(\omega_i|X_i)$.

Suppose now that a series of *observations* $y_i, i = 0, 1, \ldots, r$ of the variables Y_i is given. Each one of these observations can be considered as a *deterministic hint* relative to Ψ_i.

Define $\Theta = \Theta_0^\infty \times \Psi_0^\infty$ and extend all hints $\mathcal{T}_i, \mathcal{O}_i$ and y_i to Θ. The combination of all these hints on Θ leads to a combined hint

$$\mathcal{H} = \oplus \{\mathcal{T}_i : i = 1, 2, \ldots\} \oplus \{\mathcal{O}_i : i = 0, 1, \ldots\} \oplus_{i=1}^{r} y_i \tag{10.1}$$

englobing all information available about the system. Fix some time point k. Then the coarsening of $\mathcal{H}$ to Θ_k denoted by $\mathcal{H}(k)$ contains all the information available about the possible values of X_k; this hint determines the support and plausibility functions relative to X_k. If $k = r + 1$, then $\mathcal{H}(k)$ represents the solution to the *filter problem* relative to the uncertain dynamical system introduced above. For $k > r + 1$ it is the solution to the *prediction problem* and for $k \leq r$ it is the solution to the *smoothing problem.*

Computing the coarsening $\mathcal{H}(k)$ as defined by (10.1) is of course totally infeasible. Applying Dempster's rule on a large space as Θ is much too complicated. Also $\mathcal{H}$ contains information which is not essential for some problems. For example, it is intuitively clear that $\mathcal{T}_i$ and $\mathcal{O}_i$, for $i > r$, do not contribute to $\mathcal{H}(k)$, if $k \leq r$. Theorem 10.1 below gives a computationally much more efficient method to compute $\mathcal{H}(k)$. It is based on the fact that the family of subsets of variables $\{\{X_{i-1}, X_i\}, \{X_i, Y_i\} : i = 1, 2, \ldots\} \cup \{\{X_0, Y_0\}\}$ forms in fact a simple hypertree (see Fig. 10.1).

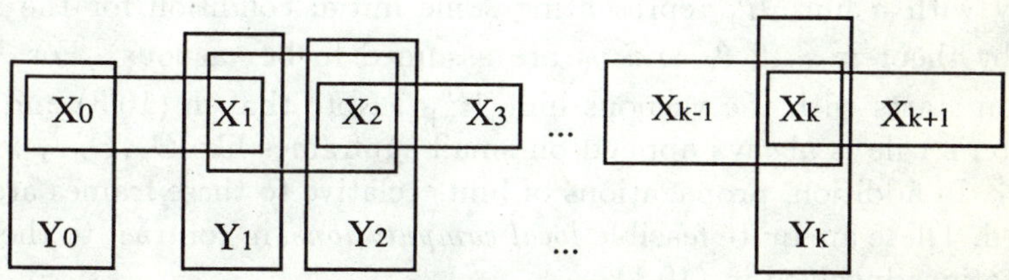

Fig. 10.1.

Define

$$\begin{aligned} \mathcal{H}_k^p &= \oplus\{\mathcal{T}_i : i \leq k\} \oplus \{\mathcal{O}_i : i \leq k\} \oplus \{y_i : i \leq k\}, \\ \mathcal{H}_k^f &= \oplus\{\mathcal{T}_i : i > k\} \oplus \{\mathcal{O}_i : i \geq k\} \oplus \{y_i : i \geq k\}, \end{aligned} \tag{10.2}$$

where by definition y_i is the vacuous hint for $i > r$.

$\mathcal{H}_k^p$ contains all available information concerning the *past* and present of the system relative to time point k, $\mathcal{H}_k^f$ contains similarly the available information concerning the *future* of the system relative to k. Let's agree on the convention that $\mathcal{H}(k-1, k)$ denotes the coarsening to X_{k-1}, X_k of an arbitrary hint $\mathcal{H}$ and that $\mathcal{H}(k, k')$ denotes its coarsening to X_k, Y_k. The following theorem is an application of theorem 8.16 to the Markov tree related to the hypertree given in Fig. 10.1. (see the proof of theorem 10.1). Of course, we are here in the context of a multivariate family of compatible frames which is, however, infinite. But as discussed in Chap. 7 and 8 this is not a problem.

Theorem 10.1. *For any* $k = 0, 1, 2 \ldots$

$$\begin{aligned} \mathcal{H}_k^p(k) =& (\mathcal{H}_{k-1}^p(k-1,k) \oplus \mathcal{T}_k)(k) \\ & \oplus (\mathcal{O}_k \oplus y_k(k,k'))(k) \\ \mathcal{H}_k^f(k) =& (\mathcal{H}_{k+1}^f(k,k+1) \oplus \mathcal{T}_{k+1})(k) \\ & \oplus (\mathcal{O}_k \oplus y_k(k,k'))(k) \end{aligned} \tag{10.3}$$

$$\begin{aligned} \mathcal{H}(k) =& (\mathcal{H}_{k-1}^p(k-1,k) \oplus \mathcal{T}_k)(k) \\ & \oplus (\mathcal{H}_{k+1}^f(k,k+1) \oplus \mathcal{T}_{k+1}))(k) \\ & \oplus (\mathcal{O}_k \oplus y_k(k,k'))(k) \end{aligned} \tag{10.4}$$

Note that (10.3) are *recursive formulas* for $\mathcal{H}_k^p(k)$ and $\mathcal{H}_k^f(k)$. In fact, $\mathcal{H}_{k-1}^p$ is carried by variables in the past relative to time point $k-1$; hence $\mathcal{H}_{k-1}^p(k-1,k)$ is the vacuous extension of $\mathcal{H}_{k-1}^p(k-1)$ to X_{k-1}, X_k (by theorem 8.12 θ_0 and y_0 are assumed to be vacuous). A similar observation applies to $\mathcal{H}_{k+1}^f(k,k+1)$. For $\mathcal{H}_k^p$ the recursion starts with the vacuous hint $\mathcal{H}_0^p$; or possibly with a hint $\mathcal{H}_0^p$ representing some initial condition for the system state (by theorem 8.12 θ_0 and y_0 are assumed to be vacuous). For $\mathcal{H}_k^f$ the recursion starts with the vacuous hint $\mathcal{H}_{r+1}^f$. Note that in (10.3) and (10.4) Dempster's rule is always applied on small subframes like $\Theta_k, \Theta_{k-1} \times \Theta_k$ or $\Theta_k \times \Psi_k$. In addition, propagations of hints relative to these frames are to be executed. These are quite feasible *local computations*, in contrast to the global computations involved in (10.1).

The first formula in (10.3) is called the *forward* propagation formula and the second one is called the *backward propagation* formula. According to (10.4), the solution of the smoothing problem at time k consists of the combination of the forward propagation up to time k (or filter-solution at time k) with the backward-propagation down to time k (a kind of backward filtering to k) and the actual observation at time k. Note the perfect symmetry of the treatment of past and future information with respect to time point k. In the prediction and the filter problem $(k > r)$ $\mathcal{H}_k^f$ is vacuous because there are no observations in the future relative to k (y_k is a vacuous hint). Hence the second and third term in (10.4) vanish. It follows that in this case $\mathcal{H}(k) = \mathcal{H}_k^p(k)$ and the prediction and filter problems are solved by the recursion (10.3).

10.2 Special Case of Stochastic Systems

Suppose that $\mathcal{T}_i = (\Omega_i, P_i, \Gamma_i, \Theta_{i-1} \times \Theta_i)$ is such that $\Gamma_i(\omega_i|\theta)$ is a singleton for all $\theta \in \Theta_{i-1}$ and all $\omega_i \in \Omega_i$ such that $X_i = \Gamma_i(\omega_i|X_{i-1})$ (X_i depends on ω_i). Similarly suppose that $\mathcal{O}_i = (\Omega_i', P_i', \Delta_i, \Theta_i \times \Psi_i)$ is such that $\Delta_i(\omega_i|X_i)$ is a singleton for all $\theta \in \Theta_i$, such that $Y_i = \Delta_i(\omega_i|X_i)$. The general model of the previous section becomes the ordinary stochastic dynamical system

$$X_{i+1} = \Gamma_i(\omega_i|X_i); \; Y_i = \Delta_i(\omega_i|X_i), \; i = 0, 1, 2, \ldots \tag{10.5}$$

In this section the general solution to the smoothing problem given by theorem 10.1 will be adapted to the model (10.5). In order to avoid some purely technical complications the state spaces Θ_i as well as the measurement spaces Ψ_i will be assumed finite.

Denote by θ the deterministic hint on Θ_{i-1}, which says that $X_{i-1} = \theta$. It is easy to see that the combined hint $(\mathcal{T}_i \oplus \theta)(i)$ is given by $(\Omega_i, P_i, \Gamma_i', \Theta_i)$, where $\Gamma_i'(\omega_i) = \Gamma_i(\omega_i|\theta)$. Similarly, the combined hint $(\mathcal{O}_i \oplus \theta)(i')$ is defined by $(\Omega_i', P_i', \Delta_i', \Psi_i)$ where $\Delta_i'(\omega_i) = \Delta_i(\omega_i|\theta)$. In general, a hint $\mathcal{H}$ on $\Theta_1 \times \Theta_2$ is called *conditional precise* relative to Θ_2 given Θ_1 if the coarsening of $\mathcal{H} \oplus \theta$ to Θ_2 is precise for all $\theta \in \Theta_1$. The general model of Sect. 10.1 thus reduces itself to the stochastic model (10.5) if $\mathcal{T}_{i+1}$ and $\mathcal{O}_i$ are all conditional precise relative to Θ_{i+1} and Ψ_i given Θ_i. Remember that for precise hints $\mathcal{H} = (\Omega, P, \Gamma, \Theta)$ support and plausibility coincide and equal probability (bpa). In particular for singletons $sp(\{\theta\}) = pl(\{\theta\}) = p(\theta) = \sum \{P(\omega) : \Gamma(\omega) = \theta\}$ is the (discrete) density of the precise hint $\mathcal{H}$ (for finite Θ).

In order to apply theorem 10.1 to system (10.5), a few theorems for conditional precise hints are needed. They are formulated for hints relative to two variables X_1 and X_2 with domains Θ_1 and Θ_2.

Theorem 10.2 *(Formula of Total Probability). If $\mathcal{H}_1$ is a precise hint on Θ_1 and $\mathcal{H}$ is a hint on $\Theta_1 \times \Theta_2$ which is conditional precise relative to Θ_2 given Θ_1, then $\mathcal{H}_2 = (\mathcal{H}_1 \oplus \mathcal{H})(2)$ is a precise hint and if $p_1(\theta_1), p_2(\theta_2)$ and $p(\theta_2|\theta_1)$ are the probability densities of the precise hints $\mathcal{H}_1, \mathcal{H}_2$ and $(\mathcal{H} \oplus \theta_1)(2)$ respectively, then*

$$p_2(\theta_2) = \sum \{p(\theta_2|\theta_1)p_1(\theta_1) : \theta_1 \in \Theta_1\} \tag{10.6}$$

for all $\theta_2 \in \Theta_2$.

Theorem 10.3. *(Bayes' Theorem). If $\mathcal{H}_1$ is a precise hint on Θ_1, $\mathcal{H}_2$ a precise hint on Θ_2 and $\mathcal{H}$ a hint on $\Theta_1 \times \Theta_2$ which is conditional precise relative to Θ_2 given Θ_1, then $(\mathcal{H}_1 \oplus \mathcal{H}_2 \oplus \mathcal{H})(1)$ is a precise hint and its probability density $p(\theta_1)$ is given by*

$$p(\theta_1) = c \sum \{p_2(\theta_2)p(\theta_2|\theta_1)p_1(\theta_1) : \theta_2 \in \Theta_2\} \tag{10.7}$$

for all $\theta_1 \in \Theta_1$.

Here and also in theorem 10.4, c is a normalization constant, defined by the requirement that the sum of $p(\theta_1)$ over all $\theta_1 \in \Theta_1$ equals one.

Theorem 10.4. *If $\mathcal{H}_2$ is a hint on Θ_2 and $\mathcal{H}$ is a hint on $\Theta_1 \times \Theta_2$ which is conditional precise relative to Θ_2 given Θ_1, then the plausibilities of singletons with respect to $(\mathcal{H}_2 \oplus \mathcal{H})(1)$ are given by*

$$pl(\theta_1) = c \sum \{pl_2(\theta_2)p(\theta_2|\theta_1) : \theta_2 \in \Theta_2\} \tag{10.8}$$

for all $\theta_1 \in \Theta_1$.

Theorem 10.5 *If $\mathcal{H}_1$ is a hint on Θ_1, $\mathcal{H}_2$ a deterministic hint focussing on $\{\theta_2\}$ and $\mathcal{H}$ a conditional precise hint relative to Θ_2 given Θ_1, then the plausibilities of the singletons with respect to $(\mathcal{H}_1 \oplus \mathcal{H}_2 \oplus \mathcal{H})(1)$ are given by*

$$pl(\theta_1) = c\; p(\theta_2|\theta_1)pl_1(\theta_1) \tag{10.9}$$

for all $\theta_1 \in \Theta_1$.

In order to develop the filtering, prediction and smoothing formulas for stochastic systems from theorem 10.1 let's define

$$\mathcal{H}_k^{fp}(k) = (\mathcal{H}_{k-1}^{p}(k-1,k) \oplus \mathcal{T}_k)(k) \tag{10.10}$$

which could be called the one-step *forward prediction (fp)* formula. It follows then from theorem 10.1 that

$$\mathcal{H}_k^{p}(k) = \mathcal{H}_k^{fp}(k) \oplus (\mathcal{O}_k \oplus y_k(k,k'))(k). \tag{10.11}$$

Note that y_k is a precise hint. In accordance with the usual assumptions for stochastic systems, an initial, precise hint $\mathcal{H}_0^p$ is given, representing an initial state distribution. It follows then by induction from theorem 10.2 and 10.3 that both $\mathcal{H}_k^p(k)$ and $\mathcal{H}_k^{fp}(k)$ are precise hints for all $k = 0, 1, 2, \ldots$ By theorems 10.2 and 10.3 the probability densities of $\mathcal{H}_k^{fp}(k)$ and $\mathcal{H}_k^p(k)$ are given by

$$\begin{aligned} p_k^{fp}(\theta) &= \sum \{p_t(\theta|\theta')p_{k-1}^{p}(\theta') : \theta' \in \Theta\}, \\ p_k^{p}(\theta) &= c\; p_0(y_k|\theta)p_k^{fp}(\theta). \end{aligned} \tag{10.12}$$

Here $p_t(\theta|\theta')$ and $p_0(\psi|\theta)$ are the probability densities of the precise hints $(\mathcal{T}_k(k-1,k) \oplus \theta')(k)$ and $(\mathcal{O}_k \oplus \theta)(k')$ respectively (time homogenity is assumed for the sake of simplification). c is simply a normalization constant. These are the well-known *filter formulas for stochastic systems.*

In an entirely symmetrical way the *backward prediction* formulas can be defined:

$$\begin{aligned} \mathcal{H}_k^{bp}(k) &= (\mathcal{H}_{k+1}^{f}(k,k+1) \oplus \mathcal{T}_{k+1})(k), \\ \mathcal{H}_k^{f}(k) &= \mathcal{H}_k^{bp}(k) \oplus (\mathcal{O}_k \oplus y_k(k,k'))(k) \end{aligned} \tag{10.13}$$

This backward recursion starts with $\mathcal{H}_r^{bp}$ as a vacuous hint and the last observation y_r and runs backwards for $k = r, r-1, r-2, \ldots$ It cannot be pretended that the hints $\mathcal{H}_k^{bp}(k)$ and $\mathcal{H}_k^f(k)$ are precise. However, by theorems 10.4 and 10.5 their respective plausibilities of singletons are given by

$$\begin{aligned} pl_k^{bp}(\theta) &= c \sum \{pl_{k+1}^f(\theta')p_t(\theta'|\theta) : \theta' \in \Theta\}, \\ pl_k^f(\theta) &= c\; p_0(y_k|\theta)pl_k^{bp}(\theta). \end{aligned} \tag{10.14}$$

These formulas allow for the recursive backward computation of the relative singleton plausibilities of the hints $\mathcal{H}_k^{bp}(k)$ and $\mathcal{H}_k^f(k)$. It is a kind of a *backward filtering scheme*. It is nicely symmetrical to the usual filtering formulae (10.12). Note however that it computes relative *singleton plausibilities* and not probability densities.

Now, by equation (10.4) we have

$$\mathcal{H}(k) = \mathcal{H}_k^{fp}(k) \oplus \mathcal{H}_k^{bp}(k) \oplus (\mathcal{O}_k \oplus y_k(k,k'))(k). \tag{10.15}$$

By theorem 4.8 $\mathcal{H}(k)$ is precise because $\mathcal{H}_k^{fp}(k)$ is precise and by theorems 10.5 and 4.8 its probability density is given by

$$p_k(\theta) = c\; p_0(y_k|\theta)p_k^{fp}(\theta)pl_k^{bp}(\theta). \tag{10.16}$$

This is the solution of the *smoothing problem for stochastic systems* as derived from theorem 10.1. For stochastic systems this problem is usually solved by classical Bayesian analysis (Pearl, 1988). Although this, of course, gives an equivalent solution, it does not exhibit the same equivalence between forward and backward filtering schemes.

10.3 Special Case of Systems with Imprecisely Known Disturbances

In this section systems with unknown, but bounded disturbances will be considered. However, in a slight generalization, the bounds are assumed to be not exactly known, but themselves uncertain. Consider then a probability space (Ω, P) where Ω is *totally ordered*. The order of the possible interpretations in Ω will reflect increasing imprecision about the possible range of the disturbances. This is described by consonant hints $\mathcal{H} = (\Omega, P, \Gamma, \Theta)$ with $\Gamma(\omega') \subseteq \Gamma(\omega'')$ whenever $\omega' \leq \omega''$ (see Sect. 2.2). Here *all* hints will depend on the same space (Ω, P) of interpretations; they will differ only in Γ (compare with Sect. 4.5). In other words, the hints will not be independent, but on the contrary fully dependent; the whole uncertainty is about which of the possible interpretation $\omega \in \Omega$ is the correct one for all hints. Dempster's rule for combining hints must be adapted accordingly: $\mathcal{H}_1 \oplus \mathcal{H}_2 = (\Omega, P, \Gamma_1 \cap \Gamma_2, \Theta)$ (see Sect. 4.5).

The dynamic system is thus described by transition relations $\mathcal{T}_i = (\Omega, P, \Gamma_i, \Theta_{i-1} \times \Theta_i)$ and observation relations $\mathcal{O}_i = (\Omega, P, \Delta_i, \Theta_i \times \Psi_i)$ which are all consonant and dependent on the same interpretation space (Ω, P). The system is then defined by the following uncertain inclusions:

$$X_{i+1} \in \Gamma_i(\omega|X_i); \; Y_i \in \Delta_i(\omega|X_i) \tag{10.17}$$

for some (unknown) $\omega \in \Omega$.

If Ω contains only one element (i.e. the hints are deterministic), then this reduces itself to the case of a system with unknown, but bounded disturbances (Kurzhanski, 1989). In the general case, the system can be qualified as a system with imprecisely known disturbances; in fact there are some formal links to fuzzy set theory and possibility theory as has been mentioned in Sect. 4.5.

A few theorems are needed in order to apply theorem 10.1 to this system:

Theorem 10.6. *If $\mathcal{H}$ is a consonant hint relative to some arbitrary product frame Θ, then both its refinement to a product frame $\Theta \times \Psi$ and its coarsening to a factor of Θ are also consonant.*

Theorem 10.7 *If $\mathcal{H}_\infty$ is a consonant hint on Θ_1 with plausibility function pl_1 and $\mathcal{H}_2$ is a consonant hint on $\Theta_1 \times \Theta_2$ with plausibility function pl_2 and both have the same interpretation space, then $\mathcal{H}_1 \oplus \mathcal{H}_2$ is also consonant on $\Theta_1 \times \Theta_2$ and its singleton plausibilities are given by*

$$pl(\theta_1, \theta_2) = c \, \min\{pl_1(\theta_1), pl_2(\theta_1, \theta_2)\}. \tag{10.18}$$

Theorem 10.8. *Let $\mathcal{H}$ be a consonant hint on $\Theta_1 \times \Theta_2$ with plausibility function pl and let $B \subseteq \Theta_2$. Then the hint $\mathcal{H}_1 = (\mathcal{H} \oplus B)(1)$ is also consonant and its singleton plausibilities are given by*

$$pl_1(\theta) = \max\{pl(\theta, \theta_2) : \theta_2 \in B\}. \tag{10.19}$$

Theorem 10.9 *If $\mathcal{H}_1$ is a consonant hint on Θ_1 with plausibility function pl_1, $\mathcal{H}$ is a consonant hint on $\Theta_1 \times \Theta_2$ with plausibility function pl and both have the same interpretation space, then $\mathcal{H}_2 = (\mathcal{H}_1 \oplus \mathcal{H})(2)$ is also consonant and its singleton plausibilities are given by*

$$pl_2(\theta_2) = c \max\{\min\{pl(\theta_2|\theta_1), pl_1(\theta_1)\} : \theta_1 \in \Theta_1\}. \tag{10.20}$$

The constant c is here always a normalization constant which guarantees that $pl(\Theta) = \max\{pl(\theta) : \theta \in \Theta\} = 1$.

According to theorems 4.11 and 10.6 all the hints $\mathcal{H}, \mathcal{H}_k^p, \mathcal{H}_k^f, \mathcal{H}_k^{fp}$ and $\mathcal{H}_k^{bp}$, as well as their restriction, are consonant. Theorem 10.1 may now be applied in the same way as in the previous section for stochastic systems. Theorems

10.8 and 10.9 together with theorem 4.11 yield the *forward filtering* formulas for the singleton plausibilites of the hints $\mathcal{H}_k^{fp}(k)$ and $\mathcal{H}_k^{p}(k)$:

$$\begin{aligned} pl_k^{fp}(\theta) &= c \max \{\min (pl_t(\theta|\theta'), pl_{k-1}^{p}(\theta')) : \theta' \in \Theta\}, \\ pl_k^{p}(\theta) &= c \min (pl_0(\theta, y_k), pl_k^{fp}(\theta)). \end{aligned} \tag{10.21}$$

Here $pl_t(\theta', \theta)$ and $pl_0(\theta, \psi)$ are the singleton plausibilities of the hints $\mathcal{T}_k$ and $\mathcal{O}_k$ respectively. Similarly, the formulas for the *backward filtering* (hints $\mathcal{H}_k^{bp}(k)$ and $\mathcal{H}_k^{f}(k)$) are

$$\begin{aligned} pl_k^{bp}(\theta) &= c \max \{\min \{pl_t(\theta, \theta'), pl_{k+1}^{f}(\theta')\} : \theta' \in \Theta\}, \\ pl_k^{f}(\theta) &= c \min \{pl_0(\theta, y_k), pl_k^{bp}(\theta)\}. \end{aligned} \tag{10.22}$$

This leads then to the *smoothing formula* for $\mathcal{H}_k(k)$ according to theorem 4.11:

$$pl_k(\theta) = c \min \{pl_0(\theta, y_k), pl_k^{fp}(\theta), pl_k^{bp}(\theta)\}. \tag{10.23}$$

Note that in the case of *systems with bounded, but unknown disturbances* all the hints involved are deterministic, i.e. simply ordinary sets. The corresponding singleton plausibility functions involved in the above filtering and smoothing formulas are then simply the *indicator functions* of the corresponding deterministic hints or sets.

The two special cases treated in sections 10.2 and 10.3 show that the mathematical theory of *hints* provides for a unifying framework in which both stochastic systems as well as systems with unknown, but bounded disturbances and a lot of other interesting systems with uncertainty can be treated. Different notions of *inclusion* may be defined for hints (see Chap. 14). This observation may lead to new, interesting *approximation schemes for the filtering and smoothing* of complex stochastic systems (especially with infinite state space). This may be one of the most fruitful applications of the theory of hints within the field of uncertain dynamical systems.

A10 Appendix: Proofs of Theorems in Chapter 10

Proof of Theorem 10.1

Fig. A10.1 shows the Markov tree associated with the hypertree of Fig. 10.1. Theorems 8.16 and 8.12 applied to the node X_t then give (10.4).

Formulas (10.3) follow equally from theorems 8.16 and 8.12 but relative to the tree with the right or left branches missing. □

Proof of Theorem 10.2

Let $\mathcal{H}_1 = (\Omega_1, P_1, \Gamma_1, \Theta_1)$ and $\mathcal{H} = (\Omega_2, P_{12}, \Gamma_{12}, \Theta_1 \times \Theta_2)$. Clearly for any $\omega_1 \in \Omega_1$ and $\omega_2 \in \Omega_2$ we have

$$((\Gamma_1(\omega_1) \times \Theta_2) \cap \Gamma_{12}(\omega_2))(2) = \Gamma_{12}(\omega_2|\theta_1) \tag{A10.1}$$

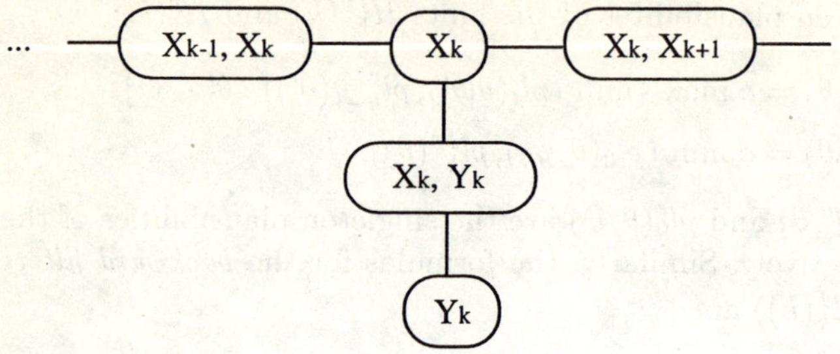

Fig. A10.1.

if $\Gamma_1(\omega_1) = \{\theta_1\}$ and hence $(\mathcal{H}_1 \oplus \mathcal{H}_2)(2)$ is precise.

If

$$u(\{\theta_2\}) = \{(\omega_1, \omega_2) : \Gamma_1(\omega_1) = \{\theta_1\}, \Gamma_{12}(\omega_2|\theta_1) = \{\theta_2\}\} \tag{A10.2}$$

then

$$\begin{aligned} p_2(\theta_2) &= P(u(\{\theta_2\})) \\ &= \sum \{p_1(\omega_1)p_{12}(\omega_2) : \Gamma_1(\omega_1) = \{\theta_1\}, \Gamma_{12}(\omega_2|\theta_1) = \{\theta_2\}\} \\ &= \sum_{\theta_1 \in \Theta_1} \Big(\sum_{\omega_1 : \Gamma_1(\omega_1) = \{\theta_1\}} p_1(\omega_1) \sum_{\omega_2 : \Gamma_{12}(\omega_2|\theta_1) = \{\theta_2\}} p_{12}(\omega_2) \Big) \\ &= \sum_{\theta_1 \in \Theta_1} p_1(\theta_1) p(\theta_2|\theta_1) \end{aligned} \tag{A10.3}$$

and (10.6) is proved. □

Proof of Theorem 10.3

Let $\mathcal{H}_i = (\Omega_i, P_i, \Gamma_i, \Theta_i)$ for $i = 1, 2$ and let $\mathcal{H} = (\Omega, P_{12}, \Gamma_{12}, \Theta_1 \times \Theta_2)$. For any $\omega_1 \in \Omega_1, \omega_2 \in \Omega_2$ and $\omega \in \Omega$ we have

$$((\Gamma_1(\omega_1) \times \Theta_2) \cap (\Theta_1 \times \Gamma_2(\omega_2)) \cap \Gamma_{12}(\omega))(1) \tag{A10.4}$$

is either empty or equals $\{\theta_1\}$ if $\Gamma_1(\omega_1) = \{\theta_1\}, \Gamma_2(\omega_2) = \{\theta_2\}$ and $\Gamma_{12}(\omega|\theta_1) = \{\theta_2\}$. Thus $(\mathcal{H}_1 \oplus \mathcal{H}_2 \oplus \mathcal{H})(1)$ is precise. If

$$u(\{\theta_1\}) = \{(\omega_1, \omega_2, \omega) : \Gamma_i(\omega_i) = \{\theta_i\}, i = 1, 2, \Gamma_{12}(\omega|\theta_1) = \{\theta_2\}\} \tag{A10.5}$$

then

$$p(\theta_1) = c \sum \{P_1(\omega_1)P_2(\omega_2)P_{12}(\omega) : (\omega_1, \omega_2, \omega) \in u(\{\theta_1\})\} \tag{A10.6}$$

where c is the constant of conditioning. If follows that

$$p(\theta_1) = c \cdot \sum_{\omega_1:\Gamma_1(\omega_1)=\{\theta_1\}} P_1(\omega_1)$$
$$\cdot \sum_{\theta_2\in\Theta_2}\Big(\Big(\sum_{\omega_2:\Gamma_2(\omega_2)=\{\theta_2\}} P_2(\omega_2)\Big)\cdot\Big(\sum_{\omega:\Gamma(\omega|\theta_1)=\{\theta_2\}} P_{12}(\omega)\Big)\Big) \quad (A10.7)$$
$$= c\cdot p_1(\theta_1)\cdot \sum_{\theta_2\in\Theta_2} p_2(\theta_2)\cdot p(\theta_2|\theta_1)$$

which proves (10.7) □

Proof of Theorem 10.4

With the notation of the previous proof let

$$v(\{\theta_1\}) = \{(\omega_2,\omega) : \exists\ \theta_2 \in \Gamma_2(\omega_2) \text{ such that } \Gamma_{12}(\omega|\theta_1) = \{\theta_2\}\}. \quad (A10.8)$$

Then

$$pl(\theta_1) = c\cdot \sum \{p_2(\omega_2)p_{12}(\omega) : (\omega_2,\omega) \in v(\{\theta_1\})\}$$
$$= c\cdot \sum_{\theta_2\in\Theta_2}\Big(\sum_{\omega_2:\theta_2\in\Gamma_2(\omega_2)} p_2(\omega_2) \sum_{\omega:\Gamma_{12}(\omega|\theta_1)=\{\theta_2\}} p_{12}(\omega)\Big) \quad (A10.9)$$
$$= c\cdot \sum_{\theta_2\in\Theta_2} pl_2(\theta_2)p(\theta_2|\theta_1).$$

□

Proof of Theorem 10.5

With the notation of the proof of theorem 10.3 let

$$v(\{\theta_1\}) = \{(\omega_1,\omega) : \theta_1 \in \Gamma_1(\omega_1), \Gamma_{12}(\omega|\theta_1) = \{\theta_2\}\}. \quad (A10.10)$$

Then

$$pl(\theta_1) = c\cdot \sum \{p_1(\omega_1)p_{12}(\omega) : (\omega_1,\omega) \in v(\{\theta_1\})\}$$
$$= c\cdot \Big(\sum_{\omega_1:\theta_1\in\Gamma_1(\omega_1)} p_1(\omega_1)\Big)\Big(\sum_{\omega:\Gamma_{12}(\omega|\theta_1)=\{\theta_2\}} p_{12}(\omega) \quad (A10.11)$$
$$= c\cdot pl_1(\theta_1)p(\theta_2|\theta_1).$$

□

Proof of Theorem 10.6

The proposition holds because both cylindrical extension as well as projection of sets maintain inclusion. □

Proof of Theorem 10.7

The consonance of $\mathcal{H}_1 \oplus \mathcal{H}_2$ is clear in view of theorems 10.6 and 4.11. The refinement of $\mathcal{H}_1$ to $\Theta_1 \times \Theta_2$ has singleton plausibilities $pl(\theta_1, \theta_2) = pl_1(\theta_1)$ for all $(\theta_1, \theta_2) \in \Theta_1 \times \Theta_2$. Then the proposition follows from theorem 4.11. □

Proof of Theorem 10.8

The hint $\mathcal{H}_1$ is clearly consonant by theorem 10.6. Then by theorem 3.7 we have

$$\begin{aligned} pl_1(\theta) &= pl_{\mathcal{H}\oplus B}(\{\theta\} \times \Theta_2) \\ &= \max\ \{pl_{\mathcal{H}\oplus B}(\theta, \theta_2) : (\theta, \theta_2) \in \{\theta\} \times \Theta_2\} \\ &= \max\ \{pl_{\mathcal{H}\oplus B}(\theta, \theta_2) : \theta_2 \in \Theta_2\} \\ &= \max\ \{pl_{\mathcal{H}\oplus B}(\theta, \theta_2) : \theta_2 \in B\} \\ &= \max\ \{pl(\theta, \theta_2) : \theta_2 \in B\}. \end{aligned} \tag{A10.12}$$

□

Proof of Theorem 10.9

By theorem 10.7 the hint $\mathcal{H}_1 \oplus \mathcal{H}$ is consonant and has singleton plausibilities

$$pl'(\theta_1, \theta_2) = c \cdot \min\ \{pl_1(\theta_1), pl(\theta, \theta_2)\}. \tag{A10.13}$$

Then by theorem 3.7 we have

$$\begin{aligned} pl_2(\theta_2) &= pl'(\Theta_1 \times \{\theta_2\}) \\ &= \max\ \{pl'(\theta_1, \theta_2) : (\theta_1, \theta_2) \in \Theta_1 \times \{\theta_2\}\} \\ &= \max\ \{pl'(\theta_1, \theta_2) : \theta_1 \in \Theta_1\} \\ &= c \cdot\ \max\ \{\min\ (pl_1(\theta_1), pl(\theta_1, \theta_2)) : \theta_1 \in \Theta_1\}. \end{aligned} \tag{A10.14}$$

□

11. Diagnostics

11.1 Relational Models

Diagnostics is the task of inferring plausible explanations for a set of observations or to decide which explanation accounts for given observations. In many cases the observations are symptoms of diseases of a medical system or faults in a technical system. Symptoms can be considered as hints for certain possible diseases and the theory of hints is therefore an obvious approach to diagnostics. The problem is then to infer the possible faults or diseases causing the observed symptoms.

An early example of a diagnostic expert system was MYCIN. The authors of this system noted that symptoms are not infallible evidence but may have exceptions. Also they remarked that symptoms are hints which bear on sets of hypotheses rather than on individual hypotheses. They concluded that probability theory was not an appropriate tool for their purposes and they developed their own ad hoc calculus of uncertainty (Shortliffe, Buchanan, 1975). Later it became clear that evidence theory could have been well adapted to the structure of MYCIN (Gordon, Shortliffe, 1984, see also the following section 11.2). Applications of evidence theory to diagnostics have also been discussed by Smets (1979, 1981). GERTIS is a newer expert system mixing evidence theory and Bayesian analysis (Yen, 1989).

The models discussed in these papers can be qualified as *relational* because they somehow link symptoms and diseases by uncertain empirical causal or evidential relations. A classical method of diagnostic analysis is based on Bayesian analysis (Peng, Reggia, 1990, Lauritzen, Spiegelhalter, 1988). In this case the relations between diseases and symptoms are described by conditional probabilities. In this section it will be shown that evidential reasoning offers an alternative to Bayesian analysis. For this purpose we first propose a general relational model for diagnostics based on assumption-based rules and we discuss a simple example. We also refer to example 5.4 in Chap. 5, which is another model, similar to the one presented in this chapter. Furthermore, a model first formulated in a Bayesian way (Lauritzen, Spiegelhalter, 1988) is reformulated and analyzed as an assumption-based knowledge base. This permits a comparison of the two approaches.

The definition of associations between diseases of faults and symptoms is a very simple way to describe the functioning of a system. There may also be more elaborate models describing the systems under different assumptions concerning the state of its components. It may then be possible to infer knowledge about the possible states of the components, if some observations on the system can be made. This leads then to a model-based approach to diagnostics as described for example by Reiter (1987). The theory of hints fits well into this approach. This will be discussed in the third section of this chapter.

Let then in a first step denote by D the set of possible diseases (or faults) whose generic elements will be denoted by d or $d_i, \ldots$ etc. M denotes the set of possible symptoms and its generic elements are denoted by $m, m_j, \ldots$ etc. It will be assumed that a disease as well as a symptom is either present or absent, i.e. degrees of diseases or symptoms (as low or high temperature for example) will not be taken into account. In this respect, as well as in a number of other respects, the model to be introduced will be a very simplistic view of a complex problem. Its main purpose is to demonstrate some essential aspects of evidential reasoning in the domain of diagnostics and to compare it with the better known Bayesian approach and for this purpose it will be sufficient.

The relation between diseases and symptoms will depend on complex interactions of internal factors in the system. To capture this idea let w denote such a particular interaction and Ω the set of all possible interactions. Then, given ω and given a certain subset D' of diseases, a certain subset $M'(\omega, D') \subseteq M$ will be present. This signifies that there is a relation between subsets of D and M for any possible ω. It will hardly be possible to specify these relations explicitly and in detail in any practical problem.This description has to be taken rather as conceptual than real. Let B_D be the Boolean cube of all Boolean vectors of length $|D|$. For a vector $x \in B_D$, let $D(x) = \{d_i \in M : x_i = 1\}$ denote its corresponding set of deseases. This defines a bijective relation between B_D and the subsets of M. Similarly let B_M denote the Boolean vectors describing in the same way the subsets of M. Then the above relation for ω can be described as a relation $\Gamma(\omega) \subseteq B_D \times B_M$. If finally a probability measure P is assumed to be given on Ω (which is assumed to be finite for simplicity's sake), then this system is described by a hint $\mathcal{S} = (\Omega, P, \Gamma, B_D \times B_M)$.

Suppose now that certain symptoms are observed. More precisely, assume that certain symptoms in a subset M^+ are found to be present, other symptoms in a subset M^- are found to be absent and the symptoms in $M - M^+ - M^-$ are not observed. Such an observation clearly defines a subset O of B_M, given by

$$\{x \in B_M : x_i = 1 \text{ for } d_i \in M^+, x_i = 0 \text{ for } d_i \in M^-\}.$$

This subset O can be considered as a deterministic hint on B_M. The solution to the problem of diagnostics in this setting is then

$$(\mathcal{S} \oplus O)|B_D. \tag{11.1}$$

From this combined hint one may get degrees of support and plausibility for any hypothesis about possible groups of diseases. Of course this is a solution only in principle. In this general form the model is hardly useful. Additional structure is needed both with respect to the form of the relations $\Gamma(\omega)$ as with respect to the probability structure on Ω, in order to get tractable models, both with regard to the specification and to the computational problems. Below and in the following section, particular cases satisfying these requirements will be introduced. A further model belonging to this class is also given in example 5.4 in Chap 5.

In chapter 10 the notion of conditional precise hints has been introduced. This notion is also of some interest in the present context. The hint $\mathcal{S}$ is conditionally precise relative to B_M given B_D, if $\mathcal{S} \oplus \{x\})|B_M$ is precise for all vectors x of B_D. Then theorem 10.3 applies. In order to interpret this theorem in the present context note that any $x \in B_D$ defines the subset D' of D containing all d_i for which $x_i = 1$. Similarly, any $y \in B_M$ defines the subset M' of M of all symptoms m_j for which $y_j = 1$. Thus, if an a priori distribution of diseases $p_0(D')$ over the power set of D is given and if an observation of symptoms M' is given, the symptoms in $M - M'$ being absent, then the a posteriori distribution on the sets of diseases can be obtained by

$$p(D'') = c \cdot p_0(D'')p(M'|D'') \tag{11.2}$$

for all $D'' \subseteq D$. Here $p(M'|D'')$ is the probability of M' (bpa) corresponding to the precise hint $(\mathcal{S} \oplus \{x\})|B_M$ if x is the vector corresponding to D''. This shows how Bayesian analysis fits into the framework of our general associative model. It represents the particular case where $\mathcal{S}$ is conditionally precise on B_M, and where furthermore an a priori distribution over the diseases is given and the observation of symptoms is complete in the sense that all possible symptoms have been observed. As the general relational model, this general Bayesian model is not tractable either; even in this case more structure has to be imposed on the model (see Peng, Reggia (1990) for an analysis of such Bayesian systems).

As a first particular model of the general associative system suppose that a disease d_i may cause a symptom m_j, given a further supplementary condition (or assumption) a_{ij} is satisfied. Assume furthermore that a symptom m_j may also be caused by an unaccounted reason b_j and suppose finally that there is a given set of diseases $(m_j) \subseteq D$ which may cause m_j and no disease outside this set can cause m_j. This can then be described by the following implications:

$$\begin{aligned} & d_i \wedge a_{ij} \rightarrow m_j, \text{ for all } d_i \in (m_j) \text{ and } m_j \in M, \\ & b_j \rightarrow m_j, \text{ for all } m_j \in M, \\ & \wedge \ \{\neg d_i : d_i \in (m_j)\} \wedge (\neg b_j) \rightarrow \neg m_j, \text{ for all } m_j \in M. \end{aligned} \tag{11.3}$$

This is a *causal* model describing how symptoms are generated by diseases. It is also an assumption-based model of the kind discussed in Chap. 5; the assumptions being the propositions a_{ij} and b_j. It can therefore be treated by the methods introduced in this chapter. A similar model has been studied by Reggia et al. (1983, 1985) in a spirit similar to Reiter's theory of diagnosis.

Note that for the first step of the analysis no probabilities have to be associated with the assumptions, which is a big advantage, because it may not be easy to assess these probabilities and the supposition of stochastic independence between these assumptions may be questionable in some cases. It is sufficient, in a first step, to compute the contradictions between assumptions and the support of hypotheses by the methods introduced in Chap. 5. This will be illustrated by a simple example. Only afterwards probabilities may possibly be introduced in order to get degrees of support of different competing hypotheses.

Example 11.1 Diagnostics of a Faulty Car. The system represented in Fig. 7.1 will be described here in the framework of the general model introduced above. The diseases d_i are represented in this example by the leaves of the tree in Fig. 7.1 and denoted as in Fig 7.1 by $b, d, e, f, g, h, i, j, k$. The observable symptoms are related in this example with the non-leaf nodes of the tree, i.e. they represent sets of possible faults:

m_1 : the car does not start
m_2 : the battery system seems to be faulty
m_3 : the starting system seems to be faulty
m_4 : the switch seems to be defective

Introducing the necessary assumptions according to the general model and defining the sets (m_j) according to the tree Fig. 7.1 leads then to the following model for the car system:

$$
\begin{array}{ll}
b \wedge a_{b1} \rightarrow m_1 & d \wedge a_{d1} \rightarrow m_1 \\
e \wedge a_{e1} \rightarrow m_1 & f \wedge a_{f1} \rightarrow m_1 \\
i \wedge a_{i1} \rightarrow m_1 & j \wedge a_{j1} \rightarrow m_1 \\
k \wedge a_{k1} \rightarrow m_1 & g \wedge a_{g1} \rightarrow m_1 \\
h \wedge a_{h1} \rightarrow m_1 & \\
e \wedge a_{e2} \rightarrow m_2 & f \wedge a_{f2} \rightarrow m_2 \\
i \wedge a_{i3} \rightarrow m_3 & j \wedge a_{j3} \rightarrow m_3 \\
k \wedge a_{k3} \rightarrow m_3 & g \wedge a_{g3} \rightarrow m_3 \\
h \wedge a_{h3} \rightarrow m_3 & \\
i \wedge a_{i4} \rightarrow m_4 & j \wedge a_{j4} \rightarrow m_4 \\
k \wedge a_{k4} \rightarrow m_4 & \\
b_1 \rightarrow m_1, \; b_2 \rightarrow m_2 & b_3 \rightarrow m_3, \; b_4 \rightarrow m_4 \\
\neg e \wedge \neg f \wedge \neg b_2 \rightarrow \neg m_2 & \\
\neg g \wedge \neg i \wedge \neg j \wedge \neg k \wedge \neg h \wedge \neg b_3 \rightarrow \neg m_3 & \\
\neg i \wedge \neg j \wedge \neg k \wedge \neg b_4 \rightarrow \neg m_4 &
\end{array}
$$

$$
\neg b \wedge \neg d \wedge \neg e \wedge \neg f \wedge \neg g \wedge \neg i \wedge \neg j \wedge \neg k \wedge \neg h \wedge \neg b_1 \rightarrow \neg m_1
$$

Suppose now that in a particular case the observations m_1 (car does not start), $\neg m_3$ (starting system does not seem faulty) and m_4 (the switch seems defective) are made. This knowledge base together with these supplementary facts can then be analyzed by the methods of Chap. 5. The results given in

Table 11.1.

Hypothesis	Minimal Quasi – Support	Degree of Support
b	$\neg b_4 \wedge a_{i3} \wedge a_{j3} \wedge a_{k3}, b_3$	0
$\neg b$	$\neg b_4 \wedge a_{i3} \wedge a_{j3} \wedge a_{k3}, b_3$	0
d	$\neg b_4 \wedge a_{i3} \wedge a_{j3} \wedge a_{k3}, b_3$	0
$\neg d$	$\neg b_4 \wedge a_{i3} \wedge a_{j3} \wedge a_{k3}, b_3$	0
e	$\neg b_4 \wedge a_{i3} \wedge a_{j3} \wedge a_{k3}, b_3$	0
$\neg e$	$\neg b_4 \wedge a_{i3} \wedge a_{j3} \wedge a_{k3}, b_3$	0
f	$\neg b_4 \wedge a_{i3} \wedge a_{j3} \wedge a_{k3}, b_3$	0
$\neg f$	$\neg b_4 \wedge a_{i3} \wedge a_{j3} \wedge a_{k3}, b_3$	0
i	$\neg b_4 \wedge a_{j3} \wedge a_{k3}, b_3$	0.0134
$\neg i$	a_{i3}, b_3	0.2960
j	$\neg b_4 \wedge a_{i3} \wedge a_{k3}, b_3$	0.0229
$\neg j$	a_{j3}, b_3	0.1954
k	$\neg b_4 \wedge a_{i3} \wedge a_{j3}, b_3$	0.0516
$\neg k$	a_{k3}, b_3	0.0948
g	$\neg b_4 \wedge a_{i3} \wedge a_{j3} \wedge a_{k3}, b_3$	0
$\neg g$	$\neg b_4 \wedge a_{i3} \wedge a_{j3} \wedge a_{k3}, a_{g3}, b_3$	0.2
h	$\neg b_4 \wedge a_{i3} \wedge a_{j3} \wedge a_{k3}, b_3$	0
$\neg h$	$\neg b_4 \wedge a_{i3} \wedge a_{j3} \wedge a_{k3}, a_{h3}, b_3$	0.0.6
m_1	$\neg b_1, \neg b_4, b_3$	0.9975
m_2	$\neg b_4 \wedge a_{i3} \wedge a_{j3} \wedge a_{k3}, b_3$	0
m_3	$\neg b_4, b_3$	0.9497
m_4	$\neg b_4, b_3$	0.9497

Tab. 11.1 have been obtained using a computer program written by U. Hänny at the Informatics Institute of the University of Fribourg.

The minimal contradictions are

$$b_3, \neg b_4 \wedge a_{i3} \wedge a_{j3} \wedge a_{k3}.$$

The assumption b_3 is clearly a contradiction because it contradicts the observation $\neg m_3$. Furthermore, a_{i3}, a_{j3} and a_{k3} are quasi-supports of the hypotheses $\neg i, \neg j$ and $\neg k$ respectively. This is clear because the observation $\neg m_3$ (the starting system is not faulty) supports the absence of a fault in the switch system (see Fig. 7.1). Also, the minimal quasi-supports of the hypothesis

$$m_1 = b \vee d \vee e \vee f \vee i \vee j \vee k \vee g \vee h$$

are

$$\neg b_1, b_3, \neg b_4$$

and those of the hypothesis

$$m_3 = i \vee j \vee k \vee g \vee h$$

are

$$b_3, \neg b_4.$$

Again, these results are easy to understand. Then degrees of support can then be computed using formula (5.35). First, we find that

$$sp'(\circ) = P(b_3) + P(\neg b_4 \wedge a_{i3} \wedge a_{j3} \wedge a_{k3})(1 - P(b_3)) = 0.335.$$

Then we have for example

$$sp(m_1) = \frac{P(\neg b_1 \vee b_3 \vee \neg b_4) - 0.335}{1 - 0.335} = 0.9975$$

and

$$sp(m_3) = \frac{P(b_3 \vee \neg b_4) - 0.335}{1 - 0.335} = 0.9497.$$

This example shows that such hierarchical systems are very easy to analyze.

Example 11.2 (Lauritzen, Spiegelhalter, 1988). We take here an example which served as a typical model of Bayesian modeling and describe it in an assumption-based knowledge base. The following text represents a small piece of fictitious medical "knowledge":

Shortness-of-breath (dyspnoea) may be due to tuberculosis, lung cancer or bronchitis, or none of them, or more than one of them. A recent visit to Asia increases the chances of tuberculosis, while smoking is known to be a risk factor for both lung cancer and bronchitis. The results of a single chest X-ray do not discriminiate between lung cancer and tuberculosis, as neither does the presence or absence of dyspnoea (Lauritzen, Spiegelhalter, 1988).

In order to describe this knowledge we introduce the following propositional symbols:

α : the person was in Asia
τ : tuberculosis is present
λ : lung cancer is present
β : bronchitis is present
σ : the person is a smoker
χ : positive X-ray
δ : dyspnoea is present

The symbols a_i represent various assumptions, their significance will be clear in the context of the rules they appear in. This is a possible representation of the knowledge expressed in the small text above:

$$\alpha \wedge a_1 \rightarrow \tau,\ a_2 \rightarrow \tau,\ \neg\alpha \wedge \neg a_2 \rightarrow \neg\tau$$

A visit to Asia can lead to tuberculosis, but other (unidentified) factors can also cause tuberculosis, and these are the only ways tuberculosis can arise.

$$\sigma \wedge a_3 \rightarrow \lambda,\ \sigma \wedge a_4 \rightarrow \beta$$
$$a_5 \rightarrow \lambda,\ a_6 \rightarrow \beta$$
$$\neg\sigma \wedge \neg a_5 \rightarrow \neg\lambda,\ \neg\sigma \wedge \neg a_6 \rightarrow \neg\beta$$

Smoking can cause lung cancer or bronchitis, but both of these diseases can be caused by other (unidentified) factors. However, they are only caused either by smoking or these factors.

$$\tau \wedge a_7 \rightarrow \chi,\ \lambda \wedge a_7 \rightarrow \chi,\ a_8 \rightarrow \chi$$
$$\neg\lambda \wedge \neg\tau \wedge \neg a_8 \rightarrow \neg\chi$$

Provided a condition is satisfied, both tuberculosis and lung cancer can cause a positive X-ray. But if this condition is not satisfied then these diseases cannot cause a positive X-ray. Of course, there may be other reasons for a positive X-ray besides lung cancer and tuberculosis.

$$\lambda \wedge a_9 \rightarrow \delta,\ \tau \wedge a_{10} \rightarrow \delta,\ \beta \wedge a_{11} \rightarrow \delta$$
$$a_{12} \rightarrow \delta,\ \neg\beta \wedge \neg\lambda \wedge \neg\tau \wedge \neg a_{12} \rightarrow \neg\delta$$

Dyspnoea may be due to lung cancer, tuberculosis or bronchitis, or none of them or more than one of them.

Note that this model explicitely represents causes or conditions for an effect by the assumptions $a_i, i = 1, \ldots, 12$. We assume the following known probabilities for the assumptions:

a_1	a_2	a_3	a_4	a_5	a_6	a_7	a_8	a_9	a_{10}	a_{11}	a_{12}
0.05	0.01	0.1	0.6	0.01	0.3	0.98	0.05	0.9	0.8	0.7	0.1

Now assume that a person who recently visited Asia arrives and complains about dyspnoea. This means that the facts α and δ are added to the knowledge base. Then the analysis can again be carried out by the methods of Chap. 5. One is particularly interested in the diseases τ, λ and β. The results of the analysis for several hypotheses involving these diseases are collected in Tab. 11.2 a). It turns out to be almost certain that one of the diseases is present. The available knowledge is, however, not sufficient to discriminate between them, even if τ and β seem probable and among them β has the largest degree of support.

The discussion with the patient reveals that he is a smoker. So σ is added to the knowledge-base and the analysis is continued. The new results are given in Tab. 11.2 b). The new information does slightly reinforce the belief that one of the diseases is present and it reinforces especially the support of β, which becomes fairly credible. But also the support of λ increases slightly and τ remains relatively unsupported.

Next an X-ray is made which turns out to be negative. This means that $\neg\chi$ is added to the knowledge-base. The new results are summerized in Tab. 11.2 c). Note that now there are contradictory assumptions. This new information now makes β nearly sure and at the same time eliminates τ and λ (their degrees of plausibility are both equal to 0.0238). Now we have a very clear diagnosis which leaves little room for doubt.

Finally, let's look at the situation if the person were a non-smoker (i.e. the observations are $\alpha, \delta, \neg\sigma$ and no X-ray has been made yet). The results are

Table 11.2 a). Person was in Asia and has dyspnoea

There is no contradiction

Hypothesis	Minimal Quasi-supports	Degree of support
τ	a1, a2	0.0595
λ	a5	0.01
β	a6	0.3
τ∨λ	¬ a12 ∧ ¬ a6 ∧ a3, a1, a2, a5	0.1276
τ∨β	¬ a12 ∧ ¬ a5 ∧ a4, a1, a2, a6	0.6936
λ∨β	a5, a6	0.307
τ∨λ∨β	¬ a12, a1, a2, a5, a6	0.9348

Table 11.2 b). Person is a smoker, was in Asia and has dyspnoea

There is no contradiction

Hypothesis	Minimal Quasi-supports	Degree of support
τ	a1, a2	0.0595
λ	a3, a5	0.109
β	a4, a6	0.72
τ∨λ	a1, a2, a3, a5	0.162
τ∨β	a1, a2, a4, a6	0.7367
λ∨β	a3, a4, a5, a6	0.7505
τ∨λ∨β	¬ a12, a1, a2, a3, a5, a6	0.9348

given in Tab. 11.2 d). Compared to the results of Tab. 11.2 b), this shows that the support of bronchitis has dramatically decreased whereas the support of tuberculosis has much increased. In addition, the presence of at least tuberculosis or bronchitis is pretty sure and lung cancer is almost not supported.

11.2 Diagnostics in a Hierarchical Hypothesis Space

In the associative model of the previous section, a hierarchical family of sets of diseases was assumed. This means that symptoms relate to natural classes

Table 11.2 c). Person is a smoker, was in Asia, has dyspnoea and the X-ray test is negative

Minimal contradictions: a5∧a7, a3∧a7, a2∧a7, a1∧a7, a8

Hypothesis	Minimal Quasi-supports	Degree of support
τ	a1, a2, a8, a3∧a7, a5∧a7	0.0014
λ	a3, a5, a8, a1∧a7, a2∧a7	0.0026
β	a4, a6, a8, a7∧¬ a12, a5∧a7 a3∧a7, a2∧a7, a1∧a7	0.966
τ∨λ	a1, a2, a3, a5, a8	0.0039
τ∨β	a1, a2, a4, a6, a8, a7∧¬ a12 a5∧a7, a3∧a7	0.9664
λ∨β	a3, a4, a5, a6, a8, a7∧¬ a12 a7∧a2, a1∧a7	0.9667
τ∨λ∨β	¬ a12, a1, a2, a3, a4, a5, a6, a8	0.9721

Table 11.2 d). Person was in Asia, has dyspnoea but is a non-smoker

There is no contradiction

Hypothesis	Minimal Quasi-supports	Degree of support
τ	a1, a2, ¬ a12∧¬ a6∧¬ a5	0.6461
λ	a5	0.01
β	a6	0.3
τ∨λ	a1, a2, a5, ¬ a12∧¬ a6	0.6555
τ∨β	¬ a12∧¬ a5, a1, a2, a6	0.9282
λ∨β	a5, a6	0.307
τ∨λ∨β	¬ a12, a1, a2, a5, a6	0.9348

of diseases or faults which can be arranged in a tree structure. In this section a similar model will be discussed and it will be shown that this tree structure greatly simplifies the computations. Fig. 7.11 may again serve as an illustrating example. This time observations (symptoms or the absence of symptoms) hint at the presence or absence of a *single* disease in the subsets represented by the nodes of a tree like Fig. 7.1. This model will this time not be solved by logical analysis as the model of the previous section, but by local propagation and combination of hints using results from Chap. 8.

In contrast to the similar model of the previous section the presence of exactly one disease is supposed now. It is a single disease, not a multiple disease model. Consider a finite set Θ of possible diseases, exactly one of which is present, and a family $\mathcal{E} = \{E_1, \ldots, E_m\}$ of subsets of Θ which has the property that either $E_i \cap E_j = \emptyset$, or $E_i \subset E_j$ or $E_j \subset E_i$, whenever $i \neq j$. Suppose that Θ and all singleton subsets $\{\theta\}$ for all $\theta \in \Theta$ are members of $\mathcal{E}$. Then $\mathcal{E}$ is called a *hierarchy* of diseases and its members can be arranged in a tree as in Fig. 7.1 which is an example of such a hierarchy. Suppose that associated with each element E of $\mathcal{E}$, except Θ, there are two simple hints $\mathcal{H}'_E$ and $\mathcal{H}''_E$, the first one with focal set E and the second one with focal set E^c. One or both of these hints may also be vacuous; that is why we may assume the existence of such hints for every element of $\mathcal{E}$, although the actual observations induce *non-vacuous* hints only for some E in $\mathcal{E}$. These two hints can then be combined on the partition $\{E, E^c\}$ of Θ to obtain the hint $\mathcal{H}_E = \mathcal{H}'_E \oplus \mathcal{H}''_E$ on Θ which is carried by $\{E, E^c\}$. Of course, the family of compatible frames to be considered here is $Part(\Theta)$: the partitions of Θ (see Chap. 7). This hint has just the three focal sets E, E^c and Θ. The problem is then to compute the hint

$$\mathcal{H} = \oplus \{\mathcal{H}_E : E \in \mathcal{E}\} \tag{11.4}$$

or at least the degrees of support and plausibility for some important hypotheses $H \subseteq \Theta$. This problem has first been formulated by Gordon, Shortliffe (1985 a and b). These authors noted that the presence of negative evidence pointing to sets E^c is disturbing, complicating the computations. They proposed therefore a simplified, approximate approach. Shafer, Logan (1987) however, pointed out that there is an exact computational method, based on results such as those of Chap. 8, which is not less efficient than the method of Gordon and Shortliffe. This approach will be presented here (see also Shafer et al., 1987).

Define $\mathcal{H}^{\downarrow}_E$ to be the combination of all hints $\mathcal{H}_B$ for all $B \in \mathcal{E}$ which are strictly included in E (strictly below the node E in the tree picture). If E is a leaf of the tree, then $\mathcal{H}^{\downarrow}_E$ is defined to be vacuous. Note that $\mathcal{H} = \mathcal{H}^{\downarrow}_{\Theta}$. Furthermore define $\mathcal{H}^{\diamond}_E$ to be the combination of all hints $\mathcal{H}_B$ for which $B \in \mathcal{E}$ is not contained in E (thus excluding $B = E$). Then clearly

$$\mathcal{H} = \mathcal{H}^{\downarrow}_E \oplus \mathcal{H}_E \oplus \mathcal{H}^{\diamond}_E \tag{11.5}$$

for any E in $\mathcal{E}$.

Denote by $S(E)$ the set of all elements of $\mathcal{E}$ which are immediately below E in the tree (the sons of E) and by $F(E)$ the node immediately above E (the father of E). We now work with partitions of Θ, which clearly form a family of compatible frames (see Chap. 7). The goal is to compute (11.4) using only relatively coarse partitions in order to reduce the computational effort.

To this end remember that the tree shown in Fig. 11.1, whose nodes are the partitions $\{E, E^c\}$ and $S(E) \cup \{E^c\}$, is a Markov tree by theorem 8.11 (this is the tree of families and dichotomies, see Chap. 8).

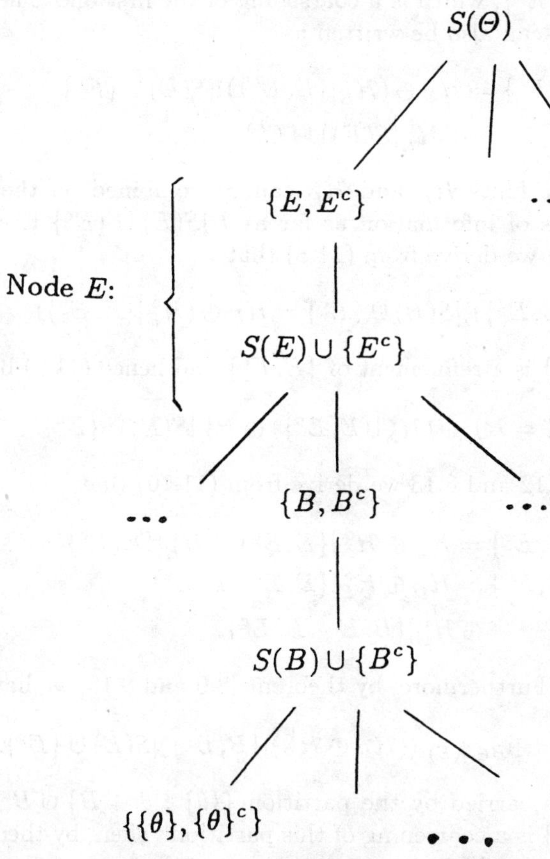

Fig. 11.1.

Then, by theorem 8.16, we have

$$
\begin{aligned}
\mathcal{H}|\{E, E^c\} &= \mathcal{H}_E \oplus (\mathcal{H}_E^{\diamond}|S(F(E)) \cup (F(E)^c))|\{E, E^c\} \\
&\quad \oplus (\mathcal{H}_E^{\downarrow}|S(E) \cup \{E^c\})|\{E, E^c\}, \\
\mathcal{H}|\{\{\theta\}, \{\theta\}^c\} &= \mathcal{H}_{\{\theta\}} \oplus (\mathcal{H}_{\{\theta\}}^{\diamond}|S(F(\{\theta\})) \cup (F(\{\theta\})^c))|\{\{\theta\}, \{\theta\}^c\}, \\
\mathcal{H}|S(E) \cup \{E^c\} &= ((\mathcal{H}_E \oplus \mathcal{H}_E^{\diamond})|\{E, E^c\})|S(E) \cup \{E^c\} \\
&\quad \oplus \{((\mathcal{H}_B \oplus \mathcal{H}_B^{\downarrow})|\{B, B^c\})|S(E) \cup \{E^c\}) : B \in S(E)\}, \\
\mathcal{H}|S(\Theta) &= \oplus\{(\mathcal{H}_B \oplus \mathcal{H}_B^{\downarrow}|\{B, B^c\})|S(\Theta) : B \in S(\Theta)\}.
\end{aligned}
\tag{11.6}
$$

The third formula of (11.6) can also be written as

$$\begin{aligned}\mathcal{H}|S(E)\cup\{E^c\} =&((\mathcal{H}_E \oplus \mathcal{H}_E^\diamond)|\{E,E^c\})|S(E)\cup\{E^c\}\\ &\oplus \mathcal{H}_E^\downarrow|S(E)\cup\{E^c\}.\end{aligned} \tag{11.7}$$

The hint $\mathcal{H}_E^\diamond$ is carried by the partition $\{E\}\cup\{\{\theta\}:\theta\in E^c\}$ and the hint $\mathcal{H}_E$ by the partition $\{E,E^c\}$, which is a coarsening of the first one. Therefore, by theorem 8.15, (11.7) can also be written as

$$\begin{aligned}\mathcal{H}|S(E)\cup\{E^c\} =&(\mathcal{H}_E \oplus (\mathcal{H}_E^\diamond|\{E,E^c\}))|S(E)\cup\{E^c\}\\ &\oplus \mathcal{H}_E^\downarrow|S(E)\cup\{E^c\}.\end{aligned} \tag{11.8}$$

This means that the hints $\mathcal{H}_E$ and $\mathcal{H}_E^\diamond$ can be combined on the partition $\{E,E^c\}$ without loss of information as far as $\mathcal{H}|S(E)\cup\{E^c\}$ is concerned. But by theorem 8.15 we derive from (11.8) that

$$\mathcal{H}_E \oplus (\mathcal{H}_E^\diamond|\{E,E^c\}))|S(E)\cup\{E^c\} = \mathcal{H}_E \oplus (\mathcal{H}_E^\diamond|\{E,E^c\}) \tag{11.9}$$

because $S(E)\cup\{E^c\}$ is a refinement of $\{E,E^c\}$ and hence (11.8) becomes

$$\mathcal{H}|S(E)\cup\{E^c\} = \mathcal{H}_E \oplus (\mathcal{H}_E^\diamond|\{E,E^c\}) \oplus \mathcal{H}_E^\downarrow|S(E)\cup\{E^c\}. \tag{11.10}$$

Then by theorems 8.12 and 8.13 we derive from (11.10) that

$$\begin{aligned}\mathcal{H}|\{E,E^c\} &= \mathcal{H}_E \oplus \mathcal{H}_E^\diamond|\{E,E^c\} \oplus \mathcal{H}_E^\downarrow|\{E,E^c\},\\ \mathcal{H}|\{B,E-B,E^c\} &= \mathcal{H}_E \oplus \mathcal{H}_E^\diamond|\{B,E-B,E^c\}\\ &\quad \oplus \mathcal{H}_E^\downarrow|\{B,E-B,E^c\}\end{aligned} \tag{11.11}$$

for every $B\in S(E)$. Furthermore, by theorems 8.9 and 8.16, we have

$$\mathcal{H}_E^\downarrow|S(E)\cup\{E^c\} = \oplus_{B\in S(E)}\ ((\mathcal{H}_B \oplus \mathcal{H}_B^\downarrow)|\{B,B^c\})|S(E)\cup\{E^c\}. \tag{11.12}$$

But note that $\mathcal{H}_B$ is carried by the partition $\{\{\theta\}:\theta\in B\}\cup B^c$ and that the partition $\{B,B^c\}$ is a coarsening of this partition. Then, by theorem 8.15, equation (11.12) can also be written as

$$\mathcal{H}_E^\downarrow|S(E)\cup\{E^c\} = \oplus_{B\in S(E)}\ (\mathcal{H}_B \oplus (\mathcal{H}_B^\downarrow|\{B,B^c\}))|S(E)\cup\{E,E^c\} \tag{11.13}$$

which says that$\mathcal{H}_B$ and $\mathcal{H}_B^\downarrow$ may be combined without loss of information on the partition $\{B,B^c\}$. This then is a recursive formula which shows how $\mathcal{H}_E^\downarrow|S(E)\cup\{E^c\}$ can be computed by going upwards in the tree $\mathcal{E}$, ending in $\mathcal{H}_\Theta^\downarrow|S(\Theta)$ and starting with the vacuous hints $\mathcal{H}_{\{\theta\}}^\downarrow$. So, this makes it possible to compute the last term in (11.8) for all E in $\mathcal{E}$. Remark finally that $\mathcal{H}|\{E,E^c\}$ can be obtained from $\mathcal{H}|S(E)\cup\{E^c\}$ simply by coarsening and that this last hint equals $\mathcal{H}_\Theta^\downarrow|S(E)\cup\{E^c\}$ because $\mathcal{H}=\mathcal{H}_\Theta^\downarrow$.

These formulas permit now the computation of $\mathcal{H}|\{E,E^c\}$ for all E in $\mathcal{E}$ in three steps.

Step 1. First

$$\mathcal{H}_E^{\downarrow}|\{E, E^c\} \text{ and } \mathcal{H}_E \oplus \mathcal{H}_E^{\downarrow}|\{E, E^c\} \tag{11.14}$$

is computed for all E in $\mathcal{E}$ using (11.13) starting at the leaves of the tree and working up towards Θ. In fact for a leaf E, the first part of (11.14) is vacuous and the second part simply equals $\mathcal{H}_E|\{E, E^c\}$. This is the base for the recursion (11.13) to compute $\mathcal{H}_E^{\downarrow}|S(E) \cup \{E^c\}$, a recursive combination of hints on the partitions $S(E) \cup \{E^c\}$. The combination with $\mathcal{H}_E$ is made on the simple partition $\{E, E^c\}$. The hints of the first part of (11.14) are stored by their respective basic probability assignments $m_E^{\downarrow}(E), m_E^{\downarrow}(E^c)$ at each node E of $\mathcal{E}$.

Step 2. Note that in the case of $E = \Theta$, (11.13) applies with $S(\Theta)$ (Θ^c is empty). Thus, at the end of step (1) $\mathcal{H}_\Theta^{\downarrow}|S(\Theta)$ is computed by the formula

$$\mathcal{H}_\Theta^{\downarrow}|S(\Theta) = \oplus_{B \in S(\Theta)}((\mathcal{H}_B)|\{B, B^c\} \oplus (\mathcal{H}_B^{\downarrow})|\{B, B^c\}))|S(\Theta). \tag{11.15}$$

This is used to derive $\mathcal{H}_\Theta^{\downarrow}|\{E, E^c\}$ for all E in $S(\Theta)$. These hints are stored by their basic probabilities $m(E), m(E^c)$ together with the hints (11.14) at the nodes E in $S(\Theta)$. Note that for a hint on a dichotomous partition $\{E, E^c\}$, $sp(E) = m(E)$ and $sp(E^c) = m(E^c)$.

Step 3. Finally the tree is traversed downwards in order to compute $\mathcal{H}|\{E, E^c\}$ for all nodes of the tree. This will be the final result. From the first part of (11.11), if $\mathcal{H}|\{E, E^c\}$ and $\mathcal{H}_E^{\downarrow}|\{E, E^c\}$ are known, $\mathcal{H}_E|\{E, E^c\} \oplus \mathcal{H}_E^{\diamond}|\{E, E^c\}$ can be determined because the partition is only bipartite (see below for details). On the other hand (11.13) can be used to compute $\mathcal{H}_E^{\downarrow}|\{B, E - B, E^c\}$ for all $B \in S(E)$, the necessary information $\mathcal{H}_B^{\downarrow}|\{B, B^c\}$ being already stored in the nodes B by step 1. Finally having found $\mathcal{H}_E|\{E, E^c\} \oplus \mathcal{H}_E^{\diamond}|\{E, E^c\}$ and $\mathcal{H}_E^{\downarrow}|\{B, E - B, E^c\}$, the second formula of (11.11) can be used to compute $\mathcal{H}|\{B, E - B, B^c\}$ and thus $\mathcal{H}|\{B, B^c\}$ for all sons B of E.

Note that $\mathcal{H}|\{E, E^c\}$ has been determined for all sons of Θ in step 2. This makes it possible to start the procedure just described and to work down the tree to the leaves. Upon termination the hints $\mathcal{H}|\{E, E^c\}$ for all $E \in \mathcal{E}$ are determined.

In the above method (11.13) is invoked several times in order to compute $\mathcal{H}_E^{\downarrow}|S(E) \cup E^c$. This involves combining *dichotomous* hints, i.e. hints with only focal sets B and B^c besides Θ. We shall show that such a hint can always be obtained as a combination of two simple hints, one focussing on B, the other on B^c. This in turn implies that these hints form in fact systems such as studied by Barnett and discussed in examples 5.3 and 5.10 in Chap. 5. Hence the simple computational formulas as obtained in example 5.10 can be applied to these combinations.

In fact let $\mathcal{H}'$ be a hint focussing on B with probability α and $\mathcal{H}''$ a hint focussing on B^c with probability β. Then $\mathcal{H}' \oplus \mathcal{H}''$ has the basic probability assignments $m(B) = \alpha(1-\beta)/(1-\alpha\beta)$ and $m(B^c) = (1-\alpha)\beta/(1-\alpha\beta)$. It follows that

$$\alpha(1-\beta)/(1-\alpha) = m(B)/(1-m(B)); \ \beta = m(B^c)/(1-m(B)) \quad (11.16)$$

and hence α and β can easily be computed from $m(B)$ and $m(B^c)$. Consider a partition $\mathcal{P}$ of Θ together with a dichotomous hint $\mathcal{H}_B$ for all $B \in \mathcal{P}$, that is a hint with focal sets B and B^c besides Θ and basic probabilities $m_B(B), m_B(B^c)$. If (11.16) is applied to (5.65) and (5.66), then it follows, for any set A which is carried by the partition $\mathcal{P}$ (i.e. A is a union of elements of $\mathcal{P}$), that

$$\begin{aligned} pl(A) &= k^{-1}(1+\sum\{m_B(B)/(1-m_B(B)) : B \subseteq A\} \\ &\quad - \prod\{m_B(B^c)/(1-m_B(B)) : B \subseteq A\}), \\ k &= 1+\sum\{m_B(B)/(1-m_B(B)) : B \in \mathcal{P}\} \\ &\quad - \prod\{m_B(B^c)/(1-m_B(B)) : B \in \mathcal{P}\}, \end{aligned} \quad (11.17)$$

where $B \subseteq A$ means all the blocks B of $\mathcal{P}$ which are in A. This can be applied to the algorithm sketched above. Here now follows a detailed description of the formulas needed in each step. Define the following basic probabilities for hints on the partition $\{E, E^c\}$:

For $\mathcal{H}_E|\{E, E^c\} : m_E(E), m_E(E^c)$
For $\mathcal{H}_E^{\downarrow}|\{E, E^c\} : m_E^{\downarrow}(E), m_E^{\downarrow}(E^c)$
For $\mathcal{H}_E|\{E, E^c\} \oplus \mathcal{H}_E^{\downarrow}\{E, E^c\} : m^+(E), m^+(E^c)$
For $\mathcal{H}_E|\{E, E^c\} \oplus \mathcal{H}_E^{\diamond}|\{E, E^c\} : m_E^{\diamond}(E), m_E^{\diamond}(E^c)$
For $\mathcal{H}|\{E, E^c\} : m(E), m(E^c)$.

Because all these hints are dichotomous, the supports sp for E and E^c equal the basic probabilities m assigned to E and E^c, in particular $sp(E) = m(E)$, $sp(E^c) = m(E^c)$.

Procedure 1:

$m_E^{\downarrow}(E), m_E^{\downarrow}(E^c)$ for all nodes E of $\mathcal{E}$ which are not leaves and not equal to Θ must be determined from $m^+(B), m^+(B^c), B \in S(E)$, using (11.13). Equation (11.17) applies here relative to the partition $S(E) \cup \{E^c\}$, where, however, the dichotomous hint relative to E^c is vacuous ($m^+(E^c) = m^+(E) = 0$), such that

$$\begin{aligned} k &= 1+\sum\{m^+(B)/(1-m^+(B)) : B \in S(E)\} \\ m_E^{\downarrow}(E) &= 1 - pl(E^c) = 1 - k^{-1} \\ m_E^{\downarrow}(E^c) &= 1 - pl(E) \\ &= k^{-1}\prod\{m^+(B^c)/(1-m+(B)) : B \in S(E)\}. \end{aligned} \quad (11.18)$$

Procedure 2:

The values of $m^+(E), m^+(E^c)$ must be determined from $m_E(E), m_E(E^c)$ and $m^{\downarrow}_E(E), m^{\downarrow}_E(E^c)$ by the combination $\mathcal{H}_E|\{E, E^c\} \oplus \mathcal{H}^{\downarrow}_E|\{E, E^c\}$ for all nodes of $\mathcal{E}$ except Θ. Note that for leaves $m^{\downarrow}_E(E) = m^{\downarrow}_E(E^c) = 0$, such that in this case $m^+(E) = m_E(E), m^+(E^c) = m_E(E^c)$. For all other nodes we obtain from theorem 4.1

$$\begin{aligned} k &= 1 - m_E(E)m^{\downarrow}_E(E^c) - m_E(E^c)m^{\downarrow}_E(E) \\ m^+(E) &= 1 - k^{-1}(1 - m_E(E))(1 - m^{\downarrow}_E(E)) \\ m^+(E^c) &= 1 - k^{-1}(1 - m_E(E^c))(1 - m^{\downarrow}_E(E^c)). \end{aligned} \tag{11.19}$$

Step 1 is executed by a repeated application of procedure 1 followed by procedure 2, starting with nodes whose sons are all leaves up to the sons of Θ.

Procedure 3:
If $E \in S(\Theta)$ then

$$\mathcal{H}|\{E, E^c\} = (\mathcal{H}^{\downarrow}_{\Theta}|\{E, E^c\} = (\mathcal{H}^{\downarrow}_{\Theta}|S(\Theta))|\{E, E^c\}$$

and hence $\mathcal{H}|\{E, E^c\}$ can be derived from $\mathcal{H}^{\downarrow}_{\Theta}|S(\Theta)$ by coarsening. But by equation (11.15) $\mathcal{H}^{\downarrow}_{\Theta}|(S(\Theta)$ can be obtained from $m^+(B), m^+(B^c)$ for all $B \in S(E)$. Hence by (11.17) we have

$$\begin{aligned} k = 1 &+ \sum \{m^+(B)/(1 - m^+(B)) : B \in S(\Theta)\} \\ &- \prod \{m^+(B^c)/(1 - m^+(B) : B \in S(\Theta)\} \end{aligned}$$

$$\begin{aligned} m(E) &= 1 - pl(E^c) \\ &= 1 - k^{-1}(1 + \sum \{m^+(B)/(1 - m^+(B)) : B \in S(\Theta), B \neq E\} \\ &\quad - \prod \{m^+(B^c)/(1 - m^+(B)) : B \in S(\Theta), B \neq E \end{aligned}$$

$$m(E^c) = 1 - pl(E) = 1 - k^{-1}(1 - m^+(E^c))/(1 - m^+(E)). \tag{11.20}$$

for all $E \in S(E)$. This procedure is applied once in step 2.

Procedure 4:

The values of $m^{\diamond}_E(E), m^{\diamond}_E(E^c)$ have to be determined from $m(E), m(E^c)$ and $m^{\downarrow}_E(E), m^{\downarrow}_E(E^c)$ for nodes E in $\mathcal{E}$ using (11.11). This involves solving the equations (compare 11.19)

$$\begin{aligned} m(E) &= 1 - k^{-1}(1 - m^{\downarrow}_E(E))(1 - m^{\diamond}_E(E)) \\ m(E^c) &= 1 - k^{-1}(1 - m^{\downarrow}_E(E^c))(1 - m^{\diamond}_E(E^c)) \end{aligned}$$

for $m^{\diamond}_E(E), m^{\diamond}_E(E^c)$:

$$
\begin{aligned}
m_E^{\diamond}(E) &= 1 - k(1-m(E))/(1-m_E^{\downarrow}(E)) \\
m_E^{\diamond}(E^c) &= 1 - k(1-m(E^c))/(1-m_E^{\downarrow}(E^c)).
\end{aligned} \tag{11.21}
$$

Furthermore, if in

$$
k = 1 - m_E^{\downarrow}(E) m_E^{\diamond}(E^c) - m_E^{\downarrow}(E^c) m_E^{\diamond}(E)
$$

(11.21) is introduced, and the resulting equation is solved for k, then it follows

$$
k = \frac{1-m(E)}{1-m_E^{\downarrow}(E)} + \frac{1-m(E^c)}{1-m_E^{\downarrow}(E^c)} + \frac{1-m(E)-m(E^c)}{1-m_E^{\downarrow}(E)-m_E^{\downarrow}(E^c)} \tag{11.22}
$$

Procedure 5:

The basic probabilities $m_E^{\downarrow}(B), m_E^{\downarrow}(B^c)$ and $m_E^{\downarrow}(B \cup E^c)$ relative to the hint $\mathcal{H}_E^{\downarrow}|S(E) \cup \{E^c\}$ must be computed for all $B \in S(E)$ from $m^+(C), m^+(C^c)$ for $C \in S(E)$ using (11.13). Once more (11.17) can be applied to the partition $S(E) \cup E^c$:

$$
\begin{aligned}
k &= 1 + \sum \{m^+(C)/(1-m^+(C)) : C \in S(E)\} \\
m_E^{\downarrow}(B) &= 1 - pl_E^{\downarrow}(B^c) \\
&= 1 - k^{-1}(1 - \sum \{m^+(C)/(1-m^+(C)) : C \in S(E), C \neq B\}) \\
m_E^{\downarrow}(B^c) &= 1 - pl_E^{\downarrow}(B) \\
&= 1 - k^{-1}(1-m^+(B^c))/(1-m^+(B)) \\
m_E^{\downarrow}(B \cup E^c) &= 1 - pl_E^{\downarrow}(B^c \cap E) \\
&= 1 - k^{-1}(1 + \sum \{m^+(C)/(1-m^+(C)) : C \in S(E), C \neq B\} \\
&\quad - \prod \{m^+(C^c)/(1-m^+(C)) : C \in S(E), C \neq B\}).
\end{aligned} \tag{11.23}
$$

Procedure 6:

The values of $m(B), m(B^c)$ are to be computed for the sons B of E from $m_E(E), m_E(E^c)$, $m_E^{\downarrow}(B), m_E^{\downarrow}(B^c)$ and $m_E^{\downarrow}(B \cup E^c)$ and $m_E^{\diamond}(E), m_E^{\diamond}(E^c)$ by the second part of (11.11). This is a simple application of Dempster's rule in the form of theorem 4.1:

$$
\begin{aligned}
k &= 1 - m_E^{\downarrow}(E) m_E^{\diamond}(E^c) - m_E^{\downarrow}(E^c) m_E^{\diamond}(E) \\
m(B) &= k^{-1}(m_E^{\diamond}(E)(m_E^{\downarrow}(B \cup E^c) - m_E^{\downarrow}(E^c)) \\
&\quad + (1 - m_E^{\diamond}(E) - m_E^{\diamond}(E^c)) m_E^{\downarrow}(B)) \\
m(B^c) &= 1 - k^{-1}(1 - m_E^{\diamond}(E^c))(1 - m_E^{\downarrow}(B^c)).
\end{aligned} \tag{11.24}
$$

In step 3 we work downwards from the sons of Θ applying repeatedly procedures 4, 5 and 6 until $m(E) = sp(E)$, $m(E^c) = sp(E^c)$ are known for all nodes. Alternatively, branches in the tree may be pruned when $sp(B)$ becomes smaller than some threshold value, because the supports of successors of a node are always at most equal to the support of the node.

Note that in applications of this model hints relative to certain nodes A of $\mathcal{E}$ will arrive sequentially and the supports of the hypotheses E and E^c for E in $\mathcal{E}$ should also be updated sequentially as new information becomes available. It is clear that in step 1 only the supports of the nodes on the path from A up to Θ change. However, in step 2 all nodes change, because one of the terms of (11.15) is altered in step 1. Thus, step 3 has to be executed in full length.

11.3 Model-Based Diagnostics

It is often possible to describe the correct functioning of a system by a mathematical model. As long as observations or measurements correspond to the predictions made by the model, the system may be assumed to be functioning correctly. When, however, a discrepancy arises between observations and model-based predictions, then an explanation for this fact has to be found. The foundation of this approach to diagnostics has been laid by Reiter (1987). The explanations or diagnosis generated by the method proposed by Reiter (1987) are in general not unique and they are not weighted by some likelihood measure which would make it possible to compare them. DeKleer, Williams (1987) introduced a priori probabilities of the failures of the components involved, which allows one to assign relative weights to the diagnostics. Unfortunately, these probabilities are not sufficient to determine a fully specified probability measure on the structure involved in the process of analysis. Therefore, these authors propose a minimum entropy technique to fix the missing probabilities.

We propose here the theory of hints as an alternative and very natural method for model based-diagnostics. The problem is best introduced by an example. Fig. 11.2 shows a logical circuit, a binary adder. Its components are logical and-gates ($A1, A2$), exclusive or-gates ($X1, X2$) and an ordinary or-gate ($O1$). An and-gate is a device whose output is the logical conjunction of its two binary inputs, an exclusive or-gate puts 1 out if exactly one of its inputs equals 1 and an or-gate has as its output the logical disjunction of its two binary inputs. If the inputs $in_1 = 1, in_2 = 0$, $in_3 = 1$ (in_3 is the carry bit from a previous addition) and the outputs $out_1 = 1$ and $out_2 = 0$ are observed (out_1 is the result of the addition and out_2 is the carry bit), then these observations are clearly in discrepancy with the expected behaviour of this system (outputs $out_1 = 0$ and $out_2 = 1$ would be correct). Therefore, some gates do not behave properly. The question is, which ones.

Reiter (1987) shows how certain selected subsets of components can be determined whose failure explains the discrepancy between observations and predictions. These then are possible diagnoses. In contrast to this, the question addressed here is, how credible or plausible is it, that any given component is

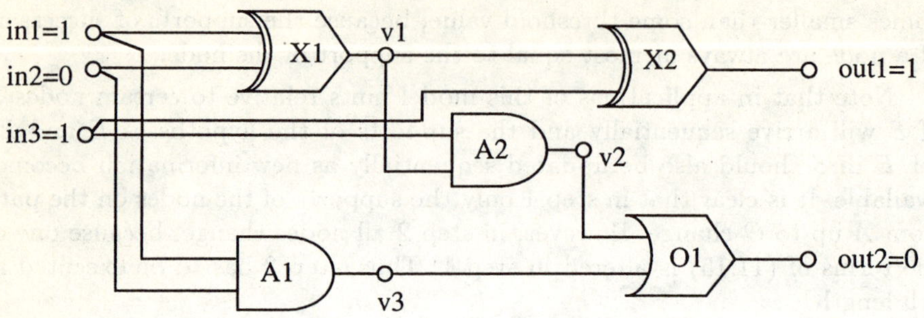

Fig. 11.2.

faulty or more generally, how credible or plausible is any conjecture about the reason for the discrepancy between observations and predictions. Thus both approaches are complementary: Reiter's method permits the creation of conjectures and the procedure presented here allows one to judge the credibility and plausibility of these hypotheses.

Suppose that there are a priori probabilities for the failure of any component of the system. To be concrete, assume that all and-gates have the same probability of failure p_a, the exclusive or-gates the probability of failure p_x and all or-gates have the probability of failure p_o, and that components fail independently from one another. Both input and output of a gate take values in the set $\Theta = \{0,1\}$ and the proper functioning of a gate is described by a relation between its inputs and output, i.e. a relation $R \subseteq \Theta^3$. For an and-gate the relation is

$$R_a = \{(0,0,0),(1,0,0),(0,1,0),(1,1,1)\}, \tag{11.25}$$

for an exclusive or-gate

$$R_x = \{(0,0,0),(1,0,1),(0,1,1),(1,1,0)\} \tag{11.26}$$

and for an or-gate

$$R_o = \{(0,0,0),(1,0,1),(0,1,1),(1,1,1)\}. \tag{11.27}$$

But these relations hold only with the probability $1 - p_a$ for each and-gate, with probability $1 - p_x$ for each exclusive or-gate and with probability $1 - p_o$ for each or-gate. In case of failure of a gate one may assume for example that each combination of inputs and output is possible, which assigns the relation $R = \Theta^3$ to the failure case. Alternatively one could assume that in a failure mode the output is surely wrong, that is, for example, the relation

$$\{(0,0,1),(1,0,1),(0,1,1),(1,1,0)\} \tag{11.28}$$

for an and-gate, or that the output is fixed to 0 for all inputs, etc. One could even allow for different failure modes corresponding to different faulty input-output relations. In any case these assumptions determine a relation between the failure modes and Θ^3. Assume for example that an and-gate has two modes only, $M_a = \{i, f\}$, where i denotes the mode "intact" and f the mode "faulty". This determines then the relation

$$\{(i,0,0,0),(i,1,0,0),(i,0,1,0),(i,1,1,1)\} \cup \{f\} \times \Theta^3 \tag{11.29}$$

in $M_a \times \Theta^3$. Such a relation can be considered as a deterministic hint on $M_a \times \Theta^3$. So with each gate is associated a variable whose values are its possible modes. For example, the exclusive-or gate X1 in Fig. 11.2 corresponds to the variable X1 taking values in $M_x = \{i, f\}$. Also, each observation of an input or output is a deterministic hint on $\Theta = \{0, 1\}$. And finally, the probabilities on the set of possible modes (i.e. $1 - p_a$ for i, p_a for f in the example) determine a precise hint on the set of possible modes (for example M_a for an and-gate).

Suppose now that the hypothesis of a particular gate being faulty is to be judged in the light of the discrepancies between observed and predicted values of certain variables. Then hints on a hypergraph corresponding to the above model have to be combined and projected to the corresponding component variable. This hypergraph as it arises for the system of Fig. 11.2 is drawn in Fig. 11.3.

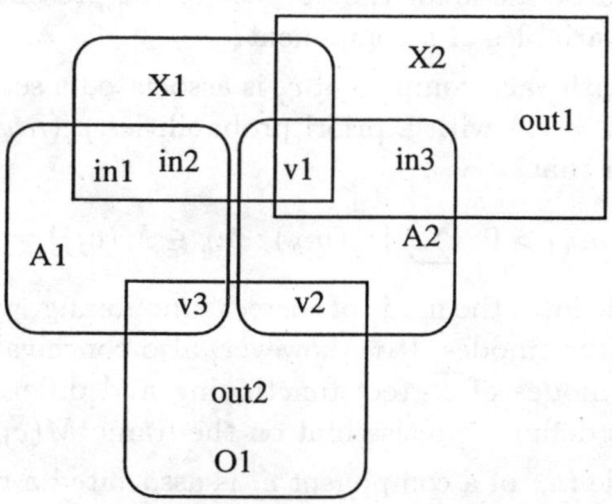

Fig. 11.3.

It has not an apparent hypertree structure, so that it seems that first a covering hypertree has to be found (see Chap. 8). It will, however, be shown below that in such problems a hypertree can be obtained in a simple way.

Composite hypotheses, like the hypothesis that all components of a certain subset of components are faulty, can also be studied in the same framework. For example, consider the hypothesis that $X2$ and $A2$ are simultaneously faulty. This is a possible explanation for the discrepancy between the observed output out_1 and out_2 and the predicted output by the model for the inputs $in_1 = 1, in_2 = 0, in_3 = 1$. There are two other possible explanations, namely $\{X1\}$ and $\{X2, O1\}$, which can for example be found by Reiter's method

(Reiter, 1987). If we want to judge the hypothesis that both $X2$ and $A2$ are faulty, i.e. compute its degree of support, then add a new hyperedge $\{X2, A2\}$ containing the two variables $X2$ and $A2$ to the hypergraph. It carries a vacuous hint. Then, as before, all hints on the new hypergraph are combined (using a covering Markov tree) and the resulting hint is projected to the hyperedge $\{X2, A2\}$. In this way any composite hypothesis about component failure can be judged.

This introductory example will be pursued below. It already points out how a general system of model-based diagnostics, based on the theory of hints, can be designed. Let there be a set C of components $\{c_1, c_2, \ldots, c_m\}$ which are linked in a network of connections to each other. These connections (variables) carry values which must satisfy certain relations defined by the nature and the mode of the components of the system. Let X denote the set of connections, each one represented by a variable from the set $X = \{x_1, x_2, \ldots, x_n\}$. The possible values of each connection or variable x_i is a set Θ_i, namely its frame. Each component c_j links a certain subset $X(c_j) \subseteq X$ of variables to which corresponds the product frame

$$\Theta(c_j) = \prod \{\Theta_k : x_k \in X(c_j)\}. \tag{11.30}$$

No distinction need be made for the purpose of the present analysis between input and output variables of a component.

Furthermore, with each component c_j is associated a set $M(c_j)$ of possible modes $\{m_0, m_1, \ldots, m_{rj}\}$ with a priori probabilities $p_j(m_h)$ for the different failure modes, such that

$$p_j(m_h) > 0; \sum \{p_j(m_h) : m_h \in M(c_j)\} = 1. \tag{11.31}$$

In general m_0 denotes the mode of correct functioning, whereas $m_1, m_2, \ldots$ denote different failure modes. It is, however, also conceivable that a component has different modes of correct functioning and different failure modes. These probabilities define a precise hint on the frame $M(c_j)$.

With each mode m_h of a component c_j is associated a relation $R_j(m_h) \subseteq \Theta(c_j)$ describing the possible values of the variables linked to the component, if the latter is in the mode m_h. This defines then a relation

$$R_j = \{m_0\} \times R_j(m_0) \cup \{m_1\} \times R_j(m_1) \cup \cdots \subseteq M(c_j) \times \Theta(c_j) \tag{11.32}$$

between the mode and the possible values of the variables in this mode. This relation can be considered as a deterministic hint relative to $M(c_j) \times \Theta(c_j)$.

Let c_j denote also the variable referring to the modes of component c_j. Then the elements of the model introduced so far define the basic hypergraph of the system

$$\{\{c_1\} \cup X(c_1), \ldots, \{c_m\} \cup X(c_m)\} \tag{11.33}$$

over the variables $C \cup X$. In addition a set of observations of values o_j for some variables x_j in a subset O of X is assumed to be available. These observations represent further hints to be added to the system. If the observations are assumed to be precise and without possible error, then these hints are deterministic and precise relative to the frame Θ_j. However, more general models of observation may be adopted. Then the hyperedges $\{x_j\}$ for $x_j \in O$ have to be added to the basic hypergraph (this is not really necessary because these are trivial hyperedges contained in the $X(c_j)$).

On this hypergraph the hints introduced above have to be combined and projected to the components. This makes it possible to judge for each component the possibility that it is faulty, or more generally, hypotheses about its fault mode. Composite hypotheses of faulty groups of components can be judged in principle by the same method, introducing an additional hyperedge representing this group of components as explained above in the example of the binary adder.

Let us illustrate this approach with two examples.

Example 11.3 Binary adder. The digital circuit represented in Fig. 11.2 will be considered here in more detail. Its corresponding hypergraph is shown in Fig. 11.3 This is clearly no hypertree. However, remark that a variable as in_1, which is linked to different gates like $X1$ and $A1$ can, without changing the nature of the problem, be split into different variables like in_{11} and in_{12} with the same value, but each linked to only one gate, one to $X1$ and the other to $A1$. If each variable with an observed value is split in this way, then the network of Fig. 11.2 changes into the network of Fig. 11.4.

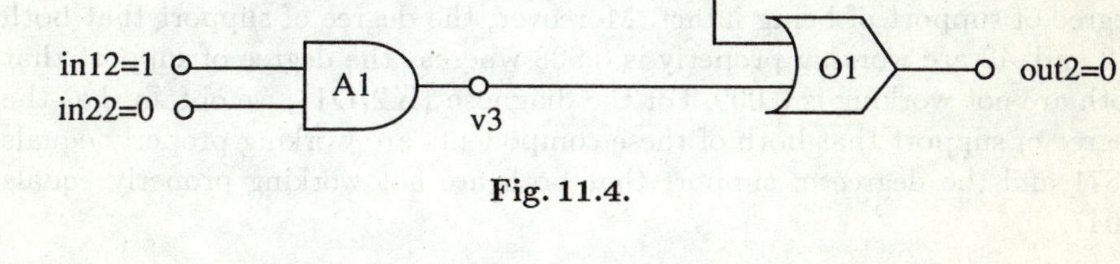

Fig. 11.4.

Both networks are clearly equivalent as far as the diagnostic problem is concerned. The hypergraph of the network of Fig. 11.4 is shown in Fig. 11.5 and it is in fact a hypertree (test this with Graham's algorithm from Chap. 8).

This trick of splitting the observed variables always simplifies the network and the related hypergraph and it always leads to a hypertree because the resulting networks are variable and relation trees and hence hypertrees by

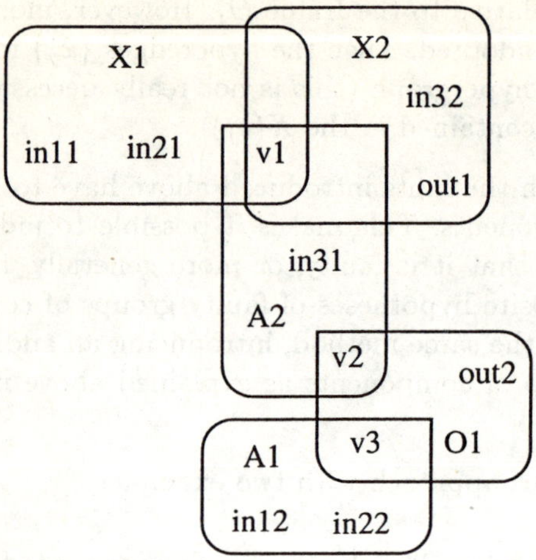

Fig. 11.5.

theorems 8.19 and 8.24 (see also the following example). It is then easy to combine and propagate the hints locally on the hypertree of Fig. 11.5.

For a numerical example, we assume that the a priori probability of failure of an and-gate, an exclusive or-gate and an or-gate are 0.01, 0.03 and 0.05 respectively. In addition, we assume that each of the five components has only two modes: intact or faulty (in the faulty mode any output is possible). The observations are those given in Fig. 11.2. Then this model can be computed using a program written by R. Hänni at the Informatics Institute of the University of Fribourg. It is found that the resulting hints are precise for each of the five components and the corresponding degrees of support are given in Tab. 11.3. These result show that the component X1 is most likely faulty and it is worth checking it carefully. The other components have a fairly high degree of support of being intact. Moreover, the degree of support that both $X2$ and $A2$ are working properly is 0.908 whereas the degree of support that both are not working is 0.009. For the diagnose $\{X2, O1\}$, we obtain that the degree of support that both of these components are working properly equals 0.871 and the degree of support that both are not working properly equals 0.047.

Table 11.3.

Component	X1	X2	A1	A2	O1
Support for faulty	0.9454	0.0829	0.01	0.0186	0.0931
Support for intact	0.0546	0.9171	0.99	0.9814	0.9069

Example 11.4 An arithmetic network. Fig. 11.6 shows a network composed of adders (A_i) and multipliers (M_j) together with some observations of input and output variables.

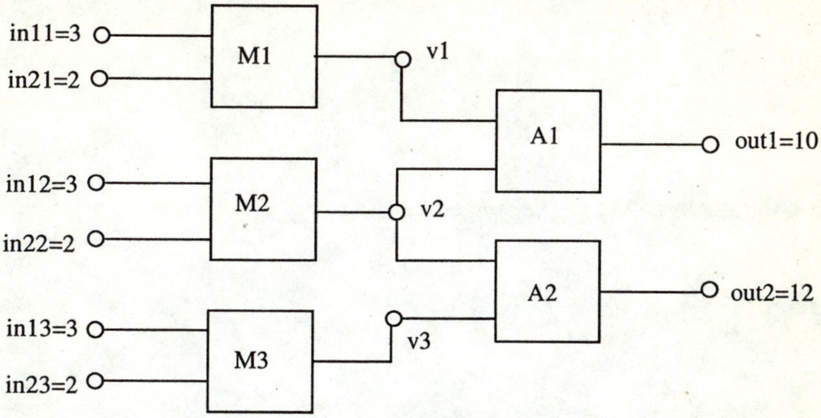

Fig. 11.6.

Again the output values do not correspond to what is expected. The observed variables are already split in Fig. 11.6. Fig. 11.7 represents the corresponding hypergraph which is a hypertree.

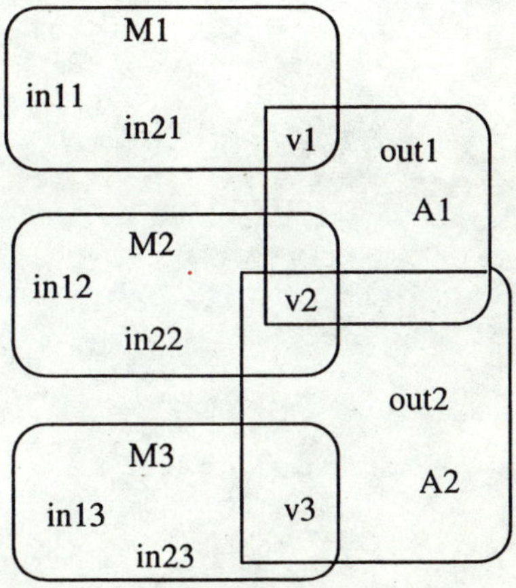

Fig. 11.7.

It can be used for local propagation of the hints.

To conclude this chapter, let us mention that the problem of detecting bad data as discussed in Sect. 1.5 can also be considered in the framework of model-based diagnostics.

12. Temporal and Spatial Reasoning

12.1 Introductory Examples

In this section a certain number of examples illustrating the type of problems to be considered in this chapter will be presented. They are all concerned with spatial or temporal reasoning under uncertainty. Then the examples are generalized and computational techniques are presented. To a large extent, this chapter reproduces results given in Monney (1991) and Kohlas, Monney (1991).

Example 12.1 Consider a certain number of events $i, j, k, h, \ldots$ and let $t_h \in \mathcal{R}$ denote the time at which the event h occurs. If $d_{ij} = t_j - t_i$ represents the duration between the events i and j, then the relation

$$d_{ik} = d_{ij} + d_{jk} \tag{12.1}$$

must be satisfied for all triple of events (i, j, k). Then $d_{ii} = 0$ and $d_{ji} = -d_{ij}$. Each constraint (12.1) is represented by a deterministic hint $\mathcal{H}(i, j, k)$ on the frame induced by the group of variables $\{d_{ik}, d_{ij}, d_{jk}\}$ (see Chap. 7). Note that the set of possible values (the frame of discernment) of each variable d_{ij} is the commutative additive group of real numbers.

Furthermore, for a certain number of variables d_{ij} we suppose that we dispose of a hint $\mathcal{H}_{ij}$ about the value of the duration d_{ij} between the events i and j. If $\mathcal{H}$ denotes the combination of all hints $\mathcal{H}(i, j, k)$ and $\mathcal{H}_{ij}$, then the problem is to compute the coarsening of $\mathcal{H}$ to a particular variable d_{rs}. This hint can be used to judge the hypothesis that the duration between the events r and s is between two given limits. This model is the starting point of the technique of time maps often used in temporal reasoning (Dean, 1985; Vere, 1983).

Another model of the same type is related to the length of a path between two nodes in a network. In fact, let $t_0, t_1, \ldots, t_m$ be a path from t_0 to t_m in a network. For a certain number of subpaths from t_i to t_j, $i < j$ suppose that we dispose of hints $\mathcal{H}_{ij}$ about their lengths d_{ij}. If we define $d_{ji} = -d_{ij}$ for $i < j$ and $d_{ii} = 0$ for all i, then the restrictions (12.1) must also be satisfied. This model is therefore a special case of the first model.

Example 12.2 Here we consider a model introduced by Allen, Kautz (1985). Let $i, j, k, \ldots$ be time intervals having durations $d_i, d_j, d_k, \ldots$ Then the ratio

$d_{ij} = d_i/d_j$ represents the length of the interval d_i taking the interval d_j as time unit. This implies that

$$d_{ik} = d_{ij} \cdot d_{jk} \tag{12.2}$$

for all triples of intervals (i, j, k) and in particular $d_{ji} = 1/d_{ij}$ and $d_{ii} = 1$. The constraints (12.2) represent again deterministic hints. Note that this time the set of possible values of each variable d_{ij} is the commutative multiplicative group of positive real numbers. If we dispose of hints about some variables d_{ij}, then we obtain a model which is similar to the one of example 12.1. It is even completely equivalent if we take logarithms of the variables d_{ij}.

Example 12.3 The geometric model of McDermott, Davis (1984) is composed of a certain number of references $f_i, f_j, f_k, \ldots$ which are defined as positively oriented orthonormal coordinate systems in the Euclidian plane. For each couple of references (f_i, f_j) the variable d_{ij} specifies the position of the reference f_j with respect to the reference f_i (see Fig. 12.1).

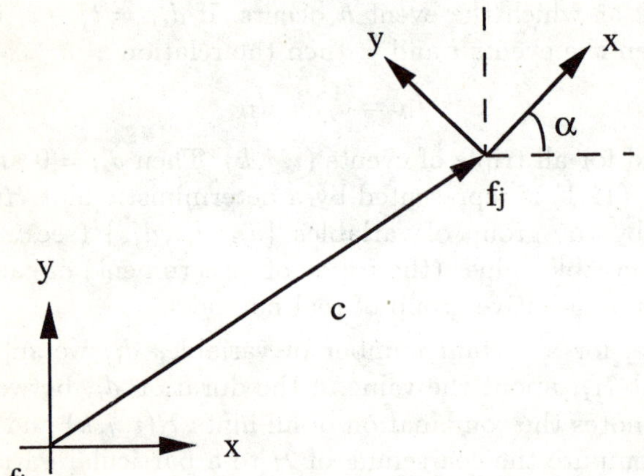

Fig. 12.1.

This position is described by the translation required to make the origin of f_i coincide with that of f_j and by the rotation required to make the x-axis of f_i coincide with that of f_j. A particular value of d_{ij} is therefore the resultant of a translation and a rotation, where these are viewed as bijective transformations of the Euclidien plane into itself. It is well known that the set of translations and rotations form a non-commutative group under the operation of composition. This group is called the group of *displacements* or also the group of *rigid motions* (Gans, 1969). So each possible value of d_{ij} is a displacement. Furthermore, it is proved that every displacement is a mapping

$\varphi : \mathcal{R}^2 \to \mathcal{R}^2$ of the form $\varphi(x) = Ax + c$ where A is a rotation matrix, i.e. A is of the form

$$\begin{pmatrix} cos(\alpha) & sin(\alpha) \\ -sin(\alpha) & cos(\alpha) \end{pmatrix}.$$

Now, if f_i, f_j, f_k are three references, it is clear that the displacement φ_{ik} expressing the position of f_k with respect to f_i is obtained by composing the displacements φ_{ij} and φ_{ik}, i.e. $\varphi_{ik} = \varphi_{jk} \circ \varphi_{ij}$ or, more explicitly, if $\varphi_{ij}(x) = A_1 x + c_1$ and $\varphi_{jk}(x) = A_2 x + c_2$, then $\varphi_{ik}(x) = A_2 A_1 x + A_2 c_1 + c_2$. This shows that if $d_{ij} = (A_1, c_1), d_{jk} = (A_2, c_2)$, then $d_{ik} = (A_2 A_1, A_2 c_1 + c_2)$. So, if we define the operation $\otimes$ in the group of displacements

$$\Theta = \{(A, c) : A \in SO(2), c \in \mathcal{R}^2\} \tag{12.3}$$

by

$$(A_1, c_1) \otimes (A_2, c_2) = (A_2 A_1, A_2 c_1 + c_2), \tag{12.4}$$

then we have $d_{ik} = d_{ij} \otimes d_{jk}$ for all i, j, k (note that $\varphi_1 \otimes \varphi_2$ corresponds to $\varphi_2 \circ \varphi_1$. Equipped with this new operation $\otimes$, it can easily be verified that the set of displacements Θ is again a group which, however, is not commutative. Since the relation $d_{ik} = d_{ij} \otimes d_{jk}$ must hold for all i, j, k, if we dispose of hints $\mathcal{H}_{ij}$ about the exact value of the variables d_{ij}, then this example is of the same nature as the two preceding ones. The focal sets of the hints $\mathcal{H}_{ij}$ are discribed by a set of possible values for the rotation angle α, typically an interval, and a set of possible values for the translation vector c. If all references are parallel, then all matrices A are identity matrices and the group $(\Theta, \otimes)$ is commutative. Of course, this group is simply the group of translations. This particular model will be studied in detail in Sect. 12.6.

Example 12.4 This example is the temporal model introduced by Allen (1983) which is often considered in temporal reasoning. It consists of time intervals represented by $i, j, k, \ldots$. Tab. 12.1 is copied from Allen (1983) and shows the possible relations between two intervals X and Y (note the symbols which are used to represent the inverse of a relation).

We dispose of hints $\mathcal{H}_{ij}$ about the position d_{ij} which actually holds between the intervals i and j. If $d_{jk} = r$, then d_{kj} must necessarily be equal to the inverse position r_i of r as defined in Tab. 12.1. Furthermore, each triple of intervals (i, j, k) must satisfy the transitivity condition

$$d_{ik} \in T(d_{ij}, d_{jk}) \tag{12.5}$$

where T is the transitivity relation given in Tab. 12.2 which is copied from Allen (1983). The constraints (12.5) represent deterministic hints and this example is therefore of the same kind as the other ones, except that here the frame of discernment of the variables d_{ij} has not the structure of a group and the constraints are not expressed by an equality relation.

Example 12.5 This example is the well known project scheduling problem. There are activities a_i which must satisfy certain precedence conditions. If the

Table 12.1.

Relation	Symbol	Symbol for inverse	Pictoral Example
X before Y	<	>	XXX YYY
X equal Y	=	=	XXX YYY
X meets Y	m	mi	XXXYYY
X overlaps Y	o	oi	XXX YYY
X during Y	d	di	XXX YYYYYY
X starts Y	s	si	XXX YYYYY
X finishes Y	f	fi	XXX YYYYY

duration of the activities is not known with certainty, then several formalisms like probability theory, fuzzy set theory or the theory of hints can be applied. If t_i and t_j denote the starting time of the activities a_i and a_j and if d_j is the duration of the activity a_j, then the condition that a_i cannot start before a_j is finished, is expressed by the constraint $t_i - t_j \geq d_j$. Therefore, if we dispose of hints about the duration of the activities, then this example is again of the same kind as the other ones, except that here the constraints are inequalities instead of equalities. This example will be discussed in detail in Sect. 12.8.

12.2 A Class of Models Based on Groups

In all examples of the previous section, except the last one, there is a finite set of indices $V = \{i, j, k, r, s, \ldots\}$. A variable d_{ij} taking its values in some fixed set Θ (its frame) is associated with every couple $(i, j) \in V^2$. Of course, the variables d_{ij} define a multivariate family of compatible frames (see Chap. 7).

Furthermore, as in examples 12.1 and 12.2 and for a special case of example 12.3, we assume that the frame Θ of the variables has the structure of a *commutative group*. However, most results in this chapter are still valid for non-commutative groups, but the theory and especially the proofs are much more simple and elegant in the case of commutative groups. We will say it explicitly each time an important result carries over to an arbitrary group and the reference for the proof, which is always Kohlas, Monney (1991), will be given. Furthermore, we will briefly come back to Allen's model (example 12.4) in this section and the project scheduling problem (example 12.5) is studied in Sect. 12.8.

Table 12.2. Transitivity table

d_{jk} / d_{ij}	<	>	d	di	o	oi	m	mi	s	si	f	fi	=
<	<	----	< o m d s	<	<	< o m d s	<	< o m d s	<	<	< o m d s	<	<
>	----	>	> oi mi d f	>	> oi mi d f	>	> oi mi d f	>	> oi mi d f	>	>	>	>
d	<	>	d	----	< o m d s	> oi mi d f	<	>	d	> oi mi d f	d	< o m d s	d
di	< o m di fi	> oi mi di si	o oi d di =	di	o di fi	oi di si	o di fi	oi di si	di fi o	di	di si oi	di	di
o	<	> oi di mi si	o d s	< o m di fi	< o m	o oi d di =	<	oi di si	o	di fi o	d s o	< o m	o
oi	< o m di fi	>	oi d f	> oi mi di si	o oi d di =	> oi mi	o di fi	>	oi d f	oi > mi	oi	oi di si	oi
m	<	> oi mi di si	o d s	<	<	o d s	<	f fi =	m	m	d s o	<	m
mi	< o m di fi	>	oi d f	>	oi d f	>	s si =	>	d f oi	>	mi	mi	mi
s	<	>	d	< o mdi fi	< o m	oi d f	<	mi	s	s si =	d	< m o	s
si	< o m di fi	>	oi d f	di	o di fi	oi	o di fi	mi	s si =	si	oi	di	si
f	<	>	d	> oi mi di si	o d s	> oi mi	m	>	d	> oi mi	f	f fi =	f
fi	<	> oi mi di si	o d s	di	o	oi di si	m	si oi di	o	di	f fi =	fi	fi
=	<	>	d	di	o	oi	m	mi	s	si	f	fi	=

So from now on, we assume that the frame Θ of the variables d_{ij} has the structure of a commutative group with neutral element e and composition law $\otimes$. The formal model considered in this section will now be introduced. For each triple $(i, j, k) \in V^3$ we impose the restriction

$$d_{ik} = d_{ij} \otimes d_{jk} \tag{12.6}$$

which is represented by a deterministic hint $\mathcal{H}(i, j, k)$ focussing on the set

$$F(i, j, k) = \{(d_{ik}, d_{ij}, d_{jk}) \in \Theta^3 : d_{ik} = d_{ij} \otimes d_{jk}\}. \tag{12.7}$$

For $i = j = k$ it follows from (12.6) that $d_{ii} = e$ and hence for $i = k$ it follows that $d_{ij} \otimes d_{ji} = e$ and thus $d_{ji} = d_{ij}^{-1}$. On the other hand, for every couple (i, j) in a subset E of V^2 we suppose to have a hint $\mathcal{H}_{ij}$ about the unknown value of the variable d_{ij}. All hints are supposed to be independent. Since in a couple (i, j) the order of the elements i and j is relevant, the existence of a hint $\mathcal{H}_{ij}$ for d_{ij} does not necessarily imply the existence of an independent hint $\mathcal{H}_{ji}$ for d_{ji}. Moreover, we suppose that these hints all have a finite number of interpretations. This defines the oriented graph $G = (V, E)$. The hints $\mathcal{H}(i, j, k)$ represent domain-dependent or structural information, whereas the hints $\mathcal{H}_{ij}$ are domain-independent. A *model* $\mathcal{M}$ is thus defined by the graph $G = (V, E)$ and by a family of hints, each one assigned to an arc $(i, j) \in E$. Synthesizing all available information means to compute the hint

$$\mathcal{H}(\mathcal{M}) = \oplus\{\mathcal{H}_{ij} : (i, j) \in E\} \oplus \{\mathcal{H}(i, j, k) : (i, j, k) \in V^3\} \tag{12.8}$$

which is called the *solution* of the model $\mathcal{M}$. Since in most cases we have to evaluate hypotheses about the value of particular variable d_{rs}, we must compute the coarsening of $\mathcal{H}(\mathcal{M})$ to the frame $\{d_{rs}\}$. To simplify notation it is denoted by $\mathcal{H}(\mathcal{M})|(r, s)$ instead of $\mathcal{H}(\mathcal{M})|\{d_{rs}\}$.

In order to compute this hint the following auxiliary results will prove useful. If $n = |V|$, let $\mathcal{V}$ denote the set of all vectors $d = (d_{ij} : (i, j) \in V^2) \in \Theta^{n \times n}$. A vector $d \in \mathcal{V}$ is called *structural* if $d_{ik} = d_{ij} \otimes d_{jk}$ for all $(i, j, k) \in V^3$ and let $\mathcal{T}$ denote the subset of all structural vectors in $\Theta^{n \times n}$.

Theorem 12.1

(1) A vector $d \in \mathcal{V}$ is structural if and only if there is a mapping $\pi : V \to \Theta$ such that $d_{ij} = \pi(j) \otimes \pi(i)^{-1}$ for all $(i, j) \in V^2$.

(2) Let $T = (V, U)$ be a tree. If $d' = (d'_{ij} : (i, j) \in U)$ is a vector in Θ^{n-1}, then there is a unique structural vector $d \in \mathcal{T}$ such that $d_{ij} = d'_{ij}$ for all $(i, j) \in U$.

The mapping π is called a *potential* associated with the vector $d \in \mathcal{T}$. This name is justified by the fact that if Θ is the additive group of real numbers, then $\mathcal{T}$ is the linear subspace of *tensions* on the complete graph over the nodes in V (Sakarovitch, 1984; Bondy, Murty, 1976). Given a mapping (potential) $\pi : V \to \Theta$, the structural vector d defined by $d_{ij} = \pi(j) \otimes \pi(i)^{-1}$ is called the *tension* associated with the potential π.

In order to solve a model $\mathcal{M}$, for all possible samples $\{\Theta_{ij} \subseteq \Theta : (i,j) \in E\}$ of focal sets of the hints $\mathcal{H}_{ij}$, the corresponding focal set of the hint $\mathcal{H}$, namely

$$\Lambda = \{d \in \mathcal{T} : d_{ij} \in \Theta_{ij} \ \forall\ (i,j) \in E\}, \tag{12.9}$$

has to be computed. The vectors in Λ are called *consistent* with respect to the sample. If Λ is empty for all samples, then $\mathcal{H}(\mathcal{M})$ does not exist and the model is called *contradictory*. Note that a model cannot be contradictory if there is no domain-independent information $\mathcal{H}_{ij}$ because in this case the neutral vector is consistent. In order to compute $\mathcal{H}(\mathcal{M})|(r,s)$, the coarsening

$$\Lambda|(r,s) = \Lambda_{rs} = \{\theta \in \Theta : \exists\ d \in \Lambda \text{ such that } d_{rs} = \theta\} \tag{12.10}$$

must be determined for every focal set Λ.

The next theorem gives some properties of the structural information which does not require Θ to be commutative.

Theorem 12.2

(1) Symmetry. If the relation

$$d_{ik} = d_{ij} \otimes d_{jk} \tag{12.11}$$

holds for the triple (i,j,k) and if d_{ki}, d_{ji}, d_{kj} are the inverses of d_{ik}, d_{ij}, d_{jk} respectively, then (12.11) also holds for every permutation of this triple.

(2) Dependency. Let (i,j,k,h) be a quadruple of elements in V. If the relation (12.12) holds for three triples among $(h,i,j), (i,j,k), (h,i,k), (h,j,k)$, then it is also satisfied for the fourth one.

We remark that all computational procedures presented in Sect. 12.3 and 12.4 can also be proved by using only this theorem which does not require Θ to be commutative, only the group structure is needed (see Kohlas, Monney, 1991).

To conclude this section, let's briefly come back to Allen's model (example 12.4). Although no group structure is present there, the symmetry property is also satisfied. In fact, it can be explicitly verified that if

$$d_{ik} \in T(d_{ij}, d_{jk}) \tag{12.12}$$

and if d_{ki}, d_{ji} and d_{kj} are the "inverses" of d_{ik}, d_{ij} and d_{jk} respectively (see Tab. 12.1), then (12.12) is also satisfied for every permutation of (i,j,k). However, the dependence property is not satisfied as the following counter-example shows. For four intervals i,j,k,h in V, let's define

$$\begin{aligned} d_{ij} &= <, d_{ik} = <, d_{ih} = < \\ d_{jh} &= >, d_{jk} = <, d_{kh} = < . \end{aligned} \tag{12.13}$$

The situation is pictured in Fig. 12.2. The three conditions $d_{ik} \in T(d_{ij}, d_{jk}), d_{ih} \in T(d_{ik}, d_{kh})$ and $d_{ih} \in T(d_{ij}, d_{jh})$ are satisfied but $d_{jh} \notin T(d_{jk}, d_{kh})$.

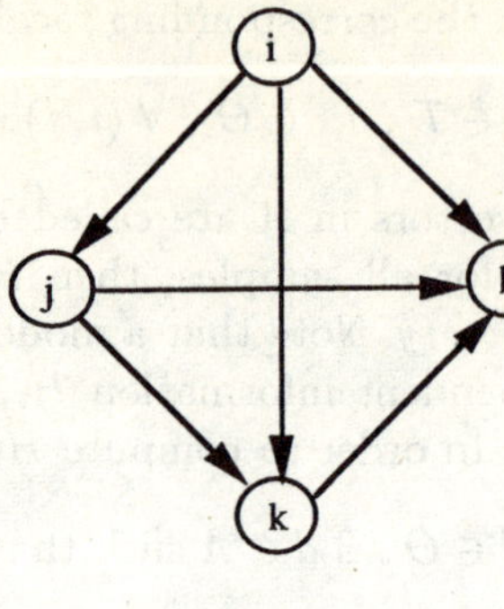

Fig. 12.2.

This implies that for Allen's model the computational results given in the following three sections cannot be proved using theorem 12.2. It is an open question whether these procedures are nevertheless correct.

12.3 Model Reductions

The computation of the solution $\mathcal{H}(\mathcal{M})$ of a model $\mathcal{M}$ is certainly not an easy task. Beside the traditional technique of local computations in Markov trees, there are two ways of reducing and hence simplifying a model (Kohlas, Monney, 1991). The presentation of these reductions is the subject of this section. Throughout it commutativity of the group Θ is assumed.

Consider a model $\mathcal{M}$ with V, E and a family of hints $\mathcal{H}_{ij}$ for $(i,j) \in E$ and suppose that $j \in V$ is a node in the graph (V, E) different from r and s whose incident arcs are exactly (i,j) and (j,k) in E. Then a new model $\mathcal{M}'$ is constructed as follows.

First, the node j is removed from V: $V' = V - \{j\}$. Then, the arcs (i,j) and (j,k) are removed from E and a *new* arc $a = (i,k)$ is added so that $E' = E - \{(i,j),(j,k)\} + \{a\}$ (see Fig. 12.3).

The arc a will be parallel to $u = (i,k)$ if $u \in E$. It remains to define the hint $\mathcal{H}_a$ for this new arc a. For two arbitrary subsets A and B of Θ, define

$$A \otimes B = \{\theta_1 \otimes \theta_2 \in \Theta : \theta_1 \in A, \theta_2 \in B\}. \tag{12.14}$$

If $\mathcal{H}_{ij} = (\Omega_1, P_1, \Gamma_1, \Theta)$ and $\mathcal{H}_{jk} = (\Omega_2, P_2, \Gamma_2, \Theta)$, then $\mathcal{H}_a$ is defined to be the following hint

$$\mathcal{H}_{ij} \otimes \mathcal{H}_{jk} = (\Omega_1 \times \Omega_2, P_1 P_2, \Gamma_1 \otimes \Gamma_2, \Theta) \tag{12.15}$$

where

$$\Gamma_1 \otimes \Gamma_2(\omega_1, \omega_2) = \Gamma_1(\omega_1) \otimes \Gamma_2(\omega_2). \tag{12.16}$$

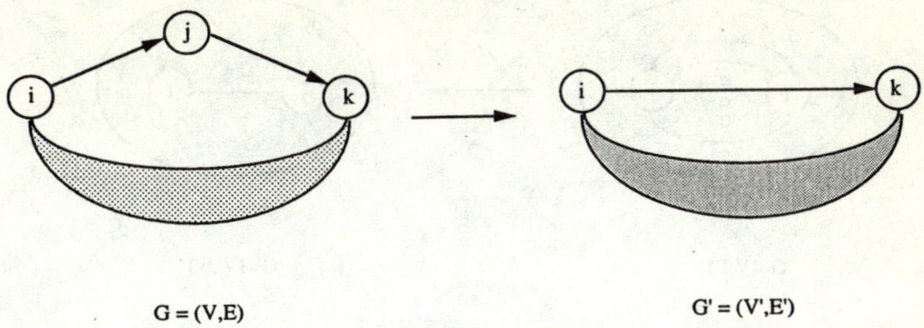

Fig. 12.3.

Note that $\mathcal{H}_a$ is nothing but the hint $(\mathcal{H}_{ij} \oplus \mathcal{H}_{jk} \oplus \mathcal{H}(i,j,k))|(i,k)$. The transformation of the model $\mathcal{M}$ into the model $\mathcal{M}'$ is called a *series reduction* with respect to node j. The following theorem asserts that the two models are equivalent as far as the variable d_{rs} is concerned.

Theorem 12.3 *If $\mathcal{M}'$ is the model obtained from the model $\mathcal{M}$ by a series reduction with respect to a node j different from r and s, then*

$$\mathcal{H}(\mathcal{M})|(r,s) = \mathcal{H}(\mathcal{M}')|(r,s). \tag{12.17}$$

This result carries over to the case where Θ is not commutative, see Kohlas, Monney, 1991. Given a hint $\mathcal{H} = (\Omega, P, \Gamma, \Theta)$, the *inverse* hint $\mathcal{H}^{-1} = (\Omega, P, \Gamma^{-1}, \Theta)$ is defined by

$$\Gamma^{-1}(\omega) = \{\theta \in \Theta : \exists\ \theta' \in \Gamma(\omega) \text{ such that } \theta = (\theta')^{-1}\}. \tag{12.18}$$

It is clear that the solution of a model remains unchanged when an arc $u = (i,j)$ is replaced by the inverse arc $v = (j,i)$ and $\mathcal{H}_v = \mathcal{H}_u^{-1}$. The orientation of the arcs is therefore not really important. This remark is useful when there are exactly two arcs a and a' incident to a node $j \in V$ because in this case it is always possible to make a series reduction with respect to j, possibly after changing the orientation of one arc.

As noted above, a model $\mathcal{M}$ may contain parallel arcs after a series reduction. These arcs can be regrouped to get a new simpler model $\mathcal{M}'$. If in general $u_1, \ldots, u_m$ denote m parallel arcs from i to j in E, then $\mathcal{M}'$ is defined by $V' = V, E' = E - \{u_1, \ldots, u_m\} + \{u\}$ where $u = (i,j)$ is a new arc from i to j. The new hint $\mathcal{H}_u$ associated with the arc u is given by $\mathcal{H}_u = \oplus_{l=1}^m \mathcal{H}_{u_l}$. Figure 12.4 illustrates the passage from $\mathcal{M}$ to $\mathcal{M}'$.

This way of reducing a model is called a *parallel reduction*. The two models $\mathcal{M}$ and $\mathcal{M}'$ are equivalent as the following theorem shows.

Theorem 12.4 *If the model $\mathcal{M}'$ is obtained from the model $\mathcal{M}$ by a parallel reduction, then*

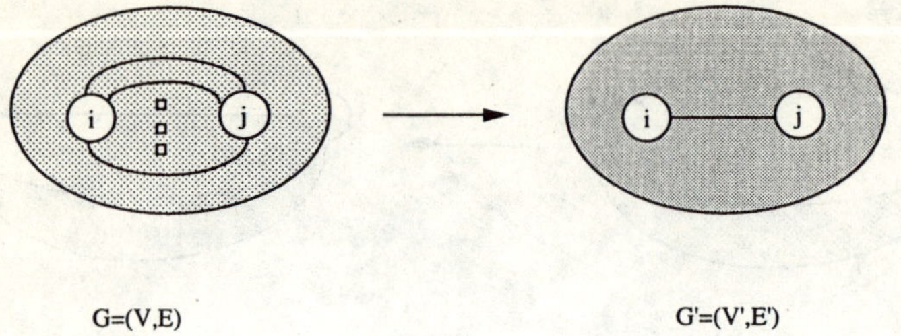

Fig. 12.4.

$$\mathcal{H}(\mathcal{M})|(r,s) = \mathcal{H}(\mathcal{M}')|(r,s) \tag{12.19}$$

for every couple $(r,s) \in V^2$.

Again, this result is also valid when the group Θ is not commutative (Kohlas, Monney, 1991).

A single series or parallel reduction is a first step in the simplification of the computation of $\mathcal{H}(\mathcal{M})|(r,s)$ because a global operation is replaced by a local computation together with a global operation on a smaller model.

The ability of making a reduction depends on the topology of the graph (V,E) of the model. Given a couple $(r,s) \in V^2$, series and parallel reductions are performed until nothing more is possible. Some graphs can be reduced to the trivial graph composed of a single arc from r to s. In this case the variable d_{rs} is called *s-p-reducible* in G. A graph is called *series-parallel* if there is a s-p-reducible variable in G. The graph G in Fig. 12.5 is series-parallel because the variable d_{34} is s-p-reducible. However, the variable d_{25} is not s-p-reducible.

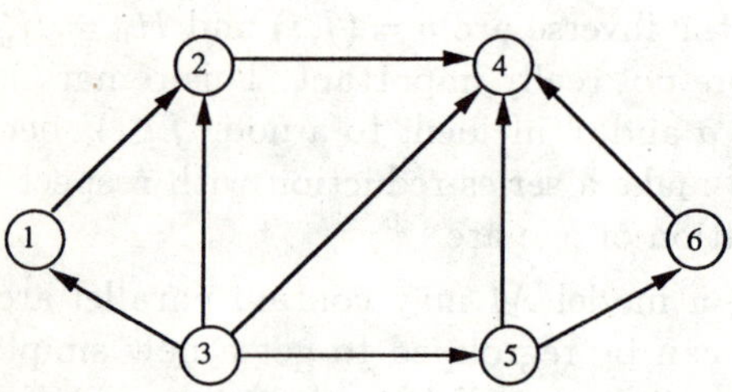

Fig. 12.5.

When the variable d_{rs} is s-p-reducible in $G = (V,E)$, then the computation of $\mathcal{H}(\mathcal{M})|(r,s)$ can be done using local computations only. The sequence in which these computations are performed can be visualized in a non-oriented bipartite tree $T = (V_1 + V_2, U)$. This tree is constructed as follows. Each time a series reduction on the arcs (i,j) and (j,k) is performed, do the following:

(1) Add new nodes to V_1 corresponding to the variables d_{ij}, d_{jk} and d_{ik} if they are not already present.
(2) Add a new node to V_2 corresponding to the relation used to make the reduction.
(3) Add three edges in U linking each of the three nodes of d_{ij}, d_{jk} and d_{ik} to the node corresponding to the relation.

Figure 12.6 shows this tree for the s-p-reducible variable d_{34} in the graph of Fig. 12.5.

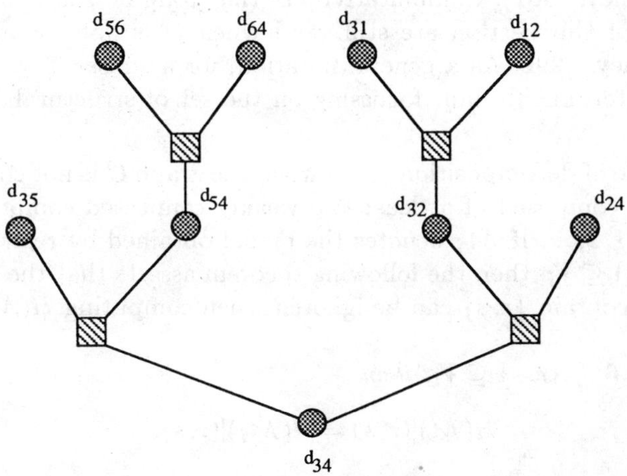

Fig. 12.6.

According to theorem 8.19 this is a Markov tree. Series and parallel reductions can be seen to correspond exactly to the local propagation procedure in the Markov tree T. Thus the s-p-reducible variables correspond nicely to Markov trees. Furthermore, it is interesting to remark that several hints can be discarded when computing $\mathcal{H}(\mathcal{M})|(r,s)$ because it is sufficient to consider only those hints which correspond to the nodes of the tree T. This is explained by the strong dependencies among the structural relations. There is a considerable redundancy in this model. The following theorem gives more information about s-p-reducible variables and series-parallel graphs.

Theorem 12.5 *If $G = (V,E)$ is a series-parallel graph and if $T(d_{rs}) = (V_1 + V_2, U)$ denotes the Markov tree corresponding to a s-p-reducible variable d_{rs} in G, then*

(1) G is weakly connected.
(2) $T(d_{ij}) = T(d_{kl}) = T$ for any s-p-reducible variables d_{ij} and d_{kl}.
(3) A variable d_{ij} is s-p-reducible if and only if it corresponds to a node in V_1.

(4) All variables d_{ij} such that $(i,j) \in E$ are s-p-reducible.

12.4 Model Decompositions

Given a model $\mathcal{M} = (V, E, I)$ and a couple of nodes $(r,s) \in V^2$, it may happen that no series or parallel reduction can be performed. Even in this situation it is sometimes possible to simplify the computation of $\mathcal{H}(\mathcal{M})|(r,s)$. The idea is to *decompose* the model into submodels which are easier to compute (Kohlas, Monney, 1991). Commutativity of the group Θ is assumed here, but all theorems of this section are still valid when Θ is not commutative, see Kohlas, Monney, 1991. As a general notation, for a subset $Z \subseteq V$, let $\mathcal{T}(Z)$ denote the deterministic hint focussing on the set of structural vectors with respect to Z.

A first type of decomposition arises when the graph G is not connected, i.e. $G = (V, E)$ is composed of at least two weakly connected components $G_j = (V_j, E_j)$, $j = 1, 2, \ldots$ If $\mathcal{M}_j$ denotes the model obtained by restricting $\mathcal{M}$ to G_j and if $\{r,s\} \subseteq V_i$, then the following theorem asserts that the components which do not contain $\{r,s\}$ can be ignored when computing $\mathcal{H}(\mathcal{M})|(r,s)$.

Theorem 12.6 *If $\{r,s\} \subseteq V_i$, then*

$$\mathcal{H}(\mathcal{M})|(r,s) = \mathcal{H}(\mathcal{M}_i)|(r,s). \tag{12.20}$$

The following theorem examines the case where r and s do not belong to the same connected component. Let $\mathcal{E}$ denote the vacuous hint on Θ.

Theorem 12.7 *Suppose that $r \in V_i$ and $s \in V_j$ with $i \neq j$. If $\mathcal{H}(\mathcal{M})$ exists then $\mathcal{H}(\mathcal{M})|(r,s) = \mathcal{E}$.*

For the rest of this chapter we suppose that $G = (V, E)$ is a connected graph. Let's introduce some graph-theoretical definitions. If $X \subseteq V$ and $E(X)$ is the set of arcs in E whose extremities are in X, then the graph $G(X) = (X, E(X))$ is called the *subgraph* of G over X. Moreover, a node $k \in V$ is called a *cut-vertex* of G if the subgraph $G(V - \{k\})$ has at least two weakly connected components. If we denote by V_j' the set of nodes of the j-th component, then define $G_j = (V_j, E_j)$ as the subgraph of G over the nodes $V_j' \cup \{k\}$. This type of decomposition is called a *series decomposition* of G (see Fig. 12.7).

If $\mathcal{M}_j$ denotes the model obtained by restricting $\mathcal{M}$ to G_j and if $\{r,s\} \subseteq V_i$, then the following theorem asserts that the subgraphs G_j not containing $\{r,s\}$ can be ignored when computing $\mathcal{H}(\mathcal{M})|(r,s)$.

Theorem 12.8 *If $G_k = (V_k, E_k), k = 1, 2, \ldots$ is a series decomposition of G and $\{r,s\} \subseteq V_i$, then*

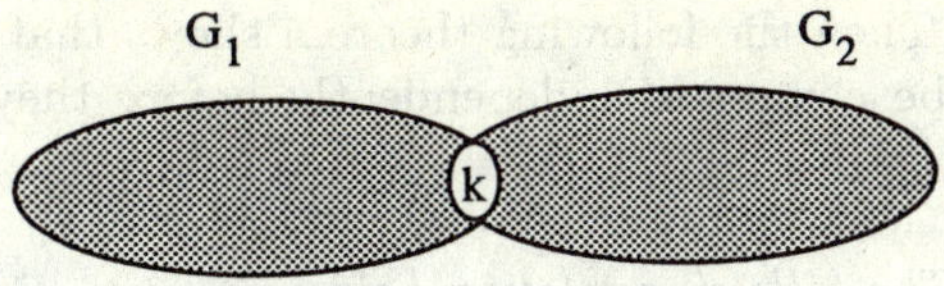

Fig. 12.7.

$$\mathcal{H}(\mathcal{M})|(r,s) = \mathcal{H}(\mathcal{M}_i)|(r,s) \tag{12.21}$$

Now consider the case where r and s do not belong to the same subgraph G_h.

Theorem 12.9 *If $G_k = (V_k, E_k), k = 1, 2, \ldots$ is a series decomposition of G and if $r \in G_i$ and $s \in G_j$ with $j \neq i$, then (compare (12.15))*

$$\mathcal{H}(\mathcal{M})|(r,s) = \mathcal{H}(\mathcal{M}_i)|(r,k) \otimes \mathcal{H}(\mathcal{M}_j)|(k,s). \tag{12.22}$$

Another, second type of decomposition arises when there are two subsets V_1 and V_2 of V such that
(1) $V = V_1 \cup V_2$, $V_1 \cap V_2 = \{r, s\}$
(2) If $G(V_i) = (V_i, E_i)$ for $i = 1, 2$, then

$$E_1 \cap E_2 = \begin{cases} (r,s) & \text{if } (r,s) \in E \\ \emptyset & \text{if } (r,s) \notin E \end{cases} \tag{12.23}$$

(see Figure 12.8)

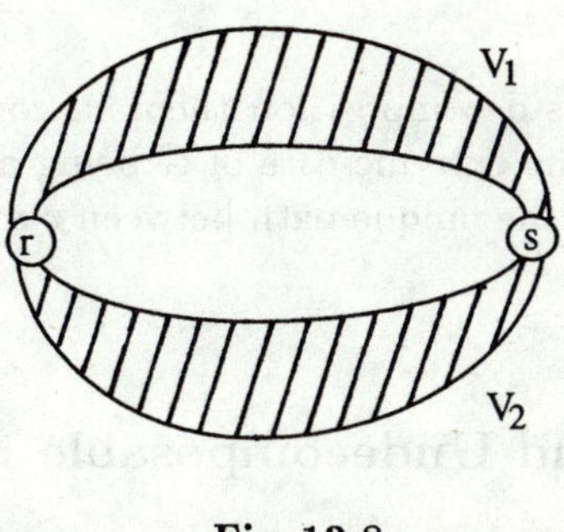

Fig. 12.8.

If the arc (r,s) is in E, then we arbitrarily remove it from E_2 (and keep it only in E_1), so that the equalities $E_1 \cap E_2 = \emptyset$ and $E_1 \cup E_2 = E$ are always satisfied. This type of decomposition is called a *parallel decomposition* of G with respect to (r,s).

Let $\mathcal{M}_i$ denote the model obtained by restricting $\mathcal{M}$ to the subgraph $G(V_i)$ for $i = 1, 2$. Then the following theorem shows that $\mathcal{H}(\mathcal{M}_1)|(r, s)$ and $\mathcal{H}(\mathcal{M}_2)|(r, s)$ can be computed independently before they are combined in order to get $\mathcal{H}(\mathcal{M})|(r, s)$.

Theorem 12.10 *The following relation holds:*

$$\mathcal{H}(\mathcal{M})|(r,s) = \mathcal{H}(\mathcal{M}_1)|(r,s) \oplus \mathcal{H}(\mathcal{M}_2)|(r,s). \tag{12.24}$$

To conclude this section let's mention that the non-structural information represented by the graph G is sometimes organized in a hierarchical way. For example, in temporal reasoning it is natural to distinguish between milestones and events of secondary importance (Allen, 1983). Figure 12.9 shows the graph of a temporal model of this type where the black nodes correspond to milestones and the white nodes are secondary events.

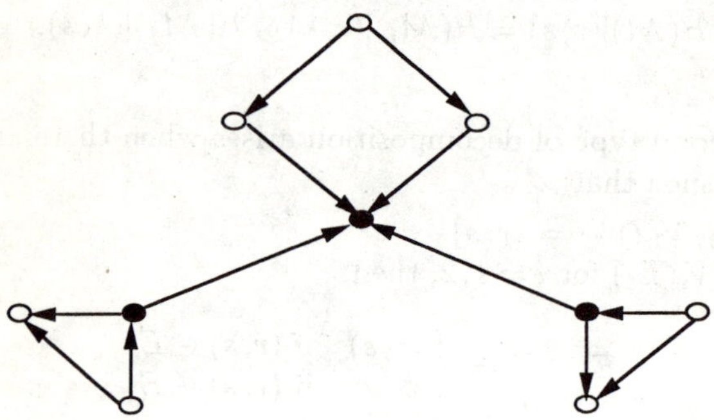

Fig. 12.9.

In this case, the series-decomposition theorem considerably reduces the computational effort. In the extreme case of G being a tree, it is sufficient to make series reductions on the unique path between r and s in order to obtain $\mathcal{H}(\mathcal{M})|(r, s)$.

12.5 Irreducible and Undecomposable Models

In this section we consider models which can neither be reduced nor decomposed. Computing such models is a rather difficult task, but it is possible to take advantage of the fact that structural vectors are tensions. Several possible approaches are presented.

The information contained in a model $\mathcal{M}$ with graph (V, E) and hints $\mathcal{H}_{ij}$ for $(i, j) \in E$ concerns the variables $d_{ij}, (i, j) \in V^2$. From another point of view we can also consider equivalently that the variables are the potentials $\pi(i), i \in V$ instead of the d_{ij}. The advantage is that one no longer has to worry about the structural relations. In this way a non-structural hint $\mathcal{H}_{ij} = (\Omega_{ij}, P_{ij}, \Gamma_{ij}, \Theta)$ about the variable d_{ij} is now transformed into the hint $\mathcal{P}_{ij} = (\Omega_{ij}, P_{ij}, \Gamma'_{ij}, \Theta^2)$ where

$$\Gamma'_{ij}(\omega) = \{(\pi(i), \pi(j)) \in \Theta^2 : \pi(j) \otimes \pi(i)^{-1} \in \Gamma_{ij}(\omega)\}. \tag{12.25}$$

To the model $\mathcal{M}$ corresponds therefore a new model $\mathcal{N}$ with variables $\pi(i), i \in V$ taking their values in the group Θ. Since the potential associated to a tension is not unique, for $r \in V$, it is useful to define the hint $\mathcal{S}_r$ as the deterministic hint focussing on $\{e\}$ such that $d_{rs} = \pi_s$. This shows that if

$$\mathcal{P}(\mathcal{N}) = \oplus \{\mathcal{P}_{ij} : (i,j) \in E\} \oplus \mathcal{S}_r, \tag{12.26}$$

then

$$\mathcal{H}(\mathcal{M})|(r,s) = \mathcal{P}(\mathcal{N})|\{s\} \tag{12.27}$$

for all $(r,s) \in V^2$. If W denotes the hypergraph associated with the hints in the right-hand side of (12.26), then the two-section graph of W is the undirected version of the graph $G = (V, E)$. Once a good covering Markov tree for W has been found, the computation of $\mathcal{P}(\mathcal{N})|\{s\}$ can be made using the local propagation procedure described in Chap. 8.

This method based on potentials is general and can also be applied to reducible and decomposable models. Furthermore, it is sometimes possible to exploit the special structure of Θ in order to construct special solution procedures (see Sect. 12.6 below). Another approach is to simply discard some hints $\mathcal{H}_{ij}$ and apply the reduction or decomposition procedures to the simplified model.

Since it is difficult to compute the exact solution $\mathcal{H}(\mathcal{M})$, it is often reasonable to use approximations. For this purpose a precise notion of inclusion between hints has to be defined (Kohlas, Monney, 1994), this subject will be discussed in detail in Chap. 14. This is done as follows. Given a hint $\mathcal{H}' = (\Omega', P', \Gamma', \Delta)$ on an arbitrary frame Δ, suppose that a supplementary source of information makes it possible to

(1) Exclude some interpretations as impossible so that only those which are in a subset Ω of Ω' remain possible.
(2) Be more precise in the determination of the focal sets corresponding to the remaining interpretations in Ω.

This leads naturally to a new hint $\mathcal{H} = (\Omega, P, \Gamma, \Delta)$ where

(1) $\Omega \subseteq \Omega'$
(2) $P(\omega) = P'(\omega)/P'(\Omega)$ for all $\omega \in \Omega$
(3) $\Gamma(\omega) \subseteq \Gamma'(\omega)$ for all $\omega \in \Omega$.

Condition (2) is based on the idea that if we learn that the unknown true interpretation is in Ω, then the probability measure P has to be conditioned on Ω. If the conditions (1) to (3) are satisfied, then $\mathcal{H}$ is said to be *finer* than $\mathcal{H}'$ and we write $\mathcal{H} \subseteq \mathcal{H}'$. It is easy to verify that this relation between hints is an order relation which is compatible with Dempster's rule, i.e. if $\mathcal{H}_1 \subseteq \mathcal{H}'_1$ and $\mathcal{H}_2 \subseteq \mathcal{H}'_2$, then $\mathcal{H}_1 \oplus \mathcal{H}_2 \subseteq \mathcal{H}_2 \oplus \mathcal{H}'_2$ (see also Sect. 15.1). If $\mathcal{H}$ is finer than $\mathcal{H}'$, we say sometimes that $\mathcal{H}'$ is an *approximation* of $\mathcal{H}$.

Now suppose that $\mathcal{M}$ is a model with a weakly connected graph $G = (V, E)$ without parallel arcs. In order to get an approximation of $\mathcal{H}(\mathcal{M})$, first a procedure inspired by Floyd's algorithm for computing shortest distances in a network is presented (Aho et al., 1983).

To start with, arbitrarily enumerate the nodes of V from $i = 1$ to $i = n$ V. From now on these nodes will be referenced by their indices. Let (Π, Q) denote the product probability space obtained from the spaces in the hints $\mathcal{H}_{jk}$ for all $(j, k) \in E$. If $\omega \in \Pi$ then the Floyd-type procedure is defined as follows:

(1) *Initialization*
For all $(j, k) \in V^2$, let

$$\Gamma'_{jk}(0,\omega) = \begin{cases} \Gamma_{jk}(\omega) & \text{if } (j,k) \in E \\ \{e\} & \text{if } j = k \\ \Theta & \text{otherwise} \end{cases} \tag{12.28}$$

(2) *Iteration*
For $i = 1, \ldots, n$ do
for all $(j, k) \in V^2$ do

$$\Gamma'_{jk}(i,\omega) = \Gamma'_{jk}(i-1,\omega) \cap \Gamma'_{ji}(i-1,\omega) \otimes \Gamma'_{ik}(i-1,\omega) \tag{12.29}$$

The result of this procedure is a family of n^2 subsets of Θ, namely :

$$\Gamma'_{jk}(\omega) = \Gamma'_{jk}(n,\omega). \tag{12.30}$$

Let $\mathcal{H}(\mathcal{M}) = (\Omega, P, \Gamma, \Theta^{n\times n})$ denote the solution of the model $\mathcal{M}$. If K is the set of contradictory interpretations, then $\Omega = \Pi - K$ and $P(\omega) = Q(\omega)/(1 - Q(K))$ for all $\omega \in \Omega$. If

$$\Gamma'(\omega) = \prod \{\Gamma'_{jk}(\omega) : (j,k) \in V^2,\} \tag{12.31}$$

then the following result holds.

Theorem 12.11 *The inclusion $\Gamma(\omega) \subseteq \Gamma'(\omega)$ holds for all $\omega \in \Omega$.*

If

$$K' = \{\omega \in \Pi : \exists\ (j,k) \in V^2 \text{ such that } \Gamma'_{jk}(\omega) = \emptyset\},$$

then the mapping Γ' is used to define a new hint

$$\mathcal{H}'(\mathcal{M}) = (\Omega', P', \Gamma', \Theta^{n\times n})$$

where $\Omega' = \Pi - K'$ and P' is a probability measure on Ω' which will be specified later. The goal is, of course, to consider the hint $\mathcal{H}'(\mathcal{M})$ as an approximation of the exact solution $\mathcal{H}(\mathcal{M})$. Theorem 12.11 clearly implies that the conditions (1) and (3) for an approximation are satisfied. Thus any probability measure P' on Ω' satisfying

$$P'(\omega) = P(\omega)P'(\Omega) \tag{12.32}$$

for all $\omega \in \Omega$ leads to an approximation of $\mathcal{H}(\mathcal{M})$. Now,

$$P'(\Omega) = 1 - P'(\Omega' - \Omega) = 1 - P'(K - K') \tag{12.33}$$

because $\Omega' - \Omega = K - K'$. Furthermore, the computation of K is sometimes relatively easy (see the next section). If K is known, then so is P. Therefore if $P'(K - K')$ is estimated by $Q(K - K')$, then the measure given by

$$P'(\omega) = \begin{cases} P(\omega)(1 - Q(K - K')) & \text{if } \omega \in \Omega \\ Q(\omega) & \text{if } \omega \in \Omega' - \Omega. \end{cases} \tag{12.34}$$

is such that $\mathcal{H}'(\mathcal{M})$ is an approximation of $\mathcal{H}(\mathcal{M})$: P' is indeed a probability on Ω' because $P'(\Omega) = P(\Omega)(1 - Q(K - K')) = 1 - Q(K - K')$ and $P'(\Omega' - \Omega) = Q(\Omega' - \Omega) = Q(K - K')$ and hence $P'(\Omega') = P'(\Omega) + P'(\Omega' - \Omega) = 1$; furthermore, condition (12.32) is satisfied because if $\omega \in \Omega$, then $P'(\omega) = P(\omega)(1 - Q(K - K')) = P(\omega)P'(\Omega)$. Of course these approximations are interesting so far as they are easy to obtain. This is actually the case here because the Floyd algorithm is fairly efficient, the only problem is the determination of the set K. One should not forget either that the Floyd-type procedure must in principle be repeated for all $\omega \in \Pi$. This is a rather large amount of computations which can possibly be simplified by the application of Monte Carlo methods.

It would be interesting to have information about the quality of the approximation obtained by this method. Given $\omega \in \Omega'$ and $(r, s) \in V^2$, a value $\theta \in \Theta$ is called *path-compatible* with respect to ω if for every path γ from r to s in G there is a vector $(d_{ij} : (i, j) \in \gamma)$ such that

(1) $d_{ij} \in \Gamma_{ij}(\omega)$ for all $(i, j) \in \gamma$
(2) $\theta = \otimes \{d_{ij} : (i, j) \in \gamma\}$.

Let $C_{rs}(\omega)$ denote the set of these path-compatible values.

Theorem 12.12 *For all $\omega \in \Omega'$ and for all $(r, s) \in V^2$, we have*

$$\Gamma'_{rs}(\omega) \subseteq C_{rs}(\omega). \tag{12.35}$$

In constraints satisfaction problems, path-compatible solutions are usually considered as good solutions (Mackworth, 1977). This result shows therefore that $\mathcal{H}'(\mathcal{M})$ could be a good approximation of $\mathcal{H}(\mathcal{M})$.

12.6 Planar Geometric Reasoning

In this section the geometric model briefly described in example 12.3 is studied more extensively. This type of model has also been considered by McDermott, Davis (1984) and Brooks (1981).

Consider a collection of n objects in the Euclidian plane $\mathcal{E}$. These objects are represented as points in $\mathcal{E}$ and will be referred to as the elements of a set $V = \{i, j, k, \ldots\}$. A positively oriented orthonormal coordinate system f_i is attached to each object $i \in V$. Their x-axes are all assumed parallel. These coordinate systems $f_i, i \in V$ serve as references to express the position $d_{ij} = (d_{ij}(1), d_{ij}(2)) \in \mathcal{R}^2$ of the object j with respect to f_i (cf. Figure 12.10).

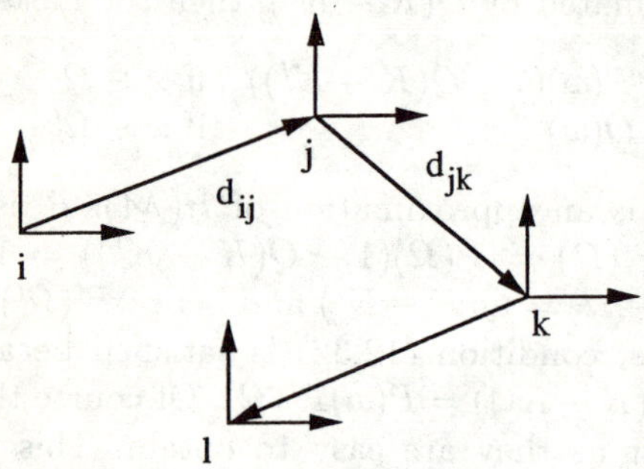

Fig. 12.10.

In this example the commutative group $(\Theta, \otimes)$ introduced in Sect. 12.2 is the group of translations $(\mathcal{R}^2, +)$. As already mentioned in Sect. 12.1, it is the fact that the references are parallel which makes Θ a commutative group. However, the reduction and decomposition results of the two preceding sections are still valid when the references are not parallel, i.e. Θ is a non-commutative group.

Before the first results are presented, let's introduce some notations. Let $\mathcal{V}$ denote the set of vectors $d = (d_{ij} : (i,j) \in V^2) \in (\mathcal{R}^2)^{n \times n}$ and let $\mathcal{T}$ denote the structural vectors of $\mathcal{V}$, i.e. those satisfying $d_{ik} = d_{ij} + d_{jk}$ for all $(i,j,k) \in V^3$. On the other hand, for $k = 1, 2$, let $\mathcal{V}(k)$ and $\mathcal{T}(k)$ denote the vectors $d(k) \in \mathcal{R}^{n \times n}$ obtained from $\mathcal{V}$ and $\mathcal{T}$ by considering only the k-th component of the couples $d_{ij} = (d_{ij}(1), d_{ij}(2))$ of the vectors d in $\mathcal{V}$ and $\mathcal{T}$ respectively. The following result is a direct consequence of the first part of theorem 12.1.

Theorem 12.13 *A vector $d(k) \in \mathcal{V}(k)$ is in $\mathcal{T}(k)$ if and only if there is a mapping*

$$\begin{aligned} \pi(k) : V &\to \mathcal{R} \\ i &\mapsto \pi_i(k) \end{aligned} \tag{12.36}$$

such that $d_{ij}(k) = \pi_j(k) - \pi_i(k)$ *for all* $(i,j) \in V^2$.

Since the existence of potentials characterizes the tensions, the set $\mathcal{T}(k) \subseteq \mathcal{R}^m, m = n^2$ is nothing but the set of tensions on the complete graph $K = (V, V^2)$ (Sakarovitch, 1984, theorem 10 (ii); see also Berge, 1973; Bondy, Murty, 1976). We have the following result.

Theorem 12.14 *For* $k \in \{1,2\}$, *the set* $\mathcal{T}(k)$ *is a linear subspace of* $\mathcal{R}^m$ *with dimension* $n-1$.

This is a well-known result from the theory of flows and tensions on a graph (Sakarovitch, 1984; Bondy, Murty, 1976: corollary 12.2 p. 216). A basis of $\mathcal{T}(k)$ can easily be obtained as follows. Let $T = (V, U)$ be a spanning tree of V. If an arc $u \in U$ is removed from T, then the remaining graph decomposes itself into exactly two connected components $T_1 = (V_1, E_1)$ and $T_2 = (V_2, E_2)$. Then a vector $\tau(u) \in \mathcal{R}^m$ is defined as follows:

$$\tau(u)_{ij} = \begin{cases} 1 & \text{if } i \in V_1, j \in V_2 \\ -1 & \text{if } i \in V_2, j \in V_1 \\ 0 & \text{otherwise} \end{cases} \tag{12.37}$$

for all $(i,j) \in V^2$. This vector is in fact a tension because it comes from the potential

$$\mu(i) = \begin{cases} 1 & \text{if } i \in V_2 \\ 0 & \text{if } i \in V_1. \end{cases} \tag{12.38}$$

Theorem 12.15 *The* $(n-1)$ *tensions in* $\tau(U) = \{\tau(u) : u \in U\}$ *form a basis of* $\mathcal{T}(k)$ *for each* $k \in \{1,2\}$. *Furthermore, if* $t \in \mathcal{T}(k)$ *and if* $t(u)$ *denotes the value of the tension* t *on the arc* $u \in U$, *then*

$$t = \sum \{t(u)\tau(u) : u \in U\}. \tag{12.39}$$

The proof of this theorem can be found in the book of Berge et al. (1962).

Consider a geometric model $\mathcal{M}$ given by the graph $G = (V, E)$ and the hints

$$\mathcal{H}_{ij} = (\Omega_{ij}, P_{ij}, \Gamma_{ij}, \mathcal{R}^2) \tag{12.40}$$

for all $(i,j) \in E$. The hint $\mathcal{H}_{ij}$ represents the non-structural information about the position of the object j with respect to the reference f_i attached to the object i. The goal is to judge certain hypotheses regarding the position of some object s with respect to some reference f_r in the light of the information encoded in the model $\mathcal{M}$. For this purpose its solution $\mathcal{H}(\mathcal{M})/(r,s)$ has to be found. This means that the projection to (r,s) of the set

$$\Lambda(\omega) = \{d \in \mathcal{T} : d_{ij} \in \Gamma_{ij}(\omega) \ \forall \ (i,j) \in E\} \tag{12.41}$$

for every combined interpretation ω in

$$\prod \{\Omega_{ij} : (i,j) \in E\} = \Pi \tag{12.42}$$

must be computed. The set $\Lambda(\omega)$ is the set of all consistent vectors associated with ω. In order to introduce more structure into the model, suppose furthermore that the focal sets $\Gamma_{ij}(\omega)$ are *convex polygons* for all $(i,j) \in E$ and for all $\omega \in \Pi$. What do series and parallel reductions look like in this special case ? In other words, how can the intersection and the sum of two convex polygons P_1 and P_2 be found ? As far as they are given by their vertices, the first task is a classical problem in computational geometry and efficient algorithms are known (Preparata, Shamos, 1985). If the polygons are given by a collection of linear inequalities, then the search for their vertices is certainly not a too difficult planar problem. For the sum of P_1 and P_2, one can proceed as follows.

A point x in a convex polygon P is a vertex if and only if there are not two different points x_1 and x_2 in P such that $x = (x_1 + x_2)/2$. If P^* denotes the vertices of P, then the following inclusion holds:

Theorem 12.16

$$(P_1 + P_2)^* \subseteq P_1^* + P_2^*. \tag{12.43}$$

If $X \subseteq \mathcal{R}^2$ then $conv(X)$ denotes the convex hull of X. Since there are good algorithms to find the convex hull of a finite set of vertices in the plane (Preparata et al., 1985), the following result gives an efficient method for computing $P_1 + P_2$.

Theorem 12.17

$$P_1 + P_2 = conv(P_1^* + P_2^*). \tag{12.44}$$

When the model is irreducible and it is not possible to perform a series or a parallel reduction (and assume that it is also indecomposable), then one can proceed as follows. Of course, the set $\Lambda(\omega)$ can be expressed as a set of linear inequalities with the variables $d_{ij}(1), d_{ij}(2)$ for all $(i,j) \in V^2$: there are the linear constraints from the conditions that d_{ij} is in the convex polygon $\Gamma_{ij}(\omega)$ and there are all the linear constraints coming from the structural relations $d_{ik} = d_{ij} + d_{jk}$ for all $(i,j,k) \in V^3$. So $\Lambda(\omega)$ is a convex polytop in $(\mathcal{R}^2)^{n\times n}$. If one is more particularly interested in the value of the variable $d_{rs} = (d_{rs}(1), d_{rs}(2))$ for a certain couple $(r,s) \in V^2$, then the projection of $\Lambda(\omega)$ to the variables $d_{rs}(1), d_{rs}(2)$ has to be computed. This projection $\Lambda(\omega)|(r,s)$ is a convex polygon which can in principle be determined by linear programming techniques. However, this is a difficult task, especially as there are a lot of constraints in the linear program. For this reason it might be advantageous to express the problem in terms of the potentials $\pi_i(1)$ and $\pi_i(2)$ so that the structural constraints need not be explicitly written. This leads to a linear program with far less constraints and variables which is easier to compute. The formal development is the following.

According to theorem 12.13, computing $\Lambda(\omega)$ is the same as computing

$$\Psi(\omega) = \{\pi \in (\mathcal{R}^2)^n : d_{ij} = \pi_j - \pi_i \in \Gamma_{ij}(\omega) \text{ for all } (i,j) \in E\}. \tag{12.45}$$

Since $\Gamma_{ij}(\omega)$ is a convex polygon, the constraint $d_{ij} \in \Gamma_{ij}(\omega)$ can be written as a family of linear inequalities of the form

$$\alpha(\omega)d_{ij}(1) + \beta(\omega)d_{ij}(2) \leq \gamma(\omega). \tag{12.46}$$

(see Fig. 12.11, where the equation of the line (D) is given by (12.46) with the inequality sign replaced by the equality sign).

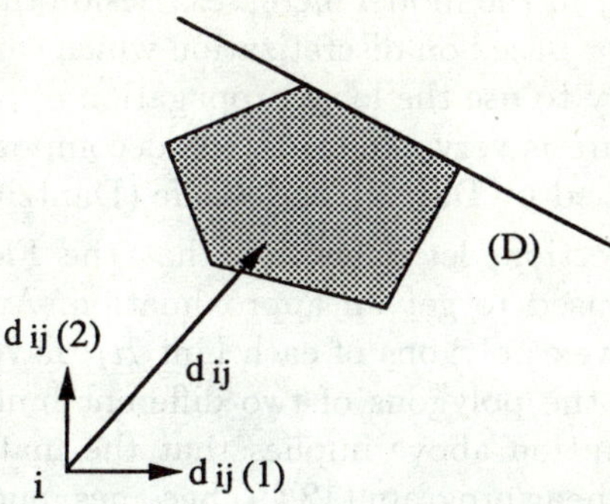

Fig. 12.11.

In terms of potentials, the inequality (12.46) reads

$$\alpha(\omega)\pi_j(1) - \alpha(\omega)\pi_i(1) + \beta(\omega)\pi_j(2) - \beta(\omega)\pi_i(2) \leq \gamma(\omega). \tag{12.47}$$

If k_{ij} is the number of edges of the polygon $\Gamma_{ij}(\omega)$, then there is a $p \times q$ matrix $A(\omega)$ and a p-vector $b(\omega)$ such that

$$\Psi(\omega) = \{\pi \in (\mathcal{R}^2)^n : A(\omega)\pi \leq b(\omega)\} \tag{12.48}$$

where $p = \sum_{(i,j)\in E} k_{ij}$ and $q = 2n$. If one is more particularly interested in the projection $\Lambda(\omega)|(r,s)$, then arbitrarily set $\pi_r = 0$ since a potential is defined up to an additive constant. So assume that the linear system $A(\omega)\pi \leq b(\omega)$ already incorporates this supplementary condition. The set $\Psi(\omega)$ is a convex polytop and its projection on the variable $\pi_s = (\pi_s(1), \pi_s(2)) = d_{rs}$ is a convex polygon which is equal to $\Lambda(\omega)|(r,s)$. The vertices of this polygon can be obtained by a parametric analysis on the objective function of the following linear program

$$\begin{aligned} A(\omega)\pi &\leq b(\omega) \\ \pi_s(1) + \mu\pi_s(2) &= z \ (Max) \end{aligned} \tag{12.49}$$

Indeed, the first phase of the simplex algorithm tells us whether $\Psi(\omega)$ is empty or not and if not, then an initial basic solution is available (Sakarovitch, 1984). From this solution a sensitivity analysis on the parameter μ leads to a set of critical values corresponding exactly to the vertices of $\Lambda(\omega)|\{r,s\}$. Refer to Simmonard (1972) for more information about the sensitivity analysis of the objective function of a linear program. By the way, most commercial linear programming software packages are able to perform such an analysis. Moreover, since the probability $P(\omega)$ is known, it is interesting to notice that (12.49) can be considered as a stochastic linear program (Kall, 1976).

Unfortunately, the method described above based on linear programming suffers from an important drawback because in principle the parametric analysis must be repeated for every combined interpretation $\omega \in \Pi$. Thus one is confronted with the problem of combinatorial growth of the cardinality of Π as the number of hints $\mathcal{H}_{ij}$ in the model increases. Beside the methods based on Monte Carlo sampling or based on discretization which can always be applied, one can possibly also try to use the local propagation of potentials in Markov trees. This last procedure is very similar to the decomposition techniques for linear programs developed by Dantzig and Wolfe (Dantzig, 1964).

To conclude this section, let's mention that the Floyd-type procedure (12.27) can always be used to get an approximation. Another possibility is to assume that the convex polygons of each hint $\mathcal{H}_{ij}$ have *parallel* edges (see Fig. 12.12). Of course, the polygons of two different hints are not necessarily parallel. The assumption above implies that the matrix $A(\omega)$ no longer depends on ω and the linear program (12.49) becomes much easier to analyze.

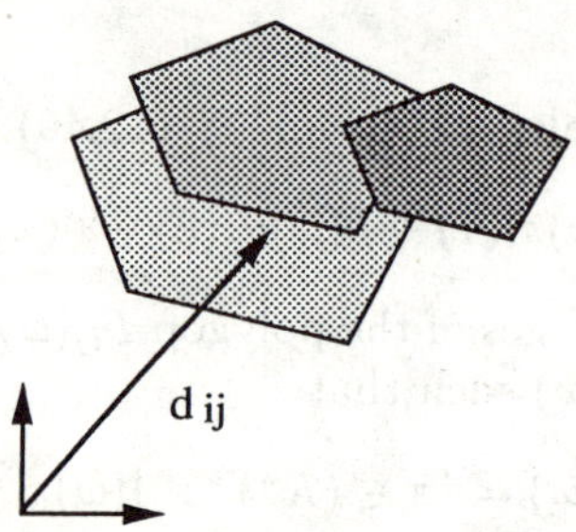

Fig. 12.12.

12.7 Rectangular Geometric Models

In this section it is assumed that the focal sets of the hints $\mathcal{H}_{ij}$ are particularly simple polygons, namely rectangles which are parallel to the axes, i.e.

$$\Gamma_{ij}(\omega) = [l_{ij}^{(1)}(\omega), u_{ij}^{(1)}(\omega)] \times [l_{ij}^{(2)}(\omega), u_{ij}^{(2)}(\omega)] \tag{12.50}$$

for all $(i,j) \in E$ and for all $\omega \in \Pi$. In this case series and parallel reductions are particularly simple because they correspond to the addition and intersection of intervals. Once all possible reductions and decompositions have been made, let's see how the sets $\Lambda(\omega)|(r,s)$ can be computed in this particular case.

The set $\Lambda(\omega)$ is composed of all vectors $d = (d(1), d(2)) \in \mathcal{V}$ such that $d(1)$ and $d(2)$ are tensions and

$$\begin{aligned} l_{ij}^{(1)}(\omega) \leq d_{ij}(1) \leq u_{ij}^{(1)}(\omega) \\ l_{ij}^{(2)}(\omega) \leq d_{ij}(2) \leq u_{ij}^{(2)}(\omega) \end{aligned} \tag{12.51}$$

for all $(i,j) \in E$. But this implies that $\Lambda(\omega) = \Lambda^{(1)}(\omega) \times \Lambda^{(2)}(\omega)$ with

$$\Lambda^{(k)}(\omega) = \{d(k) \in \mathcal{T}(k) : d_{ij}(k) \in [l_{ij}^{(k)}(\omega), u_{ij}^{(k)}(\omega)]\} \tag{12.52}$$

for $k = 1, 2$. Let $\Lambda_{rs}^{(k)}(\omega) = \Lambda^{(k)}(\omega)|(r,s)$ denote the projection of $\Lambda^{(k)}(\omega)$ to the variable d_{rs}. Since $\Lambda(\omega)$ is described by a set of linear inequalities, it is a convex set and therefore also its projection $\Lambda(\omega)|(r,s) = \Lambda_{rs}(\omega)$. But since $\Lambda_{rs}(\omega) = \Lambda_{rs}^{(1)}(\omega) \times \Lambda_{rs}^{(2)}(\omega)$ it follows that $\Lambda_{rs}^{(k)}(\omega)$ is an interval for $k = 1, 2$:

$$\Lambda_{rs}^{(k)}(\omega) = [l^{(k)}(\omega), u^{(k)}(\omega)]. \tag{12.53}$$

This shows that

$$\begin{aligned} u^{(k)}(\omega) &= \max\ \{d_{rs}(k) : d(k) \in \Lambda^{(k)}(\omega)\} \\ &= \max\ \{d_{rs}(k) : d(k) \in \mathcal{T}(k), d_{ij}(k) \in [l_{ij}^{(k)}(\omega),\ u_{ij}^{(k)}(\omega)]\}, \end{aligned} \tag{12.54}$$

i.e. $u^{(k)}(\omega)$ is the maximal possible value of a tension on the arc (r,s). This is a classical problem in graph theory (Berge, 1973). For the case where $l_{ij}^{(k)}(\omega)$ and $u_{ij}^{(k)}(\omega)$ are integers (positive or negative or null), the following algorithm of Herz can be used (except for some minor changes we follow the presentation given in Berge (1973)). In order to keep notations as simple as possible, denote by $1, 2, \ldots, m$ the arcs of the graph $G = (V, E)$ and for each arc $i = (v, w)$ let $[k_i, l_i], k_i, l_i \in \mathcal{Z} \cup \{-\infty, +\infty\}$ denote a focal interval of the hint $\mathcal{H}_{vw}$. Furthermore, assume that the arc (r,s) is in E (if this is not the case, simply consider the vacuous hint $\mathcal{H}_{rs}$ focussing on $[-\infty, +\infty]$).

First of all, given these data, it is not certain whether there is indeed at least one tension θ such that $k_i \leq \theta_i \leq l_i$ for all arcs $i = 1, \ldots, m$ (θ_i is the tension on the arc i). Finding the answer to this question is called the compatible tension problem: first find out whether such a tension exists and if the answer is yes, how we can find one. Note that if there is no compatible tension, then the interpretation ω defining the intervals $[k_i, l_i]$ is contradictory. When the intervals $[k_i, l_i]$ are relative integers (possibly infinite), then the compatible tension problem can be solved by the following algorithm.

Compatible tension algorithm (Herz, 1967; see Berge, 1973)

Start with an arbitrary tension θ with $\theta_i \in \mathcal{Z}$ for all i, compatible or not, and define the distance of θ_i from the interval $[k_i, l_i]$ by

$$d_i(\theta_i) = \begin{cases} 0 & \text{if } \theta_i \in [k_i, l_i] \\ k_i - \theta_i & \text{if } \theta_i < k_i \\ \theta_i - l_i & \text{if } \theta_i > l_i. \end{cases} \tag{12.55}$$

We shall successively reduce the quantity

$$d(\theta) = \sum_{i=1}^{m} d_i(\theta_i). \tag{12.56}$$

Note that $d(\theta)$ is finite and $d(\theta) = 0$ if and only if θ is compatible. If $d(\theta) = 0$, then a compatible tension is found and the algorithm stops. Otherwise, there is an arc i with $d_i(\theta_i) > 0$: without loss of generality suppose that $i = 1 = (b, a)$ and that $d_1(\theta_1) = k_1 - \theta_1, \theta_1 < k_1$ (the case $\theta_1 > l_1$ is treated in a symmetrical, similar way). To construct a tension θ' with $d(\theta') < d(\theta)$, *successively* label the vertices of G as follows:

Rule 1. Label vertex a, the terminal vertex of the arc 1, with the label +1.
Rule 2. Suppose that x is labelled and y is not labelled.

(a) If $\mathrm{i} = (\mathrm{x}, \mathrm{y})$ and $\theta_\mathrm{i} \leq \mathrm{k_i}$ then label y with $+\mathrm{i}$
(b) If $\mathrm{i} = (\mathrm{y}, \mathrm{x})$ and $\theta_\mathrm{i} \geq \mathrm{l_i}$ then label y with $-\mathrm{i}$.

Two cases must be distinguished:
Case 1.

The vertex b, initial vertex of the arc 1, cannot be labelled by this procedure. Then we shall define a new tension θ' such that $d(\theta) - d(\theta') > 1$, i.e. with θ' we are at least one unit closer to "compatibility". The tension θ' is constructed as follows (a tension is considered as a vector in $\mathcal{R}^m$). First let A denote the set of all labelled vertices (note that $a \in A$ and $b \notin A$) and let $\omega(A)$ denote the tension induced by the potential $\pi : V \rightarrow \mathcal{R}$ given by

$$\pi(x) = \begin{cases} 1 & \text{if } x \in A^c \\ 0 & \text{if } x \in A. \end{cases} \tag{12.57}$$

Then the tension $\theta' = \theta - \omega(A)$ is such that $d_i(\theta'_i) \leq d_i(\theta_i)$ for all i because if $i \in (A, A^c)$, then $\theta_i > k_i$ and hence $\theta'_i = \theta_i - 1 \geq k_i$ because θ_i and k_i are (relative) integers. Similarly, if $i \in (A^c, A)$, then $\theta_i < l_i$ and hence $\theta'_i = \theta_i + 1 \leq l_i$ and hence $d_i(\theta'_i) \leq d_i(\theta_i)$. Furthermore, the arc $1 = (b, a) \in (A^c, A)$ is such that $d_1(\theta_1) - d_1(\theta'_1) \geq 1$ because $\theta_1 < k_1$. Thus finally $d(\theta) - d(\theta') \geq 1$. Moreover, note that the value of the tension on the arc $1 = (b, a)$ is $\theta'_1 = \theta_1 + 1$. *The procedure is then repeated with the new tension θ' instead of θ.*

Case 2.

The vertex b can be labelled by this procedure. In this case no compatible tension exists and the algorithm stops (see next theorem).

Theorem 12.18 *In the algorithm of Herz, if the vertex b can be labelled, then there is no compatible tension.*

Of course, in case 1, it is possible to accelerate the descent toward compatability by taking $\theta' = \theta - \epsilon \cdot \omega(A)$ where $\epsilon = \min(\epsilon^+, \epsilon^-)$ and

$$\begin{aligned} \epsilon^+ &= \min \{\theta_i - k_i : i \in (A, A^c)\} \\ \epsilon^- &= \min \{l_i - \theta_i : i \in (A^c, A)\}. \end{aligned} \tag{12.58}$$

Then we have $d(\theta) - d(\theta') \geq \min(\epsilon, k_1 - \theta_1)$.

The algorithm will necessarily terminate after a finite number of steps because each time either the distance to compatibility is reduced by at least 1 and finally a compatible tension is found, or else there is no compatible tension, which is detected in case 2. Note that the improved version of the algorithm given above is still valid for values of k_i and l_i in $\mathcal{R} \cup \{-\infty, +\infty\}$, i.e. k_i, l_i don't necessarily need be (relative) integers. A new tension θ' with $d(\theta') < d(\theta)$ is found but this time the algorithm is not guaranteed to terminate after a finite number of steps because it is possible that no element in the sequence $d(\theta), d(\theta'), \ldots$ is zero. This is the same problem as with the algorithm of Ford and Fulkerson for finding a maximal flow problem when the capacities are real, not integer (or rational) numbers (Minieka, 1978). When k_i and l_i are not integers, it is better to apply another method described below.

As a consequence of this algorithm, there is the following theorem called the compatible tension theorem (Ghouila-Houri, 1960).

Theorem 12.19 *Given a graph G and (relative) integer numbers $-\infty \leq k_i \leq l_i \leq +\infty$ for $i = 1, \ldots, m$, a necessary and sufficient condition for the existence of a tension $\theta = (\theta_1, \ldots, \theta_m)$ such that $k_i \leq \theta_i \leq l_i$ for all i is that*

$$\sum_{i \in \mu^+} l_i - \sum_{i \in \mu^-} k_i \geq 0, \tag{12.59}$$

for every cycle μ and each orientation of μ.

Corollary 12.1 *There exists a tension θ such that $\theta_i \geq k_i$ for all i if and only if $\sum_{i \in \mu} k_i \leq 0$ for all circuit μ in G.*

Assuming that a compatible tension exists, the maximal value of a compatible tension on the arc $1 = (b, a)$ must be determined. In other words, what is the value of

$$\theta_1^* = \max \{\theta_1 : \theta \text{ is a tension }, k_i \leq \theta_i \leq l_i,\ i = 1, \ldots, m\}. \tag{12.60}$$

The following algorithm can be applied:

Maximum tension algorithm (Berge, 1973)

Starting from an arbitrary compatible tension θ, apply the same labelling procedure as in the compatible tension algorithm. Two cases must be distinguished.

Case 1.

If node b cannot be labelled, then the new tension $\theta' = \theta - \omega(A)$ is still compatible because $d(\theta') = 0$ since $0 \leq d(\theta') \leq d(\theta) = 0$ since θ is compatible. Furthermore, $\theta'_1 = \theta_1 + 1$ and the value of the tension on the arc 1 has increased by a value of 1. Then apply the compatible tension labelling algorithm starting with the new tension θ'.

Case 2.

If b can be labelled, then the algorithm stops and θ_1 is maximal as the following theorem asserts.

Theorem 12.20 *If node b can be labelled, then θ_1 is the maximal tension on the arc 1, i.e. $\theta_1 = \theta_1^*$.*

The next theorem is a consequence of this algorithm.

Theorem 12.21 *Given a graph G and (relative) integer numbers $-\infty \leq k_i \leq l_i \leq +\infty$ for all $i = 1, \ldots, m$, the maximum value of the tension on the arc 1 of a compatible tension equals*

$$\theta_1^* = \min \{ \sum_{i \in \mu^-} l_i - \sum_{i \in \mu^+, i \neq 1} k_i : \mu \text{ is a cycle with } 1 \in \mu^+ \}. \tag{12.61}$$

The proof of this theorem shows that if

$$B = \min \{ \sum_{i \in \mu^-} l_i - \sum_{i \in \mu^+, i \neq 1} k_i : \mu \text{ is a cycle with } 1 \in \mu^+ \}. \tag{12.62}$$

is finite, then the maximum tension algorithm stops after a finite number of iterations because each time the value of the tension on the arc 1 is increased by 1 and the tension in the arc 1 cannot exceed $B < \infty$.

Note that the maximum tension algorithm can also be used to compute the minimum value of θ_1 of a compatible tension θ because, if $\theta'_i = -\theta_i$, then

$$\begin{aligned} &\min \{\theta_1 : \theta \text{ is a tension, } k_i \leq \theta_i \leq l_i\} \\ &= - \max \{-\theta_1 : \theta \text{ is a tension, } k_i \leq \theta_i \leq l_i\} \\ &= - \max \{\theta'_1 : \theta' \text{ is a tension, } -l_i \leq \theta'_i \leq -k_i\} \end{aligned} \tag{12.63}$$

and hence the maximum tension algorithm with the intervals $[-l_i, -k_i]$ can be applied. Furthermore, according to theorem 12.21, the last line in (12.63) equals

$$\begin{aligned} &= - \min \{ (\sum_{i \in \mu^-} (-k_i) - \sum_{i \in \mu^+, i \neq 1} (-l_i)) : \mu, 1 \in \mu^+ \} \\ &= - \min \{ (\sum_{i \in \mu^+, i \neq 1} l_i - \sum_{i \in \mu^-} k_i) : \mu, 1 \in \mu^+ \} \\ &= \max \{ (\sum_{i \in \mu^-} k_i - \sum_{i \in \mu^+, i \neq 1} l_i) : \mu, 1 \in \mu^+ \}. \end{aligned} \tag{12.64}$$

Remark that the method described above can be used to solve the examples 12.1 and 12.2 when the focal sets of the non-structural hints are intervals (with (relative) integer limits).

Furthermore, in formula 12.61, if the minimum were taken over all cycles μ such that $\mu^+ = \{1\}$, then θ_i^* would be equal to the length of the shortest path from b to a in G with arc costs l_i; and in (12.64) the minimal tension in 1 would be equal to the length of the longest path from b to a in G with arc costs k_i. Therefore, computing these shortest and longest paths yield an interval which is only contained in the exact interval. However, it is possible to define networks such that the exact values correspond to the lengths of a shortest and longest path in these networks. This is another method of computing $u^{(k)}(\omega)$ and $l^{(k)}(\omega)$ which will now be presented. Unlike the method given above, this procedure can always be successfully applied even if the values k_i and l_i are not necessarily (relative) integer values.

First, according to theorems 12.6 and 12.7, one can assume without loss of generality that $G = (V, E)$ is connected. Since the inequality $d_{ij}(k) \geq l_{ij}^{(k)}(\omega)$ is equivalent to

$$-d_{ij}(k) \leq -l_{ij}^{(k)}(\omega), \tag{12.65}$$

let's consider the network $R = (V, \bar{E}, c^{(k)}(\omega))$ where

$$\bar{E} = E \cup \{a = (i,j) \in V^2 : \bar{a} = (j,i) \in E\} \tag{12.66}$$

and $c^{(k)}(\omega)$ is the mapping from $\bar{E}$ to $\mathcal{R}$ given by

$$c_a^{(k)}(\omega) = \begin{cases} u_a^{(k)}(\omega) & \text{if } a \in E \\ -l_{\bar{a}}^{(k)}(\omega) & \text{if } a \in \bar{E} - E. \end{cases} \tag{12.67}$$

If A denotes the node-to-arc incidence matrix of R, then $\Lambda^{(k)}(\omega)$ is expressed in terms of potentials by

$$\Psi_k(\omega) = \{\pi(k) \in \mathcal{R}^n : -A^T\pi(k) \leq c^{(k)}(\omega)\}. \tag{12.68}$$

Define $\pi_r(k) = 0$ so that $d_{rs}(k) = \pi_s(k)$. Then the upper limit $u^{(k)}(\omega)$ of the interval $\Lambda_{rs}^{(k)}(\omega)$ is given by the optimal value of z in the linear program

$$\begin{aligned} \pi_r(k) &= 0 \\ -A^T\pi(k) &\leq c^{(k)}(\omega) \qquad (P) \\ \pi_s(k) &= z \ (Max) \end{aligned}$$

Theorem 12.22 *The linear program (P) has a feasible solution if and only if R does not contain a circuit of negative length. If this condition is satisfied, then the maximal value of $\pi_s(k)$ (in other words $u^{(k)}(\omega)$) is the shortest distance from r to s in R.*

This result is also known as the maximum tension theorem (Sakarovitch, 1984). Any shortest path algorithm in networks which may contain circuits and possibly arcs of negative length can be used to compute $u^{(k)}(\omega)$. For such an algorithm, see Sakarovitch (1984), Aho et al. (1983) or Gondran-Minoux (1984).

A similar development can be made for $l^{(k)}(\omega)$. This time, consider the network $R' = (V, \bar{E}, c'^{(k)}(\omega))$ where

$$c'^{(k)}_a(\omega) = \begin{cases} l^{(k)}_a(\omega) & \text{if } a \in E \\ -u^{(k)}_{\bar{a}}(\omega) & \text{if } a \in \bar{E} - E. \end{cases} \tag{12.69}$$

In this case, $l^{(k)}(\omega)$ is given by the optimal value of y in the linear program

$$\begin{aligned} \pi_r(k) &= 0 \\ -A^T\pi(k) &\geq c'^{(k)}(\omega) \qquad (P') \\ \pi_s(k) &= y\ (Min) \end{aligned}$$

Theorem 12.23 *The linear program (P') has a feasible solution if and only if R' has no circuit of positive length. If this condition is satisfied, then the minimal value of $\pi_s(k)$ (in other word $l^{(k)}(\omega)$) is the longest distance from r to s in R'.*

Now let's see what the Floyd-type algorithm of Sect. 12.5 looks like in the special case considered here. The Floyd-type procedure becomes here:

(1) *Initialization*
For all $(j,k) \in V^2$, let

$$R_{jk}(0,\omega) = \begin{cases} [l^{(1)}_{jk}(\omega), u^{(1)}_{jk}(\omega)] \times [l^{(2)}_{jk}(\omega), u^{(2)}_{jk}(\omega)] & \text{if } (j,k) \in E \\ (0,0) & \text{if } j = k \\ \mathcal{R}^2 & \text{otherwise} \end{cases} \tag{12.70}$$

(2) *Iteration*
For $i = 1, \ldots, n$ do
For all $(j,k) \in V^2$ do

$$R_{jk}(i,\omega) = R_{jk}(i-1,\omega) \cap R_{ji}(i-1,\omega) + R_{ik}(i-1,\omega) \tag{12.71}$$

Remark that if

$$R_{jk}(i,\omega) = [L^{(1)}_{jk}(i,\omega), U^{(1)}_{jk}(i,\omega)] \times [L^{(2)}_{jk}(i,\omega), U^{(2)}_{jk}(i,\omega)], \tag{12.72}$$

then for $h \in \{1,2\}$ the iteration (12.71) becomes

$$L^{(h)}_{jk}(i,\omega) = Max\ \{L^{(h)}_{jk}(i-1,\omega), L^{(h)}_{ji}(i-1,\omega) + L^{(h)}_{ik}(i-1,\omega)\} \tag{12.73}$$

and

$$U_{jk}^{(h)}(i,\omega) = Min \; \{U_{jk}^{(h)}(i-1,\omega), U_{ji}^{(h)}(i-1,\omega) + U_{ik}^{(h)}(i-1,\omega)\}. \quad (12.74)$$

Let

$$R_{jk}(\omega) = [L_{jk}^{(1)}(\omega), U_{jk}^{(1)}(\omega)] \times [L_{jk}^{(2)}(\omega), U_{jk}^{(2)}(\omega)] \quad (12.75)$$

denote the final result obtained by this Floyd-type procedure. Then equation (12.73) and equation

$$L_{jk}^{(h)}(0,\omega) = l_{jk}^{(h)}(\omega) \quad (12.76)$$

imply that $L_{jk}^{(h)}(\omega)$ is the result of the application of the genuine Floyd algorithm (in the Max version) on the network $R'(E)$ obtained by restricting R' to E (Aho et al., 1983). In consequence, $L_{jk}^{(h)}(\omega)$ equals the longest distance between j and k in $R'(E)$. In a similar way, $U_{jk}^{(h)}(\omega)$ equals the shortest distance between j and k in the network $R(E)$ obtained by restricting R to E.

But by the analysis following equation (12.64), the interval obtained in this way only contains the exact interval and the inclusion is strict in general (compare with theorem 12.11).

However, it is possible to modify the initial model in order that the Floyd-type algorithm of Sect. 12.5 gives the exact intervals $\Lambda_{rs}^{(h)}(\omega)$. Indeed, consider the model $\bar{\mathcal{M}}$ with $(V, \bar{E})$ where $\bar{E}$ is defined by (12.66) and

$$\bar{\mathcal{H}}_{ij} = \begin{cases} \mathcal{H}_{ij} & \text{if } (i,j) \in E \\ -\mathcal{H}_{ij} & \text{if } (i,j) \in \bar{E} - E. \end{cases} \quad (12.77)$$

If $\bar{R}_{rs}(\omega)$ denotes the result of the Floyd-type algorithm on the model $\bar{\mathcal{M}}$, then the following result asserts the correctness of this algorithm on this model.

Theorem 12.24

(1) The combined interpretation ω is contradictory if and only if there is $(r,s) \in V^2$ such that $\bar{R}_{rs}(\omega) = \emptyset$.

(2) If ω is not contradictory, then

$$\bar{R}_{rs}(\omega) = \Lambda_{rs}(\omega) \quad (12.78)$$

for all $(r,s) \in V^2$.

Consequently, the exact solution for a model $\mathcal{M}$ can be obtained by applying the Floyd-type algorithm not to the model $\mathcal{M}$ itself, but rather to its modification $\bar{\mathcal{M}}$.

Example 12.6 Consider a model with the associated graph $G = (V, E)$ given in Fig. 12.13. (the intervals on the arcs represent $[l_{ij}^{(h)}(\omega), u_{ij}^{(h)}(\omega)]$ for $h \in \{1, 2\}$). The networks R and R' are given in Fig. 12.14 a) et b) respectively.

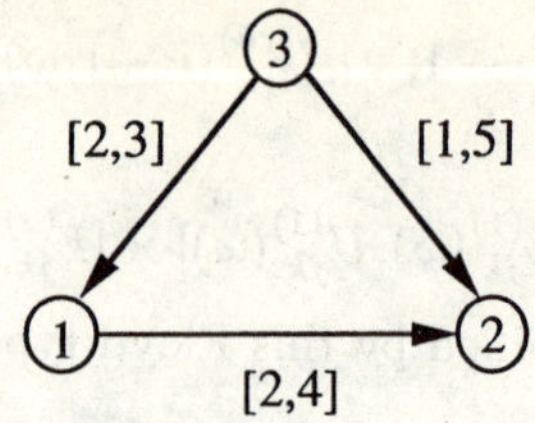

Fig. 12.13.

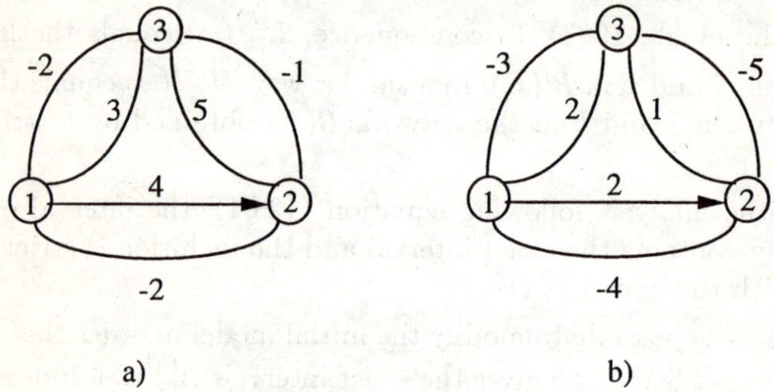

Fig. 12.14.

The result obtained by the Floyd-type algorithm on the original model gives $[L_{12}^{(h)}(\omega), U_{12}^{(h)}(\omega)] = [2,4]$ whereas the exact result $[\bar{L}_{12}^{(h)}(\omega), \bar{U}_{12}^{(h)}(\omega)]$ is $[2,3]$.

The hypothesis asserting that the focal sets of the various hints $\mathcal{H}_{ij}$ are rectangles is perhaps not sufficient to make computations feasible. Indeed, for each combined interpretation $\omega \in \Pi$ four optimal distances in different networks must be computed. It will be seen that a simpler situation arises when all hints $\mathcal{H}_{ij}$ come from the same source which can be an expert or an estimation mechanism. This source is working with an unknown degree of precision δ whose value is an element of $\Delta = \{\delta_1, \ldots, \delta_m\}$. The likelihood that δ_i is the true degree of precision is given by the probability $P(\delta_i)$. The information about the position of j with respect to f_i is then interpreted in several ways according to the degree of precision of the source. This information can thus be represented by a hint

$$\mathcal{H}_{ij} = (\Delta, P, \Gamma_{ij}, \mathcal{R}^2) \tag{12.79}$$

whose focal sets are the rectangles $\Gamma_{ij}^{(1)}(\delta) \times \Gamma_{ij}^{(2)}(\delta)$. If δ' corresponds to a higher degree of precision than δ'', then

$$\Gamma_{ij}(\delta') \subseteq \Gamma_{ij}(\delta''), \tag{12.80}$$

and $\mathcal{H}_{ij}$ is a consonant hint. So let $\mathcal{M}$ be a model of this type. In such a situation, one must not forget that the hints $\mathcal{H}_{ij}$ are dependent since they all come from the same source of information. This implies that the space of interpretations (Δ, P) is the same for all hints $\mathcal{H}_{ij}$, $(i,j) \in E$. The way in which such hints have to be combined by Dempster's rule is given by theorem 4.11.

Therefore, for the model $\mathcal{M}$

$$\Psi(\delta) = \{\pi \in (\mathcal{R}^2)^n : d_{ij} = \pi_j - \pi_i \in \Gamma_{ij}(\delta)\} \tag{12.81}$$

must be computed for all $\delta \in \Delta$. Since the number of elements of Δ is generally much smaller than that of Π, the number of optimal distances to be computed in order to find $\mathcal{H}(\mathcal{M})/(r,s)$ is considerably reduced. Furthermore, the hint $\mathcal{H}(\mathcal{M})/(r,s)$ is also consonant, which implies that the hints

$$\mathcal{H}_{ij}^{(k)} = (\Delta, p, \Gamma_{ij}^{(k)}, \mathcal{R}) \tag{12.82}$$

are also consonant.

For an arbitrary geometric model $\mathcal{M}$ the knowledge of $\mathcal{H}(\mathcal{M})/(r,s)$ allows the evaluation of the hypothesis that the object r is in a certain region H of the plane. Since such a hypothesis can be formulated for any subset $H \subseteq \mathcal{R}^2$, it is impossible to give an explicit list of all pairs of support-plausibility. Even this would not give an exhaustive picture of the information contained in $\mathcal{H}(\mathcal{M})/(r,s)$. So there is a need for methods which give a good picture of the position of the object s with respect to f_r. In that respect the following approaches can be envisaged.

(1) *Overall picture of the plausibility*

If the hint is consonant, then the plausibility function plays the role of a density function according to the first assertion of theorem 3.7. In this as well as in the general case one can always consider a partition of the plane according to a grid. Then each cell is drawn in dark grey or light grey according to its plausibility. The picture obtained in this way gives a fairly good idea of the plausibility function.

(2) *Confidence regions*

If sp and pl denote the support and plausibility functions of $\mathcal{H}(\mathcal{M})/(r,s)$ respectively, then the goal is to find a region such that $sp(H) \geq a$ and $pl(H) \geq b$ for certain fixed values a and b in $[0,1]$. These conditions mean that the hypothesis of the object s being in the region H has a sufficiently large support and the doubt about it can be neglected. This region is of course not unique. In order to get one such region, one can proceed as follows. Let $F_1, \ldots, F_n$ denote the focal sets of $\mathcal{H}(\mathcal{M})/(r,s)$ and let $m_1, \ldots, m_n$ denote their basic probability assignments. If $I = \{1, \ldots, n\}$ then the family

$$\mathcal{A}(a) = \{J \subseteq I : \sum_{i \in J} m_i \geq a\} \tag{12.83}$$

is a monotone family (i.e. $J \in \mathcal{A}(a)$ and $J' \supseteq J$ imply $J' \in \mathcal{A}(a)$) whose minimal elements are $J_1, \ldots, J_r$. Define

$$H(J_i) = \cup\{F_k : k \in J_i\}. \tag{12.84}$$

If there is at least one J_i such that $pl(H(J_i)) \geq b$, then take as H the domain $H(J_i)$ with smallest area satisfying this condition. Otherwise take a domain of small area

$$H = \cup\{F_k : k \in J\} \tag{12.85}$$

such that $pl(H) \geq b$ and $J \supseteq J_i$ for at least one index $i \in \{1, \ldots, r\}$.

(3) *Visualization windows*
Consider a geometric shape (a rectangle, a circle, etc) which can be moved in the plane. Each time, one can consider the hypothesis that the object s is within this shape. The shape is then shaded more or less darkly according to the degree of support or plausibility of the corresponding hypothesis. The shape serves as a visualization window of the support or plausibility function.

(4) *Probable and excluded regions*
A different approach can also be considered. First a region H_1 with $sp(H_1) \geq 1 - \epsilon_1$ is determined and then a second region H_2 with $pl(H_2) \leq \epsilon_2$ and $H_1 \subseteq H_2^c$ is determined. Of course, the probable region H_1 should be as small as possible and the unlikely region H_2 should be as large as possible. Note that such regions always exist (take H_1 as the union of the focal sets and H_2 as H_1^c), but finding better regions is more difficult.

12.8 Project Scheduling

Another important problem in temporal reasoning is project scheduling, especially under uncertainty. In this section this problem is analyzed from the point of view of the theory of hints and the main differences with more traditional and widely spread methods will be put forward. To fix ideas consider the following very simple example.

Example 12.7 The principal activities for constructing a house are the following:

1. Prepare the field
2. Install the crane
3. Lay the foundations
4. Install the electricity
5. Install the pipes for the evacuation of used water.

A certain number of temporal conditions between the activities must be satisfied in order to finally complete the project. These are

1. The crane cannot work without electricity
2. The crane is needed to make the foundations
3. The pipes cannot be installed and the foundations cannot be made before the field is prepared

Given these constraints and the durations of the activities, one can be interested in the earliest possible starting time of each activity or the earliest completion time of the whole project. When the duration of each activity is known, then the usual critical path method (CPM) or the metra potential method (MPM) can be applied with succes (Roy, 1970; Gondran-Minoux, 1984; Sakarovitch, 1984). However, in most real situations it is impossible to know in advance the exact duration of each activity. In the PERT method for example, it is supposed that these duration times are random variables with known beta distributions. Using some additional simplifying assumptions fairly good results are obtained. For example, it is possible to compute the probability that the project will be finished before some given time limit. Another less classical method based on fuzzy set theory has also been proposed (Dubois, Prade, 1978; Geidel, 1989).

Here we study the case where the uncertainty and imprecision about the duration of the activities is modelled by hints. Suppose that there are n activities $a_1, \ldots, a_n$ in the project and let $t_i, i = 1, \ldots, n$ denote the starting time of activity a_i and d_i its duration. In example 12.7 there are five activities and four temporal precedence conditions which can be written as

$$\begin{aligned} t_2 - t_4 &\geq d_4 \\ t_3 - t_2 &\geq d_2 \\ t_3 - t_1 &\geq d_1 \\ t_5 - t_1 &\geq d_1. \end{aligned} \tag{12.86}$$

In order to better represent the project, it is convenient to introduce two fictitious activities a_0 and a_{n+1} corresponding to the beginning and the ending of the project respectively. Of course $d_0 = 0$ and it is assumed that the project starts at time zero, i.e. $t_0 = 0$. Then the additional temporal constraints

$$\begin{aligned} t_4 - t_0 &\geq d_0 \\ t_1 - t_0 &\geq d_0 \\ t_6 - t_3 &\geq d_3 \\ t_6 - t_5 &\geq d_5. \end{aligned} \tag{12.87}$$

must be added because activities a_1 and a_4 can start immediately and the termination of activities a_3 and a_5 is a necessary and sufficient condition for the whole project to be completed. All these temporal constraints can be represented by a graph $G = (V, E)$ called a MPM network which is constructed as follows: to each activity a_i a node $i \in V$ corresponds and there is an arc $e = (i,j) \in E$ if the activity a_i must precede the activity a_j. The MPM network for example 12.7 is given in Fig. 12.15.

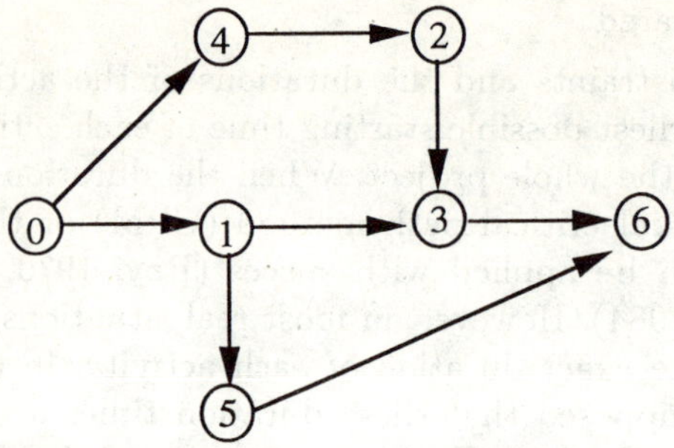

Fig. 12.15. A MPM network

The variables $t_0, \ldots, t_{n+1}, d_0, \ldots, d_n$ induce a multivariate family of compatible frames and each temporal constraint $t_j - t_i \geq d_i$ can of course be represented by a deterministic hint α_{ij} on the frame corresponding to the variables t_i, t_j, d_i. So for each arc $(i,j) \in E$ there is a deterministic hint α_{ij} representing the temporal constraint. Furthermore, there is also the deterministic hint $\mathcal{S}$ asserting that $t_0 = 0$. Unfortunately this model does not fit into the general model discussed in Sect. 12.2 because the stuctural information (12.86) and (12.87) is made up of inequalities instead of equalities of the form $d_{ik} = d_{ij} \otimes d_{kj}$

The uncertainty about the duration of the activities will also be described by hints. For each $i = 1, \ldots, n$ a hint $\mathcal{D}_i = (\Omega_i, P_i, \Gamma_i, \Theta_i)$ on the frame $\Theta_i = \{d_i\}$, representing our knowledge about the duration of the activity a_i, is assumed to be given. For example, the time required to prepare the field (activity a_1 in example 12.1) depends on two scenarios ω_1 and ω_2. In the first scenario ω_1 it is assumed that the field will be dry when its preparation starts. Then the duration of the preparation will be between 9 and 11 days. But in the scenario ω_2 it is assumed that the field will be wet and the duration will be between 10 and 15 days. These two scenarios form the set of possible interpretations Ω_1 of the hint $\mathcal{D}_1$ for the duration of the activity a_1. If the probability of ω_1 is known to be 0.7 and the probability of ω_2 is known to be 0.3, then the hint $\mathcal{D}_1$ is completely determined because $\Gamma(\omega_1) = [9, 11]$ and $\Gamma(\omega_2) = [10, 15]$. It is assumed that all these hints are independent and that their focal sets are intervals. These may be degenerate intervals consisting of only one point as for the fictitious activity a_0 where $\mathcal{D}_0$ is the deterministic hint focussing on $d_0 = 0$. These hints are assumed to be independent.

Integrating all available information means the combination of all the available hints according to Dempster's rule. So let

$$\mathcal{H} = \left(\oplus_{(i,j)\in E}\alpha_{ij}\right) \oplus \left(\oplus_{i=0}^{n}\mathcal{D}_i\right) \tag{12.88}$$

denote this combined hint. Now suppose that we are more particularly interested in the duration of the whole project, i.e. in the variable t_{n+1}. Then the hint $\mathcal{H}$ has to be marginalized to the coarser frame $\{t_{n+1}\}$. This hint is called the solution of the project scheduling problem and it will be denoted by $\mathcal{S}$.

Let $\mathcal{D}_i = (\Omega_i, P_i, \Gamma_i, \mathcal{R})$ denote the hint about the duration d_i of the activity $a_i, i = 0, 1, \ldots, n$ and let $\Gamma_i(\omega_i) = [l_i(\omega_i), L_i(\omega_i)]$. Since the hints α_{ij} are deterministic with only one interpretation, only the combined interpretations $\omega = (\omega_0, \omega_1, \ldots, \omega_n) \in \Omega = \prod_{i=0}^{n} \Omega_i$ have to be considered when computing $\mathcal{H}$. If $\Gamma(\omega)$ denotes the focal set associated with the interpretation ω and $d = (d_0, d_1, \ldots, d_n)$ and $t = (t_0, \ldots, t_{n+1})$, then

$$\Gamma(\omega) = \{(d,t) : t_0 = 0, t_j - t_i \geq d_i \ \forall\ (i,j) \in E, l_i(\omega_i) \leq d_i \leq L_i(\omega_i)\}.$$

Define

$$C(\omega) = \{d : l_i(\omega_i) \leq d_i \leq L_i(\omega_i)\} \tag{12.89}$$

and

$$B(d) = \{t : t_0 = 0, t_j - t_i \geq d_i \ \forall\ (i,j) \in E\}. \tag{12.90}$$

Then obviously

$$\Gamma(\omega) = \cup_{d\in C(\omega)}(\{d\} \times B(d)) \tag{12.91}$$

and since the projection of a union of sets is the same as the union of the projections of the sets, it follows that

$$\Gamma(\omega)|\{t_{n+1}\} = \cup_{d\in C(\omega)}(B(d)|\{t_{n+1}\}). \tag{12.92}$$

If A is the node to arc incidence matrix of the graph G, then

$$B(d) = \{t : t_0 = 0, -A^T t \geq d\}, \tag{12.93}$$

where A^T denotes the transpose of the matrix A. This shows that $B(d)$ is the set of feasible solutions of a particular linear program called minimal tension problem, namely

$$\begin{aligned} t_0 &= 0 \\ -A^T t &\geq d \qquad (P) \\ t_{n+1} &\to \text{Min} \end{aligned} \tag{12.94}$$

Then the minimal tension theorem (Sakarovitch, 1984; compare theorem 12.23) asserts in particular that the linear program (P) has a feasible solution if and only if G does not contain a circuit of positive length (with arc lengths d_i). But by construction G does not contain any circuit and hence $B(d)$ is not empty for all $d \in C(\omega)$. This shows that the set of interpretations of $\mathcal{H}$ is Ω.

Since all constraints defining $\Gamma(\omega)$ are linear, the focal set $\Gamma(\omega)$ is a convex polytope in the corresponding frame $\mathcal{R}^k$. This implies that the projection $\Gamma(\omega)|\{t_{n+1}\}$ is an interval which will be denoted by $[l(\omega), L(\omega)]$. These intervals are of course the focal sets of the hint $\mathcal{S}$, the solution of the scheduling problem. Since $B(d)$ is described by linear inequalities, the projection $B(d)|\{t_{n+1}\}$ is clearly an interval of the form $[x(d), +\infty]$ for all $d \in C(\omega)$. Using (12.92) the above discussionx shows that $L(\omega) = +\infty$ for all $\omega \in \Omega$ and it remains to compute the lower bound $l(\omega)$. But the minimal tension theorem also asserts that the minimal value of t_{n+1} in the linear program (P), namely $x(d)$, is given by the length of the longest path from 0 to $n+1$ in G with arc lengths d_i. Since the length of such a longest path cannot increase if the arc lengths are decreased, it follows from (12.92) that $l(\omega)$ is the length of the longest path from 0 to $n+1$ in G with arc lengths $l_i(\omega_i)$.

So, each focal set of $\mathcal{S}$ is of the form $[l(\omega), +\infty]$ where $l(\omega)$ is the length of a longest path in G. Since the lengths of the arcs are all non-negative and the graph contains no circuit, efficient longest path algorithms can be used to compute $l(\omega)$ (see e.g. Sakarovitch, 1984). Once the solution $\mathcal{S}$ has been computed, degrees of support of hypotheses about the duration of the project t_{n+1} can be computed. For example, we can compute the degree of support that the project will last more than a certain number of days.

An important aspect in the approach developed above is that the scenarios (combined interpretations ω) which contribute to the support of a particular hypothesis are readily available. So the results obtained can be explained, which is usually considered as a fundamental aspect in artificial intelligence.

The series and parallel reduction procedures presented in Sect. 12.3 cannot be applied directly here because the project scheduling problems considered in this section do not fit into the general model of Sect. 12.2. However, similar types of series and parallel reductions can certainly be defined. Note that such reductions are also possible when the uncertainty about the duration times is modeled by random variables (see Golenko, 1972).

A12 Appendix: Proofs of Theorems in Chapter 12

Proof of Theorem 12.1

(1). Let d be a structural vector. Select an arbitrary vertex $i' \in V$ and define an application $\pi : V \to \Theta$ by $\pi(i) = d_{i'i}$. Then the relation $d_{i'j} = d_{i'i} \otimes d_{ij}$ and the commutativity of $\otimes$ imply that

$$\pi(j) \otimes \pi(i)^{-1} = d_{i'j} \otimes d_{i'i}^{-1} = d_{ij}. \tag{A12.1}$$

Conversely, let π be a mapping such that $d_{ij} = \pi(j) \otimes \pi(i)^{-1}$ for all $(i,j) \in V^2$. Then

$$\begin{aligned} d_{ij} \otimes d_{jk} &= \pi(j) \otimes \pi(i)^{-1} \otimes \pi(k) \otimes \pi(j)^{-1} \\ &= \pi(k) \otimes \pi(i)^{-1} = d_{ik} \end{aligned} \tag{A12.2}$$

since $\otimes$ is commutative.

(2). First the existence is proved. Then an application $\pi : V \to \Theta$ is constructed as follows. Let $T' = (V, U')$ be the undirected version of the tree

$T = (V, U)$ and let $\sigma = (i_1, \ldots, i_n)$ be a tree construction sequence in nodes of T' (see Chap. 8 for this notion). Start by defining $\pi(i_1) = e$. Then assume that $\pi(i_l)$ is defined for all $l < j$ and we have to define $\pi(i_j)$. Since σ is a tree construction sequence there is $1 \leq k < j$ such that $\{i_k, i_j\} \in U'$. To simplify the notation, let $i_k = r$ and $i_j = s$. Then $\pi(r)$ is already defined and define $\pi(s)$ as follows:

$$\pi(s) = \begin{cases} \pi(r) \otimes d'_{rs} & \text{if } (r,s) \in U \\ \pi(r) \otimes (d'_{rs})^{-1} & \text{if } (s,r) \in U. \end{cases} \tag{A12.3}$$

In this way $\pi(i_l)$ is defined for all l. Now the vector d given by $d_{ij} = \pi(j) \otimes \pi(i)^{-1}$ is a structural vector with $d_{ij} = d'_{ij}$ for all $(i,j) \in U$. In order to prove uniqueness, let $d^{(1)}$ and $d^{(2)}$ be two vectors satisfying these conditions. For $(i,j) \in U^c$ it must be verified that $d^{(1)}_{ij} = d^{(2)}_{ij}$. There is exactly one chain $\mathcal{C}$ from i to j in U. If $\mathcal{C}_1$ and $\mathcal{C}_2$ denote the arcs of $\mathcal{C}$ which are "oriented from i to j" and "oriented from j to i" respectively, then

$$d^{(k)}_{ij} = \otimes\{d^{(k)}_{ij} : (i,j) \in \mathcal{C}_1\} \otimes \{(d^{(k)}_{ij})^{-1} : (i,j) \in \mathcal{C}_2\} \tag{A12.4}$$

for $k = 1, 2$. Since $d^{(1)}_{ij} = d^{(2)}_{ij}$ for all $(i,j) \in U \supseteq \mathcal{C}$ we have $d^{(1)}_{ij} = d^{(2)}_{ij}$ and the theorem is proved. □

Proof of Theorem 12.2

(1). Let's consider a permutation of (i,j,k), say (j,i,k). If we left-multiply the two members of (12.11) by d_{ji} we get $d_{ji} \otimes d_{ik} = d_{jk}$ and hence (12.11) is also satisfied for the permutation (j,i,k). The proofs for the other four permutations are completely analogous.

(2). Without loss of generality, let's suppose that (12.11) holds for the three first triples. By the associativity of $\otimes$ it follows that $d_{hk} = d_{hi} \otimes d_{ik} = d_{hi} \otimes (d_{ij} \otimes d_{jk}) = (d_{hi} \otimes d_{ij}) \otimes d_{jk} = d_{hj} \otimes d_{jk}$ and hence (12.11) also holds for (h,j,k). □

Consider an arbitrary multivariate family of compatible frames and let $\mathcal{H}_i = (\Omega_i, P_i, \Gamma_i, \Theta_I)$, $i = 1, \ldots, m$ be independent hints on the frame Θ_I and $\mathcal{I}_j = (\mathcal{O}_j, Q_j, \Upsilon_j, \Theta_J)$, $j = 1, \ldots, n$ be independent hints on the frame Θ_J with $J \subseteq I$. Let (Ω, P) denote the product probability space obtained from the hints $\mathcal{H}_i$ and let $(\mathcal{O}, Q)$ denote the product probability space obtained from the hints $\mathcal{I}_j$. Furthermore, for $\omega \in \Omega$ and $\eta \in \mathcal{O}$, define

$$\begin{aligned} \Gamma(\omega) &= \cap \{\Gamma_i(\omega_i) : i = 1, \ldots, m\} \\ \Upsilon(\eta) &= \cap \{\Upsilon(\omega_j) : j = 1, \ldots, n\}. \end{aligned} \tag{A12.5}$$

If $\mathcal{H} = \oplus \{\mathcal{H}_i : i = 1, \ldots, m\}$ and $\mathcal{I} = \oplus \{\mathcal{I}_j : j = 1, \ldots, n\}$, then the following theorem and its corollary will be useful.

Theorem A12.1 *If there is a surjective mapping $\varphi : \Omega \to \mathcal{O}$ such that*

$$\begin{aligned}&(1)\ Q(\eta) = P(\varphi^{-1}(\eta))\ \forall \eta \in \mathcal{O}\\&(2)\ \Upsilon(\eta) = \Gamma(\omega)|J\ \forall\ \eta \in \mathcal{O}, \forall\ \omega \in \varphi^{-1}(\eta),\end{aligned} \tag{A12.6}$$

then $\mathcal{H}$ exists if and only if $\mathcal{I}$ exists and in this case $\mathcal{H}|J = \mathcal{I}$.

Proof of theorem A12.1

Let

$$\begin{aligned}K_1 &= \{\omega : \Gamma(\omega) = \emptyset\} = \{\omega : \Gamma(\omega)|J = \emptyset\}\\K_2 &= \{\eta : \Upsilon(\eta) = \emptyset\}.\end{aligned} \tag{A12.7}$$

If $\omega \in K_1$, then (2) implies that $\Upsilon(\varphi(\omega)) = \Gamma(\omega)|J = \emptyset$ and hence $\varphi(\omega) \in K_2$, which proves that $K_1 \subseteq \varphi^{-1}(K_2)$. Conversely, let $\omega \in \varphi^{-1}(K_2)$. Then there is $\eta \in K_2$ such that $\varphi(\omega) = \eta$. But (2) implies that $\Gamma(\omega)|J = \Upsilon(\eta) = \emptyset$ and hence $\Gamma(\omega) = \emptyset$ and $\omega \in K_1$, which proves that $\varphi^{-1}(K_2) \subseteq K_1$ and finally $K_1 = \varphi^{-1}(K_2)$. Therefore, $\mathcal{H}$ exists if and only if $\mathcal{I}$ exists because $P(K_1) = P(\varphi^{-1}(K_2)) = Q(K_2)$ by (1).

Now suppose that $\mathcal{H}$ and $\mathcal{I}$ exist. Then define for $H \subseteq \Theta_J$ the sets

$$\begin{aligned}u_1(H) &= \{\omega : \emptyset \neq \Gamma(\omega)|J \subseteq H\}\\u_2(H) &= \{\eta : \emptyset \neq \Upsilon(\eta) \subseteq H\}.\end{aligned} \tag{A12.8}$$

It can be proved in the same way that $u_1(H) = \varphi^{-1}(u_2(H))$.

On the other hand, the relation (1) implies that $Q(H) = P(\varphi^{-1}(H))$ for all $H \subseteq \mathcal{O}$. Therefore, if sp_1 and sp_2 denote the support functions of $\mathcal{H}|J$ and $\mathcal{I}$ respectively, then

$$sp_1(H) = \frac{P(u_1(H))}{1-P(K_1)} = \frac{P(\varphi^{-1}(u_2(H)))}{1-P(\varphi^{-1}(K_2))} = \frac{Q(u_2(H))}{1-Q(K_2)} = sp_2(H) \tag{A12.9}$$

and hence $\mathcal{H}|J = \mathcal{I}$ in the sense of the equivalence relation given in Chap. 3. □

Corollary A12.1 *Let K be an arbitrary subset of J. If*

$$\begin{aligned}&(1)\ Q(\eta) = P(\varphi^{-1}(\eta))\ \forall \eta \in \mathcal{O}\\&(2)\ \Upsilon(\eta) = \Gamma(\omega)|J\ \forall\ \eta \in \mathcal{O}, \forall\ \omega \in \varphi^{-1}(\eta),\end{aligned} \tag{A12.10}$$

then $\mathcal{H}|K$ exists if and only if $\mathcal{I}|K$ exists and in this case they are equal.

Proof of Corollary A12.1

The existence conditions follow from theorem A12.1. On the other hand, since $(I, K) \dashv J$, theorem 8.12 implies that $\mathcal{I}|K = (\mathcal{H}|J)|K = \mathcal{H}|K$. □

Proof of Theorem 12.3

We also denote by $\mathcal{T}$ the deterministic hint focussing on the set T of all structural vectors with respect to V. The hint $\mathcal{T}'$ is defined similarly with respect to V'. If

$$\mathcal{L} = \oplus\{\mathcal{H}_u : u \in E - \{(i,j),(j,k)\}\}, \tag{A12.11}$$

then the hint

$$\mathcal{H}(\mathcal{M}) = \mathcal{H}_{ij} \oplus \mathcal{H}_{jk} \oplus \mathcal{L} \oplus \mathcal{T} \tag{A12.12}$$

is carried by $I = V^2$ and the hint

$$\mathcal{H}(\mathcal{M}') = (\mathcal{H}_{ij} \otimes \mathcal{H}_{jk}) \oplus \mathcal{L} \oplus \mathcal{T}' \tag{A12.13}$$

is carried by $J = (V')^2 \subseteq I$. Furthermore, if (Ω, P) and $(\mathcal{O}, Q)$ denote the product probability spaces obtained from the probability spaces $(\Omega_i, P_i), i = 1, \ldots, 4$ and $(\mathcal{O}_j, Q_j), j = 1,2,3$ of the hints on the right-hand side of (A12.9) and (A12.10) respectively, then $\mathcal{O}_1 = \Omega_1 \times \Omega_2, \mathcal{O}_2 = \Omega_3$ and we can take $\mathcal{O}_4 = \Omega_4$. Then the bijective mapping $\varphi : \Omega \to \mathcal{O}$ defined by $\varphi(\omega_1, \omega_2, \omega_3, \omega_4) = ((\omega_1, \omega_2), \omega_3, \omega_4)$ satisfies $Q = P \circ \varphi^{-1}$. Furthermore, if Γ and Υ are defined by analogy to (A12.5), we are going to prove that $\Gamma | J = \Upsilon \circ \varphi$.

In fact, let $d' \in \Gamma(\omega)|J$. Then there is $d \in \Gamma(\omega)$ such that $d' = d|J$. It follows from $d \in \Gamma(\omega)$ that $d_{ij} \in \Gamma_1(\omega_1), d_{jk} \in \Gamma_2(\omega_2)$ and $d_{ik} = d_{ij} \otimes d_{jk}$ which implies that $d'_{ik} = d_{ik} \in \Gamma_1 \otimes \Gamma_2(\omega_1, \omega_2)$ and hence $d' \in \Upsilon(\varphi(\omega))$. Conversely, let $d' \in \Upsilon(\varphi(\omega))$. We have to find a vector $d \in \Gamma(\omega)$ such that $d|J = d'$. According to the first part of theorem 12.1, there is a potential $\pi' : V' \to \Theta$ such that $d'_{hl} = \pi'(l) \otimes \pi'(h)^{-1}$ for all $(h,l) \in (V')^2$. Since $d'_{ik} \in \Gamma_1(\omega_1) \otimes \Gamma_2(\omega_2)$, there is $\theta_1 \in \Gamma_1(\omega_1)$ and $\theta_2 \in \Gamma_2(\omega_2)$ such that $d'_{ik} = \theta_1 \otimes \theta_2$. It follows that the vector $d \in \mathcal{V}$ associated with the potential $\pi : V \to \Theta$ given by

$$\pi(h) = \begin{cases} \pi'(h) & \text{if } h \in V' \\ \pi'(i) \otimes \theta_1 & \text{if } h = j \end{cases} \tag{A12.14}$$

is in $\Gamma(\omega)$ and $d|J = d'$. In fact, it is sufficient to verify that $d_{ij} \in \Gamma_1(\omega_1)$ and $d_{jk} \in \Gamma_2(\omega_2)$. But

$$d_{ij} = \pi(j) \otimes \pi(i)^{-1} = \pi'(i) \otimes \theta_1 \otimes \pi'(i)^{-1} = \theta_1 \tag{A12.15}$$

and

$$\begin{aligned} d_{jk} &= \pi(k) \otimes \pi(j)^{-1} = \pi'(k) \otimes \theta_1^{-1} \otimes \pi'(i)^{-1} \\ &= \theta_1^{-1} \otimes d'_{ik} = \theta_2. \end{aligned} \tag{A12.16}$$

Therefore there is a bijective mapping φ satisfying the conditions of corollary A12.1. Then this corollary implies theorem 12.3 because $K = \{(r,s)\} \subseteq J$ since $j \notin \{r,s\}$. □

Proof of Theorem 12.4

The associativity of Demptster's rule implies that $\mathcal{H}(\mathcal{M}) = \mathcal{H}(\mathcal{M}')$ and hence the equality (12.21) holds. □

Proof of Theorem 12.5

Since G is a series-parallel graph, there is a s-p-reducible variable d_{rs}.

(1) Series and parallel reductions clearly do not change the number of weakly connected components of G. Since a single arc between two nodes represents a single connected component, this is also true for the initial graph G.

(2) Let d_{ij} be a s-p-reducible variable. It is sufficient to verify that the nodes in V_2 corresponding to the relations in $T(d_{rs})$ and $T(d_{ij})$ are identical because these trees are bipartite and each relational node is linked to the three variable nodes which are involved in the relation. But this is the case because the relations used in the series reductions are the same, the only difference is the sequence in which they are added to the tree.

(3) If d_{ij} is s-p-reducible, then $d_{ij} \in T(d_{ij}) = T$. On the other hand, if $d_{ij} \in T$, then $d_{ij} \in T(d_{ij})$ and hence the series and parallel reductions implicitly given in the tree $T(d_{ij})$ show that d_{ij} is s-p-reducible.

(4) By definition of series and parallel reductions every variable d_{ij} with $(i,j) \in E$ is in T. Then assertion (3) implies that all these variables are s-p-reducible. □

Proof of Theorem 12.6

We apply corollary A12.1. The hint $\mathcal{H}(\mathcal{M})$ is carried by $I = V^2$ and $\mathcal{H}(\mathcal{M}_i)$ is carried by $J = (V_i)^2 \subseteq I$. Furthermore, if

$$\mathcal{T}_R = \oplus\{\mathcal{H}(i,j,k) : (i,j,k) \in V^3 - V_i^3\} \tag{A12.17}$$

then

$$\mathcal{H}(\mathcal{M}) = \oplus\{\mathcal{H}_a : a \in E\} \oplus \mathcal{T}(V_i) \oplus \mathcal{T}_R \tag{A12.18}$$

and

$$\mathcal{H}(\mathcal{M}_i) = \oplus\{\mathcal{H}_a : a \in E_i\} \oplus \mathcal{T}(V_i). \tag{A12.19}$$

If (Ω, p) and $(\mathcal{O}, q)$ denote the product probability spaces obtained from those in the hints on the right-hand side of (A12.18) and (A12.19) respectively, then the projection $\varphi : \Omega \to \mathcal{O}$ is surjective. Furthermore, condition (1) of corollary A12.1 is satisfied because $(\mathcal{O}, Q)$ is product probability space whose factors are also factors of (Ω, P).

The condition (2) is also satisfied. Indeed, let $\eta \in \mathcal{O}$ and $\omega \in \varphi^{-1}(\eta)$. Let $d' \in \Gamma(\omega)|J$. Then there is $d \in \Gamma(\omega)$ such that $d|J = d'$. But if $d \in \Gamma(\omega)$, then its restriction $d' = d|J$ to V_i^2 is clearly in $\Upsilon(\eta)$ since $\varphi(\omega) = \eta$. Conversely, let $d' \in \Upsilon(\eta)$ and let d'' be an arbitrary vector in $\Gamma(\omega)$ with $\omega \in \varphi^{-1}(\eta)$. Let π' and π'' denote potentials associated with d' and d'' respectively. Then the tension d corresponding to the potential π on V given by

$$\pi(h) = \begin{cases} \pi'(h) & \text{if } h \in V_i \\ \pi''(h) & \text{if } h \in V - V_i \end{cases} \tag{A12.20}$$

is in $\Gamma(\omega)$ because all structural constraints are satisfied (theorem 12.1 (1)) and d coincides with d'' on $(V - V_i)^2$ and d coincides with d' on V_i^2 and

$\varphi(\omega) = \eta$, i.e. the focal sets associated with ω and η are the same on the arcs of E_i. It follows that $d' = d|J \in \Gamma(\omega)|J$ and the hypotheses of corollary A12.1 are satisfied because $K = \{(r,s)\} \subseteq J$. Note that the non-connectivity is implicitly used in the verification that d is in $\Gamma(\omega)$. □

Proof of Theorem 12.7

If there are other components than G_i and G_j, then add a new arc $e = (r,s)$ in E and let $\mathcal{H}_{rs} = \mathcal{E}$. This modification does of course not change the solution $\mathcal{H}(\mathcal{M})$ and theorem 12.6 shows that the components which are different from G_i and G_j can be ignored when computing $\mathcal{H}(\mathcal{M})|(r,s)$. Hence we suppose that G has exactly two connected components G_i and G_j containing r and s respectively.

For any sample of focal sets $(\Theta_{hl} : (h,l) \in E)$ from the hints $\mathcal{H}_{hl}$ corresponding to an interpretation of $\mathcal{H}(\mathcal{M})$, it must be proved that the projection Λ_{rs} of the non-empty set

$$\Lambda = \{d \in \mathcal{T}(V) : d_{hl} \in \Theta_{hl} \ \forall\, (h,l) \in E\} \tag{A12.21}$$

onto (r,s) equals Θ. Since $\Lambda_{rs} \subseteq \Theta$, it is sufficient to verify that $\Theta \subseteq \Lambda_{rs}$. Because Λ is not empty, there is a vector $d \in \Lambda$. If d' and d'' denote the projections of d on V_i^2 and V_j^2 respectively, then of course $d' \in \mathcal{T}(V_i)$ and $d'' \in \mathcal{T}(V_j)$. The proof of assertion (2) in theorem 12.1 shows that there are potentials π' and $\tilde{\pi}$ associated with d' and d'' respectively and such that $\pi'(r) = \tilde{\pi}(s) = e$.

For any $\theta \in \Theta$ we will prove that $\theta \in \Lambda_{rs}$. The tension associated with the potential $\pi''(h) = \tilde{\pi}(h) \otimes \theta$ equals d'' because if $(u,v) \in V_j^2$ we have

$$\begin{aligned} \pi''(v) \otimes \pi''(u)^{-1} &= \tilde{\pi}(v) \otimes \theta \otimes (\tilde{\pi}(u) \otimes \theta)^{-1} \\ &= \tilde{\pi}(v) \otimes \theta \otimes \theta^{-1} \otimes \tilde{\pi}(u)^{-1} \\ &= \tilde{\pi}(v) \otimes \tilde{\pi}(u)^{-1} \\ &= d''_{uv}. \end{aligned} \tag{A12.22}$$

Then the non-connectivity implies that the potential π on V given by

$$\pi(h) = \begin{cases} \pi'(h) & \text{if } h \in V_i \\ \pi''(h) & \text{if } h \in V_j \end{cases} \tag{A12.23}$$

is well defined because $V_i \cap V_j = \emptyset$. Furthermore, the tension f associated with the potential π is in Λ because if $(u,v) \in V_i^2$ then

$$\begin{aligned} f_{uv} &= \pi(v) \otimes \pi(u)^{-1} = \pi'(v) \otimes \pi'(u)^{-1} \\ &= d'_{uv} = d_{uv} \in \Theta_{uv} \end{aligned} \tag{A12.24}$$

because $d \in \Lambda$ and if $(u,v) \in V_j^2$ then

$$\begin{aligned} f_{uv} &= \pi(v) \otimes \pi(u)^{-1} = \pi''(v) \otimes \pi''(u)^{-1} \\ &= d''_{uv} = d_{uv} \in \Theta_{uv} \end{aligned} \tag{A12.25}$$

because $d \in \Lambda$. Hence

$$\begin{aligned} f_{rs} &= \pi(s) \otimes \pi(r)^{-1} = \pi''(s) \otimes \pi'(r)^{-1} \\ &= \tilde{\pi}(s) \otimes \theta \otimes \pi'(r)^{-1} = \theta \end{aligned} \qquad (A12.26)$$

because $\tilde{\pi}(s) = \pi'(r) = e$. This shows that $\theta = f_{rs} \in \Lambda_{rs}$ and the theorem is proved. □

Proof of Theorem 12.10

The proof of this theorem is exactly the same as the one of theorem 12.6. □

In order to prove theorem 12.9 some preliminary considerations and a lemma are needed. As in the proof of theorem 12.7, suppose without loss of generality that G_i and G_j are the only two subgraphs and that $r \in G_i$ and $s \in G_j$. If Ω_{hl} denotes the set of interpreations of $\mathcal{H}_{hl}$ for all $(h,l) \in E$, then Dempster's rule implies that

$$\mathcal{H}(\mathcal{M}) = (\Omega - K, P, \Gamma, \Theta^{n \times n}). \qquad (A12.27)$$

where $\Omega = \prod\{\Omega_{hl} : (h,l) \in E\}$ and K is the set of contradictory interpretations of Ω. Similarly,if $n_h = |V_h|$ for $h \in \{i,j\}$,then

$$\mathcal{H}(\mathcal{M}_h) = (\Omega_h - K_h, P_h, \Gamma_h, \Theta^{n_h \times n_n}) \qquad (A12.28)$$

where Ω_h and K_h are defined by analogy with Ω and K. If $\varphi_h(\omega) = \omega_h$ is the projection of $\omega \in \Omega$ on Ω_h, then we have the following lemma.

Lemma A12.1 *For all $\omega \in \Omega$ we have*

$$\Gamma(\omega)|(r,s) = \Gamma_i(\omega_i)|(r,k) \otimes \Gamma_j(\omega_j)|(k,s). \qquad (A12.29)$$

Proof of Lemma A12.1

Let $\theta \in \Gamma(\omega)|(r,s)$. Then there is $d \in \mathcal{T}$ such that $d_{rs} = \theta$. If d' and d'' are the projections of d to V_i^2 and V_j^2 respectively, then $d' \in \Gamma_i(\omega_i)$ and $d'' \in \Gamma_j(\omega_j)$. But

$$\theta = d_{rs} = d_{rk} \otimes d_{ks} = d'_{rk} \otimes d''_{ks} \qquad (A12.30)$$

and hence $\theta \in \Gamma_i(\omega_i)|(r,k) \otimes \Gamma_j(\omega_j)|(k,s)$.

Conversely, let $\theta \in \Gamma_i(\omega_i)|(r,k) \otimes \Gamma_j(\omega_j)|(k,s)$. Then there is $d' \in \Gamma_i(\omega_i)$ and $d'' \in \Gamma_j(\omega_j)$ such that $d'_{rk} \otimes d''_{ks} = \theta$. We can choose potentials π' and π'' associated with d' and d'' such that $\pi'(k) = \pi''(k)$ (= e for example). Then the potential π on V given by (A12.20) is well defined (i.e. free of contradictions) and its corresponding tension d is in $\Gamma(\omega)$ because d coincides with $d' \in \Gamma_i(\omega_i)$ on V_i^2 and d coincides with $d'' \in \Gamma_j(\omega_j)$ on V_j^2 and because $\varphi_h(\omega) = \omega_h$ for $h \in \{i,j\}$. But

$$\begin{aligned}\theta = d'_{rk} \otimes d''_{ks} &= (\pi(k) \otimes \pi(r)^{-1}) \otimes (\pi(s) \otimes \pi(k)^{-1}) \\ &= \pi(s) \otimes \pi(r)^{-1} = d_{rs}\end{aligned} \tag{A12.31}$$

and therefore $\theta \in \Gamma(\omega)|(r,s)$. □

Proof of Theorem 12.9

First we prove that $\Omega - K = (\Omega_i - K_i) \times (\Omega_j - K_j)$. Indeed, let $\omega = (\omega_i, \omega_j) \in \Omega = \Omega_i \times \Omega_j$. Then we have $\omega \in \Omega - K$ if and only if $\Gamma(\omega)|(r,s) \neq \emptyset$ which holds by lemma A12.1 if and only if $\Gamma_i(\omega_i)|(r,k) \neq \emptyset$ and $\Gamma_j(\omega_j)|(k,s) \neq \emptyset$ which in turn is true if and only if $\omega_i \in \Omega_i - K_i$ and $\omega_j \in \Omega_j - K_j$. According to the very definition of $\otimes$ for hints (see (12.16)) it follows that the hints on the left-hand and right-hand side of (12.22) have the same sets of interpretations. This implies in particular that the existence of one entails the existence of the other.

Again by lemma A12.1 it remains to prove that the interpretations have the same probability, i.e.

$$\frac{P(\omega)}{1 - P(K)} = \frac{P_i(\omega_i)}{1 - P_i(K_i)} \cdot \frac{P_j(\omega_j)}{1 - P_j(K_j)} \tag{A12.32}$$

where P, P_i and P_j are the product measures associated with the hints $\mathcal{H}_a$ for $a \in E$, E_i and E_j respectively. Since $E = E_i + E_j$ implies $P(\omega) = P_i(\omega_i)P_j(\omega_j)$, $\Omega - K = (\Omega_i - K_i) \times (\Omega_j - K_j)$ implies $1 - P(K) = (1 - P_i(K_i))(1 - P_j(K_j))$ and hence (A12.32) is satisfied which proves the theorem. □

In order to avoid unnecessary repetitions, if h is an element of $\{1,2\}$, then the hints $\mathcal{H}(\mathcal{M})$ and $\mathcal{H}(\mathcal{M}_h)$ are defined as in (A12.27) and (A12.28) respectively. Similarly, if $\varphi_h(\omega) = \omega_h$ denotes the projection of $\omega \in \Omega$ on Ω_h, then we have the following lemma.

Lemma A12.2 *For all* $\omega \in \Omega$,

$$\Gamma(\omega)|(r,s) = \Gamma_1(\omega_1)|(r,s) \cap \Gamma_2(\omega_2)|(r,s). \tag{A12.33}$$

Proof of Lemma A12.2

Let $\theta \in \Gamma(\omega)|(r,s)$. Then there is $d \in \Gamma(\omega)$ such that $d_{rs} = \theta$. But the projections d' and d'' of d to V_1^2 and V_2^2 are in $\Gamma_1(\omega_1)$ and $\Gamma_2(\omega_2)$ respectively. Since $V_1 \cap V_2 = \{r,s\}$, it follows that

$$\theta = d_{rs} = d'_{rs} = d''_{rs} \tag{A12.34}$$

is in $\Gamma_1(\omega_1)|(r,s) \cap \Gamma_2(\omega_2)|(r,s)$. Conversely, let $\theta \in \Gamma_1(\omega_1)|(r,s) \cap \Gamma_2(\omega_2)|(r,s)$. Then there is $d' \in \Gamma_1(\omega_1)$ and $d'' \in \Gamma_2(\omega_2)$ such that $d'_{rs} = d''_{rs} = \theta$. On the other hand, let π' and π'' be potentials associated with d' and d'' respectively and such that $\pi'(r) = \pi''(r)$. Since $d'_{rs} = d''_{rs}$, we must necessarily have $\pi'(s) = \pi''(s)$. But this implies that the potential π on V given by

$$\pi(h) = \begin{cases} \pi'(h) & \text{if } h \in V_1 \\ \pi''(h) & \text{if } h \in V_2 \end{cases} \tag{A12.35}$$

is well defined because $V_1 \cap V_2 = \{r, s\}$. But then the tension d associated with π is in $\Gamma(\omega)$ because d coincides with d' on V_1^2 and d coincides with d'' on V_2^2. Furthermore

$$d_{rs} = \pi(s) \otimes \pi(r)^{-1} = d'_{rs} = \theta \tag{A12.36}$$

and therefore $\theta \in \Gamma(\omega)|(r, s)$, which proves the lemma. □

Proof of Theorem 12.10

If we proceed as in the proof of theorem 12.9, it can easily be seen that lemma A12.2 implies that

$$\Omega - K = (\Omega_1 - K_1) \times (\Omega_2 - K_2). \tag{A12.37}$$

This implies that the set of interpretations on the right-hand side of (12.24) equals $(\Omega_1 - K_1) \times (\Omega_2 - K_2)$ and hence the two sides of (12.24) have the same set of interpretations. Again by lemma A12.2 it remains to prove that these interpretations have the same probability, i.e. equality (A12.32) holds if we replace i by 1 and j by 2. But (A12.37) implies that this equality is satisfied and the theorem is proved. □

Proof of Theorem 12.11

Let $d \in \Gamma(\omega)$. We prove by recurrence on i that $d_{jk} \in \Gamma'_{jk}(i, \omega)$ for all $(j, k) \in V^2$ and for all $i = 0, 1 \ldots, n$. The assertion is trivially satisfied for $i = 0$. On the other hand, if it is true for $i - 1$, then $d_{jk} \in \Gamma'_{jk}(i-1, \omega)$, $d_{ji} \in \Gamma'_{ji}(i-1, \omega)$ and $d_{ik} \in \Gamma'_{ik}(i-1, \omega)$. Since

$$d_{jk} = d_{ji} \otimes d_{ik} \in \Gamma'_{ji}(i-1, \omega) \otimes \Gamma'_{ik}(i-1, \omega), \tag{A12.38}$$

it follows that $d_{jk} \in \Gamma'_{jk}(i, \omega)$. Hence $d_{jk} \in \Gamma'_{jk}(\omega)$ for all $(j, k) \in V^2$ and therefore $d \in \Gamma'(\omega)$. □

Proof of Theorem 12.12

If $C_{rs}(k, \omega)$ denotes the set of compatible values with respect to the paths from r to s whose intermediate nodes are in the set $\{1, \ldots, k\}$, then we prove by recurrence on k that $\Gamma'_{rs}(k, \omega) \subseteq C_{rs}(k, \omega)$. The assertion is trivially satisfied for $k = 0$. To show the step from $(k - 1)$ to k let $\theta \in \Gamma'_{rs}(k, \omega)$ and let γ be a path between r and s whose intermediate nodes are smaller or equal to k. If γ does not pass by k, then the induction hypothesis implies that θ is compatible with γ.

Suppose that γ goes through k only once. The case where it goes more than once is completely similar. Then the path γ can be decomposed into two subpaths γ_1 and γ_2 which do not pass by k and which link r to k and k to s respectively. Since by the Floyd-type algorithms there is $\theta' \in \Gamma'_{rk}(k-1,\omega)$ and $\theta'' \in \Gamma'_{ks}(k-1,\omega)$ such that $\theta = \theta' \otimes \theta''$, then the induction hypothesis implies that θ is compatible with γ. □

Proof of Theorem 12.16

Let $x \in (P_1 + P_2)^*$ and select $x_1 \in P_1$ and $x_2 \in P_2$ such that $x = x_1 + x_2$. We prove that $x_1 \in P_1^*$ et $x_2 \in P_2^*$. If this were not true, then $x_1 \notin P_1^*$ or $x_2 \notin P_2^*$. Suppose that at least $x_1 \notin P_1^*$. Thus there is $y_1 \neq y_2$ in P_1 such that $x_1 = (y_1 + y_2)/2$. Then

$$\begin{aligned} x = x_1 + x_2 &= (y_1 + y_2)/2 + (x_2 + x_2)/2 \\ &= (y_1 + x_2)/2 + (y_2 + x_2)/2 \end{aligned} \tag{A12.39}$$

and hence x is not a vertex because $y_i + x_2 \in P_1 + P_2$ for $i = 1, 2$. But this is a contradiction. The case that $x_2 \notin P_2^*$ is treated similarly. □

Proof of Theorem 12.17

Since the convex hull procedure preserves inclusion, theorem 12.16 implies that

$$\begin{aligned} P_1 + P_2 = conv\ [(P_1 + P_2)^*] &\subseteq conv\ (P_1^* + P_2^*) \\ &\subseteq conv\ (P_1 + P_2) = P_1 + P_2 \end{aligned} \tag{A12.40}$$

because $P_1 + P_2$ is a convex polygon and $conv(P^*) = P$ for any convex polygon. □

To prove theorem 12.18 the following lemma is needed. Given an orientation γ of a cycle μ in G, let μ^+ denote the arcs in μ which are oriented like γ and let μ^- denote those which are not oriented like γ.

Lemma A12.3 *A necessary condition for the existence of a compatible tension is that*

$$\sum_{i \in \mu^-} l_i - \sum_{i \in \mu^+} k_i \geq 0 \tag{A12.41}$$

for every cycle μ and each orientation γ of μ.

Proof of Lemma A12.3

Since cycles and tensions are orthogonal spaces (Bondy, Murty, 1976), if a compatible tension θ exists, then for every cycle μ and each orientation γ we have

$$\begin{aligned} 0 = \langle [\mu], \theta \rangle &= \langle [\mu]^+, \theta \rangle - \langle [\mu]^-, \theta \rangle \\ &\geq \sum_{i \in \mu^+} k_i - \sum_{i \in \mu^-} l_i, \end{aligned} \tag{A12.42}$$

where $[\mu]$ is the representative vector of the cycle μ (i.e.

$$[\mu]_i = \begin{cases} 1 & \text{if } i \in \mu^+ \\ -1 & \text{if } i \in \mu^- \\ 0 & \text{otherwise} \end{cases} \tag{A12.43}$$

and $[\mu]^+, [\mu]^-$ are defined by

$$[\mu]_i^+ = \begin{cases} 1 & \text{if } i \in \mu^+ \\ 0 & \text{otherwise,} \end{cases} \tag{A12.44}$$

$$[\mu]_i^- = \begin{cases} 1 & \text{if } i \in \mu^- \\ 0 & \text{otherwise.} \end{cases} \tag{A12.45}$$

□

Proof of Theorem 12.18

If the labelling procedure labels b, then there is a chain of nodes $\omega = [a, a_1, a_2, \ldots, b]$ in G such that each vertex has been labelled from the preceding one. So, if $i \in \omega^+$ (i.e. the arc i is traversed in the sequence ω), then $\theta_i \leq k_i$ and similarly if $i \in \omega^-$ then $\theta_i \geq l_i$. Then $\mu = [a, a_1, a_2, \ldots, b, a]$ is a cycle and since $\theta_1 < k_1$ it follows that

$$0 = \langle \theta, [\mu] \rangle = \sum_{i \in \mu^+} \theta_i - \sum_{i \in \mu^-} \theta_i < \sum_{i \in \mu^+} k_i - \sum_{i \in \mu^-} l_i. \tag{A12.46}$$

Then by lemma A12.3 it follows that there is no compatible tension. □

Proof of Theorem 12.19

The necessity is given by lemma A12.3 and if (12.59) is satisfied, then case 2 in the algorithm of Herz never applies because this would lead to a violation of (12.59): see the proof of lemma A12.3. This shows that a compatible tension exists. □

Proof of Corollary 12.1

This is a direct consequence of the compatible tension theorem by taking $l_i = +\infty$ for all i. □

Proof of Theorem 12.20

To prove this theorem the following result is needed. For any compatible tension θ and for any cycle μ such that $1 \in \mu^+$, since cycles and tensions are orthogonal (Bondy, Murty, 1976), we have

$$0 = \langle \theta, [\mu] \rangle \geq \theta_1 + \sum_{i \in \mu^+, i \neq 1} k_i - \sum_{i \in \mu^-} l_i, \tag{A12.47}$$

or, equivalently,

$$\theta_1 \leq \sum_{i \in \mu^-} l_i - \sum_{i \in \mu^+, i \neq 1} k_i. \tag{A12.48}$$

Since node b can be labelled there is a chain ω from a to b in G such that $\theta_i = k_i$ for all $i \in \omega^+$ and $\theta_i = l_i$ for all $i \in \omega^-$. This implies that for the cycle $\mu = \omega + (b,a)$ we have

$$\theta_1 = \sum_{i\in\mu^-} \theta_i - \sum_{i\in\mu^+, i\neq 1} \theta_i = \sum_{i\in\mu^-} l_i - \sum_{i\in\mu^+, i\neq 1} k_i. \qquad (A12.49)$$

Hence by (A12.48) θ_1 is indeed maximal. □

Proof of Theorem 12.21

Let B denote the right-hand side of (12.61). Then (A12.48) implies that $\theta_1^* \leq B$. Now suppose that θ is a compatible tension such that $\theta_1 < B$. We must prove that θ_1 is not maximal. Indeed, if $\theta_1 < B$, then case 2 in the labelling algorithm cannot occur because otherwise there would be a cycle μ such that $1 \in \mu^+$ and $\theta_1 = \sum_{i\in\mu^-} l_i - \sum_{i\in\mu^+, i\neq 1} k_i$ (see (A12.49)), which is a contradiction to $\theta_1 < B$. Hence case 1 applies and the new tension θ' satisfies $\theta_1' > \theta_1$ which proves that θ_1 is not maximal. □

Proof of Theorem 12.24

(1). This is a sufficient condition because $\Lambda_{rs}(\omega) \subseteq \bar{R}_{rs}(\omega)$ according to theorem 12.11.

The condition is also necessary. Indeed, if ω is contradictory, then there is $h \in \{1,2\}$ such that $\Lambda^{(h)}(\omega) = \emptyset$. But this implies that $\Psi_h(\omega) = \emptyset$ and therefore the linear program (P) has no feasible solution.

On the other hand, let $\bar{R}_{jk}(\omega)$ denote the right-hand side of (12.75) with L and U replaced by $\bar{L}$ and $\bar{U}$ respectively. Since $-[a,b] = [-b,-a]$, the application of the Floyd-type algorithm to the model $\bar{\mathcal{M}}$ means in particular that the genuine Floyd algorithm to the network R for the computation of the upper bounds $\bar{U}_{rs}^{(h)}(\omega)$ must be applied. But the first part of theorem 12.22 implies that R contains at least one circuit γ of negative length and hence $\bar{U}_{ii}^{(h)}(\omega) < 0$ for all nodes on this circuit. But $\bar{L}_{ii}^{(h)}(0,\omega) = 0$ and the iteration step of the Floyd-type algorithm implies that $\bar{L}_{ii}^{(h)}(\omega) \geq 0$ for all $i \in V$. Since the upper bound is smaller than the lower bound, it follows that the h-side of $\bar{R}_{ii}(\omega)$ is empty and hence $\bar{R}_{ii}(\omega)$ is also empty for all $i \in \gamma$.

(2). It must be proved that

$$[\bar{L}_{rs}^{(h)}(\omega), \bar{U}_{rs}^{(h)}(\omega)] = [l^{(h)}(\omega), u^{(h)}(\omega)] \qquad (A12.50)$$

for $h = 1,2$. But it has been seen in the proof of (1) that $\bar{U}_{rs}^{(h)}(\omega)$ is the result of the application of the genuine Floyd algorithm to the network R. On the other hand, since ω is not contradictory, it follows that the linear program (P) has a feasible solution. Hence $u^{(h}(\omega)$ equals the shortest distance from r to s in R according to the second part of theorem 12.22. Since the genuine Floyd algorithm computes shortest distances, it follows indeed that $\bar{U}_{rs}^{(h)}(\omega) = u^{(h)}(\omega)$. Using theorem 12.23, we can easily prove, in a similar way, that $\bar{L}_{rs}^{(h)}(\omega) = l^{(h)}(\omega)$. □

Part III

A General Theory of Hints

13. The General Model of a Hint

13.1 Families of Implied and Possible Hypotheses

Up to now, with the exception of certain examples, hints were always assumed to allow only a set of possible interpretations. In this third part of the book, this assumption will now be dropped. Whereas for many practical applications the assumption of a finite number of possible interpretations may be sufficient, part II has already shown that there are problems where a more general notion of a hint is needed. Moreover, it will be shown in this part of the book that the generalization of the notion of a hint leads to a rich mathematical theory which has some interest of its own.

So far only few attempts to develop a general theory of evidence have been reported in the literature. One of them is by Shafer (1979). In this theory support (or belief) functions are introduced axiomatically as Choquet capacities which are monotone of order ∞, i.e. the result of theorem 3.5 is essentially taken as the starting point. Using a result by Choquet an allocation of probability for support functions can be derived (see Sect. 13.3). This concept provides a probabilistic interpretation of belief functions. However, with this theory the description of Dempster's rule of combination of belief functions is not clear at all. This is the big problem with this approach.

A straightforward generalization of the notion of hints, as introduced in part II of this book, is provided by random sets, as proposed by Nguyen (1978) and Hestir et al. (1991). This model allows in particular for a clear and straightforward definition of Dempster's rule and also for an immediate generalization of the notions of degrees of support and plausibility, although problems of measurability arise. However, it has been shown that the theory based on random sets is not general enough (Kohlas, 1993). It generates only a special class of support functions, but not the whole class of belief functions as considered by Shafer (1979). Something more general than random sets is needed. This something are random filters as will be explained in this chapter.

The meaning of this notion will be explained in this section. Chapters 13 to 15 are devoted to the study of a theory of hints based on this notion. Random sets appear in this theory as an important special case, which as such will however not be pursued specially in this book, except in the form of closed random intervals (Chap. 16). This latter case is of course of particular importance for practical applications and deserves thus particular attention.

The very notion of a filter points to the question of the relation between hints and the topology on the frames of discernment of the hints. The study of this interesting and important question remains largely to be done. Part III of this book is thus only a short introduction to a subject on which most of the work still has to be carried out.

Just as in Chap. 2, the starting point is a precise question, whose answer is unknown, but for which a hint is available. Let Θ be an arbitrary set (finite or not) whose elements θ represent the possible answers to a question. One of the elements of Θ is assumed to represent the true, but unknown answer. The set Θ is called the *frame of discernment* and Ω denotes as usual the *set of possible interpretations* of the information contained in the hint to be represented. Ω will now be arbitrary, not necessarily a finite set. One of the elements $\omega \in \Omega$ must be the correct interpretation, but it is unknown which one. Again, not all possible interpretations are equally likely. Thus a probability space $(\Omega, \mathcal{A}, P)$ with a σ-algebra $\mathcal{A}$ and a known probability measure P on $\mathcal{A}$ is introduced. Each possible interpretation ω somehow restricts the possible answers within Θ. In the simplest case, one can assume as in part I that if ω is the correct interpretation, then the correct answer θ must be within some subset $\Gamma(\omega)$ of Θ, the *focal set* of the interpretation. However, a slightly more general concept proves now to be appropriate.

Assuming that ω is the correct interpretation, one can ask which are the subsets of Θ (considered as propositions) which are *implied* by this interpretation, i.e. which must be necessarily true. Let $\mathcal{S}$ denote the family of these subsets. Then, any superset of an implied subset is also implied and if two subsets are implied their intersection is surely also implied. In other terms, $\mathcal{S}$ must be a *filter*:

(1) $H \in \mathcal{S}$ and $H \subseteq H'$ imply $H' \in \mathcal{S}$
(2) $H_1 \in \mathcal{S}$, $H_2 \in \mathcal{S}$ imply $H_1 \cap H_2 \in \mathcal{S}$.
(3) Θ always belongs to $\mathcal{S}$, $\emptyset$ never belongs to $\mathcal{S}$.

In a dual manner one can also consider the family $\mathcal{P}$ of subsets of Θ which are *possible* under the interpretation ω: a subset $H \subseteq \Theta$ is possible if and only if its complement H^c is not implied, $H^c \notin \mathcal{S}$. It follows easily that

(1') $H \in \mathcal{P}$ and $H \subseteq H'$ imply $H' \in \mathcal{P}$
(2') $H_1 \in \mathcal{P}$, $H_2 \in \mathcal{P}$ imply $H_1 \cup H_2 \in \mathcal{P}$.
(3') Θ always belongs to $\mathcal{P}$, $\emptyset$ never belongs to $\mathcal{P}$.

Furthermore, if $H \in \mathcal{S}$, then $H^c \notin \mathcal{S}$ and thus $\mathcal{S} \subseteq \mathcal{P}$. A pair of such dual families $\mathcal{R} = (\mathcal{S}, \mathcal{P})$ will be called a *restriction*. The idea is, that each interpretation ω imposes a restriction on the hypotheses which must be necessarily true or which are possible. To each restriction $\mathcal{R}$ one assigns two zero-one-valued set functions

$$sp_{\mathcal{R}}(H) = \begin{cases} 1 & \text{if } H \in \mathcal{S} \\ 0 & \text{otherwise} \end{cases}$$

and

$$pl_{\mathcal{R}}(H) = \begin{cases} 1 & \text{if } H \in \mathcal{P} \\ 0 & \text{otherwise.} \end{cases}$$

A restriction $\mathcal{R}$ is called *vacuous*, if $\mathcal{S}$ contains only Θ (in which case $\mathcal{P}$ contains all subsets of Θ except the empty set). A vacuous restriction does not at all restrict the possible answers. It is used to represent the situation that under some interpretations a hint contains no information about the question considered.

The set $R = \cap\{H : H \in \mathcal{S}\}$ is called the *base* of the restriction $\mathcal{R}$. It is easy to see that R equals the set of all possible points $\theta \in \Theta$, i.e. those for which $\{\theta\} \in \mathcal{P}$. One might wonder whether $\mathcal{S}$ should not be closed under arbitrary intersections and thus $R \in \mathcal{S}$. This will not be assumed here because it would restrict the theory too much (as will be seen in theorem 13.5). However, a restriction $\mathcal{R}$ with $R \in \mathcal{S}$ will be called *set-based* because in this case $\mathcal{S} = \{H \subseteq \Theta : R \subseteq H\}$ and $\mathcal{P} = \{H \subseteq \Theta : R \cap H \neq \emptyset\}$. In the literature, a filter which is closed under arbitrary intersections is called a *principal filter.* So the filter associated with a set-based restriction is principal. This shows that the case of interpretations with focal sets is of course contained in the general theory developed here. For a set-based restriction we write $\mathcal{R} = R$. Similarly, if (2) and (2') hold for countable interesections or unions respectively, the restriction will be called a *σ-restriction.* A hint will now be defined as in Chap. 2, but using the notion of a restriction instead of a focal set. It will be assumed that every possible interpretation $\omega \in \Omega$ has a restriction $\Gamma(\omega)$ assigned to it, called its *focal restriction*, describing its implied and possible propositions. Γ is a mapping from Ω into the set $\mathcal{R}(\Theta)$ of restrictions on Θ. A *hint* $\mathcal{H}$ is thus finally a quintupel $\mathcal{H} = (\Omega, \mathcal{A}, P, \Gamma, \Theta)$ of elements as described above.

A hint $\mathcal{H} = (\Omega, \mathcal{A}, P, \Gamma, \Theta)$ is called *set-focussed* when its restrictions $\Gamma(\omega)$ are set-based for all $\omega \in \Omega$. The bases of $\Gamma(\omega)$ are then called *focal sets.* If Θ is a finite set, then all restrictions and thus all hints are set-focussed. But even in the general case many important classes of hints are set-focussed. Indeed, if $\Gamma(\omega)$ equals either some fixed subset $F \subseteq \Theta$ or Θ itself for all ω, then the hint is called *simple.* If $\Gamma(\omega)$ equals the vacuous restriction for all $\omega \in \Omega$, then the hint is called *vacuous*; it represents full ignorance about the question at hand. If $\Gamma(\omega)$ equals the same restriction $\mathcal{R}$ for all ω, then the hint is called *deterministic.* If $\mathcal{H}$ is a set-focussed hint whose focal sets $\Gamma(\omega)$ all contain only one single point $\theta(\omega) \in \Theta$, then the hint is called *precise.* A precise hint corresponds essentially to a random variable (under reserve of the appropriate measurability condition, see Chap. 16.).

Restrictions are fundamental to the theory. In many respects they behave like ordinary subsets of Θ. Thus, as a preparation for the subsequent discussion, some elementary operations with restrictions are defined:

(1) *Intersection, conjunction; union, disjunction:* If $\mathcal{R}_1$ and $\mathcal{R}_2$ are two restrictions known to hold on Θ, then their conjunction forms a new restriction $\mathcal{R} = \mathcal{R}_1 \cap \mathcal{R}_2$ defined by $\mathcal{S} = \{H_1 \cap H_2 : H_1 \in \mathcal{S}_1, H_2 \in \mathcal{S}_2\}$. It is easily verified, that $\mathcal{S}$ is a filter unless $\emptyset \in \mathcal{S}$. In the latter case $\mathcal{R}_1$ and $\mathcal{R}_2$ are considered

to be contradictory and $\mathcal{R}_1 \cap \mathcal{R}_2$ does not exist. If $\mathcal{R}_1 = R_1$ and $\mathcal{R}_2 = R_2$, then $\mathcal{R}_1 \cap \mathcal{R}_2 = R_1 \cap R_2$ and the base of $\mathcal{R}_1 \cap \mathcal{R}_2$ is always $R_1 \cap R_2$. In a similar way $\mathcal{R} = \mathcal{R}_1 \cup \mathcal{R}_2$ is defined by $\mathcal{S} = \{H_1 \cup H_2 : H_1 \in \mathcal{S}_1, H_2 \in \mathcal{S}_2\}$. If $\mathcal{R}_1 = R_1$ and $\mathcal{R}_2 = R_2$, then $\mathcal{R}_1 \cup \mathcal{R}_2 = R_1 \cup R_2$ and the base of $\mathcal{R}_1 \cup \mathcal{R}_2$ is always $R_1 \cup R_2$. Both intersection and union of restrictions may be defined in this way for arbitrary families of restrictions (not only finite ones).

(2) *Extension of restrictions.* If $\mathcal{R}$ is a restriction on a set Θ_1, then it may become necessary to extend it to a product space $\Theta_1 \times \Theta_2$. As $\mathcal{R}$ contains no information or restriction relative to the new dimension Θ_2, this extension must be the *cylindrical extension* of all implied sets. The restriction $\mathcal{R} \times \Theta_2$ on $\Theta_1 \times \Theta_2$ is defined by the filter $\mathcal{S} \times \Theta_2 = \{H \subseteq \Theta_1 \times \Theta_2 :$ there is an $A \in \mathcal{S}$ such that $A \times \Theta_2 \subseteq H\}$. If $H \in \mathcal{S} \times \Theta_2$, then denote by $pr_1(H)$ the projection of H to Θ_1. The family of possible sets of the restriction $\mathcal{R} \times \Theta_2$ is then given by the following theorem.

Theorem 13.1 *The family of possible sets of the extended restriction* $\mathcal{R} \times \Theta_2$ *is* $\mathcal{P} \times \Theta_2 = \{H \subseteq \Theta_1 \times \Theta_2 : pr_1(H) \in \mathcal{P}\}$.

If $\mathcal{R}_1$ and $\mathcal{R}_2$ are two restrictions on Θ_1 and Θ_2 respectively, then the restriction $(\mathcal{R}_1 \times \Theta_2) \cap (\Theta_1 \times \mathcal{R}_2)$ can be formed. The following theorem gives the implied and possible sets of this restriction.

Theorem 13.2 *Let* $\mathcal{R} = (\mathcal{R}_1 \times \Theta_2) \cap (\Theta_1 \times \mathcal{R}_2)$ *be a restriction on* $\Theta_1 \times \Theta_2$. *Then its implied sets are those in the filter* $\mathcal{S} = \{H \subseteq \Theta_1 \times \Theta_2 :$ there is $H_1 \in \mathcal{S}_1, H_2 \in \mathcal{S}_2$ such that $H_1 \times H_2 \subseteq H\}$ *and its possible sets are those in the set* $\mathcal{P} = \{H \subseteq \Theta_1 \times \Theta_2 : (H_1 \times H_2) \cap H \neq \emptyset$ for all $H_1 \in \mathcal{S}_1, H_2 \in \mathcal{S}_2\}$.

In view of this theorem we denote the restriction $(\mathcal{R}_1 \times \Theta_2) \cap (\Theta_1 \times \mathcal{R}_2)$ by $\mathcal{R}_1 \times \mathcal{R}_2$ and their families of implied and possible subsets by $\mathcal{S}_1 \times \mathcal{S}_2$ and $\mathcal{P}_1 \times \mathcal{P}_2$.

13.2 Support and Plausibility Functions

As in part I, hints will be used to judge hypotheses about the true, but unknown answer to the question considered. A hypothesis is again simply a subset H of Θ representing some proposition. To what extent does a hint $\mathcal{H} = (\Omega, \mathcal{A}, P, \Gamma, \Theta)$ support a hypothesis, to what extent does it contradict it? To what degree does a hypothesis appear possible or plausible in the light of a hint? The answer to these questions is given in the same spirit as in Chap. 3. Let $\mathcal{S}(\omega)$ and $\mathcal{P}(\omega)$ denote the families of implied and possible subsets of the restriction $\Gamma(\omega)$. In order to study the questions posed above, the set

$$u(H) = \{\omega \in \Omega : H \in \mathcal{S}(\omega)\} \tag{13.1}$$

of all interpretations which imply or support H will be considered (compare with (3.1)). If an interpretation $\omega \in u(H)$ proves to be the correct one, then the hypothesis is assured. Furthermore, the set

$$v(H) = \{\omega \in \Omega : H \in \mathcal{P}(\omega)\} \tag{13.2}$$

of all interpretations under which H is possible will also be considered (compare with (3.4)). If the correct interpretation belongs to $v(H)$, then the hypothesis is not excluded, is *possible*, even if not necessarily assured. The following theorem lists some simple properties of the mappings u and v from the power set $\mathcal{P}(\Theta)$ into the power set $\mathcal{P}(\Omega)$. This theorem generalizes theorems 3.1 and 3.3 of Chap. 3.

Theorem 13.3

(1) $u(\emptyset) = v(\emptyset) = \emptyset$
(2) $u(\Theta) = v(\Theta) = \Omega$
(3) $u(H) = v(H^c)^c$
(4) $v(H) = u(H^c)^c$
(5) $u(\cap\{H_i : i \in \mathcal{C}\}) = \cap\{u(H_i) : i \in \mathcal{C}\}$, *where* $\mathcal{C}$ *is finite in general, countable for hints with* σ*-restrictions* $\Gamma(\omega)$, $\omega \in \Omega$ *and arbitrary for set-focussed hints.*
(6) $u(\cup\{H_i : i \in \mathcal{C}\}) \supseteq \cup\{u(H_i) : i \in \mathcal{C}\}$ *for an arbitrary* $\mathcal{C}$.
(7) $v(\cup\{H_i : i \in \mathcal{C}\}) = \cup\{v(H_i) : i \in \mathcal{C}\}$, *where* $\mathcal{C}$ *is finite in general, countable for hints with* σ*-restrictions* $\Gamma(\omega)$, $\omega \in \Omega$ *and arbitrary for set-focussed hints.*
(8) $v(\cup\{H_i : i \in \mathcal{C}\}) \subseteq \cup\{v(H_i) : i \in \mathcal{C}\}$ *for an arbitrary* $\mathcal{C}$.
(9) $u(H') \subseteq u(H'')$ *if* $H' \subseteq H''$.
(10) $v(H') \subseteq v(H'')$ *if* $H' \subseteq H''$.

In view of (5) u is called a $\cap$ - homomorphism and in view of (7) v is called a $\cup$ - homomorphism.

If $u(H) \in \mathcal{A}$, then $P(u(H))$ is the probability that the hint $\mathcal{H}$ supports, or allows one to deduce or to prove the hypothesis H. Thus

$$sp(H) = P(u(H)) \tag{13.3}$$

is called the *degree of support* of the hypothesis H.

Similarly, if $v(H) \in \mathcal{A}$, then $P(v(H))$ is the probability that the hypothesis H is possible in the light of the hint $\mathcal{H}$. The value

$$pl(H) = P(v(H)) \tag{13.4}$$

will be called the *degree of plausibility* of H.

The functions sp and pl are mappings from some subset of the power set $\mathcal{P}(\Theta)$ to $[0, 1]$. These functions will be called *support* and *plausibility functions* respectively. Note that the functions $sp_{\mathcal{R}}$ and $pl_{\mathcal{R}}$ introduced above are the support and plausibility functions of the deterministic hint which focus on the restriction $\mathcal{R}$. Support and plausibility functions play a similar role relative to hints as that of distribution functions relative to random variables.

Note that there is a strong link between degree of support and degree of plausibility. In fact, according to theorem 13.3 (4) we have

$$pl(H) = P(v(H)) = P(u(H^c)^c) = 1 - sp(H^c) \tag{13.5}$$

whenever the corresponding probabilities are defined. Similarly, by theorem 13.3 (3)

$$sp(H) = P(u(H)) = P(v(H^c)^c) = 1 - pl(H^c). \tag{13.6}$$

Let $\mathcal{E}_s$ be the class of all subsets H of Θ for which $u(H) \in \mathcal{A}$, i.e. for which the degree of support is defined. The sets of $\mathcal{E}_s$ are called *s-measurable* and $\mathcal{E}_s$ is the domain of the set-function sp. Similarly let $\mathcal{E}_p$ be the class of all subsets $H \subseteq \Theta$ for which $v(H) \in \mathcal{A}$, i.e. for which the degree of plausibility is defined. The sets of $\mathcal{E}_p$ are called *p-measurable* and $\mathcal{E}_p$ is the domain of the set-function pl.

Theorem 13.4

(1) $\mathcal{E}_s$ is a multiplicative class (i.e. closed under finite intersections) or a σ-multiplicative class (closed under countable intersections) depending on whether the focuses $\Gamma(\omega)$ are restrictions or σ-restrictions for all $\omega \in \Omega$.

(2) $\mathcal{E}_p$ is an additive class (i.e. closed under finite unions) or a σ-additive class (closed under countable unions) depending on whether the focuses $\Gamma(\omega)$ are restrictions or σ-restrictions for all $\omega \in \Omega$.

(3) $\mathcal{E}_p = \{H \subseteq \Theta : H^c \in \mathcal{E}_s\}$, $\mathcal{E}_s = \{H \subseteq \Theta : H^c \in \mathcal{E}_p\}$ and $\emptyset, \Theta$ belong to both $\mathcal{E}_s$ and $\mathcal{E}_p$.

The families $\mathcal{E}_s$ and $\mathcal{E}_p$ are called *dual* classes of s- and p-measurable sets because by theorem 13.4 (3) $\mathcal{E}_p = \{H^c : H \in \mathcal{E}_s\}$ and $\mathcal{E}_s = \{H^c : H \in \mathcal{E}_p\}$. If Ω is a finite set, then *all* subsets of Θ are s- and p-measurable. In general however, $\mathcal{E}_s$ and $\mathcal{E}_p$ are strict subclasses of the power set of Θ. Note that if $\mathcal{E}_s$ and $\mathcal{E}_p$ are algebras, then they coincide. Let's illustrate theorem 13.4 by a simple example. If $(\Omega, \mathcal{A}, P)$ is a probability space and $B \subseteq \Omega$ a subset which does not belong to $\mathcal{A}$, $\Gamma(\omega) = F \subseteq \Theta$ for all $\omega \in B, \Gamma(\omega) = \Theta$ otherwise, then $\mathcal{E}_s$ contains all subsets of Θ, which do not contain F plus the set Θ. We have $u(H) = \emptyset$ for all $H \in \mathcal{E}_s, H \neq \Theta$ and thus $sp(H) = 0$, unless $H = \Theta$. $\mathcal{E}_p$ contains all subsets of Θ which are not contained in F^c plus $\emptyset$.

In particular, note that set-focussed hints have σ-restrictions and hence their s- and p-measurable classes $\mathcal{E}_s$ and $\mathcal{E}_p$ are σ-multiplicative and σ-additive respectively. It is possible to characterize support and plausibility functions. The following main theorem does this.

Theorem 13.5 *(Characterization of support and plausibility functions). Let $\mathcal{E}_s$ and $\mathcal{E}_p$ be dual multiplicative and additive classes of subsets of a set Θ.*

Then the set-functions $sp : \mathcal{E}_s \to [0,1]$ *and* $pl : \mathcal{E}_p \to [0,1]$ *are the support- and plausibility functions of a hint* $\mathcal{H} = (\Omega, \mathcal{A}, P, \Gamma, \Theta)$ *if and only if the following conditions hold:*

(1) $sp(\emptyset) = pl(\emptyset) = 0$ *and* $sp(\Theta) = pl(\Theta) = 1$.

(2) sp is monotone of order ∞, *i.e.*

$$sp(E) \geq \sum \{(-1)^{|I|+1} sp(\cap_{i \in I} E_i) : \emptyset \neq I \subseteq \{1, \ldots, n\}\} \tag{13.7}$$

for all $n \geq 1$ *and sets* $E, E_i \in \mathcal{E}_s$ *such that* $E \supseteq E_i$; *and pl is alternating of order* ∞, *i.e.*

$$pl(E) \leq \sum \{(-1)^{|I|+1} pl(\cup_{i \in I} E_i) : \emptyset \neq I \subseteq \{1, \ldots, n\}\} \tag{13.8}$$

for all $n \geq 1$ *and sets* $E, E_i \in \mathcal{E}_p$ *such that* $E \subseteq E_i$.

Furthermore, if $\mathcal{E}_s$ *and* $\mathcal{E}_p$ *are dual* σ*-multiplicative and* σ*-additive classes of subsets of a set* Θ, *then the set-functions* $sp : \mathcal{E}_s \to [0,1]$ *and* $pl : \mathcal{E}_p \to [0,1]$ *are the support- and plausibility functions of a hint* $\mathcal{H} = (\Omega, \mathcal{A}, P, \Gamma, \Theta)$ *with* σ*-restrictions as focusses* $\Gamma(\omega)$ *if and only if in addition to (1), (2) the following condition holds:*

(3) sp and pl are continuous, i.e. if $E_1 \supseteq E_2 \supseteq \ldots$ *is a monotone decreasing sequence of sets of* $\mathcal{E}_s$, *then*

$$sp(\cap_{i=1}^{\infty} E_i) = \lim_{i \to \infty} sp(E_i) \tag{13.9}$$

and if $E_1 \subseteq E_2 \subseteq \ldots$ *is a monotone increasing sequence of sets of* $\mathcal{E}_p$, *then*

$$pl(\cup_{i=1}^{\infty} E_i) = \lim_{i \to \infty} pl(E_i). \tag{13.10}$$

This theorem shows why the theory of hints must be based on random filters rather than random sets. In the latter case (13.7) and (13.8) hold of course. But for random sets $\mathcal{E}_s$ and $\mathcal{E}_p$ are σ-multiplicative and σ-additive respectively, such that their support and plausibility functions are continuous. Thus not all support and plausibility functions satisfying (13.7) and (13.8) can be obtained from random sets, and even not all functions satisfying (13.9) and (13.10) can be obtained from random sets, random σ-filters being sufficient for these conditions.

13.3 Allocation and Allowment of Probability

Consider a hint $\mathcal{H} = (\Omega, \mathcal{A}, P, \Gamma, \Theta)$. By the application $P \circ u$ from $\mathcal{E}_s$ to $\mathcal{A}$, a probability is allocated to each set $E \in \mathcal{E}_s$. Shafer (1979) proposes to replace the probability space $(\Omega, \mathcal{A}, P)$ by a probability algebra, i.e. a complete Boolean algebra with a positive and completely additive probability measure. This probability algebra can be obtained from $(\Omega, \mathcal{A}, P)$ as a quotient as shown below. This is a convenient technical means to study hints, support and plausibility function. For G. Shafer, however, it plays a much more important role because he uses probability allocations to define and understand the meaning of belief and belief functions.

Define two sets A_1 and A_2 in $\mathcal{A}$ to be equivalent ($A_1 \sim A_2$) if their symmetric difference $A_1 \Delta A_2 = (A_1 - A_2) \cup (A_2 - A_1)$ has probability zero. The relation $\sim$ is reflexive, symmetric and transitive. It induces thus a decomposition of $\mathcal{A}$ into a family $\mathcal{M}$ of equivalence classes. Let $\mathcal{B}_0 \subseteq \mathcal{M}$ be a family of classes and $\mathcal{B} \subseteq \mathcal{A}$ the family of all finite intersections of representatives of $\mathcal{B}_0$. Set $\beta = \inf_{B \in \mathcal{B}} P(B)$ and select a sequence $(B_n)_{n \geq 1}$ of elements of $\mathcal{B}$ such that $\lim_{n \to \infty} P(B_n) = \beta$. Then, if $B = \cap_{n=1}^{\infty} B_n$, it follows that $P(B) = \lim_{n \to \infty} P(\cap_{i=1}^{n} B_i) = \beta$ because $\beta \leq P(\cap_{i=1}^{n} B_i) \leq P(B_n)$. Furthermore, all the sets B obtained by the above construction are equivalent and define a class in $\mathcal{M}$ which is denoted by $\wedge \mathcal{B}_0$. In a similar way $\vee \mathcal{B}_0$ can be defined. If $\alpha \in \mathcal{M}$, then the family of all complements of representatives $A \in \alpha$ forms a class, denoted by α^c, the complement of α. With these operations $\mathcal{M}$ becomes a complete Boolean algebra, i.e. $\wedge \mathcal{B}_0$ and $\vee \mathcal{B}_0$ are defined and are in $\mathcal{M}$ for arbitrary classes $\mathcal{B}_0 \subseteq \mathcal{M}$. Note that by the above construction any class $\mathcal{B}_0$ contains a countable subclass $\mathcal{C}_0$ such that $\wedge \mathcal{B}_0 = \wedge \mathcal{C}_0$ or $\vee \mathcal{B}_0 = \vee \mathcal{C}_0$.

The null class (the class containing $\emptyset$ as representative) is denoted by Λ, the unity class (the class containing Ω) is denoted by $\mathcal{V}$. For $\alpha, \beta \in \mathcal{M}, \alpha \neq \beta$ we write $\alpha < \beta$ if the representatives of α are contained in the representatives of β (up to subsets of probability zero).

Define a probability measure μ on $\mathcal{M}$ by $\mu(\alpha) = P(A)$, if $A \in \alpha \in \mathcal{M}$. Then $\mu(\Lambda) = 0, \mu(\mathcal{V}) = 1, 0 < \mu(\alpha) < 1$ for all $\alpha \in \mathcal{M}, \alpha \neq \Lambda, \alpha \neq \mathcal{V}$; μ is a positive measure. Also $\mu(\alpha) < \mu(\beta)$ if $\alpha < \beta$. Now let $\mathcal{B}$ be a family of pairwise disjoint elements of $\mathcal{M}$, i.e. if $\beta', \beta'' \in \mathcal{B}$, then $\beta' \wedge \beta'' = \Lambda$. This means that we may also choose pairwise disjoint representatives of elements of $\mathcal{B}$. Thus there exists a sequence $(B_i)_{i \geq 1}$ of pairwise disjoint sets such that $B = \cup_{i=1}^{\infty} B_i$ is a representative of $\vee \mathcal{B}$. It follows that $\mu(\vee \mathcal{B}) = \sum_{i=1}^{\infty} \mu(\beta_i)$, if β_i is the class of B_i. Then $\mathcal{B}$ must be countable, because otherwise two different sequences of B_i could be combined to give another value for $\mu(\vee \mathcal{B})$ which, however, is uniquely defined. Hence it is possible to write $\mu(\vee \mathcal{B}) = \sum_{\beta \in \mathcal{B}} \mu(\beta)$ whenever $\mathcal{B}$ is a family of pairwise disjoint elements; μ is therefore completely additive. We refer to Halmos (1963) for more details about complete probability spaces.

Let π be the mapping from $\mathcal{A}$ to $\mathcal{M}$ which maps any set A of $\mathcal{A}$ to the class α in $\mathcal{M}$ containing A. Clearly $\pi(\cap_{i=1}^{\infty} A_i) = \wedge_{i=1}^{\infty} \pi(A_i)$ and π is a $\cap$-homomorphism which maintains intersections of countable families. In the same way π is also a $\cup$-homomorphism and π maintains also complements. Thus, $\rho = \pi \circ u$ is also a $\cap$-homomorphism from $\mathcal{E}_s$ to $\mathcal{M}$ and $\xi = \pi \circ v$ is a $\cup$-homomorphism from $\mathcal{E}_p$ to $\mathcal{M}$. Note that if $\mathcal{E}_s$ is σ-multiplicative and $\mathcal{E}_p$ is σ-additive, then countable intersections and unions are maintained, $\rho(\cap_{i=1}^{\infty} E_i) = \wedge_{i=1}^{\infty} \rho(E_i)$ and $\xi(\cup_{i=1}^{\infty} \xi(E_i) = \vee_{i=1}^{\infty} \xi(E_i)$. Furthermore $sp(E) = \mu(\rho(E))$ and $pl(E) = \mu(\xi(E))$. A $\cap$-homomorphism $\rho : \mathcal{E}_s \to \mathcal{M}$ is called an *allocation of probability* and a $\cup$-homomorphism $\xi : \mathcal{E}_p \to \mathcal{M}$ is called an *allowment of probability*. G. Shafer (1979) discusses in detail the use of these concepts to interpret belief and plausibility functions. Note that the duality relations

$$\xi(E) = \rho(E^c)^c; \ \rho(E) = \xi(E^c)^c \tag{13.11}$$

hold.

Let's collect here a few elementary results about the measure μ, the allocation of probability ρ and the allowment of probability ξ.

For any family $\mathcal{B}$ of elements of $\mathcal{M}$ clearly $\mu(\wedge\mathcal{B}) \leq \inf_{\alpha\in\mathcal{B}} \mu(\alpha)$ and $\mu(\vee\mathcal{B}) \geq \sup_{\alpha\in\mathcal{B}} \mu(\alpha)$. There are, however, important cases where there is equality. A family $\mathcal{B} \subseteq \mathcal{M}$ is called a *downward net*, if for any $\beta', \beta'' \in \mathcal{B}$ there is also a $\beta \in \mathcal{B}$ such that $\beta \leq \beta' \wedge \beta''$. An upward net is defined similarly: a family $\mathcal{B} \subseteq \mathcal{M}$ is called an *upward net*, if for any $\beta', \beta'' \in \mathcal{B}$ there is also a $\beta \in \mathcal{B}$ such that $\beta \geq \beta' \vee \beta''$.

Theorem 13.6

(i) If $\mathcal{B}$ is a downward net, then

$$\mu(\wedge\mathcal{B}) = \inf_{\alpha\in\mathcal{B}} \mu(\alpha). \tag{13.12}$$

(ii) If $\mathcal{B}$ is an upward net, then

$$\mu(\vee\mathcal{B}) = \sup_{\alpha\in\mathcal{B}} \mu(\alpha). \tag{13.13}$$

Furthermore, the inclusion-exclusion formulas from probability theory are of course also valid in $(\mathcal{M}, \mu)$.

Theorem 13.7 *For any collection $\alpha_i, i = 1, 2, \ldots, n, n \geq 1$ of elements of $\mathcal{M}$ we have*

$$\mu(\vee\{\alpha_i : i = 1, \ldots, n\}) = \sum\{(-1)^{|I|+1}\mu(\wedge_{i\in I}\alpha_i) : \emptyset \neq I \subseteq \{1, \ldots, n\}\} \tag{13.14}$$

$$\mu(\wedge\{\alpha_i : i = 1, \ldots, n\}) = \sum\{(-1)^{|I|+1}\mu(\vee_{i\in I}\alpha_i) : \emptyset \neq I \subseteq \{1, \ldots, n\}\}. \tag{13.15}$$

Allocations of probability and allowments of probability can be used in the same way as hints to generate monotone and alternating functions of order ∞ on $\mathcal{E}_s$ and $\mathcal{E}_p$.

Theorem 13.8

(i) If $\rho : \mathcal{E}_s \to \mathcal{M}$ is an allocation of probability with respect to a multiplicative class $\mathcal{E}_s$ of subsets of Θ and a probability algebra $\mathcal{M}$, then $\mu \circ \rho : \mathcal{E}_s \to [0,1]$ is monotone of order ∞ and $\mu(\rho(\emptyset)) = 0, \mu(\rho(\Theta)) = 1$.

(ii) If $\xi : \mathcal{E}_p \to \mathcal{M}$ *is an allowment of probability with respect to an additive class* $\mathcal{E}_p$ *of subsets of* Θ *and a probability algebra* $\mathcal{M}$*, then* $\mu \circ \xi : \mathcal{E}_p \to [0,1]$ *is alternating of order* ∞ *and* $\mu(\xi(\emptyset)) = 0, \mu(\xi(\Theta)) = 1$*.*

(iii) If ρ *is an allocation of probability relative to a* σ*-multiplicative class* $\mathcal{E}_s$ *and* ξ *is an allowment of probability relative to a* σ*-additive class* $\mathcal{E}_p$*, then* $\mu \circ \rho : \mathcal{E}_s \to [0,1]$ *and* $\mu \circ \xi : \mathcal{E}_p \to [0,1]$ *are continuous, i.e. if* $E_1 \supseteq E_2 \supseteq \cdots$, $E_i \in \mathcal{E}_s$*, then* $\mu(\rho(\cap_{i=1}^{\infty} E_i)) = lim_{i\to\infty}\mu(\rho(E_i))$ *and if* $E_1 \supseteq E_2, \ldots$, $E_i \in \mathcal{E}_p$*, then* $\mu(\xi(\cup_{i=1}^{\infty} E_i)) = lim_{i\to\infty}\mu(\xi(E_i))$*.*

13.4 Extensions of Support and Plausibility Functions

Does it make sense to define the degree of support for a hypothesis $H \subseteq \Theta$ outside the class $\mathcal{E}_s$ of s-measurable subsets? If $u(H) \subseteq \Omega$ is not measurable, then the model of the hint $\mathcal{H}$ does not contain the necessary information to determine the probability of the set of interpretations supporting H. But any measurable set of interpetations $A \subseteq \Omega$ which is contained in $u(H)$ is a support for H. Hence one may say that the unknown support for H must be at least $P(A)$, for any $A \subseteq u(H)$ and $A \in \mathcal{A}$. Thus, in the absence of further information, the support of H could be defined as

$$sp_e(H) = \sup\{P(A) : A \subseteq u(H), A \in \mathcal{A}\} = P_*(u(H)) \tag{13.16}$$

where P_* is the interior probability to P. This is an extension of the support function sp onto the whole power set $\mathcal{P}(\Theta)$ because of course the restriction of sp_e to $\mathcal{E}_s$ equals sp. We call sp_e the *vacuous extension* of sp to underline that no information not contained in the hint $(\Omega, \mathcal{A}, P, \Gamma, \Theta)$ has been added.

By duality, we may also extend the plausibility functions pl from $\mathcal{E}_p$ to $\mathcal{P}(\Theta)$:

$$pl_e(H) = 1 - sp_e(H^c). \tag{13.17}$$

This is similarly called the vacuous extension of pl. This name is justified by the following proposition:

Theorem 13.9 *If* P^* *is the exterior probability to* P*, then*

$$pl_e(H) = inf\{P(A) : A \supseteq v(H), A \in \mathcal{A}\} = P^*(v(H)). \tag{13.18}$$

This theorem generalizes in fact results on the connection between interior and exterior probability measures and support and plausibility functions found by Ruspini (1987) and Fagin, Halpern (1989), see also Shafer (1990).

Furthermore, it turns out that sp_e and pl_e satisfy also the conditions of theorem 13.5.

Theorem 13.10 *Let sp_e and pl_e be the extended support- and plausibility functions of a hint $\mathcal{H} = (\Omega, \mathcal{A}, P, \Gamma, \Theta)$. Then*

(1) sp_e and pl_e are monotone and alternating of order ∞ respectively on $\mathcal{P}(\Theta)$.

(2) If the hint $\mathcal{H}$ focuses on σ-restrictions, then sp_e and pl_e are also continuous.

There can be several distinct hints defining the same classes $\mathcal{E}_s$ and $\mathcal{E}_p$ of s- and p-measurable subsets and the same support and plausibility functions sp and pl as will be shown in the following chapter. Such hints are called *equivalent* (compare also with Sect. 3.3). Equivalent hints have, however, not necessarily the same *extended* support and plausibility functions ! In fact, let's take up the simple example introduced above. If $(\Omega, \mathcal{A}, P)$ is a probability space, B' and B'' two different, non-measurable subsets of Ω, then $\Gamma'(\omega) = F$ for $\omega \in B'$, $\Gamma'(\omega) = \Theta$ otherwise and $\Gamma''(\omega) = F$ for $\omega \in B''$, $\Gamma''(\omega) = \Theta$ otherwise, define two equivalent hints. Their respective extended support functions do clearly not coincide if B' and B'' have different interior probabilities. Furthermore, for both of these hints the hint constructed in the proof of theorem 13.5 has as set of interpretations the set of filters in the class of sets H not containing F. The whole probability mass can then be concentrated on the vacuous restriction Θ. It is a vacuous hint. This hint is again equivalent to the hints defined above. For this hint clearly $sp_e(H) = 0$ for all $H \neq \Theta$ and $pl_e(H) = 1$ for all $H \neq \emptyset$. Support and plausibility functions together with their associated hints have in fact a rich mathematical structure which is studied in the following chapter.

A13 Appendix: Proofs of Theorems in Chapter 13

Proof of Theorem 13.1

The fact that $H \in \mathcal{P} \times \Theta_2$ means that H^c contains no set $A \times \Theta_2$ with $A \in \mathcal{S}$. Hence H intersects with every set $A \times \Theta_2$ with $A \in \mathcal{S}$, and $pr_1(H)$ intersects with every set $A \in \mathcal{S}$. Thus $(pr_1(H))^c \notin \mathcal{S}$ and hence $pr_1(H) \in \mathcal{P}$. This argument also goes in the other direction. □

Proof of Theorem 13.2

If $H \in \mathcal{S}$ then $H = H_1 \cap H_2$, $H_1 \in \mathcal{S}_1 \times \Theta_2, H_2 \in \Theta_1 \times \mathcal{S}_2$. But then there are $A_1 \in \mathcal{S}_1, A_2 \in \mathcal{S}_2$ such that $A_1 \times \Theta_2 \subseteq H_1$ and $\Theta_1 \times A_2 \subseteq H_2$ and hence $A_1 \times A_2 \subseteq H$. Inversely, with $H_1 \in \mathcal{S}_1$ and $H_2 \in \mathcal{S}_2$, $H_1 \times H_2 = (H_1 \times \Theta_2) \cap (\Theta_1 \times H_2) \subseteq H$ implies $H \in \mathcal{S}$ because $H_1 \times \Theta_2 \in \mathcal{S}_1 \times \Theta_2$ and $\Theta_1 \times H_2 \in \Theta_1 \times \mathcal{S}_2$.

$H \in \mathcal{P}$ iff $H^c \notin \mathcal{S}$ iff $H_1 \times H_2 \not\subseteq H^c$ for all $H_1 \in \mathcal{S}_1, H_2 \in \mathcal{S}_2$. But this last condition is equivalent to $(H_1 \times H_2) \cap H \neq \emptyset$ for all $H_1 \in \mathcal{S}_1, H_2 \in \mathcal{S}_2$. □

Proof of Theorem 13.3

(1) and (2) are trivial.
(3) and (4): by definition $v(H)^c = \{\omega \in \Omega : H^c \in \mathcal{S}(\omega)\} = u(H^c)$ proves (4). (3) follows by applying (4) to H^c.
(5): If $\omega \in u(H_i)$ for all $i \in \mathcal{C}$, then $H_i \in \mathcal{S}(\omega)$, thus $\cap\{H_i : i \in \mathcal{C}\} \in \mathcal{S}(\omega)$ and therefore $\omega \in u(\cap\{H_i : i \in \mathcal{C}\})$. Inversely, $\omega \in u(\cap\{H_i : i \in \mathcal{C}\})$ implies $\cap\{H_i : i \in \mathcal{C}\} \in \mathcal{S}(\omega)$, hence $H_i \in \mathcal{S}(\omega)$ and $\omega \in u(H_i)$ for all $i \in \mathcal{C}$.
(6): If $\omega \in u(H_i)$ for some $i \in \mathcal{C}$, then $H_i \in \mathcal{S}(\omega)$, thus $\cup\{H_i : i \in \mathcal{C}\} \in \mathcal{S}(\omega)$ and $\omega \in u(\cup\{H_i : i \in \mathcal{C}\})$.
(7) and (8) are proved using (3),(4),(5) and (6) together with de Morgan laws.
(9) and (10) follow immediately from the definitions of u and v.

□

Proof of Theorem 13.4

Assertions (1) and (2) are direct consequences of theorem 13.3 (5) and (7) and the fact that $\mathcal{A}$ is a σ-algebra. (3): $H \in \mathcal{E}_s$ is equivalent to $u(H) \in \mathcal{A}$, which is equivalent to $v(H^c) \in \mathcal{A}$ (theorem 13.3 (3) and (4)) which finally is equivalent to $H^c \in \mathcal{E}_p$. $\emptyset, \Theta$ belong to $\mathcal{E}_s$ and $\mathcal{E}_p$ because of theorem 13.3 (1) and (2). □

Proof of Theorem 13.5

Let's first show that the conditions of the theorem are necessary. Thus let $\mathcal{H} = (\Omega, \mathcal{A}, P, \Gamma, \Theta)$ be a hint and *sp* and *pl* its support and plausibility functions. (1) follows from theorem 13.3 (1) and (2). In order to prove (2) for the support function, the well-known inclusion-exclusion formula of probability theory, together with theorem 13.3 (5), (6) and (9) is used:

$$\begin{aligned} sp(H) &= P(u(E)) \geq P(u(\cup\{E_i : i = 1, 2, \ldots, n\}) \\ &\geq P(\cup\{u(E_i) : i = 1, 2, \ldots, n\}) \\ &= \sum\{(-1)^{|I|+1} P(\cap_{i \in I} u(E_i)) : \emptyset \neq I \subseteq \{1, \ldots, n\}\} \\ &= \sum\{(-1)^{|I|+1} P(u(\cap_{i \in I} E_i)) : \emptyset \neq I \subseteq \{1, \ldots, n\}\} \\ &= \sum\{(-1)^{|I|+1} sp(\cap_{i \in I} E_i) : \emptyset \neq I \subseteq \{1, \ldots, n\}\}. \end{aligned}$$

Condition (2) for the plausibility function is proved in the same way or by using (2) for the support function together with (13.5).

Suppose that $\mathcal{H}$ has σ-restrictions as focuses $\Gamma(\omega)$. $E_1 \supseteq E_2 \supseteq \ldots$ implies $u(E_1) \supseteq u(E_2) \supseteq \ldots$ (theorem 13.3 (9)). By the continuity of probabilities and theorem 13.3 (5)

$$\begin{aligned} sp(\cap_{i=1}^{\infty} E_i) &= P(u(\cap_{i=1}^{\infty} E_i)) \\ &= P(\cap_{i=1}^{\infty} u(E_i)) \\ &= \lim_{i \to \infty} P(u(E_i)) = \lim_{i \to \infty} sp(E_i) \end{aligned} \tag{A13.1}$$

Equation (13.10) is obtained in a similar way. This proves the necessity of (3) for hints with σ-restrictions as focal restrictions.

Next sufficiency is proved. The proof of the sufficiency of (1) and (2) depends on an integral representation theorem of Choquet (1953, 1969), as has been noted by G. Shafer (1979). The construction will be outlined here for further reference. Let l be the space of all real-valued functions $f : \mathcal{E}_s \to \mathcal{R}$. l is a locally convex Hausdorff linear vector space with a topology induced by ordinary pointwise convergence of functions $f \in l$. Within l let $\mathcal{X}$ denote the set of all functions satisfying conditions (1) and (2) of the theorem. $\mathcal{X}$ is a convex and compact set. Its extreme points ext($\mathcal{X}$) are exactly the set of all functions satisfying (1) and (2) and taking only binary values zero or one (Choquet, 1953, p. 261). These functions are in one-to-one correspondence with filters in $\mathcal{E}_s$, which in turn correspond to restrictions on Θ (add all $H \subseteq \Theta$ which contain a set from the filter in $\mathcal{E}_s$ to the family $\mathcal{S}$ of implied propositions). The set ext($\mathcal{X}$) is compact and thus equal to its closure; closure(ext($\mathcal{X}$)) = ext($\mathcal{X}$).

According to Choquet (1953, p. 237 or 1969, vol. II, p. 117), for every $f \in \mathcal{X}$ there is a probability measure P on ext($\mathcal{X}$) such that

$$f(E) = \int_{ext(\mathcal{X})} h(E)dP(h) = P(\{h \in ext(\mathcal{X}) : h(E) = 1\}). \qquad (A13.2)$$

for every $E \in \mathcal{E}_s$.

Let now $\Omega = ext(\mathcal{X}) = \mathcal{R}(\mathcal{E}_s)$ be the set of all restrictions on $\mathcal{E}_s$ and define $\Gamma(\omega) = \omega$, where ω is considered as a restriction on Θ. Then $\mathcal{H} = (\Omega, \mathcal{A}, P, \Gamma, \Theta)$ is a hint whose support function on $\mathcal{E}_s$ equals $f \in \mathcal{X}$. This construction explains the use of restrictions rather than focal sets in the definition of hints.

To show the sufficiency of conditions (1) and (2) relative to alternating functions g use the duality relation to define $f(E) = 1 - g(E^c)$ which is monotone. The hint constructed with f as support function has then g as plausibility function.

If condition (3) is added, then the same construction is carried out relative to the convex subset $\mathcal{X}'$ of functions satisfying (1), (2) and (3). This time ext($\mathcal{X}'$) is in one-to-one relation with σ-filters on $\mathcal{E}_s$ and $\Gamma(\omega)$ will therefore be σ-restrictions. This completes the proof of the sufficiency of the conditions. □

Proof of Theorem 13.6

There is a countable subfamily $\mathcal{C}$ of $\mathcal{B}$ such that $\wedge\mathcal{B} = \wedge\mathcal{C}$. If $\gamma_1, \gamma_2, \ldots$ are the elements of $\mathcal{C}$, then define $\gamma' = \gamma_1$ and select elements $\gamma_i' \in \mathcal{B}$ such that $\gamma_2' \leq \gamma_1' \wedge \gamma_2$, $\gamma_3' \leq \gamma_2' \wedge \gamma_3 \ldots$ etc. It follows $\gamma_1' \geq \gamma_2' \geq \ldots$ and if $\mathcal{C}'$ denotes the family of the γ_i', then still $\wedge\mathcal{B} = \wedge\mathcal{C}'$. However, $\mu(\wedge\mathcal{B}) = \mu(\wedge\mathcal{C}') = \lim_{i\to\infty} \mu(\gamma_i') \geq \inf_{\alpha\in\mathcal{B}} \mu(\alpha)$ and hence $\mu(\wedge\mathcal{B}) = \inf_{\alpha\in\mathcal{B}} \mu(\alpha)$. This proves (i). (ii) is proved in an analogous way. □

Proof of Theorem 13.7

This is a direct consequence of the inclusion-exclusion formula in $(\Omega, \mathcal{A}, P)$. □

Proof of Theorem 13.8

Assertions (1) and (2) are proved in exactly the same way as (2) of theorem 13.5. Assertion (3) follows from $\rho(\cap_{i=1}^{\infty} E_i) = \wedge_{i=1}^{\infty} \rho(E_i)$, $\xi(\cup_{i=1}^{\infty} E_i) = \vee_{i=1}^{\infty} \xi(E_i)$ and the continuity of μ. □

Proof of Theorem 13.9

From the definitions (13.16) and (13.17) and theorem 13.3 (4) it follows that

$$\begin{aligned} pl_e(H) &= 1 - sp_e(H^c) \\ &= 1 - \sup \{P(A) : A \in \mathcal{A}, A \subseteq u(H^c)\} \\ &= 1 - \sup \{P(A) : A \in \mathcal{A}, u(H^c)^c \subseteq A^c\} \\ &= \inf \{P(A^c) : A \in \mathcal{A}, v(H) \subseteq A^c\} \\ &= \inf \{P(A) : A \in \mathcal{A}, v(H) \subseteq A\}. \end{aligned}$$

□

Proof of Theorem 13.10

A direct proof of the monotonicity and alternance of sp_e and pl_e seems to be surprisingly difficult. Therefore the following proof depends on the construction of an extension of the allocation of probability ρ for sp to an allocation of probability ρ_e for sp_e. (1) follows then from theorem 13.8.

Define $A(H) = \{\alpha \in \mathcal{M} : \exists\, A \in \alpha \text{ such that } A \subseteq u(H)\}$. If $\alpha', \alpha'' \in A(H)$, then also $\alpha = \alpha' \vee \alpha'' \in A(H)$ and $A(H)$ is an upward net. Therefore, by theorem 13.6, we have

$$\sup_{\alpha \in A(H)} \mu(\alpha) = \mu(\vee A(H)). \tag{A13.3}$$

The mapping $\rho : \mathcal{E}_s \to \mathcal{M}$ is extended to a mapping $\rho_e : \mathcal{P}(\Theta) \to \mathcal{M}$ by defining $\rho_e(H) = \vee A(H)$. First we show that $sp_e = \mu \circ \rho_e$. Select a sequence $\alpha_i \in A(H)$ such that $\mu(\alpha_i) \to \sup_{\alpha \in A(H)} \mu(\alpha)$. There is then a representative $A_i \in \alpha_i$ such that $A_i \subseteq u(H)$ and for $i \to \infty$

$$P(A_i) = \mu(\alpha_i) \to \sup_{\alpha \in A(H)} \mu(\alpha) \leq sp_e(H). \tag{A13.4}$$

On the other hand, there is also a sequence of sets $A_i \in \mathcal{A}$ such that $A_i \subseteq u(H)$ and

$$P(A_i) \to P_*(u(H)) = sp_e(H). \tag{A13.5}$$

For this same sequence $\mu(\pi(A_i)) \to sp_e(H)$ where $\pi(A_i) \in A(H)$ and hence $sp_e(H) \leq \sup_{\alpha \in A(H)} \mu(\alpha)$. Therefore $sp_e(H) = \sup_{\alpha \in A(H)} \mu(\alpha) = \mu(\vee A(H)) = \mu(\rho_e(H))$.

Let's show next that ρ_e is a $\cap$ - homomorphism. Fix $H_1, H_2 \subseteq \Theta$ and consider $\vee A(H_1)$ and $\vee A(H_2)$. Let A_1 and A_2 be two sets in $\mathcal{A}$ representing $\vee A(H_1)$ and $\vee A(H_2)$ respectively. Then $A = A_1 \cap A_2$ is a representative of $(\vee A(H_1)) \wedge (\vee A(H_2))$. It will be shown that it is also a representative of $\vee A(H_1 \cap H_2)$. From $P(A_i) = sp_e(H_i)$, $A_i \subseteq u(H_i)$ for $i = 1, 2$ it follows that $A = A_1 \cap A_2 \subseteq u(H_1) \cap u(H_2) = u(H_1 \cap H_2)$ and thus $P(A) \leq sp_e(H_1 \cap H_2)$. In fact, equality must hold here. According to Halmos (1964) every subset $H \subseteq \Omega$ has a so-called measurable kernel, i.e. a subset $H' \in \mathcal{A}$ such $H' \subseteq H$ and $P(G) = 0$ for every $G \in \mathcal{A}$ satisfying $G \subseteq H - H'$. Then Halmos proves that $P(H') = P^*(H)$. Therefore we have to prove that $A_1 \cap A_2$ is a measurable kernel of $u(H_1) \cap u(H_2) = u(H_1 \cap H_2)$ because in this case $P(A) = sp_e(H_1 \cap H_2)$. Suppose that this is not true, i.e. there is a set G in $\mathcal{A}$ such that $G \subseteq u(H_1) \cap u(H_2) - A_1 \cap A_2$ and $P(G) > 0$. If we define $M_1 = G \cap A_1$, $M_2 = G \cap A_2$ and $M_3 = G - (M_1 + M_2)$ then $G = M_1 + M_2 + M_3$ and hence there must be at least one M_i such that $P(M_i) > 0$. But this is a contradiction to the definition A_1 or A_2: for example, if $P(M_1) > 0$, then $A_2 + M_1 \subseteq u(H_2)$ but $P(A_2 + M_1) > P(A_2) = sp_e(H_2)$ which is impossible. The other cases $P(M_2) > 0$ and $P(M_3) > 0$ are treated similarly. Hence $P(A) = sp_e(H_1 \cap H_2)$ and A is a representative of $\vee A(H_1 \cap H_2)$. This implies

$$\rho_e(H_1 \cap H_2) = \vee A(H_1 \cap H_2) = (\vee A(H_1)) \wedge (\vee A(H_2)) = \rho_e(H_1) \wedge \rho_e(H_2)$$

and ρ_e is a $\cap$ - homomorphism. This implies (1) for sp_e by theorem 13.8 and by duality (1) follows also for pl_e.

In order to prove (2) fix a sequence $E_1 \supseteq E_2 \supseteq \ldots$ and an $\epsilon > 0$ and select $A'_i \in \mathcal{A}, A'_i \subseteq u(E_i)$ such that $sp_e(E_i) - P(A'_i) \leq \epsilon/2$. Define $A_i = \cup_{j \geq i} A'_j \supseteq A'_i$. Note that still $A_i \subseteq u(E_i)$ and thus still $sp_e(E_i) - P(A_i) \leq \epsilon/2$. The sequence A_i is decreasing, $A_1 \supseteq A_2 \supseteq \ldots$ Let

$$A = \cap_{i=1}^{\infty} A_i \subseteq \cap_{i=1}^{\infty} u(E_i). \tag{A13.6}$$

We can derive from theorem 13.3 (5) that for hints with σ-restrictions $A \subseteq u(\cap_{i=1}^{\infty} E_i)$. Furthermore $P(A_i) \to P(A)$. Fix n such that $P(A_i) - P(A) < \epsilon/2$ for all $i > n$. Now $sp_e(\cap_{i=1}^{\infty} E_j) > P(A)$ and

$$sp_e(E_i) - sp_e(\cap_{j=1}^{\infty} E_j) < P(A_i) + \epsilon/2 - P(A) < \epsilon/2 + \epsilon/2 = \epsilon. \tag{A13.7}$$

Since $\epsilon > 0$ is arbitrary it follows that $sp_e(E_i) \to sp_e(\cap_{j=1}^{\infty} E_j)$ for $i \to \infty$. The continuity of pl_e follows by duality. □

14. Structure of Support and Plausibility

14.1 Refining Hints

In chapter 8 we have seen how the frame of discernment of a hint may be coarsened and how in this way information contained in a hint may be lost. This may be a deliberate action with the goal of simplifying a discussion. In this section the coarsening or rather the inverse process of refining the information contained in a hint is discussed in a new comprehensive way. Unlike in Chap. 6 where frames were refined and coarsened , here we refine and coarsen hints. A general notion of inclusion of hints is defined. This notion has a practical interest as means to construct approximations in order to simplify computations. But this inclusion of hints is also an important theoretical tool which allows one to study the fine structure of hints and their corresponding support and plausibility functions. This aspect is discussed in the subsequent section 14.2.

A hint $\mathcal{H} = (\Omega, \mathcal{A}, P, \Gamma, \Theta)$ can be refined in several respects by adding supplementary information to it:

(1) The restrictions $\Gamma(\omega)$ associated with the interpretations ω may become more precise. A restriction $(\mathcal{S}', \mathcal{P}')$ is said to be more precise than (or included in) a restriction $(\mathcal{S}, \mathcal{P})$ iff $\mathcal{S}' \supseteq \mathcal{S}$ (or equivalently $\mathcal{P}' \subseteq \mathcal{P}$), i.e. if it implies more propositions (or equivalently less propositions are possible). Then we write $(\mathcal{S}', \mathcal{P}') \subseteq (\mathcal{S}, \mathcal{P})$.

(2) Some interpretations which were originally considered possible may become known as impossible. The new set of possible interpretations Ω' becomes a subset of Ω. This implies that the original probability P must be conditioned on Ω'. This leads to a new probability space $(\Omega', \mathcal{A}', P')$ of possible interpretations, where $\mathcal{A}' = \mathcal{A} \cap \Omega'$ and $P'(A') = P^*(A')/P^*(\Omega')$ for all $A' \in \mathcal{A}'$, provided that $P^*(\Omega') > 0$. Note that Ω' is not necessarily measurable; P' is still a probability measure on $\mathcal{A}'$ (Neveu, 1964).

(3) The probability measure P' on the set of possible interpretations Ω' may be extended from the σ-algebra $\mathcal{A}'$ to a probability measure P'' on a larger σ-algebra $\mathcal{A}'' \subseteq \mathcal{P}(\Omega')$ containing $\mathcal{A}'$. Let's note that in this case

$$P'_*(A) \leq P''(A) \leq P'^*(A) \tag{14.1}$$

for all $A \in \mathcal{A}''$.

Thus, if we define $\Omega'' = \Omega'$, then combining all three refining steps in the above sequence leads to a new, refined hint $\mathcal{H}'' = (\Omega'', \mathcal{A}'', P'', \Gamma'', \Theta)$, such that $\Omega'' \subseteq \Omega, \mathcal{A}'' \supseteq \mathcal{A}', P''$ is an extension to $\mathcal{A}''$ of P' and $\Gamma''(\omega) \subseteq \Gamma(\omega)$ for all $\omega \in \Omega''$. In this case we write $\mathcal{H}'' \subseteq \mathcal{H}$ and say that $\mathcal{H}''$ is *included* in or is *finer* than $\mathcal{H}$ (and $\mathcal{H}$ is *coarser* then $\mathcal{H}''$). Of course, often not all three refining steps are present; in particular it is usual that only step (1) or steps (1) and (3) are present. These particular cases correspond to Yager's (1985) definition of inclusion. See also Dubois, Prade (1990) for a discussion of approximation of support and plausibility functions. The following simple example illustrates this notion of inclusion.

Example 14.1 Extension and marginal. Consider a refining $\tau : \Theta \to \Lambda$ in a family of compatible frames. If $\mathcal{H} = (\Omega, \mathcal{A}, P, \Gamma, \Theta)$ is a hint on Θ then the *extension* of $\mathcal{H}$ to the frame Λ is defined as follows (compare with Sect. 8.2). First, the probability space $(\Omega, \mathcal{A}, P)$ remains unchanged and for each $\omega \in \Omega$ let $\Gamma \downarrow (\omega)$ denote the restriction in Λ induced by the filter

$$\mathcal{S}(\omega) \downarrow = \{A \subseteq \Lambda : \exists\ S \in \mathcal{S}(\omega) \text{ such that } A \supseteq \tau(S)\}, \tag{14.2}$$

where $\Gamma(\omega) = (\mathcal{S}(\omega), \mathcal{P}(\omega))$ is the restriction on Θ associated with ω. It is easily verified that $\mathcal{S}(\omega) \downarrow$ is indeed a filter. The extension of $\mathcal{H}$ is then defined to be the hint $\mathcal{H} \downarrow \Lambda = (\Omega, \mathcal{A}, P, \Gamma \downarrow, \Lambda)$ and this definition coincides with the usual definition for a set-focussed hint. Similarly, if $\mathcal{H} = (\Omega, \mathcal{A}, P, \Gamma, \Lambda)$ is a hint on Λ, then the *marginal* of $\mathcal{H}$ to the frame Θ is defined as follows. Again, the probability space $(\Omega, \mathcal{A}, P)$ remains unchanged and for each $\omega \in \Omega$ let $\Gamma \uparrow (\omega)$ denote the restriction in Θ induced by the filter $\mathcal{S}(\omega) \uparrow = v(\mathcal{S}(\omega))$ if $\Gamma(\omega) = (\mathcal{S}(\omega), \mathcal{P}(\omega))$. Again it can easily be shown that $\mathcal{S}(\omega) \uparrow$ is indeed a filter. For example, let's show that if $H \in v(\mathcal{S}(\omega))$ and $H' \supseteq H$, then $H' \in v(\mathcal{S}(\omega))$. We prove that $\tau(H') \in \mathcal{S}(\omega)$ and $v(\tau(H')) = H'$. The last equality holds by theorem 7.2 (11), and $\tau(H') \in \mathcal{S}(\omega)$ because there is an $S \in \mathcal{S}(\omega)$ such that $H = v(S)$ and thus $\tau(H') \supseteq \tau(H) = \tau(v(S)) \supseteq S$ which proves that $\tau(H') \in \mathcal{S}(\omega)$ by the first property of a filter. The marginal of $\mathcal{H}$ is then defined to be the hint $\mathcal{H} \uparrow \Theta = (\Omega, \mathcal{A}, P, \Gamma \uparrow, \Theta)$ and this definition coincides with the usual one for set-focussed hints.

Now we prove that a hint is always finer than the extension of its marginal, i.e. $\mathcal{H} \subseteq (\mathcal{H} \uparrow \Theta) \downarrow \Lambda$ for any hint $\mathcal{H} = (\Omega, \mathcal{A}, P, \Gamma, \Lambda)$ and any refining $\tau : \Theta \to \Lambda$. If $\Gamma(\omega) = (\mathcal{S}(\omega), \mathcal{P}(\omega))$, then it must be proved that $\mathcal{S}(\omega) \supseteq (\mathcal{S}(\omega) \uparrow) \downarrow$. So let $S \in (\mathcal{S}(\omega) \uparrow) \downarrow$. Then there is $X \in v(\mathcal{S}(\omega))$ such that $S \supseteq \tau(X)$ and hence there is $Y \in \mathcal{S}(\omega)$ such that $X = v(Y)$. But this implies that $S \supseteq \tau(v(Y)) \supseteq Y$ and hence $S \in \mathcal{S}(\omega)$ because $Y \in \mathcal{S}(\omega)$ and $S \supseteq Y$.

Example 14.2 Aggregation of interpretations. Let $\mathcal{H} = (\Omega, \mathcal{A}, P, \Gamma, \Theta)$ be a hint on a frame Θ and let $\mathcal{Q}$ be a partition of Ω. The hint may be simplified by grouping the interpretations which belong the a common block $Q \in \mathcal{Q}$. For $\omega \in \Omega$, if $Q(\omega)$ is the block of $\mathcal{Q}$ containing ω, then define

$$\Gamma'(\omega) = \cup\ \{\Gamma(\omega') : \omega' \in Q(\omega)\} \tag{14.3}$$

(see Sect. 13.1 for the definition of a union of restrictions). The hint $\mathcal{H}' = (\Omega, \mathcal{A}, P, \Gamma', \Theta)$ is then coarser than the original hint because $\Gamma'(\omega) \supseteq \Gamma(\omega)$ for all $\omega \in \Omega$. Indeed, with obvious notations, if $S \in \mathcal{S}'(\omega)$ then there is $S(\omega') \in \mathcal{S}(\omega')$ for all $\omega' \in Q(\omega)$ such that $S = \cup \{S(\omega') : \omega' \in Q(\omega)\}$. Then $S \supseteq S(\omega)$ because $\omega \in Q(\omega)$ and hence $S \in \mathcal{S}(\omega)$ because $S(\omega) \in \mathcal{S}(\omega)$ and $\mathcal{S}(\omega)$ is a filter.

Example 14.3 Conditioning on interpretations. Let $\mathcal{H}$ be a simple hint with two possible interpretations, $\Omega = \{\omega_0, \omega_1\}$, representing an unreliable source. The interpretation ω_0 means that the source correctly reports the information which is looked for, that is $\Gamma(\omega_0) = \{\theta_0\}$ for some $\theta_0 \in \Theta$, and ω_1 is the interpretation that the source is not reliable, such that $\Gamma(\omega_1) = \Theta$. Now, if it is learned that the source is indeed reliable, then ω_1 is excluded and ω_0 remains the only possible interpretation. This defines a deterministic hint $\mathcal{H}'$ pointing to $\{\omega_0\}$. According to step (2) above we have $\mathcal{H}' \subseteq \mathcal{H}$. But suppose that in a second case it becomes known that the source is not reliable which excludes ω_0. Then this defines another deterministic hint $\mathcal{H}''$ which is in fact vacuous. Again by step (2) above $\mathcal{H}'' \subseteq \mathcal{H}$. This may seem to be counterintuitive: in fact $\mathcal{H}$ may support rather strongly $\{\theta_0\}$, leaving only a small doubt that θ_0 is the correct answer. Then $\mathcal{H}'$ is intuitively finer than $\mathcal{H}$ because it asserts that θ_0 is surely the correct answer. In contrast $\mathcal{H}''$ is void of any information about Θ. How can this information be finer than the original one ? The point is that in a "non-monotone" theory like the theory of hints, a new, added information may sometimes invert a seemingly clear situation into a situation of confusion or ignorance, as is the case in this example. Even if the new information is less informative, it is still a refining of the original one because new information has been taken into account, namely that nothing is known.

Similar examples will be found later in Chap. 15 in relation with Dempster's rule. This feature of evidence theory may be disturbing at first sight and it surely complicates life. But it should be seen as a realistic and indispensable property of reasoning under uncertainty, that new information may sometimes introduce doubt about what one used to believe as being rather sure.

This inclusion relation between hints satisfies the following properties:

Theorem 14.1

(1) Reflexivity: for any hint $\mathcal{H}$, we have $\mathcal{H} \subseteq \mathcal{H}$.
(2) Antisymmetry: If $\mathcal{H}_1 \subseteq \mathcal{H}_2$ and $\mathcal{H}_2 \subseteq \mathcal{H}_1$, then $\mathcal{H}_1 = \mathcal{H}_2$.

This notion of inclusion of hints leads to the following comparison of the corresponding support and plausibility functions:

Theorem 14.2 *Let $\mathcal{H}'' = (\Omega'', \mathcal{A}'', P'', \Gamma'', \Theta)$ and $\mathcal{H} = (\Omega, \mathcal{A}, P, \Gamma, \Theta)$ be two hints such that $\mathcal{H}'' \subseteq \mathcal{H}$ and let sp''_e, pl''_e and sp_e, pl_e denote their respective extended support and plausibility functions. If $k = P^*(\Omega'')$, then*

$$\begin{aligned} sp_e(H) &\leq k \cdot sp''_e(H) + (1-k) \\ pl_e(H) &\geq k \cdot pl''_e(H) \end{aligned} \tag{14.4}$$

for all $H \subseteq \Theta$. *If only refining step (1) and possibly (3) are present, then* $k = 1$ *and* $[sp''_e(H), pl''_e(H)] \subseteq [sp_e(H), pl_e(H)]$.

To any hint $\mathcal{H} = (\Omega, \mathcal{A}, P, \Gamma, \Theta)$ a particular vacuous hint

$$\mathcal{V} = (\Omega, \mathcal{A}, P, \Gamma_{vac}, \Theta)$$

can be associated, where $\Gamma_{vac}(\omega)$ is the vacuous restriction for all ω. Clearly $\Gamma(\omega) \subseteq \Gamma_{vac}(\omega)$ for all ω and therefore we have always $\mathcal{H} \sqsubseteq \mathcal{V}$. But this does not imply that *any* vacuous hint is coarser than $\mathcal{H}$ (compare example 14.3).

14.2 The Structure of Support and Plausibility Functions

This section continues the discussion started in Sect. 3.3 on the equivalence of hints. This notion becomes more important in the present general theory of hints. In particular it leads to a study of the finer structure of support and plausibility functions on general frames of discernement Θ.

A hint generates a support function sp on some multiplicative class $\mathcal{E}_s$ satisfying conditions (1), (2) and possibly (3) of theorem 13.5. Inversely, for every function on $\mathcal{E}_s$ satisfying (1), (2) and possibly (3) there is a hint which generates this support function. However, this hint is not determined uniquely by sp as was noted at the end of Chap. 13. There is a whole family of hints with a given support function sp on $\mathcal{E}_s$, but with different extensions sp_e to the whole power set of Θ. The goal of this section is to study such families of hints. Extensions of support functions were already studied by Shafer (1979). This section puts some of the results obtained by Shafer into the perspective of hints.

As a preparation, note that with any restriction $\mathcal{R} = (\mathcal{S}, \mathcal{P})$ in $\mathcal{R}(\Theta)$ a filter $\varphi(\mathcal{R}) = \mathcal{S} \cap \mathcal{E}_s$ on $\mathcal{E}_s$ can be associated. The mapping φ from $\mathcal{R}(\Theta)$ to $\mathcal{F}(\mathcal{E}_s)$, the family of filters on $\mathcal{E}_s$, is onto because for any filter $\mathcal{F} \in \mathcal{F}(\mathcal{E}_s)$ the restriction $\mathcal{R}_{\mathcal{E}_s}(\mathcal{F}) \in \mathcal{R}(\Theta)$ defined by its class of implied propositions $\mathcal{S}_{\mathcal{E}_s} = \{H \subseteq \Theta :$ there is an $E \in \mathcal{F}$ such that $E \subseteq H\}$ is in $\varphi^{-1}(\mathcal{F})$. This shows that $\{\varphi^{-1}(\mathcal{F}) : \mathcal{F} \in \mathcal{F}(\mathcal{E}_s)\}$ is a partition of $\mathcal{R}(\Theta)$. Moreover, $\mathcal{R}_{\mathcal{E}_s}(\mathcal{F})$ is the coarsest restriction in $\varphi^{-1}(\mathcal{F})$: if $\mathcal{R}' \in \varphi^{-1}(\mathcal{F})$ then $\mathcal{R}' \sqsubseteq \mathcal{R}_{\mathcal{E}_s}(\mathcal{F})$.

Now let sp be a function satisfying conditions (1) and (2) of theorem 13.5 relative to a multiplicative family $\mathcal{E}_s$. Let $\mathcal{F}(\mathcal{E}_s)$ be the family of filters on $\mathcal{E}_s$. In the proof of theorem 13.5 it is stated that there is a σ-algebra $\mathcal{A}''$ on $\mathcal{F}(\mathcal{E}_s)$ and a probability measure P'' on $\mathcal{A}''$ such that the hint $\mathcal{H}'' = (\mathcal{F}(\mathcal{E}_s), \mathcal{A}'', P'', \Gamma'', \Theta)$ has the support function sp on $\mathcal{E}_s$, which means that $\mathcal{E}_s \subseteq \mathcal{E}''_s$ and $sp''(H) = sp(H)$ for $H \in \mathcal{E}_s$ if $\mathcal{E}''_s$ is the family of measurable hypothese of $\mathcal{H}''$ and sp'' is its support function. Note that $(\mathcal{A}'', P'')$ is not necessarily unique. The mapping Γ'' maps any filter $\mathcal{F}$ in $\mathcal{F}(\mathcal{E}_s)$ to the restriction $\mathcal{R}_{\mathcal{E}_s}(\mathcal{F})$ on Θ. Of course, P'' depends on sp and the notation P''_{sp} would be more appropriate but we stick to P'' to avoid heavy notations. Note that using φ the probability space $(\mathcal{F}(\mathcal{E}_s), \mathcal{A}'', P'')$ induces a probability space $(\mathcal{R}(\Theta), \mathcal{A}', P')$ where $\mathcal{A}'$ is the σ-algebra of all sets $\varphi^{-1}(A'')$ for $A'' \in \mathcal{A}''$ and $P'(A') = P''(\varphi(A'))$ for $A' \in \mathcal{A}'$. If we define $\Gamma_{\mathcal{E}_s}(\mathcal{R}) = \mathcal{R}_{\mathcal{E}_s}(\varphi(\mathcal{R}))$, then the following theorem states that the hint $\mathcal{H}' = (\mathcal{R}(\Theta), \mathcal{A}', P', \Gamma_{\mathcal{E}_s}, \Theta)$ also generates the support function sp on $\mathcal{E}_s$.

Theorem 14.3 *If $\mathcal{E}'_s$ denotes the family of measurable hypotheses of the hint $\mathcal{H}' = (\mathcal{R}(\Theta), \mathcal{A}', P', \Gamma_{\mathcal{E}_s}, \Theta)$ and sp' is its support function, then $\mathcal{E}_s \subseteq \mathcal{E}'_s$ and $sp'(H) = sp(H)$ for all $H \in \mathcal{E}_s$.*

Let $u_{\mathcal{E}_s}(H)$ be the function (13.1) defined with respect to $\Gamma_{\mathcal{E}_s}$ and let $u_{\mathcal{E}_s}(\mathcal{E}_s)$ be the images of $\mathcal{E}_s$ with respect to $u_{\mathcal{E}_s}$. By theorem 13.3 (5) $u_{\mathcal{E}_s}(\mathcal{E}_s)$ is a multiplicative class. The set $u_{\mathcal{E}_s}(\mathcal{E}_s)$ is contained in $\mathcal{A}'$. Now, let $\mathcal{A}_{\mathcal{E}_s}$ be the smallest σ-algebra containing $u_{\mathcal{E}_s}(\mathcal{E}_s)$; $\mathcal{A}_{\mathcal{E}_s}$ is a subalgebra of $\mathcal{A}'$. Finally, let P_{sp} be the restriction of P' to $\mathcal{A}_{\mathcal{E}_s}$. Then the set of measurable hypotheses with respect to the hint $(\mathcal{R}(\Theta), \mathcal{A}_{\mathcal{E}_s}, P_{sp}, \Gamma_{\mathcal{E}_s}, \Theta)$ contains $\mathcal{E}_s$ and its support function restricted to $\mathcal{E}_s$ coincides with sp, i.e. $P_{sp}(u_{\mathcal{E}_s}(H)) = sp(H)$ for all $H \in \mathcal{E}_s$. It is important to remark that both $\Gamma_{\mathcal{E}_s}$ and $\mathcal{A}_{\mathcal{E}_s}$ are uniquely determined by $\mathcal{E}_s$, whereas the probability P_{sp} is derived from P'' which is not unique in general. Therefore, any hint

$$\mathcal{C}^{\mathcal{E}_s}_{sp} = (\mathcal{R}(\Theta), \mathcal{A}_{\mathcal{E}_s}, P, \Gamma_{\mathcal{E}_s}, \Theta) \tag{14.5}$$

such that $P(u_{\mathcal{E}_s}(H)) = sp(H)$ for all $H \in \mathcal{E}_s$ is called a *canonical hint* with respect to sp on $\mathcal{E}_s$. Note that any probability P_{sp} as defined above satisfies this condition and hence there is always at least one canonical hint.

Loosely speaking, the following theorem states that if we extend a support function to a larger multiplicative class then we obtain a more informative canonical hint.

Theorem 14.4 *Let sp be a support function on $\mathcal{E}_s$ and sp' a support function on $\mathcal{E}'_s$ such that $\mathcal{E}_s \subseteq \mathcal{E}'_s$ and sp' restricted to $\mathcal{E}_s$ coincides with sp. Then*

(1) There exist canonical hints $\mathcal{C}^{\mathcal{E}_s}_{sp}$ and $\mathcal{C}^{\mathcal{E}'_s}_{sp'}$ such that $\mathcal{C}^{\mathcal{E}'_s}_{sp'} \subseteq \mathcal{C}^{\mathcal{E}_s}_{sp}$.

(2) Let sp_e and sp'_e denote the extended support functions of sp and sp' with respect to the canonical hints $\mathcal{C}^{\mathcal{E}_s}_{sp}$ and $\mathcal{C}^{\mathcal{E}'_s}_{sp'}$ given above. If pl_e and pl'_e are the corresponding extended plausibility functions, i.e. $pl_e(H) = 1 - sp_e(H^c)$ and $pl'_e(H) = 1 - sp'_e(H^c)$, then

$$[sp'_e(H), pl'_e(H)] \subseteq [sp_e(H), pl_e(H)] \tag{14.6}$$

for all $H \subseteq \Theta$.

Theorem 14.5 *Let sp be a support function on a multiplicative class $\mathcal{E}_s$. Then there is a canonical hint $\mathcal{C}_{sp}^{\mathcal{E}_s}$ with respect to sp on $\mathcal{E}_s$ such that its extended support function sp_e satisfies*

$$\begin{aligned} sp_e(H) = sup\ \{\sum\{(-1)^{|I|+1} sp(\cap_{i\in I} E_i) : \emptyset \neq I \subseteq \{1,\ldots,n\}\} : \\ E_i \subseteq H, E_i \in \mathcal{E}_s, i = 1,\ldots,n; n = 1,\ldots\} \end{aligned} \tag{14.7}$$

for all $H \subseteq \Theta$. If we define $\mathcal{E}_p = \{H \subseteq \Theta : H^c \in \mathcal{E}_s\}$ and $pl(H) = 1 - sp(H^c)$ for all $H \in \mathcal{E}_p$, then the extended plausibility function pl_e of $\mathcal{C}_{sp}^{\mathcal{E}_s}$ satisfies

$$\begin{aligned} pl_e(H) = inf\ \{\sum\{(-1)^{|I|+1} pl(\cup_{i\in I} E_i) : \emptyset \neq I \subseteq \{1,\ldots,n\}\} : \\ E_i \supseteq H, E_i \in \mathcal{E}_p, i = 1,\ldots,n; n = 1,\ldots\} \end{aligned} \tag{14.8}$$

for all $H \subseteq \Theta$.

Since two equivalent hints with the same class of s-measurable sets may generate two different extended support functions (see the example at the end of Chap. 13), it is not excluded that two different canonical hints for *sp* on $\mathcal{E}_s$ generate different extensions. As a consequence, there might exist a canonical hint whose extension is different from the right-hand side of (14.7). Also, note that the restriction to $\mathcal{E}_s$ of any such extension (e.g. sp_m) coincides with *sp*.

Theorem 14.6 *Let sp be a support function on $\mathcal{E}_s$ and sp′ a support function on $\mathcal{P}(\Theta)$ such that $sp'(H) = sp(H)$ for all $H \in \mathcal{E}_s$. Then there is a canonical hint $\mathcal{C}_{sp}^{\mathcal{E}_s}$ of sp whose extended support function sp_e satisfies*

$$sp_e(H) \leq sp'(H) \tag{14.9}$$

for all $H \subseteq \Theta$.

Theorem 14.7 *Let sp be a support function on an algebra $\mathcal{E}$. Then there is a canonical hint $\mathcal{C}_{sp}^{\mathcal{E}}$ whose extended support function sp_e satisfies*

$$sp_e(H) = sup\ \{sp(E) : E \in \mathcal{E}, E \subseteq H\} \tag{14.10}$$

for all $H \subseteq \Theta$. Furthermore, if pl_e is the extended plausibility function with respect to $\mathcal{C}_{sp}^{\mathcal{E}}$ and $pl(H) = 1 - sp(H^c)$ for all $H \in \mathcal{E}$, then for all $H \subseteq \Theta$ we have

$$pl_e(H) = inf\ \{pl(E) : E \in \mathcal{E}, E \supseteq H\}. \tag{14.11}$$

Theorem 14.8 *Let sp be a continuous support function on a σ-algebra $\mathcal{E}$. Then there is a canonical hint of sp whose extended support function sp_e satisfies the following properties:*

(1) sp_e is continuous, i.e.

$$sp_e(\cap_{i=1}^{\infty} E_i) = \lim_{i\to\infty} sp_e(E_i) \tag{14.12}$$

for every collection of sets $E_1 \supseteq E_2 \supseteq \ldots$

(2)

$$sp_e(H) = sup\,\{\lim_{i\to\infty} sp(E_i) : E_1 \supseteq E_2 \supseteq \ldots \in \mathcal{E}, \cap_{i=1}^{\infty} E_i \subseteq H\}. \quad (14.13)$$

If we define $pl(E) = 1 - sp(E^c)$ *then the extended plausibility function* pl_e *satisfies*

$$pl_e(H) = inf\,\{\lim_{i\to\infty} pl(E_i) : E_1 \subseteq E_2 \supseteq \ldots \in \mathcal{E}, \cup_{i=1}^{\infty} E_i \supseteq H\}. \quad (14.14)$$

A14 Appendix: Proofs of Theorems in Chapter 14

Proof of Theorem 14.1

These two assertions follow directly from the definition of inclusion. □

Proof of Theorem 14.2

Let $v''(H)$ and $v(H)$ be the subsets of interpretations of Ω'' and Ω respectively under which H is possible. Then clearly $v''(H) \subseteq v(H) \cap \Omega''$ by the refining step (1).

Now, for any $H \subseteq \Theta$,

$$pl_e(H) = P^*(v(H)) \geq P^*(v(H) \cap \Omega'') \geq P^*(v''(H)) = P^*(v''(H) \cap \Omega'').$$

Let $P'(A) = P^*(A)/P^*(\Omega'')$ for $A \in \mathcal{A} \cap \Omega''$ and $P'^*(A)$ denote the outer probability measure with respect to P'. Then it follows easily that $P'^*(v''(H) \cap \Omega'') = P^*(v''(H) \cap \Omega'')/P^*(\Omega'')$ and hence

$$pl_e(H) \geq P'^*(v''(H) \cap \Omega'')P^*(\Omega''). \quad (A14.1)$$

If $P''^*(A)$ is the outer measure with respect to the probability measure P'' on $\mathcal{A}''$, then clearly $P'^*(A) \geq P''^*(A)$ for any $A \subseteq \Omega''$. Thus

$$\begin{aligned} pl_e(H) &\geq P''^*(v''(H) \cap \Omega'')P^*(\Omega'') = P''^*(v''(H))P^*(\Omega'') \\ &= pl''_e(H)P^*(\Omega'') = k \cdot pl''_e(H). \end{aligned} \quad (A14.2)$$

This proves the first part of equation (14.4).

By (13.17) we have

$$\begin{aligned} sp_e(H) &= 1 - pl_e(H^c) \leq 1 - k \cdot pl''_e(H^c) \\ &= 1 - k \cdot (1 - sp''_e(H)) = k \cdot sp''_e(H) + (1 - k). \end{aligned} \quad (A14.3)$$

This proves the second part of equation (14.4). This completes the proof of the theorem because the last assertion follows from (14.4). □

Proof of Theorem 14.3

By definition we have

$$u''(H) = \{\mathcal{F} : H \in \mathcal{R}_{\mathcal{E}_s}(\mathcal{F})\} \tag{A14.4}$$

and hence

$$\begin{aligned} \varphi^{-1}(u''(H)) &= \{\mathcal{R} : \varphi(\mathcal{R}) \in u''(H)\} \\ &= \{\mathcal{R} : H \in \mathcal{R}_{\mathcal{E}_s}(\varphi(\mathcal{R}))\} \\ &= \{\mathcal{R} : H \in \Gamma_{\mathcal{E}_s}(\mathcal{R})\} = u'(H), \end{aligned} \tag{A14.5}$$

which implies that

$$\varphi(u'(H)) = u''(H) \tag{A14.6}$$

because φ is onto. If $H \in \mathcal{E}''_s$ then $u''(H) \in \mathcal{A}''$ and hence $u'(H) \in \mathcal{A}'$ by equation (A14.6), which shows that $H \in \mathcal{E}'_s$ and finally $\mathcal{E}''_s \subseteq \mathcal{E}'_s$. Since $\mathcal{E}_s \subseteq \mathcal{E}''_s$ it follows that $\mathcal{E}_s \subseteq \mathcal{E}'_s$.

Now if $H \in \mathcal{E}_s$ then $H \in \mathcal{E}'_s$ and

$$\begin{aligned} sp'(H) = P'(u'(H)) &= P''(\varphi(u'(H))) \\ &= P''(u''(H)) = sp''(H) = sp(H). \end{aligned} \tag{A14.7}$$

□

Proof of Theorem 14.4

First we prove assertion (1). Let

$$\mathcal{C}^{\mathcal{E}'_s}_{sp'} = (\mathcal{R}(\Theta), \mathcal{A}_{\mathcal{E}'_s}, P_{sp'}, \Gamma_{\mathcal{E}'_s}, \Theta) \tag{A14.8}$$

be the canonical hint with respect to sp' on $\mathcal{E}'_s$ which is derived from a probability P''. Furthermore, let

$$\mathcal{C}^{\mathcal{E}_s}_{sp} = (\mathcal{R}(\Theta), \mathcal{A}_{\mathcal{E}_s}, P, \Gamma_{\mathcal{E}_s}, \Theta) \tag{A14.9}$$

be a hint which will be proved to be a canonical hint with respect to sp on $\mathcal{E}_s$ as soon as P is properly defined (see below). In order to prove the inclusion

$$\mathcal{C}^{\mathcal{E}'_s}_{sp'} \subseteq \mathcal{C}^{\mathcal{E}_s}_{sp} \tag{A14.10}$$

we need to show points (1) and (3) of the inclusion of hints. In other words, we need to show that

(1) $\Gamma_{\mathcal{E}'_s}(\mathcal{R}) \subseteq \Gamma_{\mathcal{E}_s}(\mathcal{R})$
(2) $\mathcal{A}_{\mathcal{E}'_s} \supseteq \mathcal{A}_{\mathcal{E}_s}$
(3) $P_{sp'}$ restricted to $\mathcal{A}_{\mathcal{E}_s}$ coincides with P.

Furthermore, for $\mathcal{C}^{\mathcal{E}_s}_{sp}$ to be a canonical hint it must be proved that

(4) $P(u_{\mathcal{E}_s}(H)) = sp(H)$ for all $H \in \mathcal{E}_s$.

We start by proving (1). If we represent a restriction by its filters of implied propositions, then

$$\begin{aligned} \Gamma_{\mathcal{E}'_s}(\mathcal{F}) &= \mathcal{R}_{\mathcal{E}'_s}(\mathcal{F} \cap \mathcal{E}'_s) \\ &= \{H \subseteq \Theta : \exists\, K \in \mathcal{F} \cap \mathcal{E}'_s \text{ such that } H \supseteq K\} \end{aligned} \tag{A14.11}$$

and similarly

$$\Gamma_{\mathcal{E}_s}(\mathcal{F}) = \{H \subseteq \Theta : \exists\, K \in \mathcal{F} \cap \mathcal{E}_s \text{ such that } H \supseteq K\}. \tag{A14.12}$$

It must be shown that $\Gamma_{\mathcal{E}_s}(\mathcal{F}) \subseteq \Gamma_{\mathcal{E}'_s}(\mathcal{F})$, which is true because $\mathcal{E}'_s \subseteq \mathcal{E}_s$.

Now we prove (2). Since $\mathcal{A}_{\mathcal{E}_s}$ is the smallest σ-algebra containing $u_{\mathcal{E}_s}(\mathcal{E}_s)$, it is sufficient to show that $\mathcal{A}_{\mathcal{E}'_s} \supseteq u_{\mathcal{E}_s}(\mathcal{E}_s)$ and since $\mathcal{A}_{\mathcal{E}'_s} \supseteq u_{\mathcal{E}'_s}(\mathcal{E}'_s)$ it remains to show that $u_{\mathcal{E}'_s}(\mathcal{E}'_s) \supseteq u_{\mathcal{E}_s}(\mathcal{E}_s)$. So let $A = u_{\mathcal{E}_s}(H)$ with $H \in \mathcal{E}_s$. Since $\mathcal{E}_s \subseteq \mathcal{E}'_s$ we have $H \in \mathcal{E}'_s$ and it remains to show that $u_{\mathcal{E}_s}(H) = u_{\mathcal{E}'_s}(H)$. But

$$\begin{aligned} u_{\mathcal{E}_s}(H) &= \{\mathcal{F} : H \in \Gamma_{\mathcal{E}_s}(\mathcal{F})\} \\ &= \{\mathcal{F} : H \in \mathcal{R}_{\mathcal{E}_s}(\mathcal{F} \cap \mathcal{E}_s)\} \\ &= \{\mathcal{F} : \exists\, K \in \mathcal{F} \cap \mathcal{E}_s \text{ such that } H \supseteq K\} \end{aligned} \tag{A14.13}$$

and similarly

$$u_{\mathcal{E}'_s}(H) = \{\mathcal{F} : \exists\, K \in \mathcal{F} \cap \mathcal{E}'_s \text{ such that} H \supseteq K\}. \tag{A14.14}$$

If $\mathcal{F} \in u_{\mathcal{E}_s}(H)$ then there is $K \in \mathcal{F} \cap \mathcal{E}_s$ such that $H \supseteq K$ and hence there is $K \in \mathcal{F} \cap \mathcal{E}'_s$ such that $H \supseteq K$ because $\mathcal{E}_s \subseteq \mathcal{E}'_s$, which shows that $\mathcal{F} \in u_{\mathcal{E}'_s}(H)$ and $u_{\mathcal{E}_s}(H) \subseteq u_{\mathcal{E}'_s}(H)$. Inversely, if $\mathcal{F} \in u_{\mathcal{E}'_s}(H)$ then there is $K \in \mathcal{F} \cap \mathcal{E}'_s$ such that $H \supseteq K$ and hence $H \in \mathcal{F}$ since $H \supseteq K$ and K is in the filter $\mathcal{F}$. So $H \in \mathcal{F} \cap \mathcal{E}_s$, which proves that $\mathcal{F} \in u_{\mathcal{E}_s}(H)$ and $u_{\mathcal{E}'_s}(H) \subseteq u_{\mathcal{E}_s}(H)$ and finally $u_{\mathcal{E}_s}(H) = u_{\mathcal{E}'_s}(H)$.

Now we define P as the restriction of $P_{sp'}$ to $\mathcal{A}_{\mathcal{E}_s}$, which implies that (3) is automatically satisfied. It remains to prove (4). If $H \in \mathcal{E}_s$ then $u_{\mathcal{E}_s}(H) = u_{\mathcal{E}'_s}(H)$ implies that

$$\begin{aligned} P(u_{\mathcal{E}_s}(H)) &= (P_{sp'}|\mathcal{A}_{\mathcal{E}_s})(u_{\mathcal{E}_s}(H)) \\ &= P_{sp'}(u_{\mathcal{E}_s}(H)) \\ &= P_{sp'}(u_{\mathcal{E}'_s}(H)) \\ &= sp'(H) = sp(H) \end{aligned} \tag{A14.15}$$

where the second last equality holds because $H \in \mathcal{E}'_s$ since $\mathcal{E}_s \subseteq \mathcal{E}'_s$ and the last equality holds because sp' restricted to $\mathcal{E}_s$ coincides with sp.

The assertion (2) follows from theorem 14.2 □

Proof of Theorem 14.5

Let $sp_m(H)$ denote the right-hand side of (14.7). It will be shown below that sp_m is a support function, i.e. it satisfies the conditions (1) and (2) of theorem 13.5. Let $H \in \mathcal{E}_s$ and let $E_1, \ldots, E_n$ be a collection of sets in $\mathcal{E}_s$ with $E_i \subseteq H$ for all $i = 1, \ldots, n$. Since sp is a support function on $\mathcal{E}_s$ it follows that

$$sp(H) \geq \sum\{(-1)^{|I|+1} sp(\cap_{i\in I} E_i) : \emptyset \neq I \subseteq \{1, \ldots, n\}\} \tag{A14.16}$$

which implies that $sp(H) \geq sp_m(H)$. On the other hand, we clearly have $sp_m(H) \geq sp(H)$ if $H \in \mathcal{E}_s$, which proves that $sp_m(H) = sp(H)$ for all $H \in \mathcal{E}_s$. Then, according to theorem 14.4, there exist canonical hints $\mathcal{C}_{sp}^{\mathcal{E}_s}$ and $\mathcal{C}_{sp_m}^{\mathcal{P}(\Theta)}$ such that $\mathcal{C}_{sp_m}^{\mathcal{P}(\Theta)} \subseteq \mathcal{C}_{sp}^{\mathcal{E}_s}$. Let sp_e denote the extended support function with respect to $\mathcal{C}_{sp}^{\mathcal{E}_s}$. Since the extended support function with respect to $\mathcal{C}_{sp_m}^{\mathcal{P}(\Theta)}$ equals sp_m, theorem 14.4 (2) implies that $sp_e(H) \leq sp_m(H)$ for all $H \subseteq \Theta$. On the other hand, if $\mathcal{C}_{sp}^{\mathcal{E}_s} = (\mathcal{R}(\Theta), \mathcal{A}_{\mathcal{E}_s}, P, \Gamma_{\mathcal{E}_s}, \Theta)$, then for all $H \subseteq \Theta$ we have

$$\begin{aligned}
sp_e(H) &= P_*(u_{\mathcal{E}_s}(H)) \\
&\geq sup\ \{P(\cup_{i=1}^n u_{\mathcal{E}_s}(E_i)) : E_i \subseteq H, E_i \in \mathcal{E}_s, i = 1, \ldots, n; n = 1, \ldots\}\} \\
&= sup\ \{\sum\{(-1)^{|I|+1} P(\cap_{i \in I} u_{\mathcal{E}_s}(E_i)) : \emptyset \neq I \subseteq \{1, \ldots, n\}\} : \\
&\quad E_i \subseteq H, E_i \in \mathcal{E}_s, i = 1, \ldots, n; n = 1, \ldots\}\} \\
&= sup\ \{\sum\{(-1)^{|I|+1} sp(\cap_{i \in I} E_i) : \emptyset \neq I \subseteq \{1, \ldots, n\}\} : \\
&\quad E_i \subseteq H, E_i \in \mathcal{E}_s, i = 1, \ldots, n; n = 1, \ldots\}\} \\
&= sp_m(H)
\end{aligned} \tag{A14.17}$$

because

$$\cup_{i=1}^n u_{\mathcal{E}_s}(E_i) \subseteq u_{\mathcal{E}_s}(\cup_{i=1}^n E_i) \subseteq u_{\mathcal{E}_s}(H). \tag{A14.18}$$

So finally $sp_e(H) = sp_m(H)$. It remains to prove that sp_m indeed is a support function, i.e. it satisfies conditions (1) and (2) of theorem 13.5.

Let $\rho : \mathcal{E}_s \to \mathcal{M}$ be the allocation of probability corresponding to $\mathcal{C}_{sp}^{\mathcal{E}_s}$ and μ the associated completely additive probability measure on $\mathcal{M}$. Then, according to Sect. 13.3, for all $E \in \mathcal{E}_s$ we have

$$sp_{\mathcal{C}_{sp}^{\mathcal{E}_s}}(E) = \mu(\rho(E)) \tag{A14.19}$$

and hence $sp(E) = \mu(\rho(E))$ because $\mathcal{C}_{sp}^{\mathcal{E}_s}$ is a canonical hint for sp.

Define $\bar{\rho}(H) = \vee\{\rho(E) : E \in \mathcal{E}_s, E \subseteq H\}$ for all $H \subseteq \Theta$. We first show that $\bar{\rho}$ is an allocation of probability, i.e. a $\cap$-intersection homomorphism on the power set $\mathcal{P}(\Theta)$ of Θ. Clearly $\bar{\rho}(H_1) \geq \bar{\rho}(H_1 \cap H_2)$, $\bar{\rho}(H_2) \geq \bar{\rho}(H_1 \cap H_2)$ and thus $\bar{\rho}(H_1) \cap \bar{\rho}(H_2) \geq \bar{\rho}(H_1 \cap H_2)$. On the other hand, let $E_i \in \mathcal{E}_s$, $E_i \subseteq H_i$ for $i = 1, 2$. Then

$$\rho(E_1) \wedge \rho(E_2) = \rho(E_1 \cap E_2);\ E_1 \cap E_2 \in \mathcal{E}_s, E_1 \cap E_2 \subseteq H_1 \cap H_2. \tag{A14.20}$$

From this it follows that

$$\rho(E_1) \wedge \rho(E_2) \leq \bar{\rho}(H_1 \cap H_2) \tag{A14.21}$$

and this in turn implies that $\bar{\rho}(H_1) \wedge \bar{\rho}(H_2) \leq \bar{\rho}(H_1 \cap H_2)$ and thus $\bar{\rho}(H_1) \wedge \bar{\rho}(H_2) = \bar{\rho}(H_1 \cap H_2)$.

By theorem 13.8 $\mu(\bar{\rho}(H))$ satisfies conditions (1) and (2) of theorem 13.5. It remains to show that $sp_m(H) = \mu(\bar{\rho}(H))$.

$$\begin{aligned}sp_m(H) &= \sup\{\textstyle\sum\{(-1)^{|I|+1}sp(\cap_{i\in I}E_i) : \emptyset\neq I\subseteq\{1,\ldots,n\}\} :\\ &\qquad E_i\subseteq H, E_i\in\mathcal{E}_s, i=1,\ldots,n; n=1,\ldots\}\\ &= \sup\{\textstyle\sum\{(-1)^{|I|+1}\mu(\wedge_{i\in I}\rho(E_i)) : \emptyset\neq I\subseteq\{1,\ldots,n\}\} : \ldots\}\\ &= \sup\{\mu(\vee_{i=1}^n\rho(E_i)) : E_i\in\mathcal{E}_s, E_i\subseteq H, i=1,\ldots,n; n=1,\ldots\},\end{aligned} \tag{A14.22}$$

where the last equality is justified by theorem 13.7.

The family $\vee_{i=1}^n\rho(E_i)$ for $E_i\in\mathcal{E}_s$, $E_i\subseteq H$ forms an upward net because $(\vee_{i=1}^n\rho(E_i))\vee(\vee_{i=n+1}^m\rho(E_i))\leq\vee_{i=1}^m\rho(E_i)$. Thus by theorem 13.6 the last term in (A14.22) equals

$$\begin{aligned}&\mu(\vee\{\vee_{i=1}^n\rho(E_i) : E_i\in\mathcal{E}_s, E_i\subseteq H, i=1,\ldots,n; n=1,\ldots\})\\ &= \mu(\vee\{\rho(E) : E\in\mathcal{E}_s, E\subseteq H\}) = \mu(\bar{\rho}(H)).\end{aligned} \tag{A14.23}$$

Equation (14.8) follows from (14.7) using (13.17) and theorem 13.3 (3) and (4) together with de Morgan laws. □

Proof of Theorem 14.6

Since sp' restricted to $\mathcal{E}_s$ coincides with sp, theorem 14.4 (1) implies that there exist canonical hints $\mathcal{C}_{sp}^{\mathcal{E}_s}$ and $\mathcal{C}_{sp'}^{\mathcal{P}(\Theta)}$ such that $\mathcal{C}_{sp'}^{\mathcal{P}(\Theta)}\subseteq\mathcal{C}_{sp}^{\mathcal{E}_s}$. Since the extension of sp' is sp' itself, theorem 14.4 (2) implies that $sp_e(H)\leq sp'_e(H)$ for all $H\subseteq\Theta$. □

Proof of Theorem 14.7

Define

$$\begin{aligned}sp_m(H) = sup\{\textstyle\sum\{(-1)^{|I|+1}sp(\cap_{i\in I}E_i) : \emptyset\neq I\subseteq\{1,\ldots,n\}\} :\\ E_i\subseteq H, E_i\in\mathcal{E}, i=1,\ldots,n; n=1,\ldots\}\end{aligned} \tag{A14.24}$$

for all $H\subseteq\Theta$. Then by theorem 14.5 there is a canonical hint $\mathcal{C}_{sp}^{\mathcal{E}}$ such that its extended support function sp_e satisfies $sp_e = sp_m$. So it must be proved that $sp_m(H) = sup\,\{sp(E) : E\in\mathcal{E}, E\subseteq H\}$. It is clear that $sp_m(H)\geq sup\,\{sp(E) : E\in\mathcal{E}, E\subseteq H\}$ and hence it remains to prove the reverse inequality. Let ρ and μ be the allocation of probability and the probability measure generated from $\mathcal{C}_{sp}^{\mathcal{E}}$ and recall that $sp(H)=\mu(\rho(H))$ for all $H\in\mathcal{E}$. Also note that

$$\begin{aligned}\rho(\cup_{i=1}^n E_i) &= \pi(u(\cup_{i=1}^n E_i))\geq\pi(\cup_{i=1}^n u(E_i))\\ &= \vee_{i=1}^n\pi(u(E_i)) = \vee_{i=1}^n\pi(u(E_i)) = \vee_{i=1}^n\rho(E_i).\end{aligned} \tag{A14.25}$$

This equation together with equation (A14.22) imply that

$$\begin{aligned}sp_m(H) &= sup\,\{\mu(\vee_{i=1}^n\rho(E_i)) : E_i\in\mathcal{E}, E_i\subseteq H, i=1,\ldots,n; n=1,\ldots\}\\ &\leq sup\,\{\mu(\rho(\cup_{i=1}^n E_i)) : E_i\in\mathcal{E}, E_i\subseteq H, i=1,\ldots,n; n=1,\ldots\}\\ &\leq sup\,\{\mu(\rho(E)) : E\in\mathcal{E}, E\subseteq H\}\\ &= sup\,\{sp(E) : E\in\mathcal{E}, E\subseteq H\},\end{aligned} \tag{A14.26}$$

which implies that $sp_m = sp_e$. Equation (14.11) can easily be proved in a similar way using the allowment of probability ξ generated by $\mathcal{C}^{\mathcal{E}}_{sp}$. □

Proof of Theorem 14.8

Let $sp_{cm}(H)$ denote the right-hand side of equation (14.13) and let $\rho : \mathcal{E} \to \mathcal{M}$ denote the allocation of probability corresponding to sp and let μ be the associated probability measure on $\mathcal{M}$. Note that $\rho(\cap_{i=1}^{\infty} E_i) = \wedge_{i=1}^{\infty} \rho(E_i)$ because $\mathcal{E}$ is a σ-algebra. Let's define $\bar{\rho} : \mathcal{P}(\Theta) \to \mathcal{M}$ by

$$\bar{\rho}(H) = \vee \{\wedge_{i=1}^{\infty} \rho(E_i) : E_1 \supseteq E_2 \supseteq \ldots \in \mathcal{E}, \cap_{i=1}^{\infty} E_i \subseteq H\}. \qquad (A14.27)$$

First we show that $sp_{cm} = \mu \circ \bar{\rho}$. In fact,

$$\begin{aligned} sp_{cm}(H) &= \sup \{lim_{i \to \infty} sp(E_i) : E_1 \supseteq E_2 \supseteq \ldots \in \mathcal{E}, \cap_{i=1}^{\infty} E_i \subseteq H\} \\ &= \sup \{sp(\cap_{i=1}^{\infty} E_i) : E_1 \supseteq E_2 \supseteq \ldots \in \mathcal{E}, \cap_{i=1}^{\infty} E_i \subseteq H\} \\ &= \sup \{\mu(\rho(\cap_{i=1}^{\infty} E_i)) : E_1 \supseteq E_2 \supseteq \ldots \in \mathcal{E}, \cap_{i=1}^{\infty} E_i \subseteq H\}. \end{aligned} \qquad (A14.28)$$

Now let's show that the family $\rho(\cap_{i=1}^{\infty} E_i)$, where

$$E_1 \supseteq E_2 \supseteq \ldots, E_i \in \mathcal{E}_s, \cap_{i=1}^{\infty} E_i \subseteq H \qquad (A14.29)$$

forms an upward net. Let $\rho(\cap_{i=1}^{\infty} E_i)$ and $\rho(\cap_{i=1}^{\infty} E'_i)$ be two elements of the family. Since $\rho(A) \vee \rho(B) \leq \rho(A \cup B)$, it follows that

$$\begin{aligned} \rho(\cap_{i=1}^{\infty} E_i) &\leq \rho(\cap_{i=1}^{\infty} E_i) \vee \rho(\cap_{i=1}^{\infty} E'_i) \\ &\leq \rho((\cap_{i=1}^{\infty} E_i) \cup (\cap_{i=1}^{\infty} E'_i)) \\ &= \rho(\cap_{i=1}^{\infty} (E_i \cup E'_i)) \end{aligned} \qquad (A14.30)$$

and similarly $\rho(\cap_{i=1}^{\infty} E'_i) \leq \rho(\cap_{i=1}^{\infty} (E_i \cup E'_i))$. So finally

$$\rho(\cap_{i=1}^{\infty} E_i) \vee \rho(\cap_{i=1}^{\infty} E'_i) \leq \rho(\cap_{i=1}^{\infty} (E_i \cup E'_i)) \qquad (A14.31)$$

and $\rho(\cap_{i=1}^{\infty} (E_i \cup E'_i))$ is in the family because $E_i \cup E'_i$ is in $\mathcal{E}$ since $\mathcal{E}$ is a σ-algebra. Then by theorem 13.6 (ii) the last term of (A14.28) equals

$$\mu(\vee\{\rho(\cap_{i=1}^{\infty} E_i) : \ldots\}) = \mu(\bar{\rho}(H)). \qquad (A14.32)$$

We are going to show below that $\bar{\rho}(\cap_{i=1}^{\infty} H_i) = \wedge_{i=1}^{\infty} \bar{\rho}(H_i)$ for any sequence $H_1, H_2, \ldots$ in $\mathcal{P}(\Theta)$. Then by theorem 13.8 we can conclude that sp_{cm} is a support function and that $sp_{cm}(\cap_{i=1}^{\infty} H_i) = \lim_{i \to \infty} sp_{cm}(H_i)$, i.e. sp_{cm} is continuous.

Clearly we have $sp_m(H) = sp(H)$ for $H \in \mathcal{E}$. Since sp_{cm} is a support function, theorem 14.3 implies that there are canonical hints $\mathcal{C}^{\mathcal{E}}_{sp}$ and $\mathcal{C}^{\mathcal{P}(\Theta)}_{sp_{cm}}$ whose extensions sp_e and sp_{cm} satisfy $sp_e(H) \leq sp_{cm}(H)$ for all $H \subseteq \Theta$. Since sp is continuous, note that in $\mathcal{C}^{\mathcal{E}}_{sp} = (\mathcal{R}(\Theta), \mathcal{A}_{\mathcal{E}}, P, \Gamma_{\mathcal{E}}, \Theta)$ the set of restrictions $\mathcal{R}(\Theta)$ is this time the set of all σ-restrictions $\mathcal{R}_\sigma(\Theta)$ and $\Gamma_{\mathcal{E}}(\mathcal{R})$ is also a σ-restriction. This implies that $u_{\mathcal{E}}(\cap_{i=1}^{\infty} E_i) = \cap_{i=1}^{\infty} u_{\mathcal{E}}(E_i)$ according to theorem 13.3 (5). Then for all $H \subseteq \Theta$ we have

$$\begin{aligned} sp_e(H) &= sup\ \{P(A) : A \in \mathcal{A}_{\mathcal{E}}, A \subseteq u_{\mathcal{E}}(H)\} \\ &\geq sup\ \{P(u_{\mathcal{E}}(E)) : E \in \mathcal{E}, E \subseteq H\} \\ &\geq sup\ \{P(u_{\mathcal{E}}(\cap_{i=1}^{\infty} E_i)) : E_1 \supseteq E_2 \supseteq \ldots \in \mathcal{E}, \cap_{i=1}^{\infty} E_i \subseteq H\} \\ &= sup\ \{P(\cap_{i=1}^{\infty} u_{\mathcal{E}}(E_i)) : E_1 \supseteq E_2 \supseteq \ldots \in \mathcal{E}, \cap_{i=1}^{\infty} E_i \subseteq H\} \\ &= sup\ \{\lim_{i\to\infty} P(u_{\mathcal{E}}(E_i)) : E_1 \supseteq E_2 \supseteq \ldots \in \mathcal{E}, \cap_{i=1}^{\infty} E_i \subseteq H\} \\ &= sup\ \{\lim_{i\to\infty} sp(E_i) : E_1 \supseteq E_2 \supseteq \ldots \in \mathcal{E}, \cap_{i=1}^{\infty} E_i \subseteq H\} \\ &= sp_{cm}(H). \end{aligned} \tag{A14.33}$$

So finally $sp_e = sp_{cm}$ and assertions (1) and (2) of the theorem are proved. It remains to show that

$$\bar{\rho}(\cap_{i=1}^{\infty} H_i) = \wedge_{i=1}^{\infty} \bar{\rho}(H_i). \tag{A14.34}$$

For $H \subseteq \Theta$, define

$$\tilde{\rho}(H) = \vee\{\wedge_{B\in\mathcal{B}} \rho(B) : \mathcal{B} \text{ is acountable subset of } \mathcal{E}, \cap\mathcal{B} \subseteq H\}. \tag{A14.35}$$

Then we have clearly $\bar{\rho}(H) \leq \tilde{\rho}(H)$. On the other hand, we also have $\bar{\rho}(H) \geq \tilde{\rho}(H)$ because for any sequence $B_1, B_2, \ldots$ in $\mathcal{E}$ there is a decreasing sequence $E_1, E_2, \ldots$ defined by

$$E_i = B_1 \cap B_2 \cap \ldots \cap B_i \tag{A14.36}$$

which satisfies both $\cap_{i=1}^{\infty} E_i = \cap_{i=1}^{\infty} B_i$ and $\wedge_{i=1}^{\infty} \rho(E_i) = \wedge_{i=1}^{\infty} \rho(B_i)$. So $\bar{\rho} = \tilde{\rho}$ and hence it must be proved that

$$\tilde{\rho}(\cap_{i=1}^{\infty} A_i) = \wedge_{i=1}^{\infty} \tilde{\rho}(A_i) \tag{A14.37}$$

for every sequence $A_1, A_2, \ldots$ of subsets of Θ. But the equation (A14.37) is proved on page 835 of Shafer (1979).

The assertion of the theorem concerning the extended plausibility function pl_e follows from duality. □

15. Dempster's Rule in the General Case

15.1 Combination of Hints

Let $\mathcal{H}_1$ and $\mathcal{H}_2$ be two hints relative to the same frame Θ defined by $(\Omega_1, \mathcal{A}_1, P_1, \Gamma_1, \Theta)$ and $(\Omega_2, \mathcal{A}_2, P_2, \Gamma_2, \Theta)$. The basic idea for the combination of these hints is as before in Chap. 3 that in each hint there must be exactly one correct interpretation $\omega_i, i = 1, 2$ such that, looking at the two hints together, ω_1 and ω_2 must be simultaneously correct interpretations. (ω_1, ω_2) is then the combined correct interpretation. Therefore, in order to combine the two hints $\mathcal{H}_1$ and $\mathcal{H}_2$ into one new combined hint, form first the product space of the combined interpretations from the two hints $(\Omega_1 \times \Omega_2, \mathcal{A}_1 \otimes \mathcal{A}_2, P')$ where P' is any probability measure on $\mathcal{A}_1 \otimes \mathcal{A}_2$ reflecting the common likelihood of combined interpretations and having P_1, P_2 as marginals. The two hints are called *independent* if the interpretations of the two hints are stochastically independent. Then P' is the product measure of P_1 and P_2. This is the case which will be pursued here.

If the combined interpretation (ω_1, ω_2) is the correct one, then the restriction

$$\Gamma(\omega_1, \omega_2) = \Gamma_1(\omega_1) \cap \Gamma_2(\omega_2) \tag{15.1}$$

must necessarily hold, unless $\Gamma_1(\omega_1) \cap \Gamma_2(\omega_2)$ does not exist. In the latter case, the interpretations ω_1 and ω_2 are called *contradictory.*

Let $\mathcal{S}(\omega_1, \omega_2)$ and $\mathcal{P}(\omega_1, \omega_2)$ be the families of implied and possible subsets relative to the restriction $\Gamma(\omega_1, \omega_2)$. We introduce here the convention that for contradictory interpretations ω_1 and ω_2 the set $\mathcal{S}(\omega_1, \omega_2)$ contains all subsets of Θ whereas $\mathcal{P}(\omega_1, \omega_2)$ is empty. Define

$$\begin{aligned} u'(H) &= \{(\omega_1, \omega_2) \in \Omega_1 \times \Omega_2 : H \in \mathcal{S}(\omega_1, \omega_2)\}, \\ v'(H) &= \{(\omega_1, \omega_2) \in \Omega_1 \times \Omega_2 : H \in \mathcal{P}(\omega_1, \omega_2)\}. \end{aligned} \tag{15.2}$$

Theorem 13.3, except (1) and (2), clearly applies to u' and v'; (1) is replaced by $v'(\emptyset) = \emptyset$ and (2) by $u'(\Theta) = \Omega_1 \times \Omega_2$. $u'(\emptyset)$ represents the set of contradictory interpretations pairs. Such a pair cannot be the correct one (contradictions are not possible). Therefore contradictory interpretations must be eliminated and the probability must be conditioned on the event $v'(\Theta)$ that there in no contradiction. Provided that $u'(\emptyset)$ is measurable, i.e. $u'(\emptyset) \in \mathcal{A}_1 \otimes \mathcal{A}_2$, and that $P'(v'(\Theta)) > 0$, the new combined hint $\mathcal{H}_1 \oplus \mathcal{H}_2 = (\Omega, \mathcal{A}, P, \Gamma, \Theta)$ can be formed, where $\Omega = v'(\Theta)$, $\mathcal{A} = v'(\Theta) \cap \mathcal{A}_1 \otimes \mathcal{A}_2$, $P(A) = P'(A)/P'(v'(\Theta))$ for all $A \in v'(\Theta) \cap \mathcal{A}_1 \otimes \mathcal{A}_2$ and Γ is defined by (15.1) (and restricted to Ω). This way of combining hints is called, as before, *Dempster's rule.*

Let u and v be defined by (13.1) and (13.2) relative to the hint $\mathcal{H}_1 \oplus \mathcal{H}_2$. Then $u(H) = u'(H) \cap \Omega = u'(H) - u'(\emptyset)$ and $v(H) = v'(H)$.

Dempster's rule can be extended to the case where $u'(\emptyset)$ is not measurable. In this case, provided that $P'^*(v'(\Theta)) > 0$, the probability space $(\Omega, \mathcal{A}, P)$ can be considered where

$$P(A) = P'^*(A \cap v'(\Theta))/P'^*(v'(\Theta))$$

and Ω, $\mathcal{A}$ are defined as above. It is this space, in which the contradictory interpretations have been eliminated, that is used to construct the combined hint $\mathcal{H}_1 \oplus \mathcal{H}_2 = (\Omega, \mathcal{A}, P, \Gamma, \Theta)$. As before, we have $u(H) = u'(H) \cap \Omega = u'(H) - u'(\emptyset)$ and $v(H) = v'(H)$.

Let $\mathcal{E}_s$ and $\mathcal{E}_p$ be the classes of s- and p-measurable sets relative to the hint $\mathcal{H}_1 \oplus \mathcal{H}_2$. Denote by $\mathcal{E}'_s$ and $\mathcal{E}'_p$ the classes of sets H such that $u'(H)$ and $v'(H)$ are measurable with respect to $\mathcal{A}_1 \otimes \mathcal{A}_2$. From $u'(H) \in \mathcal{A}_1 \otimes \mathcal{A}_2$ it follows that $u(H) \in \Omega \cap \mathcal{A}_1 \otimes \mathcal{A}_2$ and thus $\mathcal{E}'_s \subseteq \mathcal{E}_s$. Similarly, because $v'(H) \subseteq \Omega$, $v'(H) \in \mathcal{A}_1 \otimes \mathcal{A}_2$ implies $v(H) \in \Omega \cap \mathcal{A}_1 \otimes \mathcal{A}_2$ or $\mathcal{E}'_p \subseteq \mathcal{E}_p$. If $u'(\emptyset)$ is measurable, then $\mathcal{E}'_s = \mathcal{E}_s$ and $\mathcal{E}'_p = \mathcal{E}_p$.

There is an interesting integral representation of the support- and plausibility functions of the hint obtained by combining two set-focussed hints:

Theorem 15.1 *Let $\mathcal{H}_1 \oplus \mathcal{H}_2 = (\Omega, \mathcal{A}, P, \Gamma, \Theta)$ be a combined hint derived from set-focussed hints $\mathcal{H}_i = (\Omega_i, \mathcal{A}_i, P_i, \Gamma_i, \Theta)$, $i = 1, 2$ and let sp, pl and sp_i, pl_i denote the support and plausibility functions of $\mathcal{H}_1 \oplus \mathcal{H}_2$ and $\mathcal{H}_i$ respectively. Let $T_i(\omega_i)$ be the bases of $\Gamma_i(\omega_i)$, $\Gamma_i(\omega_i) = T_i(\omega_i)$. Then for all $H \in \mathcal{E}'_s$*

$$\begin{aligned} sp(H) &= c \int sp_1(T_2^c(\omega_2) \cup H)\, dP_2(\omega_2) - (c-1) \\ &= c \int sp_2(T_1^c(\omega_1) \cup H)\, dP_1(\omega_1) - (c-1) \end{aligned} \tag{15.3}$$

and for all $H \in \mathcal{E}'_p$

$$\begin{aligned} pl(H) &= c \int pl_1(T_2(\omega_2) \cap H)\, dP_2(\omega_2) \\ &= c \int pl_2(T_1(\omega_1) \cap H)\, dP_1(\omega_1). \end{aligned} \tag{15.4}$$

This theorem is a generalization of theorem 4.2 of Chap. 4.

If $u'(\emptyset)$ is measurable, then $\Theta \in \mathcal{E}'_p$ and

$$pl(\Theta) = c \int_{\Omega_1} pl_2(T_1(\omega_1))\, dP_1(\omega_1) = 1. \tag{15.5}$$

This makes it possible to determine the constant c. In the general case c equals $1/P'^*(v'(\Theta))$ as can be seen in the proof of the theorem.

One can define an unnormalized "pseudo" plausibility function $pl'(H) = P'(v'(H))$ on $\mathcal{E}'_p$. pl' can be extended to the power set $\mathcal{P}(\Theta)$ of Θ in exactly the same way as an ordinary plausibility function by $pl'_e(H) = P'^*(v'(H))$. It is true that pl'_e determines already the extension of the plausibility function (and therefore also of the support function) of the combined hint $\mathcal{H}_1 \oplus \mathcal{H}_2$ as the following theorem shows.

Theorem 15.2 *If $pl'(\Theta) > 0$, then for all $H \subseteq \Theta$*

$$pl_e(H) = pl'_e(H)/pl'_e(\Theta). \tag{15.6}$$

An important special case of Dempster's rule arises if one of the two hints $\mathcal{H}_1$ and $\mathcal{H}_2$, which are to be combined, is a simple hint whose interpretations focus with probability 1 on a set-based restriction $\mathcal{R}$ with base R. If this is the case for say $\mathcal{H}_2$, then we say that $\mathcal{H}_1$ is to be *conditioned* on the condition R and the result is written as $\mathcal{H}_1 \oplus R$. R is used here as a substitute for $\mathcal{H}_2$. Let $\mathcal{H} = (\Omega, \mathcal{A}, P, \Gamma, \Theta)$ be an arbitrary hint with dual classes $\mathcal{E}_s$ and $\mathcal{E}_p$ of $s-$ and $p-$measurable sets and applications u, v defined by (13.1) and (13.2). Then $\mathcal{H} \oplus R$ has focal restrictions $\Gamma(\omega) \cap R$ for all $\omega \in v(R)$. If $v''(H)$ is defined by (13.2) relative to the hint $\mathcal{H} \oplus R$, then $v''(H) = v(R \cap H)$. $u(R^c)$ contains all interpretations of $\mathcal{H}$ which are in contradiction to the condition R and these are almost all interpretations of $\mathcal{H}$ which are contradictory to $\mathcal{H}_2$.

The following theorem shows how the support and plausibility functions of a conditioned hint $\mathcal{H}$ can be obtained from those of the hint $\mathcal{H}$.

Theorem 15.3 *Denote the support- and plausibility functions of $\mathcal{H} \oplus R$ by $sp(\cdot|R)$ and $pl(\cdot|R)$ and the extended support- and plausibility functions by $sp_e(\cdot|R)$ and $pl_e(\cdot|R)$. Let sp and pl (sp_e and pl_e) be the (extended) support and plausibility functions of $\mathcal{H}$. Then, if R is p-measurable*

$$\begin{aligned} sp(H|R) &= (sp(R^c \cup H) - sp(R^c))/(1 - sp(R^c)), \text{ for all } H \in \mathcal{E}_s, \\ pl(H|R) &= pl(R \cap H)/pl(R), \text{ for all } H \in \mathcal{E}_p. \end{aligned} \tag{15.7}$$

In the general case

$$\begin{aligned} sp_e(H|R) &= (sp_e(R^c \cup H) - sp_e(R^c))/(1 - sp_e(R^c)), \text{ for all } H \subseteq \Theta, \\ pl_e(E|R) &= pl_e(R \cap H)/pl_e(R), \text{ for all } H \subseteq \Theta. \end{aligned} \tag{15.8}$$

This theorem is a generalization of theorem 4.7. Note as a corollary that $\mathcal{H} \oplus \Theta = \mathcal{H}$: the vacuous hint Θ is a neutral element for the operation $\oplus$. The next theorem states that inclusion is maintained under Dempster's rule.

Theorem 15.4 *Let $\mathcal{H}_1, \mathcal{H}_2, \mathcal{H}_1'', \mathcal{H}_2''$ be four hints such that $\mathcal{H}_1'' \subseteq \mathcal{H}_1$ and $\mathcal{H}_2'' \subseteq \mathcal{H}_2$. Then $\mathcal{H}_1'' \oplus \mathcal{H}_2'' \subseteq \mathcal{H}_1 \oplus \mathcal{H}_2$.*

In the case of theorem 15.4, the constant k appearing in theorem 14.2 equals $P^*(\Omega'')$, where Ω'' contains all combined interpretations (ω_1, ω_2) which are not contradictory under $\mathcal{H}_1'' \oplus \mathcal{H}_2''$. Some combined interpretations, which are not contradictory under $\mathcal{H}_1 \oplus \mathcal{H}_2$, may however be contradictory under $\mathcal{H}_1'' \oplus \mathcal{H}_2''$. This accounts for the possible difference between Ω and Ω''. If the situation is such that $\Omega'' = \Omega$, then $k = 1$ and $[sp_e''(H), pl_e''(H)] \subseteq [sp_e(H), pl_e(H)]$.

Let $\mathcal{V}$ be a vacuous hint. Then theorem 15.4 implies that $\mathcal{H}_1 \oplus \mathcal{H}_2 \subseteq \mathcal{H}_1 \oplus \mathcal{V}$. Similarily $\mathcal{H}_1 \oplus \mathcal{H}_2 \subseteq \mathcal{V} \oplus \mathcal{H}_2$. As the combination of a hint with a vacuous hint does not add new information to the hint, this result shows that a combined hint $\mathcal{H}_1 \oplus \mathcal{H}_2$ is always finer than each of the two hints $\mathcal{H}_1$ and $\mathcal{H}_2$ alone. And in particular, if sp is the support function of $\mathcal{H}_1 \oplus \mathcal{H}_2$, then we have $[sp_e(H), pl_e(H)] \subseteq [sp_{1e}(H), pl_{1e}(H)]$ and $[sp_e(H), pl_e(H)] \subseteq [sp_{2e}(H), pl_{2e}(H)]$ if $\mathcal{H}_1$ and $\mathcal{H}_2$ have no contradictory interpretations.

15.2 Hints on Product Spaces

Up to now we have considered hints relative to the same frame of discernment Θ. Hints may however be relative to different frames or questions which may or may not be related as has been discussed in Chap. 7. Here we will consider two hints $\mathcal{H}_1$ and $\mathcal{H}_2$ relative to two completely independent frames Θ_1 and Θ_2. Let $\mathcal{H}_1$ be defined by $(\Omega_1, \mathcal{A}_1, P_1, \Gamma_1, \Theta_1)$ and $\mathcal{H}_2$ by $(\Omega_2, \mathcal{A}_2, P_2, \Gamma_2, \Theta_2)$. Denote the respective support- and plausibility functions by sp_i and pl_i, $i = 1,2$ and define u_i and v_i by (13.1) and (13.2) relative to the hint $\mathcal{H}_i$. These two hints may be considered on the product space $\Theta_1 \times \Theta_2$, where they can be combined to answer combined questions, relating to both frames Θ_1 and Θ_2 simultaneously. In order to study this combination we must first show how a hint relative to a frame Θ_1 can be extended to a hint relative to a new, extended frame $\Theta_1 \times \Theta_2$ in the general case.

This extension is obtained in the spirit of Sect. 8.2 by extending the restrictions $\Gamma_1(\omega_1)$ in Θ_1 to $\Theta_1 \times \Theta_2$ (see Sect. 13.1). The extension of a hint $\mathcal{H}_1 = (\Omega_1, \mathcal{A}_1, P_1, \Gamma_1, \Theta_1)$ to the product space $\Theta_1 \times \Theta_2$ is defined by $\mathcal{H}_1 | \Theta_1 \times \Theta_2 = (\Omega_1, \mathcal{A}_1, P_1, \Gamma_1 \times \Theta_2, \Theta_1 \times \Theta_2)$, where $(\Gamma_1 \times \Theta_2)(\omega_1) = \Gamma_1(\omega_1) \times \{\Theta_2\}$. This is called the *extension* because no information is added to the hint. This can of course be formally expressed in terms of family of compatible frames and refinings.

Theorem 15.5 *For any subset H of $\Theta_1 \times \Theta_2$ denote by $C_1(H)$ the base of the largest cylinder with base in Θ_1 contained in H and by $C_1'(H)$ the base of the smallest cylinder set with base in Θ_1 containing H. Let sp_1, pl_1 and sp_1', pl_1'*

be the support and plausibility functions of the hints $\mathcal{H}_1$ and $\mathcal{H}_1|\Theta_1 \times \Theta_2$ respectively. Then

(i) A set $H \subseteq \Theta_1 \times \Theta_2$ is s-measurable with respect to the extended hint $\mathcal{H}_1|\Theta_1 \times \Theta_2$ iff $C_1(H) \in \mathcal{E}_{s1}$ and in this case $sp'_1(H) = sp_1(C_1(H))$. More generally, for any $H \subseteq \Theta_1 \times \Theta_2$ we have

$$sp'_{1e}(H) = sp_{1e}(C_1(H)). \tag{15.9}$$

(ii) A set $H \subseteq \Theta_1 \times \Theta_2$ is p-measurable with respect to the extended hint $\mathcal{H}_1|\Theta_1 \times \Theta_2$, iff $C'_1(H) \in \mathcal{E}_{p1}$ and in this case $pl'_1(H) = pl_1(C'_1(H))$. More generally, for any $H \subseteq \Theta_1 \times \Theta_2$, we have

$$pl'_{1e}(H) = pl_{1e}(C'_1(H)). \tag{15.10}$$

Let $\mathcal{H}_1 \oplus \mathcal{H}_2 = (\Omega_1 \times \Omega_2, \mathcal{A}_1 \otimes \mathcal{A}_2, P, \Gamma, \Theta_1 \times \Theta_2)$ denote the combined hint of $\mathcal{H}_1|\Theta_1 \times \Theta_2$ and $\mathcal{H}_2|\Theta_1 \times \Theta_2$ on $\Theta_1 \times \Theta_2$. Note that this hint focuses on restrictions $\Gamma(\omega_1, \omega_2) = (\Gamma_1(\omega_1) \times \Theta_2) \cap (\Theta_1 \times \Gamma_2(\omega_2)) = \Gamma_1(\omega_1) \times \Gamma_2(\omega_2)$ and that there are no contradictory combined interpretations. Let sp and pl denote the support and plausibility functions of $\mathcal{H}_1 \oplus \mathcal{H}_2$ on $\Theta_1 \times \Theta_2$, sp_e, pl_e their extensions to the power set of $\Theta_1 \times \Theta_2$, and $\mathcal{E}_s$ and $\mathcal{E}_p$ the dual multiplicative and addiditve classes of the s- and p-measurable sets. Then we have the following result.

Theorem 15.6

(i) If $A \in \mathcal{E}_{s1}$, $B \in \mathcal{E}_{s2}$, then $A \times B \in \mathcal{E}_s$ and

$$sp(A \times B) = sp_1(A) sp_2(B) \tag{15.11}$$

(ii) If $A \in \mathcal{E}_{p1}$, $B \in \mathcal{E}_{p2}$, then $A \times B \in \mathcal{E}_p$ and

$$pl(A \times B) = pl_1(A) pl_2(B). \tag{15.12}$$

(iii) More generally, for any $A \subseteq \Theta_1$ and any $B \subseteq \Theta_2$:

$$\begin{aligned} sp_e(A \times B) &= sp_{1e}(A) sp_{2e}(B) \\ pl_e(A \times B) &= pl_{1e}(A) pl_{2e}(B). \end{aligned} \tag{15.13}$$

A15 Appendix: Proofs of Theorems in Chapter 15

Proof of Theorem 15.1

Consider $v'(H) = v(H) \in \mathcal{A}_1 \otimes \mathcal{A}_2$. For all $\omega_1 \in \Omega_1$, the ω_1-section of the set $v'(H)$ is the subset $v'(H|\omega_1) = \{\omega_2 \in \Omega_2 : T_1(\omega_1) \cap T_2(\omega_2) \cap H \neq \emptyset\} = v_2(T_1(\omega_1) \cap H)$ of Ω_2. Similarly, the ω_2-section of $v'(H)$ is the subset $v'(H|\omega_2) = v_1(T_2(\omega_2) \cap H)$ of Ω_1. The ω_1- and ω_2-sections of a set of

$\mathcal{A}_1 \otimes \mathcal{A}_2$ belong to $\mathcal{A}_1$ and $\mathcal{A}_2$ respectively (Halmos, 1950; Neveu, 1964). Thus $v_1(T_2(\omega_2) \cap H) \in \mathcal{A}_1$ for all $\omega_2 \in \Omega_2$ and $v_2(T_1(\omega_1) \cap H) \in \mathcal{A}_2$ for all $\omega_1 \in \Omega_1$, whenever $v'(H) \in \mathcal{A}_1 \otimes \mathcal{A}_2$. Let $i(\omega_1, \omega_2)$ be the indicator function of $v'(H)$. i is measurable with respect to $\mathcal{A}_1 \otimes \mathcal{A}_2$, the ω_1-section i_{ω_1} (defined on Ω_2 by $i_{\omega_1}(\omega_2) = i(\omega_1, \omega_2)$) is measurable with respect to $\mathcal{A}_2$ and

$$y(\omega_1) = \int_{\Omega_2} i_{\omega_1}(\omega_2)\, dP_2(\omega_2) \tag{A15.1}$$

is defined for all $\omega_1 \in \Omega_1$, positive and measurable with respect to $\mathcal{A}_1$ (Neveu, 1964, Prop. III 2.1). Now $i_{\omega_1}(\omega_2) = 1$, iff $\omega_2 \in v_2(T_1(\omega_1) \cap H)$ and hence $y(\omega_1) = P_2(v_2(T_1(\omega_1) \cap H)) = pl_2(T_1(\omega_1) \cap H)$. Furthermore, by Fubini's theorem (Neveu, 1964, Prop. III 2.1)

$$\begin{aligned} P'(v'(H)) &= \int_{\Omega_1 \times \Omega_2} i(\omega_1, \omega_2)\, dP_1(\omega_1)\, dP_2(\omega_2) \\ &= \int_{\Omega_1} dP_1(\omega_1) \int_{\Omega_2} i_{\omega_1}(\omega_2)\, dP_2(\omega_2) \\ &= \int_{\Omega_1} pl_2(T_1(\omega_1) \cap H)\, dP_1(\omega_1). \end{aligned} \tag{A15.2}$$

On $\mathcal{A}_1 \otimes \mathcal{A}_2$ we have $P' = P'^*$ and thus $pl(H) = P'(v'(H))/P'^*(v'(\Theta))$. It follows for any $H \in \mathcal{E}'_p$ that

$$pl(H) = c \int_{\Omega_1} pl_2(T_1(\omega_1) \cap H)\, dP_1(\omega_1). \tag{A15.3}$$

The second formula in (15.4) follows by symmetry.

If $H \in \mathcal{E}'_s$, then

$$\begin{aligned} sp(H) = 1 - pl(H^c) &= 1 - c \int_{\Omega_1} pl_2(T_1(\omega_1) \cap H^c)\, dP_1(\omega_1) \\ &= 1 - c \int_{\Omega_1} (1 - sp_2(T_1^c(\omega_1) \cup H)\, dP_1(\omega_1) \\ &= c \int_{\Omega_1} sp_2(T_1^c(\omega_1) \cup H)\, dP_1(\omega_1) - (c-1). \end{aligned} \tag{A15.4}$$

The second formula in (15.3) follows in the same way. □

Proof of Theorem 15.2

If $pl'_e(\Theta) = P'^*(v'(\Theta)) = P'^*(\Omega) > 0$, then $\mathcal{H}_1 \oplus \mathcal{H}_2$ can be defined. In this case

$$\begin{aligned} pl_e(H) &= \inf\,\{P(A) : A \in \Omega \cap \mathcal{A}_1 \otimes \mathcal{A}_2,\ v(H) \subseteq A\} \\ &= \inf\,\{P'^*(A)/P'^*(\Omega) : A \in \Omega \cap \mathcal{A}_1 \otimes \mathcal{A}_2,\ v'(H) \subseteq A\} \\ &= \inf\,\{P'^*(A \cap \Omega) : A \in \mathcal{A}_1 \otimes \mathcal{A}_2,\ v'(H) \subseteq A\}/P'^*(\Omega). \end{aligned}$$

For any $\epsilon > 0$ an $A \in \mathcal{A}_1 \otimes \mathcal{A}_2, A \supseteq v'(H)$ can be found such that $P'(A) - \epsilon < pl'_e(H) = P'^*(v'(H)) \leq P'(A)$. However, we have $P'^*(A \cap \Omega) \leq P'(A)$ because $A \cap \Omega \subseteq A$ and $P'^*(A \cap \Omega) \geq pl'_e(H)$ because $A \cap \Omega \supseteq v'(H)$ and thus

$$P'^*(A \cap \Omega) - \epsilon \leq pl'_e(H) \leq P'^*(A \cap \Omega). \tag{A15.5}$$

This implies that

$$pl'_e(H) = \inf \{P'^*(A \cap \Omega) : A \in \mathcal{A}_1 \otimes \mathcal{A}_2,\ v'(H) \subseteq A\} \tag{A15.6}$$

and hence indeed $pl_e(H) = pl'_e(H)/pl'_e(\Theta)$. □

Proof of Theorem 15.3

We have for $H \in \mathcal{E}_p$

$$pl(H|R) = P(v(R \cap H))/P(v(R)) = pl(R \cap H)/pl(R). \tag{A15.7}$$

For $H \in \mathcal{E}_s$ it follows that

$$\begin{aligned} sp(H|R) &= 1 - pl(H^c|R) = 1 - pl(R \cap H^c)/pl(R) \\ &= (pl(R) - pl(R \cap H^c))/pl(R) \\ &= (1 - sp(R^c)) - (1 - sp(R^c \cup H))/(1 - sp(R^c)) \\ &= (sp(R^c \cup H) - sp(R^c))/(1 - sp(R^c)). \end{aligned} \tag{A15.8}$$

This proves (15.7). In the general case,

$$\begin{aligned} pl_e(H|R) &= inf\ \{P^*(A)/P^*(v(R)) : A \in v(R) \cap \mathcal{A}, A \supseteq v(R \cap H)\} \\ &= P^*(v(R \cap H))/P^*(v(R)) = pl_e(R \cap H)/pl_e(R). \end{aligned} \tag{A15.9}$$

The formula for $sp_e(H|R)$ is derived as above from duality. This proves (15.8). □

Proof of Theorem 15.4

$\Gamma''_1(\omega_1) \subseteq \Gamma_1(\omega_1)$ and $\Gamma''_2(\omega_2) \subseteq \Gamma_2(\omega_2)$ imply

$$\Gamma''_1(\omega_1) \cap \Gamma''_2(\omega_2) \subseteq \Gamma_1(\omega_1) \cap \Gamma_2(\omega_2).$$

This, together with $\Omega''_1 \subseteq \Omega_1$ and $\Omega''_2 \subseteq \Omega_2$ implies $\Omega'' \subseteq \Omega$. Also $\mathcal{A}''_1 \supseteq \mathcal{A}_1 \cap \Omega''_1$ and $\mathcal{A}''_2 \supseteq \mathcal{A}_2 \cap \Omega''_2$ imply that

$$\begin{aligned} \mathcal{A}'' &= \mathcal{A}''_1 \otimes \mathcal{A}''_2 \cap \Omega'' \supseteq (\mathcal{A}_1 \cap \Omega''_1) \otimes (\mathcal{A}_2 \cap \Omega''_2) \cap \Omega'' \\ &= (\mathcal{A}_1 \otimes \mathcal{A}_2) \cap (\Omega''_1 \times \Omega''_2) \cap \Omega'' \\ &= \mathcal{A}_1 \otimes \mathcal{A}_2 \cap \Omega'' = (\mathcal{A}_1 \otimes \mathcal{A}_2 \cap \Omega) \cap \Omega'' = \mathcal{A} \cap \Omega''. \end{aligned} \tag{A15.10}$$

It remains to show that

$$P''(A) = P^*(A)/P^*(\Omega'') \tag{A15.11}$$

for any $A \in \mathcal{A} \cap \Omega''$. Let Q'', Q denote the product measures of P''_1 and P''_2 and P_1 and P_2 on the product spaces $(\Omega''_1 \times \Omega''_2, \mathcal{A}''_1 \otimes \mathcal{A}''_2)$ and $(\Omega_1 \times \Omega_2, \mathcal{A}_1 \otimes \mathcal{A}_2)$

respectively. Then by definition $P''(A) = Q''^*(A)/Q''^*(\Omega'')$ for any $A \in \mathcal{A} \cap \Omega''$. It is thus sufficient to show that $Q''^*(A) = k \cdot P^*(A)$ for some constant k independent of A.

To begin with, let's suppose that the sets $\Omega_1'', \Omega_2'', \Omega''$ and Ω are measurable with respect to $\mathcal{A}_1, \mathcal{A}_2, \mathcal{A}$ and $\mathcal{A}_1 \otimes \mathcal{A}_2$ respectively. Then $P''(A) = Q''(A)/Q''(\Omega'')$ and $P^*(A)/P^*(\Omega'') = P(A)/P(\Omega'')$ for $A \in \mathcal{A} \cap \Omega''$ and we must prove that $Q^*(A) = k \cdot P(A)$. Let X_A denote the indicator function of A. Then

$$Q''(A) = \int P_1''(d\omega_1)P_2''(d\omega_2)X_A. \tag{A15.12}$$

Because X_A is a measurable function with respect to $\mathcal{A}$, it is sufficient to take the restrictions of the probability measures P_1'' and P_2'' to $\mathcal{A}_1$ and $\mathcal{A}_2$. But there these probabilities are conditional probabilities such that

$$\begin{aligned} Q''(A) &= \int P_1(d\omega_1)P_2(d\omega_2)X_A/P_1(\Omega_1'')P_2(\Omega_2'') \\ &= Q(A)/P_1(\Omega_1'')P_2(\Omega_2'') = P(A)(Q(\Omega)/P_1(\Omega_1'')P_2(\Omega_2'')). \end{aligned} \tag{A15.13}$$

This proves the theorem in the case of measurable sets $\Omega_1'', \Omega_2'', \Omega''$ and Ω. If Ω is not measurable, then there is a measurable set $\bar{\Omega}$, containing Ω, such that $Q^*(\Omega) = Q(\bar{\Omega})$. If $A \in \mathcal{A} \cap \Omega$, then $\bar{A} = A \cap \bar{\Omega}$ is measurable, contains A, and $Q^*(A) = Q(\bar{A})$. Thus $P(\bar{A}) = P(A)$ for all $A \in \mathcal{A} \cap \Omega$ and Ω may be replaced by $\bar{\Omega}$ and $\mathcal{A} \cap \Omega$ by $\mathcal{A} \cap \bar{\Omega}$ without changing the relevant probability values. In this way the case where some or all sets $\Omega_1'', \Omega_2'', \Omega''$ and Ω are not measurable can be reduced to the former case. This proves the theorem. □

Proof of Theorem 15.5

(i) Define u_1' by (13.1) with respect to the extended hint $\mathcal{H}_1|\Theta_1 \times \Theta_2$:

$$\begin{aligned} u_1'(H) &= \{\omega \in \Omega_1 : H \in \mathcal{S}_1(\omega) \times \Theta_2\} \\ &= \{\omega \in \Omega_1 : C_1(H) \times \Theta_2 \in \mathcal{S}_1(\omega) \times \Theta_2\} \\ &= \{\omega \in \Omega_1 : C_1(H) \in \mathcal{S}_1(\omega)\} = u_1(C_1(H)). \end{aligned} \tag{A15.14}$$

Thus H is s-measurable if and only if $C_1(H) \in \mathcal{E}_{s1}$ and in this case $sp_1'(H) = sp_1(C_1(H))$. Furthermore

$$\begin{aligned} sp_{1e}'(H) &= \sup\ \{P(A) : A \in \mathcal{A}_1, A \subseteq u_1'(H)\} \\ &= \sup\ \{P(A) : A \in \mathcal{A}_1, A \subseteq u_1(C_1(H))\} = sp_{1e}(C_1(H)). \end{aligned} \tag{A15.15}$$

(ii) is proved similarly using

$$\begin{aligned} v_1'(H) &= \{\omega \in \Omega_1 : H \in \mathcal{P}_1(\omega) \times \Theta_2\} \\ &= \{\omega \in \Omega_1 : pr_1(H) \in \mathcal{P}_1(\omega)\} \\ &= \{\omega \in \Omega_1 : C_1'(H) \in \mathcal{P}_1(\omega)\} = v_1(C_1'(H)). \end{aligned} \tag{A15.16}$$

□

Proof of Theorem 15.6

(i) Define u according to (13.1) relative to the hint $\mathcal{H}_1 \oplus \mathcal{H}_2$, such that

$$\begin{aligned} u(A \times B) &= \{(\omega_1, \omega_2) \in \Omega_1 \times \Omega_2 : A \times B \in \mathcal{S}_1(\omega_1) \times \mathcal{S}_2(\omega_2)\} \\ &= \{\omega_1 \in \Omega_1 : A \in \mathcal{S}_1(\omega_1)\} \times \{\omega_2 \in \Omega_2 : B \in \mathcal{S}_2(\omega_2)\} \\ &= u_1(A) \times u_2(B) \end{aligned} \tag{A15.17}$$

(see theorem 13.2). Hence, if $A \in \mathcal{E}_{s1}$, $B \in \mathcal{E}_{s2}$, then $A \times B$ is s-measurable and

$$\begin{aligned} sp(A \times B) &= P(u(A \times B)) = P(u_1(A) \times u_2(B)) \\ &= P_1(u_1(A))P_2(u_2(B)) = sp_1(A)sp_2(B). \end{aligned} \tag{A15.18}$$

(ii) Define v according to (13.2) relative to the hint $\mathcal{H}_1 \oplus \mathcal{H}_2$, such that

$$v(A \times B) = \{(\omega_1, \omega_2) \in \Omega_1 \times \Omega_2 : A \times B \in \mathcal{P}_1(\omega_1) \times \mathcal{P}_2(\omega_2)\} \tag{A15.19}$$

Now, $A \times B \in \mathcal{P}_1(\omega_1) \times \mathcal{P}_2(\omega_2)$ iff $(A \times B)^c = (A^c \times \Theta_2) \cup (\Theta_1 \times B^c) \notin \mathcal{S}_1(\omega_1) \times \mathcal{S}_2(\omega_2)$. This is the case iff both $A^c \times \Theta_2 \notin \mathcal{S}_1(\omega_1) \times \mathcal{S}_2(\omega_2)$ and $\Theta_1 \times B^c \notin \mathcal{S}_1(\omega_1) \times \mathcal{S}_2(\omega_2)$ (see theorem 13.2). This in turn is equivalent to $A^c \notin \mathcal{S}_1(\omega_1)$ and $B^c \notin \mathcal{S}_2(\omega_2)$ or to $A \in \mathcal{P}_1(\omega_1)$ and $B \in \mathcal{P}_2(\omega_2)$. Thus $v(A \times B) = v_1(A) \times v_2(B)$ and (ii) follows in the same way as (i).

(iii) First we prove that $sp_e(A \times B) = sp_e(A)sp_e(B)$. We have

$$sp_e(A \times B) = P_*(u(A \times B)) = P_*(u_1(A) \times u_2(B)) \tag{A15.20}$$

and

$$\begin{aligned} sp_{1e}(A) &= P_{1*}(u_1(A)) \\ sp_{2e}(B) &= P_{2*}(u_2(B)). \end{aligned} \tag{A15.21}$$

So we must prove that if A and B are arbitrary subsets of Ω_1 and Ω_2, then

$$P_*(A \times B) = P_{1*}(A) \times P_{2*}(B). \tag{A15.22}$$

According to Halmos (1950), there exist $A^* \in \mathcal{A}_1$ and $B^* \in \mathcal{A}_2$ such that $P_{1*}(A) = P_1(A^*)$ and $P_{2*}(B) = P_2(B^*)$ and $A^* \subseteq A$, $B^* \subseteq B$. Then $A^* \times B^* \subseteq A \times B$ and hence

$$\begin{aligned} P_*(A \times B) &\geq P(A^* \times B^*) = P_1(A^*)P_2(B^*) \\ &= P_{1*}(A)P_{2*}(B). \end{aligned} \tag{A15.23}$$

Now let's prove the reverse inequality. Again, there is $E^* \in \mathcal{A}_1 \otimes \mathcal{A}_2$ such that $P(E^*) = P_*(A \times B)$ and $E^* \subseteq A \times B$. Since $E^* \cup A^* \times B^* \subseteq A \times B$ we have $P(E^* \cup A^* \times B^*) = P_*(A \times B)$ and so we can assume that $E^* \supseteq A^* \times B^*$. Then by Fubini's theorem we have

$$\begin{aligned} P_*(A \times B) = P(E^*) &= \int P_2(E^*_x)dP_1(x) \\ &= \int 1_{A^*} P_2(E^*_x)dP_1(x), \end{aligned} \tag{A15.24}$$

where the last equality is justified because the set of points x for which $P_2(E_x^*)$ is different from $1_{A^*}P_2(E_x^*)$ is contained in $A - A^*$ which itself is contained in a set $X \in \mathcal{A}_1$ with $P_1(X) = 0$. Since on A^* we have $E_x^* \subseteq B$ it follows that

$$\begin{aligned}\int_{A^*} P_2(E_x^*)dP_1(x) &\leq \int_{A^*} P_{2*}(B)dP_1(x) \\ &= P_{2*}(B)P_1(A^*) = P_{2*}(B)P_{1*}(A).\end{aligned} \tag{A15.25}$$

This proves finally that

$$P_*(A \times B) = P_{1*}(A)P_{2*}(B). \tag{A15.26}$$

Now let's prove that $pl_e(A \times B) = pl_{1e}(A)pl_{2e}(B)$. Since $v(A \times B) = v_1(A) \times v_2(B)$, we have to show that

$$P^*(A \times B) = P_1^*(A)P_2^*(B) \tag{A15.27}$$

for arbitrary sets A and B included in Ω_1 and Ω_2. But this equality can be proved in the same way as we proved (A15.22) (see also exercise 9 p. 150 of Halmos, 1950). □

16. Closed Random Intervals

16.1 Simple Closed Random Intervals

An important practical problem is modeling evidence about real valued variables. Although the general theory developed in the previous chapter applies to this case, it is necessary, for practical reasons, to limit the models to some simple structures. Such a simple structure is given by set-focused hints whose focal sets are closed intervals of real numbers. Random intervals describe two essential aspects of evidence about numbers: an interval expresses the imprecision of the knowledge about a number and the randomness captures the uncertainty about the number, the possibility of different cases. Due to its simplicity this is a very useful model for practical purposes. This chapter is therefore devoted to a study of this class of hints relative to real numbers. Several results in this chapter can be found in Dempster (1968 b), but here the discussion is extended in several respects.

In the first section of this chapter hints with only a finite number of distinct closed intervals as focal sets will be studied. This class of hints is called simple closed random intervals. In the following section, hints with an arbitrary, in general infinite number of closed intervals as focals sets will be introduced. It will be shown that an interesting subclass of those hints is the class obtained as limits of simple closed random intervals. This class will be called closed random intervals. A special subclass of closed random intervals are random points. Hints with single point focal sets are called precise. In section 16.4 random points will be studied especially in relation to the classical notion of random variables as a particular subclass of the closed random intervals.

Let then $(\Omega, \mathcal{A}, P)$ be a probability space and $A_1, \ldots, A_n$ a finite partition of Ω by measurable sets. Note that the intervals given below are the only non-empty closed intervals in $\mathcal{R}$:

$$\begin{aligned}
&[x, y], \text{ for } -\infty < x \leq y < +\infty \\
&(-\infty, y], \text{ for } y < +\infty \\
&[x, +\infty), \text{ for } -\infty < x \\
&(-\infty, +\infty).
\end{aligned} \tag{16.1}$$

To keep the notation simple, any interval of the form (16.1) will be denoted by $[x, y]$. For all $j = 1, \ldots, n$ let $[x_j, y_j]$ be an interval of the form (16.1).

Then we define a mapping Γ from Ω to the set of non-empty closed real intervals by $\Gamma(\omega) = [x_j, y_j]$ if $\omega \in A_j$. These elements define then a hint $\mathcal{H} = (\Omega, \mathcal{A}, P, \Gamma, \mathcal{R})$ relative to the set of real numbers. The n intervals $[x_j, y_j]$ are the focal sets of the hint $\mathcal{H}$.

Although Ω is not assumed to be a finite set here, it is clear that the hint $\mathcal{H}$ is equivalent to a hint with a finite set of interpretations. In this sense, the theory of the hints considered in this section corresponds essentially to the elementary theory exposed in Part I. Thus, if H is any subset of real numbers, then we define as in Sect. 3.1

$$\begin{aligned} u(H) &= \{\omega \in \Omega : \Gamma(\omega) \subseteq H\} = \sum \{A_j : [x_j, y_j] \subseteq H\}, \\ v(H) &= \{\omega \in \Omega : \Gamma(\omega) \cap H \neq \emptyset\} = \sum \{A_j : [x_j, y_j] \cap H \neq \emptyset\}, \\ w(H) &= \{\omega \in \Omega : \Gamma(\omega) \supseteq H\} = \sum \{A_j : [x_j, y_j] \supseteq H\}. \end{aligned} \tag{16.2}$$

These are all $\mathcal{A}$-measurable sets, such that $sp(H) = P(u(H))$, $pl(H) = P(v(H))$ and $q(H) = P(w(H))$ are defined for any subset H of real numbers.

Let $\mathcal{R}^2 = (\mathcal{R} \cup \{-\infty, +\infty\}) \times (\mathcal{R} \cup \{-\infty, +\infty\})$ denote the extended plane. Then the focal intervals $[x_j, y_j]$ of the hint $\mathcal{H}$ determine n points (x_j, y_j) in the extended plane which have assigned probability masses $m_j = P(A_j)$. This in turn defines a two-dimensional discrete random variable $(X(\omega), Y(\omega))$ with (x_j, y_j), $j = 1, 2, \ldots, n$ as possible values. X and Y are univariate discrete random variables. For $X \subseteq \mathcal{R}^2$ define $P'(X) = \sum \{m_j : (x_j, y_j) \in X\}$ and for $(x, y) \in \mathcal{P}^u = \{(x, y) \in \mathcal{R}^2 : x \leq y\}$ define $\Gamma'(x, y) = [x, y]$. Then the hint $\mathcal{H}' = (\mathcal{P}^u, \mathcal{P}(\mathcal{P}^u), P', \Gamma', \mathcal{R})$ is equivalent to $\mathcal{H}$ because for $H \subseteq \mathcal{R}$

$$\begin{aligned} sp'(H) &= P'(\{(x, y) : [x, y] \subseteq H\}) \\ &= \sum \{m_j : [x_j, y_j] \subseteq H\} = P(u(H)) = sp(H). \end{aligned} \tag{16.3}$$

Such a class of equivalent hints, as represented by a hint like $\mathcal{H}'$, is called a *simple closed random interval* (simple cri for short). The collection of values $m_j, j = 1, \ldots, n$ is called the basic probability assignment (bpa) of the simple cri.

Let $\mathcal{H}' = (\Omega', \mathcal{A}', P', \Gamma', \mathcal{R})$ and $\mathcal{H}'' = (\Omega'', \mathcal{A}'', P'', \Gamma'', \mathcal{R})$ be two simple closed random intervals. Then, if $\omega' \in A'_j$ and $\omega'' \in A''_k$, the mapping

$$\Gamma(\omega', \omega'') = [x'_j, y'_j] \cap [x''_k, y''_k] \tag{16.4}$$

is the multivalued mapping of the combined hint $\mathcal{H}' \oplus \mathcal{H}''$. Furthermore, the set

$$\begin{aligned} u(\emptyset) &= \{(\omega', \omega'') : \Gamma(\omega', \omega'') = \emptyset\} \\ &= \sum \{A'_j \times A''_k : [x'_j, y'_j] \cap [x''_k, y''_k] = \emptyset\} \end{aligned} \tag{16.5}$$

is measurable, i.e. it is in $\mathcal{A}' \otimes \mathcal{A}''$. Hence the two hints are combinable if

$$k = \sum \{m'_j m''_k : [x'_j, y'_j] \cap [x''_k, y''_k] = \emptyset\} \neq 1. \tag{16.6}$$

The combined hint $\mathcal{H}' \oplus \mathcal{H}''$ is then again a simple closed random interval with $\Omega = \Omega' \times \Omega'' - u(\emptyset)$ and the partition

$$A_h = \sum \{A'_j \times A''_k : [x'_j, y'_j] \cap [x''_k, y''_k] = [x_h, y_h]\} \tag{16.7}$$

corresponding to the closed intervals $[x_h, y_h]$. Its bpa is given by

$$m_h = \sum \{m'_j m''_k : [x'_j, y'_j] \cap [x''_k, y''_k] = [x_h, y_h]\}/(1-k). \tag{16.8}$$

Theorem 4.4 regarding commonality functions applies here and gives

$$q(H) = q'(H)q''(H)/(1-k) \tag{16.9}$$

for any nonempty subset H of real numbers.

Simple closed random intervals are both conceptually simple and easy to treat numerically as the above discussion shows. In the following section more general closed random intervals will be introduced, but in such a way that they can be approximated by simple cris. This will lead to a richer theory without the loss of the simplicity of numerical computations.

16.2 Closed Random Intervals

Let $(\Omega, \mathcal{A}, P)$ be as before a probability space and consider a mapping Γ from Ω into the family of closed intervals in the extended line $\mathcal{R}$

$$\Gamma(\omega) = [X(\omega), Y(\omega)] \subseteq \mathcal{R}, \tag{16.10}$$

where X and Y represent mappings from Ω into $\mathcal{R} \cup \{-\infty, +\infty\}$ such that $X(\omega) \leq Y(\omega)$ for all $\omega \in \Omega$. It seems reasonable to assume here that both applications X and Y are *measurable* with respect to the Borel algebra on $\mathcal{R} \cup \{-\infty, +\infty\}$ and $\mathcal{A}$. This implies that the sets

$$\begin{aligned}
u[x,y] &= \{\omega : [X(\omega), Y(\omega)] \subseteq [x,y]\} \\
&= \{\omega : x \leq X(\omega), Y(\omega) \leq y\} \\
&= \{\omega : X(\omega) \geq x\} \cap \{\omega : Y(\omega) \leq y\}, \\
v[x,y] &= \{\omega : [X(\omega), Y(\omega)] \cap [x,y] \neq \emptyset\} \\
&= \{\omega : x \leq X(\omega) \leq y\} \cup \{\omega : x \leq Y(\omega) \leq y\}, \\
w[x,y] &= \{\omega : [X(\omega), Y(\omega)] \supseteq [x,y]\} \\
&= \{\omega : X(\omega) \leq x, y \leq Y(\omega)\} \\
&= \{\omega : X(\omega) \leq x\} \cap \{\omega : Y(\omega) \geq y\}
\end{aligned} \tag{16.11}$$

are all measurable for any numbers $-\infty \leq x \leq y \leq +\infty$ of the extended real line. This is also true if the closed interval $[x, y]$ is replaced by the half-open intervals $(x, y]$ or $[x, y)$ or by the open interval (x, y). Inversely, if we impose these sets to be measurable, then X and Y must necessarily be measurable. This is a first justification for assuming X and Y to be measurable, because one wants at least to be able to obtain degrees of support and plausibilities for these intervals (without to resort to inner and outer measures). More importantly, this measurability condition also assures that the corresponding hint can be approximated by simple cris as will be shown below.

So, let's consider hints $\mathcal{H} = (\Omega, \mathcal{A}, P, \Gamma, \mathcal{R})$, where $\Gamma(\omega) = [X(\omega), Y(\omega)]$, X, Y are measurable and $X(\omega) \leq Y(\omega)$ for all $\omega \in \Omega$. Since X and Y are measurable, i.e. *random variables*, the bivariate random variable (X, Y) can be considered (it is indeed a measurable mapping). According to Dempster (1968 b), its joint distribution Q on the two-dimensional Borel σ-algebra $\mathcal{B}_2$ is completely characterized by the commulative distribution function

$$Q(x, y) = P(X \leq x, Y \geq y) = P([X, Y] \supseteq [x, y]) \tag{16.12}$$

for all numbers $x \leq y$ (Dempster actually asserts this when X and Y are finite, but the assertion is still valid in the more general case considered here). Since $Q(B) = 0$ for any Borel set B included in the lower half-plane $\{x > y\}$, Q is actually a probability measure on the σ-algebra formed by the Borel sets included in the upper half-plane $\{x \leq y\} = \mathcal{P}^u$. This defines a new hint $\mathcal{H}' = (\mathcal{P}^u, \mathcal{B}_2, Q, \Gamma', \mathcal{R})$ where $\Gamma'(x, y) = [x, y]$.

Now we prove that $\mathcal{E}'_s \subseteq \mathcal{E}_s$ (and hence $\mathcal{E}'_p \subseteq \mathcal{E}_p$ by duality). In fact, let $A \subseteq \mathcal{R}$ such that $A \in \mathcal{E}'_s$, i.e. $u'(A) = \{(x, y) \in \mathcal{P}^u : [x, y] \subseteq A\}$ is in $\mathcal{B}_2^u$. Then we have

$$\begin{aligned} u(A) &= \{\omega : [X(\omega), Y(\omega)] \subseteq A\} \\ &= \{\omega : (X(\omega), Y(\omega)) \in \{(x, y) \in \mathcal{P}^u : [x, y] \subseteq A\}\} \\ &= (X, Y)^{-1}(\{(x, y) \in \mathcal{P}^u : [x, y] \subseteq A\}) \\ &= (X, Y)^{-1}(u'(A)) \end{aligned} \tag{16.13}$$

and hence $u(A) \in \mathcal{A}$, i.e. $A \in \mathcal{E}_s$, since (X, Y) is a measurable mapping. Furthermore, since $Q = P \circ (X, Y)^{-1}$ it follows that

$$\begin{aligned} sp(A) = P(u(A)) &= P((X, Y)^{-1}(u'(A))) \\ &= Q(u'(A)) = sp'(A) \end{aligned} \tag{16.14}$$

for all $A \in \mathcal{E}'_s$ and hence $\mathcal{H}$ and $\mathcal{H}'$ are *equivalent* with respect to $\mathcal{E}'_s$. Such an equivalence class of hints as represented by $\mathcal{H}'$ is called a *closed random interval* (cri). It is easily verified that $\mathcal{E}'_s$ contains both closed intervals as well as their complements. Dempster (1968 b) assumes without proof that all Borel sets of $\mathcal{R}$ are s-measurable and p-measurable, i.e they are in $\mathcal{E}_s$ and $\mathcal{E}_p$. Whereas this may be intuitive, it seems not be easy to prove. Note that $Q(x, y)$ is nothing else than the commonality of closed intervals, that is $Q(x, y) = q[x, y] = P(w[x, y])$. From its definition it follows that $Q(x, y)$ is continuous from above for x and from below for y. This allows to obtain the commonality for all kind of intervals from $Q(x,y)$; for example $q(x,y] = Q(x^-,y)$, $q[x,y) = Q(x,y^+)$ and $q(x,y) = Q(x^-,y^+)$, where $Q(x^-,y) = \lim\ Q(x',x)$ for $x' \uparrow x$ and $Q(x,y^+) = \lim\ Q(x,y')$ for $y' \downarrow y$, etc.

Another way to determine the distribution of (X,Y) is by the function

$$\begin{aligned} S(x,y) &= P(\{[X,Y] \subset [x,y]\}) = P(\{x < X, Y < y\}) \\ &= P(\{x < X\} \cap \{Y < y\}) = 1 - P(\{X \leq x\} \cup \{y \leq Y\}) \\ &= 1 - (P\{X \leq x\} + P\{y \leq Y\} - P\{X \leq x, y \leq Y\}) \\ &= 1 - (F(x) + G(y) - Q(x,y)), \end{aligned} \tag{16.15}$$

where $F(x) = P\{X \leq x\}$ and $G(y) = P\{y \leq Y\}$ are the marginal distribution functions of X and Y respectively. These functions can be obtained either from $S(x,y)$ or from $Q(x,y)$ by integration, for example

$$F(x) = \int_x^{+\infty} Q(x,dy), \ G(y) = \int_{-\infty}^{y} Q(dx,y). \tag{16.16}$$

Note that $S(x,x) = 0$, such that

$$F(x) + G(x) - Q(x,x) = 1. \tag{16.17}$$

The distribution function $S(x,y)$ equals the degree of support of the open interval (x,y), $S(x,y) = sp((x,y)) = P((u(x,y)))$. It is continuous from below for x and from above for y. $S(x,y)$ allows therefore also for the determination of the supports of half-open and closed intervals, $sp(x,y] = S(x,y^+)$, $sp[x,y) = S(x^-,y)$ and $sp[x,y] = S(x^-,y^+)$.

The distribution of a closed random interval is thus determined either by its commonality on closed intervals or by its support on open intervals. We may also consider the degrees of plausibility of intervals, for example

$$\begin{aligned} pl([x,y]) &= P(v([x,y])) = 1 - P\{\{X > y\} \cup \{Y < x\}\} \\ &= 1 - (P\{X > y\} + P\{Y < x\}) \\ &= 1 - ((1 - F(y)) + (1 - G(x)) \\ &= F(y) + G(x) - 1. \end{aligned} \tag{16.18}$$

Similar formulas can be derived for half-open and open intervals. We remark that the knowledge of the plausibility function on intervals - in contrast to the support function - is not sufficient to determine the distribution of the closed random set; it determines only the marginal distributions F and G, $pl[-\infty,y] = F(y)$ and $pl[x,+\infty] = G(x)$.

Simple closed random intervals lead to discrete distributions for (X,Y) as a special case of closed random intervals. Another important special case of random variables (X,Y) are those with absolutely continuous distributions. Then a density function $q_{xy}(x,y)$ is defined on $\mathcal{R}^2$ with $q_{xy}(x,y) = 0$ for $y < x$. It follows that

$$Q(x,y) = \int_{-\infty}^{x} du \int_{y}^{+\infty} dv q_{xy}(u,v); \; S(x,y) = \int_{x}^{+\infty} du \int_{-\infty}^{y} dv q_{xy}(u,v). \tag{16.19}$$

In this case all intervals $[x,y], (x,y], [x,y)$ and (x,y) have the same commonality, support as well as the same plausibility. Define

$$q_x(x,y) = \int_{y}^{+\infty} q_{xy}(x,v)dv; \; q_y(x,y) = \int_{-\infty}^{x} q_{xy}(u,y)du. \tag{16.20}$$

Then we have

$$Q(x,y) = \int_{-\infty}^{x} q_x(u,y)du = \int_{y}^{+\infty} q_y(x,v)dv. \tag{16.21}$$

From this and (16.20) it follows that

$$\begin{aligned}(\partial/\partial x)Q(x,y) &= q_x(x,y); \; (\partial/\partial y)Q(x,y) = -q_y(x,y),\\ (\partial^2/\partial x\partial y)Q(x,y) &= (\partial/\partial x)(-q_y(x,y))\\ &= -q_{xy}(x,y) = (\partial/\partial y)q_x(x,y) = (\partial^2/\partial y\partial x)Q(x,y).\end{aligned} \tag{16.22}$$

Furthermore, in the case of absolutely continuous distribution, the marginal distributions $F(x)$ and $G(y)$ have also density functions $f(x)$ and $g(y)$, namley

$$\begin{aligned}f(x) &= \int_{x}^{+\infty} q_{xy}(x,v)dv\\ g(y) &= \int_{-\infty}^{y} q_{xy}(u,v)du.\end{aligned} \tag{16.23}$$

It follows that

$$\begin{aligned}(d/dx)F(x) &= f(x) = (\partial/\partial x)Q(x,y)|_{x=y}\\ (d/dy)G(y) &= -g(y) = (\partial/\partial y)Q(x,y)|_{x=y}.\end{aligned} \tag{16.24}$$

From (16.15) we obtain finally the following results:

$$\begin{aligned}&(\partial/\partial x)S(x,y) = -f(x) + q_x(x,y); \; (\partial/\partial y)S(x,y) = g(y) - q_y(x,y)\\ &(\partial^2/\partial x\partial y)S(x,y) = -q_{xy}(x,y) = (\partial^2/\partial y\partial x)S(x,y).\end{aligned} \tag{16.25}$$

Let now $\mathcal{H}' = (\Omega', \mathcal{A}', P', \Gamma', \mathcal{R})$ and $\mathcal{H}' = (\Omega'', \mathcal{A}'', P'', \Gamma'', \mathcal{R})$ be two closed and independent random intervals. As in the case of simple cris,

$$\Gamma(\omega',\omega'') = [X'(\omega'),Y'(\omega')] \cap [X''(\omega''),Y''(\omega'')] \tag{16.26}$$

is the multivalued mapping associated to $\mathcal{H}' \oplus \mathcal{H}''$. This intersection is again an interval $[X(\omega',\omega''),Y(\omega',\omega'')]$, if it is not empty. In fact, $X(\omega',\omega'') = \max\{X'(\omega'),X''(\omega'')\}$, $Y(\omega',\omega'') = \min\{Y'(\omega'),Y''(\omega'')\}$ and

$$\begin{aligned}u(\emptyset) &= \{(\omega',\omega'') : \Gamma(\omega',\omega'') = \emptyset\}\\ &= \{Y' < X''\} \cup \{Y'' < X'\}.\end{aligned} \tag{16.27}$$

This shows that $u(\emptyset)$ is measurable. If P denotes the product probability on $(\Omega' \times \Omega'', \mathcal{A}' \otimes \mathcal{A}'')$, then let $k = P(u(\emptyset))$. Consider

$$\begin{aligned} Q(x,y) &= P\{[X,Y] \supseteq [x,y]\} \\ &= P\{[X',Y'] \supseteq [x,y], [X'',Y''] \supseteq [x,y]\} \\ &= P'\{[X',Y'] \supseteq [x,y]\}P''\{[X'',Y''] \supseteq [x,y]\} \\ &= Q'(x,y)Q''(x,y). \end{aligned} \tag{16.28}$$

After conditioning on non-contradictory interpretations

$$\Omega = \{(\omega',\omega'') : \Gamma(\omega',\omega'') \neq \emptyset\},$$

the distribution function $Q(x,y)$ of the random variables (X,Y) becomes

$$Q(x,y) = Q'(x,y)Q''(x,y)/(1-k). \tag{16.29}$$

This corresponds to theorem 4.4 for closed random intervals. This discussion shows that the combination of cris by Dempster's rule leads again to a cri. If $[X',Y']$ and $[X'',Y'']$ are independent cris, then $[X',Y'] \oplus [X'',Y'']$ denotes the cri obtained by Dempster's rule from $[X',Y']$ and $[X'',Y'']$.

If both cris $\mathcal{H}'$ and $\mathcal{H}''$ have absolutely continuous distributions, then the combined cri has also an absolutely continuous distribution and its density function $q_{xy}(x,y)$ can be obtained from (16.29) and (16.22):

$$\begin{aligned} q_x(x,y) &= (q'_x(x,y)Q''(x,y) + Q'(x,y)q''_x(x,y))/(1-k) \\ q_y(x,y) &= (q'_y(x,y)Q''(x,y) + Q'(x,y)q''_y(x,y)/(1-k) \\ q_{xy}(x,y) &= (q'_{xy}(x,y)Q''(x,y) + q'_x(x,y)q''_y(x,y) \\ &\quad + q'_y(x,y)q''_x(x,y) + Q'(x,y)q''_{xy}(x,y))/(1-k). \end{aligned} \tag{16.30}$$

All these results and formulas generalize of course to the combination of more than two cris.

Next we study closed random intervals as limits of simple closed random intervals. In particular we show that any cri is a limit of coarser simple cris and inversely, that that any limit of simple cris is a cri. Let $\mathcal{H} = (\Omega, \mathcal{A}, P, \Gamma, \mathcal{R})$ be a closed random interval represented by the pair of random variables $(X(\omega), Y(\omega))$, $X(\omega) \leq Y(\omega)$ for all $\omega \in \Omega$. Fix an integer n and partition the real line $\mathcal{R} \cup \{-\infty, +\infty\}$ first into the intervals $[-\infty, -n), [+n, +\infty]$ and $[-n, +n)$ and the latter into subintervals $[k/2n, (k+1)/2n)$, $k = -n2^n, \ldots, 0, \ldots, n2^n - 1$. Let denote the indicator function of the set B by

$$i_B(x) = \begin{cases} 1 & \text{if } x \in B \\ 0 & \text{otherwise.} \end{cases}$$

Then define the simple random variable X_n by

$$\begin{aligned} X_n(\omega) = & -\infty \cdot i_{[-\infty,-n)}(X(\omega)) \\ & + \sum_{k=-n2^n}^{n2^n-1} k/2^n \cdot i_{[k/2^n,(k+1)/2^n)}(X(\omega)) \\ & + n \cdot i_{[n,+\infty]}(X(\omega)). \end{aligned} \tag{16.31}$$

Clearly $X_n(\omega) \leq X(\omega)$ for all ω and $X_n(\omega) > X(\omega) - 1/2^n$ for $\omega \in \{-n \leq X(\omega) < n\}$ and $X_n(\omega) = n$ for $\omega \in \{X(\omega) \geq n\}$. Hence for any ω there is an $N(\omega)$ such that $|X_n(\omega) - X(\omega)| \leq 1/2^n$ for all $n > N(\omega)$ (note that $X = +\infty$ is excluded in a cri). This shows that $X_n(\omega) \uparrow X(\omega)$ for all $\omega \in \Omega$.

Similarly, decompose the line into the intervals $[-\infty, -n]$, $(+n, +\infty]$ and $(-n, +n]$ and the latter into the intervals $(k/2^n, (k+1)/2^n]$ for $k = -n2^n, \ldots, 0, \ldots, n2^n - 1$. Define then the simple random variable Y_n by

$$\begin{aligned} Y_n(\omega) = & -\infty \cdot i_{[-\infty,-n]}(Y(\omega)) \\ & + \sum_{k=-n2^n}^{n2^n-1} k/2^n \cdot i_{(k/2^n,(k+1)/2^n]}(Y(\omega)) \\ & + n \cdot i_{(n,+\infty]}(Y(\omega)). \end{aligned} \tag{16.32}$$

This time we have $Y_n(\omega) \geq Y(\omega)$ for all ω, $Y_n(\omega) < Y(\omega) + 1/2^n$ for $\omega \in \{-n < Y(\omega) \leq n\}$, $Y_n(\omega) = n$ for $\omega \in \{Y(\omega) \leq -n\}$. For any ω there is an $N(\omega)$ such that $|Y_n(\omega) - Y(\omega)| \leq 1/2^n$ for all $n > N(\omega)$ (note that $Y = -\infty$ is excluded in a cri). This shows that $Y_n(\omega) \downarrow Y(\omega)$ for all $\omega \in \Omega$.

Together this shows that $X_n \leq X \leq Y \leq Y_n$ or $[X_n, Y_n] \supseteq [X, Y]$ and $[X_n, Y_n]$ is a simple cri which is coarser than the cri $[X, Y]$ in the sense of inclusion of hints. We say that the limit of $[X_n, Y_n]$ is $[X, Y]$ and we write $\lim\, [X_n, Y_n] = [X, Y]$ if the sequences of random variables $X_n \leq Y_n$ converge to $X \leq Y$ for all ω (for the subsequent discussion it would be sufficient to assume this convergence for almost all ω only). It is well known that such limits of simple random variables are measurable functions, that is again random variables (see for example Neveu, 1964). Thus the above analysis shows that we can always approximate - in the above sense - a cri $[X, Y]$ by a sequence of simple cris $[X_n, Y_n]$ containing $[X, Y]$.

Inversely, if $[X_n, Y_n]$, $n = 1, \ldots, n$ is any sequence of simple cris such that $\lim X_n(\omega) = X(\omega)$ and $\lim Y_n(\omega) = Y(\omega)$ for all ω, then X and Y are also measurable and hence $[X, Y]$ is also a cri (this follows from the fact that X_n and Y_n are simple random variables, i.e. they can be represented as a finite linear combination of measurable indicator functions, see for example Neveu, 1964.

Next the convergence of distributions, or more precisely commonality, support and plausibility functions are studied. The following theorem gives the main result in this respect.

Theorem 16.1 *(Weak Convergence Theorem). Let $[X_n, Y_n]$ be a sequence of cris converging to a cri $[X, Y]$. Then*

(1) If Q and Q_n are the distribution functions of (X, Y) and (X_n, Y_n) respectively, then the sequence $Q_n(x, y)$ converges to $Q(x, y)$ at every point of continuity $(x, y), -\infty < x < y < \infty$ of $Q(x, y)$. The sequences $Q_n(-\infty, y)$ and $Q_n(x, +\infty)$ converge to $Q(-\infty, y)$ and $Q(x, +\infty)$ at every point of continuity $-\infty < y$ and $x < +\infty$ of $Q(-\infty, y)$ and $Q(x, +\infty)$ respectively.

(2) If sp and sp_n are the support functions of open intervals for the cris $[X_n, Y_n]$ and $[X, Y]$ respectively, then the sequence $sp_n(x, y)$ converges to $sp(x, y)$ at every point of continuity $(x, y), -\infty < x < y < \infty$ of $sp(x, y)$. The sequences $sp_n(-\infty, y)$ and $sp_n(x, +\infty)$ converge to $sp(-\infty, y)$ and $sp(x, +\infty)$ at every point of continuity $-\infty < y$ and $x < +\infty$ of $sp(-\infty, y)$ and $sp(x, +\infty)$ respectively.

(3) If pl and pl_n are the plausibility functions of closed intervals for the cris $[X_n, Y_n]$ and $[X, Y]$ respectively, then the sequence $pl_n[x, y]$ converges to $pl[x, y]$ at every point of continuity $(x, y), -\infty < x < y < \infty$ of $pl[x, y]$. The sequences $pl_n[-\infty, y]$ and $pl_n[x, +\infty]$ converge to $pl[-\infty, y]$ and $pl[x, +\infty]$ at every point of continuity y and x of $pl[-\infty, y]$ and $pl[x, +\infty]$ respectively.

Similar convergence theorems also hold for open, half-open and closed intervals. And in fact at points of continuity of $Q(x, y)$ and using self-explanatory notations we have $Q(x, y) = Q[x, y) = Q(x, y] = Q[x, y]$ and simiarily for sp and pl. This theorem shows in particular how cris can be approximated by simple cris in distribution. This subject will be pursued in the next section.

16.3 Monotone Convergence of Cris and Dempster's Rule

Let $(\Omega, \mathcal{A}, P)$ be a probability space and $[X(\omega), Y(\omega)]$ and $[X'(\omega), Y'(\omega)]$ two cris defined on Ω such that $[X(\omega), Y(\omega)] \subseteq [X'(\omega), Y'(\omega)]$ for all ω. Then looked at these two cris as hints $\mathcal{H} = (\Omega, \mathcal{A}, P, \Gamma, \mathcal{R})$ and $\mathcal{H}' = (\Omega, \mathcal{A}, P, \Gamma', \mathcal{R})$ with $\Gamma(\omega) = [X(\omega), Y(\omega)]$ and $\Gamma'(\omega) = [X'(\omega), Y'(\omega)]$ we have clearly the inclusion relation $\mathcal{H} \subseteq \mathcal{H}'$ (see section 14.1). Therefore, by theorem 14.2 the inclusion

$$[sp(x, y), pl(x, y)] \subseteq [sp'(x, y), pl'(x, y)] \tag{16.33}$$

is valid for $x \leq y$.

Then consider a sequence $[X_n, Y_n]$ of cris converging to a cri $[X, Y]$ such that $[X_{n+1}, Y_{n+1}] \subseteq [X_n, Y_n]$ for all $n = 1, \ldots$. This will be denoted by $[X_n, Y_n] \downarrow [X, Y]$. Then by the above observation and the weak convergence theorem 16.1

$$\begin{aligned} sp_n(x, y) &\uparrow sp(x, y) \text{ for } n \to \infty \\ pl_n(x, y) &\downarrow pl(x, y) \text{ for } n \to \infty \end{aligned} \tag{16.34}$$

at every point (x, y) of continuity of $sp(x, y)$ and $pl(x, y)$ respectively. A similar result holds for a sequence with $[X_{n+1}, Y_{n+1}] \supseteq [X_n, Y_n]$ for $n = 1, \ldots$.

More interesting is the study of the approximation of combined cris by coarser ones, for example simple ones. Consider two independent cris $[X'(\omega'), Y'(\omega')]$ and $[X''(\omega''), Y''(\omega'')]$. Let

$$[X'_n(\omega'), Y'_n(\omega')]$$
$$[X''_n(\omega''), Y''_n(\omega'')]$$

be two sequences of cris which approximate

$$[X'(\omega'), Y'(\omega')]$$
$$[X''(\omega''), Y''(\omega'')]$$

monotonically from the exterior, that is

$$[X'_{n+1}, Y'_{n+1}] \subseteq [X'_n, Y'_n]$$
$$[X''_{n+1}, Y''_{n+1}] \subseteq [X''_n, Y''_n]$$

for all $n = 1, \ldots$ and $[X'_n, Y'_n]$ and $X''_n, Y''_n]$ converge to $[X', Y']$ and $[X'', Y'']$ respectively (such sequences always exist according to Sect. 16.2, we can even choose sequences of simple cris). We study the combined cri

$$[X_n, Y_n] = [X'_n, Y'_n] \oplus [X''_n, Y''_n]$$

for $n \to \infty$. Let

$$\Omega_n = \{[X'_n, Y'_n] \cap [X''_n, Y''_n] \neq \emptyset\} \subseteq \Omega' \times \Omega''. \tag{16.35}$$

Then we have $X_n = \max(X'_n, X''_n)$ and $Y_n = \min(Y'_n, Y''_n)$ on Ω_n. Clearly $\Omega_{n+1} \subseteq \Omega_n$ for $n = 1, \ldots$ such that $\Omega = \cap \Omega_n \subseteq \Omega_n$ for all n. It follows that $[X_n, Y_n]$ converges to $[X, Y] = [\max(X', X''), \min(Y', Y'')]$ on Ω and $X_n \uparrow X$, $Y_n \downarrow Y$. In this sense one may say that $[X'_n, Y'_n] \oplus [X''_n, Y''_n]$ approximates $[X', Y'] \oplus [X'', Y'']$ from the exterior. If these combined cris are looked at as hints, since Dempster's rule preserves inclusion of hints, we have

$$\begin{aligned}[X', Y'] \oplus [X'', Y''] \subseteq \ldots [X'_{n+1}, Y'_{n+1}] \oplus [X''_{n+1}, Y''_{n+1}] \\ \subseteq [X'_n, Y'_n] \oplus [X''_n, Y''_n] \subseteq \ldots\end{aligned} \tag{16.36}$$

where $\subseteq$ denotes the inclusion of hints.

The following theorem shows that if a combined cri is approximated monotonically from the exterior by a sequence of combined cris, then its support and plausibility functions may also be approximated by the corresponding support and plausibility functions of the coarser cris.

Theorem 16.2 *Let $[X'_n, Y'_n] \downarrow [X', Y']$ and $[X''_n, Y''_n] \downarrow [X'', Y'']$. If $sp_n(x, y)$ and $pl_n(x, y)$ denote the support and plausibility functions of open intervals relative to the combined cri $[X'_n, Y'_n] \oplus [X''_n, Y''_n]$ and $sp(x, y)$, $pl(x, y)$ denote the support and plausibility functions of open intervals relative to $[X', Y'] \oplus [X'', Y'']$, then*

$$\lim_{n\to\infty} sp_n(x,y) = sp(x,y), \quad \lim_{n\to\infty} pl_n(x,y) = pl(x,y) \tag{16.37}$$

at every point of continuity (x, y) of $sp(x, y)$ and $pl(x, y)$ respectively.

The same result also holds with respect to the support and plausibility of half-open and closed intervals, $sp[x, y)$, $sp(x, y]$, $pl[x, y)$, $pl(x, y]$ and $pl[x, y]$. As a corollary, it follows from this theorem and from (16.36) using theorem 14.2 that

$$\frac{sp_n(x,y)}{k_n} - \frac{1-k_n}{k_n} \leq sp(x,y) \leq pl(x,y) \leq \frac{pl_n(x,y)}{k_n}. \tag{16.38}$$

with $k_n = P(X \leq Y)/P(X_n \leq Y_n) \uparrow 1$ for $n \to \infty$. This shows that if the support and the plausibility of a combined cri is approximated from the exterior, then in general the inclusion $[sp(x, y), pl(x, y)] \subseteq [sp_n(x, y), pl_n(x, y)]$ does not hold.

16.4 Random Variables

Degenerate cris with $X(\omega) = Y(\omega)$, $(-\infty < X = Y < +\infty)$ for all $\omega \in \Omega$ are of particular interest. They correspond of course to random variables. Viewed as hints they represent precise hints. How do they fit into the present general theory of hints? It has already been shown in Chap. 3 and 4 that precise hints have some particular properties (see for example theorems 4.8 and 4.9). But these results where derived only for simple degenerate cris (in the present terminology). Do they carry over to the general case?

With obvious notations, first note that

$$\begin{aligned} sp(x, y] &= P(x < X \leq y) = F(y) - F(x) = pl(x, y], \\ sp[x, y) &= P(x \leq Y < y) = G(x) - G(y) = pl[x, y). \end{aligned} \tag{16.39}$$

This extends theorem 3.8 saying that the support and plausibility of a precise hint are equal and define a probability measure on the frame of discernement. Here the supports of half-open intervals define distribution functions which specify a probability measure on the Borel sets of the real line as is well known.

For numerical or other purposes, random variables (rvs) are often approximated by simple random variables (having only a finite number of possible values). This is justified by the fact that any rv is the limit of a sequence of simple rvs and by the weak convergence theorem of distribution functions. In the light of the discussion in the previous sections, it seems however also appropriate to approximate rvs by simple cris from the exterior. This idea will be pursued to some extend in the sequel.

Thus, let $Z : \Omega \to \mathcal{R}$ be a rv with $(\Omega, \mathcal{A}, P)$ as probability space. A possible exterior approximation of Z by a simple cri can be obtained from a finite set of points $x_1 < x_2 < \ldots < x_n$ on the real line as follows. Define

$$Z'(\omega) = \begin{cases} -\infty, & \text{if } Z(\omega) \leq x_1 \\ x_i & \text{if } x_i < Z(\omega) \leq x_{i+1} \\ x_n & \text{if } x_n < Z(\omega) \end{cases}$$

$$Z''(\omega) = \begin{cases} x_1, & \text{if } Z(\omega) \leq x_1 \\ x_{i+1} & \text{if } x_i < Z(\omega) \leq x_{i+1} \\ +\infty & \text{if } x_n < Z(\omega). \end{cases}$$

Then clearly $Z \in [Z', Z'']$, which corresponds to an inclusion of cris if Z is seen as a degenerate cri or as inclusion of hints, if Z is viewed as a precise hint. The cri $[Z', Z'']$ is a special kind of a simple cri, because the intervals form a partition of the real line (if one discards the limit points of the intervals which belong always to the two neighboring intervals; this small inconvenience could be avoided by considering random half-open intervals). Such a simple cri will be called a *partition* cri. A simple cri $[Z', Z'']$ with $Z \in [Z', Z'']$ is called a *cover* cri of Z.

Let sp, pl denote now the support and plausibility functions of the simple cri $[Z', Z'']$, whereas $F(x)$ represents the distribution function of the rv Z. From the inclusion of Z in $[Z', Z'']$ it from theorem 14.2 that

$$sp(x, y] \leq F(y) - F(x) \leq pl(x, y]. \tag{16.40}$$

Consider a sequence of sets of points $x_{i,1} < x_{i,2} < \ldots x_{i,n_i}$ for $i = 1, \ldots$ such that $x_{i,1} \to -\infty$ and $x_{i,n_i} \to +\infty$ for $i \to \infty$ and such that max $(x_{i,j+1} - x_{i,j})$ converges to zero for $i \to \infty$, then the corresponding sequence of cris $[Z'_i, Z''_i]$ converges to Z. From theorem 16.1 it follows then that $sp_i(x, y]$ and $pl_i(x, y]$ converge both to $F(y) - F(x)$ at every point of continuity (x, y) of the function $F(y) - F(x)$, and $sp_i(-\infty, y]$ and $pl_i(-\infty, y]$ converge to $F(y)$ at every point of continuity y of $F(y)$.

These observations form the base for the use of cover cris to approximate rvs. The following example illustrates this method.

Example 16.1 Stochastic Interval Arithmetics. Consider two independent rvs $Z_1 : \Omega_1 \to \mathcal{R}$ and $Z_2 : \Omega_2 \to \mathcal{R}$ and suppose the sum $Z = Z_1 + Z_2 : \Omega_1 \times \Omega_2 \to \mathcal{R}$ has to be computed. The solution of this problem is of course given by the convolution interval

$$F(y) = \int_{-\infty}^{+\infty} F_1(y - x) dF_2(x) \tag{16.41}$$

which gives the distribution function of the sum in terms of the distribution functions of the summands. But the numerical computation of this convolution is not always easy.

Let $[Z'_i, Z''_i]$ be the simple partition cover of Z_i corresponding to the points $x_{i,1} < x_{i,2} < \ldots x_{i,n_i}$ for $i = 1, 2$. Clearly, $Z_i \in [Z'_i, Z''_i]$ implies

$$Z = Z_1 + Z_2 \in [Z'_1 + Z'_2, Z''_1 + Z''_2] = [Z', Z''].$$

This last cover cri $[Z', Z'']$ for Z is of course still a simple cri, but no more a partition cri in general. But the supports and plausibilities of $[Z', Z'']$ are easy to compute because $[Z', Z'']$ has a finite number of focal intervals.

These are then lower and upper bounds for $F(y) - F(x)$ because $Z \in [Z', Z'']$ (compare (16.40)). Furthermore, as the partition of the cover cris for Z_i become finer and finer, $[Z', Z'']$ converges to $Z = Z_1 + Z_2$, therefore these supports and plausibility approximate the distribution of Z better and better.

Other arithmetic operations between rvs such as substraction, multiplication, division, max and min can be treated similarily. Essentially this method corresponds to a stochastic version of interval arithmetics, the latter being a well known method in numerical analysis. The approximation of rvs by cover cris is also of potential interest in domains such as stochastic filtering of nonlinear systems, where difficult numerical problems have to be solved. The basic reason is that covers are maintained by Dempster's rule (theorem 15.4) and that filtering of dynamical systems is essentially an application of Dempster's rule (see Chap. 10).

In chapter 4, the combination of an arbitrary hint with a precise hint has been considered. This subject will pursued here in the context of cris. So, let $[X, Y]$, $X, Y : \Omega_1 \rightarrow \mathcal{R}$ be a cri and $Z : \Omega_2 \rightarrow \mathcal{R}$ an independent rv. Let P be the product measure on the space $(\Omega_1 \times \Omega_2, \mathcal{A}_1 \otimes \mathcal{A}_2)$. Then $[X, Y] \oplus Z$ is defined unless $P(X \leq Z \leq Y) = 0$. And this combination must be again a rv. Define $\Omega = \{(\omega_1, \omega_2) : Z(\omega_2) \in [X(\omega_1), Y(\omega_1)]\}$. If $P(\Omega) > 0$, then $[X, Y] \oplus Z$ is defined on Ω as the rv $Z'(\omega_1, \omega_2) = Z(\omega_2)$, the corresponding σ-algebra is $\Omega \cap \mathcal{A}_1 \otimes \mathcal{A}_2$ and the probability measure is defined by $P'(\cdot) = P(\cdot)/P(\Omega)$. What is the distribution function of Z' ? The following theorem answers this question:

Theorem 16.3. *Let Z be a rv with distribution function F, $[X, Y]$ a cri with $pl(x) = pl[x, x]$ as its plausibility function of singletons. If Z and $[X, Y]$ are independent, then the distribution function F' of the rv $[X, Y] \oplus Z$, if it exists, is given by*

$$F'(y) = c \cdot \int_{-\infty}^{y} pl(x) dF(x), \; c^{-1} = \int_{-\infty}^{+\infty} pl(x) dF(x). \tag{16.42}$$

This is the cri-version of theorem 4.8. Like there, only the plausibility function of singletons (the likelihood function) of the cri is needed to compute the distribution of the combination of the cri with a rv. This is essentially Bayes theorem (see Chap. 10, where the relation to the classical Bayes theorem should become clear).

A16 Appendix: Proofs of Theorems in Chapter 16

Proof of Theorem 16.1

(1) Let's start with the case of the distribution (or commonality) function $Q(x,y)$. If (x,y), $-\infty < x < y < +\infty$ is a point of continuity of $Q(x,y)$, select x', y' such that $x' < x < y < y'$. Define the sets

$$\begin{aligned} A' &= \{[X,Y] \supseteq [x',y']\} \\ A_n &= \{[X_n,Y_n] \supseteq [x,y]\} \\ A'_n &= \{|X_n - X| \leq x - x', |Y_n - Y| \leq y' - y\} \\ B'_n &= \{|X_n - X| > x - x'\} \\ C'_n &= \{|Y_n - Y| > y' - y\}. \end{aligned}$$

Then, either $X_n \leq x, y \leq Y_n$ or else either $X_n > x$ or $Y_n < y$. Then for all n we have $\Omega = A'_n \cup B'_n \cup C'_n$ and hence $A' = A' \cap \Omega \subseteq (A'_n \cap A') \cup B'_n \cup C'_n$. But obviously we have $A'_n \cap A' \subseteq A_n$ and hence $A' \subseteq A_n \cup B'_n \cup C'_n$. Therefore

$$Q(x',y') = P(A') \leq P(A_n) + P(B'_n) + P(C'_n)$$

for all n. The convergence of $[X_n, Y_n]$ to $[X,Y]$, which is a convergence almost sure of X_n and Y_n to X and Y, implies convergence in probability for these variables and therefore $P(B_n) \to 0$ and $P(C'_n) \to 0$ for $n \to \infty$. Hence

$$Q(x',y') \leq \liminf_{n\to\infty} Q_n(x,y). \tag{A16.1}$$

Now select x'', y'' such that $x < x'' \leq y'' < y$ and define

$$\begin{aligned} A'' &= \{[X,Y] \supseteq [x'',y'']\} \\ A''_n &= \{|X_n - X| \leq x'' - x, |Y_n - Y| \leq y - y''\} \\ B''_n &= \{|X_n - X| > x'' - x\} \\ C''_n &= \{|Y_n - Y| > y - y''\}. \end{aligned}$$

As above it can be concluded that $A_n \subseteq A''_n \cup B''_n \cup C''_n$ and it follows that

$$P(A_n) \leq P(A'') + P(B''_n) + P(C''_n)$$

or, as $P(B''_n) \to 0$ and $P(C''_n) \to 0$ for $n \to \infty$

$$Q(x'',y''') \geq \limsup_{n\to\infty} Q_n(x,y). \tag{A16.2}$$

Equations (A16.1) and (A16.2) together show that

$$Q(x',y') \leq \liminf Q_n(x,y) \leq \limsup Q_n(x,y) \leq Q(x'',y''). \tag{A16.3}$$

Because (x,y) is a point of continuity of Q, then, if (x',y') and (x'',y'') both converge to (x,y), $Q(x',y')$ and $Q(x'',y'')$ converge both to $Q(x,y)$ and it follows therefore from (A16.3) that

$$\lim Q_n(x,y) = Q(x,y).$$

Next, suppose that $x = -\infty$ and let y be a point of continuity of $Q(-\infty, y)$. Let $y < y'$ and define

$$\begin{aligned} A' &= \{X = -\infty, y' \leq Y\} \\ A_n &= \{Y_n = -\infty, y \leq Y_n\} \\ C'_n &= \{|Y_n - Y| > y' - y\}. \end{aligned}$$

Then $A' \subseteq A_n \cup C'_n$ and as above it follows that

$$Q(-\infty, y') \leq \liminf_{n\to\infty} Q_n(-\infty, y). \tag{A16.4}$$

On the other hand, let $-\infty < y'' < y$ and

$$A'' = \{X = -\infty, y'' \leq Y\},$$

whereas A_n and C''_n remain defined as above. This time $A_n \subseteq A'' \cup C''_n$ and therefore

$$Q(-\infty, y'') \geq \limsup_{n\to\infty} Q_n(-\infty, y). \tag{A16.5}$$

Now, if y' and y'' converge both to y, because y is a point of continuity of $Q(-\infty, y)$, it follows from (A16.4) and A(16.5) that

$$\lim_{n\to\infty} Q_n(-\infty, y) = Q(-\infty, y).$$

The case $x < y = +\infty$ is proved in an analogous way.

The remaining assertions (2) and (3) for the support function sp for open intervals and of plausibility functions for closed intervals, as well as the other cases for half-open intervals are proved similarily. □

Proof of Thoeorem 16.2

The support $sp_n(x,y)$ of the open interval (x,y) is defined as

$$sp_n(x,y) = \frac{P\{x < X_n \leq Y_n < y\}}{P\{X_n \leq Y_n\}} = \frac{\tilde{s}p_n(x,y)}{P(\Omega_n)},$$

where P is the product measure of P' and P'' on $\Omega'_n \times \Omega''_n$ and $\tilde{s}p_n(x,y) = P(\{x < X_n \leq Y_n < y\})$. As $\Omega = \cap \Omega_n$, $\Omega_{n+1} \subseteq \Omega_n$, it follows that $P(\Omega_n) \downarrow P(\Omega)$ for $n \to \infty$. Define $\tilde{X}_n = X_n$ if $\omega \in \Omega_n$ and $\tilde{X}_n = -\infty$ otherwise and $\tilde{Y}_n = Y_n$ if $\omega \in \Omega_n$ and $\tilde{Y}_n = +\infty$ otherwise. Then

$$\begin{aligned} \tilde{s}p_n(x,y) = P&\{x < X_n \leq Y_n < y\} \\ P&\{x < X_n, Y_n < y, \omega \in \Omega_n\} \\ P&\{x < \tilde{X}_n, \tilde{Y}_n < y\} \end{aligned}$$

and $(\tilde{X}_n, \tilde{Y}_n) \to (\tilde{X}, \tilde{Y})$ where $\tilde{X} = X$ if $\omega \in \Omega$ and $\tilde{X} = -\infty$ otherwise and $\tilde{Y} = Y$ if $\omega \in \Omega$ and $\tilde{Y} = +\infty$ otherwise. From theorem 16.1 it follows that $\tilde{s}p_n(x,y)$ converges to $\tilde{s}p(x,y) = P\{x < X \leq Y < y\}$ at every point of continuity of $\tilde{s}p(x,y)$. But

$$\frac{P\{x < X \leq Y < y\}}{P(\Omega)} = sp(x,y).$$

This proves the theorem for $sp(x,y)$. The other cases are proved simiarily. □

Proof of Theorem 16.3

Approximate Z by a sequence of partition covers $[Z'_n, Z''_n]$ defined by a sequence of points $x_{-n,n} < \ldots < x_{k,n} < x_{k+1,n} < \ldots x_{n,n}$ such that $x_{-n,n} \rightarrow -\infty$, $x_{n,n} \rightarrow +\infty$ and $\max_k (x_{k,n} - x_{k-1,n}) \rightarrow 0$ for $n \rightarrow \infty$. Under these condition $[Z'_n, Z''_n]$ converges clearly to Z for each $\omega_2 \in \Omega_2$. Consider now $[X,Y] \oplus [Z'_n, Z''_n]$. We have

$$[X,Y] \cap [Z'_n, Z''_n] = \begin{cases} [x_{k,n}, x_{k+1,n}] & \text{if } X \leq Z'_n = x_{k,n}, Z''_n = x_{k+1,n} \leq Y \\ [X, x_{k+1,n}] & \text{if } Z'_n = x_{k,n} \leq X \leq Z''_n = x_{k+1,n} \leq Y \\ [x_{k,n}, Y] & \text{if } X \leq Z'_n = x_{k,n} \leq Z''_n = x_{k+1,n} \\ \emptyset & \text{otherwise} \end{cases}$$

Here $x_{-n-1,n} = -\infty$, $x_{n+1,n} = +\infty$ are defined as possible values of Z'_n and Z''_n respectively.

It follows now that

$$\begin{aligned} \tilde{s}p_n(x,y] &= P\{\emptyset \neq [X,Y] \cap [Z'_n, Z''_n] \subseteq (x,y]\} \\ &= \sum_{k=k'}^{k''} P\{x_{k,n} < Z \leq x_{k+1,n}, [X,Y] \cap [x_{k,n}, x_{k+1,n}] \neq \emptyset\} \\ &\quad + P\{x_{k'-1,n} < Z \leq x_{k',n}, x < X \leq x_{k',n}\} \\ &\quad + P\{x_{k'',n} < Z \leq x_{k''+1,n}, x_{k'',n} \leq Y \leq y\}, \end{aligned}$$

where k' denotes the smallest value of k such that $x_{k,n} > x$ and k'' the greatest value of k such that $x_{k,n} \leq y$. We obtain

$$\begin{aligned} \tilde{s}p_n(x,y] &= \sum_{k=k'}^{k''} (F(x_{k+1,n}) - F(x_{k,n})) pl[x_{k,n}, x_{k+1,n}] \\ &\quad + (F(x_{k',n}) - F(x_{k'-1,n}))(F_X(x_{k',n}) - F_X(x)) \\ &\quad + (F(x_{k''+1,n}) - F(x_{k'',n}))(G_Y(x_{k'',n}) - G_Y(y^+)). \end{aligned} \tag{A16.6}$$

For any $x \in [x_{k,n}, x_{k+1,n}]$ we have

$$pl[x_{k,n}, x_{k+1,n}] \geq pl[x,x].$$

Note that

$$\begin{aligned} pl[x,x] &= P_1(X \leq x \leq Y) = 1 - P_1\{X > x\} - P_1\{Y < x\} \\ &= 1 - (1 - F_X(x)) - (1 - G_Y(x)) = F_X(x) + G_Y(x) - 1. \end{aligned}$$

Define $pl_n(x) = pl[x_{k,n}, x_{k+1,n}]$ if $x \in [x_{k,n}, x_{k+1,n}]$ and $pl(x) = pl[x, x]$. Then, if $x \in [x_{k,n}, x_{k+1,n}]$,

$$\begin{aligned} pl_n(x) &= P\{X \leq x \leq Y\} + P\{x_{k,n} \leq Y < x\} + P\{x < X \leq x_{k+1,n}\} \\ &= pl(x) + (G_Y(x_{k,n}) - G_Y(x)) + (F_X(x_{k+1,n}) - F_X(x)). \end{aligned} \tag{A16.7}$$

Now, F_X is continuous from the right, whereas G_Y is continuous from the left. Therefore, for any sequences $x_{k+1,n} \downarrow x$ and $x_{k,n} \uparrow x$ the last two terms in (A16.7) vanish in the limit, which implies that $pl_n(x) \downarrow pl(x) \geq 0$ for $n \to \infty$. This in turn means that the first term in (A16.6) converges to the Lebesgue-Stieltjes integral

$$\int_x^y pl(x)dF(x) = \int i_{(x,y]} pl(x)dF(x). \tag{A16.8}$$

The last two terms in (A16.6) vanish for $n \to \infty$ because their first factors vanish at the continuity points x and y of F.

Now, by theorem 16.2 $\tilde{s}p_n(x, y] \to \tilde{s}p(x, y]$ at every point of continuity of sp. Thus,

$$\tilde{s}p(x, y] = \int_x^y pl(x)dF(x) \tag{A16.9}$$

at every point of continuity x and y of F and

$$F'(y) = sp(-\infty, y] = c \cdot \int_{-\infty}^y pl(x)dF(x) \tag{A16.10}$$

$$1 = F'(+\infty) = c \cdot \int_{-\infty}^{+\infty} pl(x)dF(x), \tag{A16.11}$$

which proves the theorem. □

References

Abraham, J.A. (1979): An Improved Algorithm for Network Reliability. *IEEE Trans. on Reliability*, **R-28**, 58–61.

Aho, A.V., Hopcroft, J.E., Ullmann, J.D. (1983): *Data Structures and Algorithms.* Addison-Wesley, Reading, Mass.

Allen, J.F. (1983): Maintaining Knowledge about Temporal Intervals. *Comm. of the ACM,* **26**, 832-843.

Allen, J.F., Kautz, H.A. (1985): A Model of Naive Temporal Reasoning. In Hoobs, J., Moore, R. (Eds.): Formal Theories of the Commonsense World. Ablex, Norwood, NJ, 251-268.

Almond R., Kong, A. (1991): Optimality Issues in Constructing a Markov Tree From Graphical Models. Research Report A-3, Harvard Univ., Dept. of Statistics.

Arnborg S., Corneil, D., Proskurowski, A. (1987): Complexity of Finding Embeddings in a k-tree. SIAM J. of Algebraic and Discrete Methods, **8**, 277–284.

Ball, M. (1986): Computational Complexitiy of Network Reliability Analysis: An Overview. *IEEE Trans. on Reliability*, **R-35**, 230–239.

Barlow, R.E., Proschan, R. (1975): *Statistical Theory of Reliability and Life testing.* New York.

Barnett, J.A. (1981): Computational Methods for a Mathematical Theory of Evidence. In: Proc. *IJCAI-81*, Vancouver, BC, 868–875.

Beeri, C., Fagin, R., Maier, D., Yannakakis, M. (1983): On the Desirability of Acyclic Database Schemes. *J. of ACM*, **30**, no. 3, 479-513.

Berge, C. (1973): *Graphs and Hypergraphs.* North-Holland.

Berge, C. (1985): *Graphs.* Second Revised Edition. North-Holland.

Berge, C., Ghouila-Houri, A. (1962): *Programmes, Jeux et Réseaux de Transports.* Dunod, Paris.

Bernoulli, J. (1713): *Ars Conjectandi.* Reprinted in 1968 by Culture et Civilisation, 115 Avenue Gabriel Lebon, Brussels.

Bertele, U., Brioschi, F. (1972): *Nonserial Dynamic Programming.* Academic Press.

Besnard, P., Kohlas, J. (1994): Evidence Theory Based on General Consequence Relations. *Internal Working Paper no. 94-08, Institute of Informatics, University of Fribourg, Switzerland.*

Birkhoff, G. (1967): *Lattice theory.* Third Edition. American Mathematical Society.

Birkhoff, G., Bartee T.C. (1970): *Modern Applied Algebra.* Mc-Graw Hill, New York.

Bodily, S.E. (1985): *Modern decision making.* New York.

Bondy, J.A., Murty, U.S. (1976): *Graph Theory with Applications.* Elsevier.

Bossu, G., Siegel, P. (1985): Saturation, Nonmonotonic Reasoning, and the Closed-World Assumption. *Artificial Intelligence* **25**, 23–67.

Brooks, R.A. (1981): Symbolic Reasoning Among 3-D Models and 2-D Images. *Artificial Intelligence*, **32**, 281-331.

Buneman, P. (1974): A Characterization of Rigid Circuit Graphs. *Discrete Math.*, **9**, 205–212.

Bunke, H. (1975): Statistical inference: fiducial and structural vs. likelihood. *Math. Operationsforschung und Statistik*, **6**, 667-676

Cardona, L. (1993): Die Behandlung von Inferenznetzen mit unzuverlässigen Regeln anhand der Theorie der Hinweise. *Doctoral Dissertation*, Institute for Automation and O.R., Univ. of Fribourg, Switzerland.

Cannings, C., Thompson, E.A., Skolnick, M.H. (1978): Probability Functions on Complex Pedigrees. *Advances in Applied Probability*, **10**, 26–61.

Chang, C.E., Lee R.C.T. (1973): *Symbolic Logic and Mechanical Theorem Proving.* Academic Press.

Choquet, G. (1953): Theory of capacities. *Ann. Inst. Fourier* (Grenoble) **5**, 131–295

Choquet, G. (1969): *Lectures on Analysis.* Benjamin Cummings. New York.

Dantzig, G.B. (1964): *Linear Programming and Extensions.* Princeton University Press.

Dawid, A.P., Stone, M. (1982): The Functional-Model Basis of Fiducial Inference. *Ann. Math. Stat.*, **10**, 1954-1967

Dean, T. (1985): Temporal Imaginery: An Approach to Reasoning about Time for Planning and Problem Solving. Research Report 433, Yale University, New Haven, CT.

De Kleer, J. (1986 a): An Assumption-based TMS. *Artificial Intelligence*, **28**, 127–162.

De Kleer, J. (1986 b): Extending the ATMS. *Artificial Intelligence*, **28**, 163–196.

De Kleer, J., Williams, B.C. (1987): Diagnosing Multiple Faults. *Artificial Intelligence*, **32**, 97–130.

Dempster, A.P. (1965): New Methods for Reasoning Towards Posterior Distributions Based on Sample Data. *Annals of Math. Statistics*, **37**, 355–374.

Dempster, A.P. (1967): Upper and Lower Probabilities Induced by a Multivalued Mapping. *Ann. Math. Statist.* **38**, 325–339.

Dempster, A.P. (1968 a): A Generalization of Bayesian Inferences. *J. Royal Stat. Society*, B, **30**, 205-247.

Dempster, A.P. (1968 b): Upper and Lower Probabilities Generated by a Random Closed Interval. *Annals of Math. Stat.*, **39**, 957–966.

Dempster, A.P. (1988): Construction and Local Computation Aspects of Network Belief Functions. In: Oliver, R.M., Smith, J.Q. (ed.): *Influence Diagrams, Belief Nets and Decision Analysis.* 121–141. Wiley.

Dirac, G.A. (1961): On Rigid Circuit Graphs. Abh. Math. Sem. Univ. Hamburg, **25**, 71–76. MR24 #57

Dubois, D., Prade, H. (1978): Algorithmes de plus courts chemnins pour traiter des donnée floues. *RAIRO*, **12**, no. 2, 213-227.

Dubois, D., Prade, H. (1985): *Théorie des possibilités. Applications à la représentation des connaissances en informatique.* Masson, Paris

Dubois, D., Prade, H. (1986): A set-theoretic view of belief functions. Logical operations and approximations by fuzzy sets. *Int. J. General Systems*, **12**, 193-226.

Dubois, D., Prade, H. (1987): Properties of measures of information in evidence and possibility theories. *Fuzzy Sets and Systems*, **24**, 161–182

Dubois, D., Prade, H. (1990): Consonant approximations of belief functions. *Int. J. of Approximate Reasoning*, **4**, 419–449

Fagin, R. (1983): Degrees of Acyclicity for Hypergraphs and Relational Database Schemes. *J. of ACM*, **30**, no. 3, 514-550. Fagin, R., Halpern, J.Y. (1989): Uncertainty, Belief, and Probability. *IJCAI-89: 1161-1167.*

Feller, W. (1968): *An Introduction to Probability Theory and its Applications.* Third edition, vol. 1, Wiley

Fraser, D.A.S. (1968): *The structure of inference.* Wiley, New York.

Fraser, D.A.S. (1988): *Structural inference.* Entry in: Encyclopedia of Statistical Sciences. Vol. 9, ed. by S. Kotz and N.L. Johnson, Wiley, New York, 20-27.

Fulkerson D.R., Gross, O.A. (1965): Incidence Matrices and Interval Graphs. *Pacific J. Math.*, **15**, 835-855. MR32 #383812.

Gans, D. (1969): *Transformations and Geometries.* Appleton-Century-Crofts Eds.

Gavril, F. (1974): The Intersection Graphs of Subtrees in Trees are Exactly the Chordal Graphs. *J. Combin. Theory*, **B 16**, 47–56. MR48 #10868.

Geidel, J. (1989): Zeitplanung von Projeckten mit Unsharfen Daten. Diplomarbeit. Institut für Wirtschaftstheorie und O.R., Univ. Karlsruhe.

Ghouila-Houri, A. (1960): Sur l'Existence d'un Flot ou d'une Tension Prenant ses Valeurs dans un Groupe Abélien. *Comptes Rendus de l'Académie des Sciences*, **250**, 3931-3932.

Ghouila-Houri, A. (1964): Flots et Tensions dans un Graphe. *Annales de l'Ecole Normale*, **81**, 267-339.

Golenko, D.I. (1972): *Statistische Methoden der Netzplantechnik.* Teubner.

Golumbic, M.C. (1980): *Algorithmic Graph Theory and Perfect Graphs.* Academic Press.

Gondran, M., Minoux, M. (1984): *Graphs and Algorithms.* Wiley.

Gordon, J, Shortliffe, E.H. (1984): The Dempster-Shafer theory of evidence. In Buchanan, B.G., Shortliffe, E.H. (1984): Rule-based expert systems: the MYCIN experiments at the Stanford heuristic programming project. Addison Weseley, Reading, 272–292

Graham, M.H. (1979): On the universal relation. Tech. Rep. Univ. of Toronto, Canada.

Grätzer G. (1978): *General lattice theory.* Academic Press, New York.

Hacking, I. (1975): *The emergence of probability.* Cambridge

Hall, M. (1967): *Combinatorial Theory.* Blaisdell

Halmos, P.R. (1950): *Measure Theory.* Van Nostrand.

Halmos, P.R. (1963): *Lectures on Boolean algebras.* Van Nostrand-Reinhold, London.

Hässig, K. (1979): *Graphentheoretische Methoden des Operations Research.* Teubner.

Heckermann, D.E. (1986): Probabilistic Interpretations for MYCIN's Certainty Factors. In Kanal, L., Lemmer J., Eds. (1986): Uncertainty in artificial intelligence. North Holland

Heidtmann, K.D. (1989): Smaller Sums of Disjoint Products by Subproduct Inversion. *IEEE Trans. on Reliability*, **R-38**, 305–311.

Hestir, K., Nguyen, H.T., Rogers, G.S. (1991): A Random Set Formalism for Evidential Reasoning. In: Goodman, I.R., Gupta, M.M., Nguyen, Rogers G.S. (eds.): *Conditional Logic in Expert Systems*, North-Holland, 309–344.

Hsia, Y.T.; Shenoy, P.P. (1989): MacEvidence: a Visual Evidential Language For Knowledge-Based Systems. The Univsersity of Kansas, School of Business, Working Paper 211.

Inoue, K. (1991 a): An Abductive Procedure for the CMS/ATMS. In: Martin, J.P., Reinfrank, M. (eds.): *Proc. of the 1990 Workshop on Truth Maintenance Systems*, Lecture Notes in AI, Springer.

Inoue, K. (1991 b): Consequence-Finding Based on Ordered Resolution. *Proc. IJCAI-91*, 158–164.

Inoue, K. (1992): Linear Resolution for Consequence Finding. *Artificial Intelligence*, **56**, 301–353.

Jensen, F.V. (1988): Junction Trees and Decomposable Hypergraphs. JUDEX Research Report, Aalborg Univ., Denmark.

Kall, P. (1976): *Stochastic Linear Programming.* Springer Verlag.

Kean, A., Ksiknis, G. (1990): An Incremental Method for Generating Prime Implicants/Implicates. *J. Symbolic Computation*, **9**, 185–206.

Kjaerulff U. (1990): Triangulation of Graphs. Algorithms Giving Small Total State Space. Research Reoport R 90-09. Institute for Electronic Systems, Univ. of Aalborg, Denmark.

Knuth D.E. (1969): *The Art of Computer Programming. Volume 1, Fundamental Algorithms.* Addison-Wesley.

Kohlas, J. (1987): *Zuverlässigkeit und Verfügbarkeit.* Teubner.

Kohlas, J. (1991 a): The Reliability of Reasoning With Unreliable Arguments. *Annals of Oparations Research*, **32**, 67–113.

Kohlas, J. (1991 b): Describing Uncertainty in Dynamical Systems by Uncertain Restrictions. In: Di Masi, G.B., Gombani, A., Kurzhanski, A.B. (Eds.): *Modeling, Estimation and Control of Systems with Uncertainty.* Birkhäuser, Boston, 210-223.

Kohlas, J. (1993): Support and Plausibility Functions Induced by Filter-Valued Mappings. *Int. J. of General Systems*, **21**, 343–363.

Kohlas, J. (1993 a): Symbolic Evidence, Arguments, Supports and Valuation Networks. In: Clark, M., Kruse, R., Moral, S. (eds.): *Quantitative Approaches to Reasoning and Uncertainty.* Lecture Notes in Computer Science 747, Springer, 186–198.

Kohlas, J. (1994): Mathematical Foundations of Evidence Theory. *Internal Working Paper no. 94-09, Institute of Informatics, University of Fribourg, Switzerland.*

Kohlas, J., Monney, P.A. (1990): Modeling and Reasoning with Hints. Working Paper 174. Institute for Automation and O.R. University of Fribourg, Switzerland.

Kohlas, J., Monney, P.A. (1991): Propagaging Belief Functions Through Constraint Systems. *Int. Journal of Approximate Reasoning*, **5**, 433-461.

Kohlas, J., Monney, P.A. (1994): Representation of Evidence by Hints. In: Yager, R.R., Kacprzyk J., Fedrizzi, M. (eds.): *Advances in the Dempster-Shafer Theory of Evidence*, 473–492, Wiley.

Kong, A. (1986): Multivariate Belief Functions and Graphical Models. Ph. D., Dept. of Statistics, Harvard Univ.

Kruse, R, Schwecke, E., Heinsohn, J. (1991): *Uncertainty and Vagueness in Knowledge Based Systems.* Springer.

Kurzhanski, A.B. (1989): The treatment of uncertainty in system dynamic models. *International Institute for Applied Systems Analysis, Laxenburg, Austria. Internal Paper.*

Lambert, J.H. (1764): *Neues Organon.* Reprinted in 1965 as the first two volumes of Lambert's Philosophische Schriften by Georg Olms Verlagsbuchandlung, Hildesheim.

Lange, K., Boehnke, M. (1983): Extensions to Pedigree Analysis. Optimal Calculation of Mendelian Likelihoods. *Hum. Hered.*, **33**, 291–301.

Laskey, K.B., Lehner, P.E. (1989): Assumptions, beliefs and probabilities. Artificial Intelligence, **41**, 65–77.

Lauritzen, S.L., Speed, T.P., Vijayan, K. (1984): Decomposable Graphs and Hypergraphs. *J. Australian Math. Society (Series A)*, **36**, 12–29.

Lauritzen, S.L., Spiegelhalter, D.J. (1988): Local Computations With Probabilities on Graphical Structures and Their Application to Expert Systems. *J. Royal Stat. Soc.*, Series B, **50**, 157–224.

Leimer, H.G. (1985): Strongly Decomposable Graphs and Hypergraphs. Doctoral Thesis, Johannes Gutenberg Universität, Mainz.

Leimer, H.G. (1989): Triangulated Graphs With Marked Vertices. *Annals of Discrete Math.*, **41**, 311–324.

Mackworth, A.K. (1977): Consistency in Networks of Relations. *Artificial Intelligence*, **8**, 99-108.

Mac Lane, S., Birkhoff, G. (1971): *Algèbre. Vol. 2: Les Grands Théorèmes.* Gauthier-Vilars, Paris.

Maier, D. (1983): *The Theory of Relational Data Bases.* Computer Science Press.

McDermott, D., Davis, E. (1984): Planning Routes Through Uncertain Teritory. *Artificial Intelligence*, **22**, 107-156.

Mellouli, K. (1988): On the Propagation of Beliefs in Networks Using the Dempster -Shafer Theory of Evidence. Ph.D., School of Business, The University of Kansas, Lawrence.

Minieka, E. (1978): *Optimization Algorithms for Networks and Graphs.* Dekker.

Monney, P.A. (1991): Le raisonnement temporel et géométrique sous incertitude. Une approche avec la théorie mathématique des indications. Thèse de Doctorat. Institut de Mathématiques, Univ. de Fribourg.

Neveu, J. (1964): *Bases Mathématiques du Calcul des Probabilités.* Masson. Paris.

Ngair, T.H., Provan, G. (1993): A Lattice-Theoretic Analysis of ATMS Problem Solving. In: *Proceedings of the ECSQARU 1993 Conference, Granada, Spain.* Lecture Notes in Computer Science, vol. 747, pp 282–289, Springer.

Nguyen, H.T. (1978): On Random Sets and Belief Functions. *J. Math. Analysis and Applications*, **65**, 531–542.

Oliver, R.M., Smith, J.Q. (Eds.) (1990): *Influence Diagrams. Beliefs and Decision Analysis.* Wiley.

Olmsted, S.M. (1983): On Representing and Solving Decision Problems. *Ph. D. Thesis*, Department of Engineering-Economic Systems. Stanford University.

Pearl, J. (1988): *Probabilistic Reasoning in Intelligent Systems. Networks of Plausible Inference.* Morgan and Kaufman.

Peng, Y., Reggia, J. (1990): *Aductive Inference Model for Diagnostic Problem Solving.* Springer, New York.

Provan, G.M. (1990): A Logic-based Analysis of Dempster-Shafer Theory. *Int. J. Approximate Reasoning*, **4**, 451–495.

Plante, A. (1979): On the validation of fidudical techniques. *The Canadian J. of Statistics*, **7**, no. 2, 217-226.

Preparata, F.P., Shamos, M.I. (1985): *Computational Geometry. An Introduction.* Springer Verlag, New York.

Reggia, J., Nau, D.S., Wang, P.Y. (1983): Diagnostic Expert Systems Based on a Set Covering Model. *Int. J. Man-Machine Studies*, **19**, 437–460.

Reggia, J., Nau, D.S., Wang, P.Y. (1985): A Formal Model of Diagnostic Inference. *Information Sciences*, **37**, 227–256.

Reiter, A. (1987): A Theory of Diagnosis from First Principles. *Artificial Intelligence*, **32**, 57–95.

Reiter, R., De Kleer, J. (1987): Foundations of Assumption-based Truth Maintenance Systems. *Preliminary Report.* Proc. Amer. Assoc. A.I. 183–188.

Robinson, J.A. (1965): A Machine Oriented Logic Based on the Resolution Principle. *Journal of the ACM*, **12**, 23–41.

Rose, D.J. (1970): Triangulated Graphs and the Elimination Process. *J. Math. Analysis and Applications*, **32**, 597–609. MR42 #5840.

Rose, D.J. (1972): A Graph-Theoretic Study of the Numerical Solution of Sparse Positive Definite Systems of Linear Equations. In Read R.C. (Ed.) (1972): Graph Theory and Computing, Academic Press, pp. 183-217. MR49 #6579.

Rose, D.J. (1974): On Simple Characterizations of k-Trees. *Discrete Math.*, **7**, 317–322. MR49 #101.

Rose D.J., Tarjan, R.E. (1975): Algorithmic Aspects of Vertex Elimination. *Proc. 7th Annu. ACM Symp. Theory Computing*, 245–254. MR56 #7320.

Rose D.J., Tarjan, R.E. (1978): Algorithmic Aspects of Vertex Elimination of Directed Graphs. SIAM J. Appl. Math., **34**, 176–179.

Rose D.J., Tarjan, R.E., Leuker, G.S. (1976): Algorithmic Aspects of Vertex Elimination on Graphs. *SIAM J. Computing*, **5**, 266–283.

Rota, G.C. (1964): On the foundations of combinatorial theory. Part I, Theory of Möbius functions. Zeitschrift für Wahrscheinlichkeitstheory, **2**, 340–368

Roy, B. (1970): *Algèbre Moderne et Théorie des Graphes.* Tommes 1 et 2. Dunod, Paris.

Ruspini, E.H. (1987): The Logical Foundations of Evidential Reasoning. *Tech. Note 408, AI Center, SRI International, Menlo Park, CA.*

Sakarovitch, M. (1984): *Optimisation Combinatoire.* Tomes 1 et 2. Hermann, Paris.

Schaller, J.P. (1991): *Multiple criteria decision aid under incomplete information.* Ph.D. thesis, University of Fribourg, Faculty of Economics, Switzerland

Shafer, G. (1976 a): *A Mathematical Theory of Evidence.* Princeton Univ. Press

Shafer, G. (1976 b): A Theory of Statistical Evidence. Harper and Hooker (Eds.): *Foundations of Probability Theory, Statistical Inference, and Statistical Theories of Science*, Vol. II, Reidel, Dordrecht-Holland, 365-436.

Shafer, G. (1978): Non-additive probabilities in the work of Bernoulli and Lambert. *Archives for the history of science*, **19**, 309–370

Shafer, G. (1979): Allocations of probability. *Annals of Probability*, **7**, 827–839

Shafer, G. (1981): Constructive Probability. *Synthese*, **48**, 1-60.

Shafer, G. (1982): Belief Function and Parametric Models. *J. Royal Stat. Society*, B, **44**, 322-352.

Shafer, G. (1987): Belief functions and possibility measures. In Bezdek, J.C., Ed. (1987): *Analysis of fuzzy information*, vol. 1, Mathematics and Logic, CRC Press, 51–84

Shafer, G. (1990): Perspectives on the Theory and Practice of Belief Functions. *Int. J. of Approximate Reasoning*, **4**, 323–362.

Shafer, G. (1991): An Axiomatic Study of Computation in Hypertrees. Working Paper 232, School of Business, The University of Kansas.

Shafer, G., Logan, R. (1987): Implementing Dempster's Rule for a Hierarchical Evidence. *Artificial Intelligence*, **33**, 271–298.

Shafer, G., Shenoy, P.P., Meloulli, K. (1987): Propagating Belief Functions in Qualitative Markov Trees. *Int. J. of Approximate Reasoning*, **1**, 349–400.

Shenoy, P.P. (1991 a): Valuation-Based Systems: A Framework for Managing Uncertainty in Expert Systems. Working Papter 226, School of Business, The University of Kansas.

Shenoy, P.P. (1991 b): Independence in Valuation-Based Systems. The University of Kansas, School of Business, Working Paper 236.

Shenoy, P.P., Shafer, G. (1990): Axioms for Probability and Belief Functions Propagation. In: Shachter R.D. et al. (Eds.) (1990): Uncertainty in Artificial Intelligence 4. North-Holland.

Shortliffe, E.H., Buchanan, B.G. (1975): A model of inexact reasoning in medicine. Mathematical Biosciences, **23**, 351–379.

Siegel, P. (1987): *Représentation et utilisation de la connaissance en calcul propositionnel.* Ph.D. thesis, Université de Marseille II.

Simmonard, M. (1972): *Programmation Linéaire.* Tomes 1 et 2. Dunod, Paris.

Smets, P. (1979): Modèle Quantitatif du Diagnostique Médical. *Bulletin de l'Académie de Médecine de Belgique*, **134**, 330-343.

Smets, P. (1981): Medical Diagnosis: Fuzzy Sets and Degrees of Belief. *Fuzzy Sets and Systems*, **5**, 259–266.

Smets, P. (1988): Belief functions. In Smets, P., Ed. (1988): *Non standard logics for automated reasoning*, Academic Press, London, 253–286.

Tarjan R.E., Yannakakis, M. (1984): Simple Linear Time Algorithms to Test Chordality of Graphs, Test Acyclicity of Hypergraphs, and Selectively Reduce Acyclic Hypergraphs. *SIAM J. Computing*, **13**, 566–579.

Thoma, H.M. (1989): Factorization of belief functions. Ph.D. thesis, Harvard Univ., Dept. of Statistics.

Thomas, A. (1986): Optimal Computation of Probability Functions for Pedigree Analysis. *MA J. Math. Appl. Med. and Bio.*, **3**, 167–178.

Vere, S. (1983): Planning in Time: Windows and Durations for Activities and Goals. *IEEE Trans. Patt. Anal. Mach. Intel.* **5** 246-267

Walley, P. (1987): Belief function representations of statistical evidence. *The Annals of Statistics*, **15**, no. 4, 1439-1465.

Walter, J.R. (1972): Representations of Rigid Cycle Graphs. Ph.D. Thesis, Wayne State Univ.

Wen, W.X. (1991): Optimal Decomposition of Belief Networks. In: Bonissone P. et al (Eds.) (1991): Uncertainty in Artificial Intelligence 6. North-Holland.

Yager, R.R. (1985): The Entailment Principle for Dempster-Shafer Granule. *Tech. Report MII-512, Iona College, New Rochelle, NY.*

Yannakakis, M. (1981): Computing the Minimum Fill-in is NP-complete. *SIAM J. of Algebraic and Discrete Methods*, Vol. 2, No. 1, 77–79.

Yu, C.T., Ozsoyoglu, M.Z. (1979): An algorighm for tree-query membership of a distributed query. In *Proc. 1979 IEEE Compsac, IEEE, New York,* 306-312.

Index

Vol. 330: P.M. Kort, Optimal Dynamic Investment Policies of a Value Maximizing Firm. VII, 185 pages. 1989.

Vol. 331: A. Lewandowski, A.P. Wierzbicki (Eds.), Aspiration Based Decision Support Systems. X, 400 pages. 1989.

Vol. 332: T.R. Gulledge, Jr., L.A. Litteral (Eds.), Cost Analysis Applications of Economics and Operations Research. Proceedings. VII, 422 pages. 1989.

Vol. 333: N. Dellaert, Production to Order. VII, 158 pages. 1989.

Vol. 334: H.-W. Lorenz, Nonlinear Dynamical Economics and Chaotic Motion. XI, 248 pages. 1989.

Vol. 335: A.G. Lockett, G. Islei (Eds.), Improving Decision Making in Organisations. Proceedings. IX, 606 pages. 1989.

Vol. 336: T. Puu, Nonlinear Economic Dynamics. VII, 119 pages. 1989.

Vol. 337: A. Lewandowski, I. Stanchev (Eds.), Methodology and Software for Interactive Decision Support. VIII, 309 pages. 1989.

Vol. 338: J.K. Ho, R.P. Sundarraj, DECOMP: an Implementation of Dantzig-Wolfe Decomposition for Linear Programming. VI, 206 pages.

Vol. 339: J. Terceiro Lomba, Estimation of Dynamic Econometric Models with Errors in Variables. VIII, 116 pages. 1990.

Vol. 340: T. Vasko, R. Ayres, L. Fontvieille (Eds.), Life Cycles and Long Waves. XIV, 293 pages. 1990.

Vol. 341: G.R. Uhlich, Descriptive Theories of Bargaining. IX, 165 pages. 1990.

Vol. 342: K. Okuguchi, F. Szidarovszky, The Theory of Oligopoly with Multi-Product Firms. V, 167 pages. 1990.

Vol. 343: C. Chiarella, The Elements of a Nonlinear Theory of Economic Dynamics. IX, 149 pages. 1990.

Vol. 344: K. Neumann, Stochastic Project Networks. XI, 237 pages. 1990.

Vol. 345: A. Cambini, E. Castagnoli, L. Martein, P Mazzoleni, S. Schaible (Eds.), Generalized Convexity and Fractional Programming with Economic Applications. Proceedings, 1988. VII, 361 pages. 1990.

Vol. 346: R. von Randow (Ed.), Integer Programming and Related Areas. A Classified Bibliography 1984–1987. XIII, 514 pages. 1990.

Vol. 347: D. Ríos Insua, Sensitivity Analysis in Multiobjective Decision Making. XI, 193 pages. 1990.

Vol. 348: H. Störmer, Binary Functions and their Applications. VIII, 151 pages. 1990.

Vol. 349: G.A. Pfann, Dynamic Modelling of Stochastic Demand for Manufacturing Employment. VI, 158 pages. 1990.

Vol. 350: W.-B. Zhang, Economic Dynamics. X, 232 pages. 1990.

Vol. 351: A. Lewandowski, V. Volkovich (Eds.), Multiobjective Problems of Mathematical Programming. Proceedings, 1988. VII, 315 pages. 1991.

Vol. 352: O. van Hilten, Optimal Firm Behaviour in the Context of Technological Progress and a Business Cycle. XII, 229 pages. 1991.

Vol. 353: G. Ricci (Ed.), Decision Processes in Economics. Proceedings, 1989. III, 209 pages 1991.

Vol. 354: M. Ivaldi, A Structural Analysis of Expectation Formation. XII, 230 pages. 1991.

Vol. 355: M. Salomon. Deterministic Lotsizing Models for Production Planning. VII, 158 pages. 1991.

Vol. 356: P. Korhonen, A. Lewandowski, J . Wallenius (Eds.), Multiple Criteria Decision Support. Proceedings, 1989. XII, 393 pages. 1991.

Vol. 357: P. Zörnig, Degeneracy Graphs and Simplex Cycling. XV, 194 pages. 1991.

Vol. 358: P. Knottnerus, Linear Models with Correlated Disturbances. VIII, 196 pages. 1991.

Vol. 359: E. de Jong, Exchange Rate Determination and Optimal Economic Policy Under Various Exchange Rate Regimes. VII, 270 pages. 1991.

Vol. 360: P. Stalder, Regime Translations, Spillovers and Buffer Stocks. VI, 193 pages . 1991.

Vol. 361: C. F. Daganzo, Logistics Systems Analysis. X, 321 pages. 1991.

Vol. 362: F. Gehrels, Essays In Macroeconomics of an Open Economy. VII, 183 pages. 1991.

Vol. 363: C. Puppe, Distorted Probabilities and Choice under Risk. VIII, 100 pages . 1991

Vol. 364: B. Horvath, Are Policy Variables Exogenous? XII, 162 pages. 1991.

Vol. 365: G. A. Heuer, U. Leopold-Wildburger. Balanced Silverman Games on General Discrete Sets. V, 140 pages. 1991.

Vol. 366: J. Gruber (Ed.), Econometric Decision Models. Proceedings, 1989. VIII, 636 pages. 1991.

Vol. 367: M. Grauer, D. B. Pressmar (Eds.), Parallel Computing and Mathematical Optimization. Proceedings. V, 208 pages. 1991.

Vol. 368: M. Fedrizzi, J. Kacprzyk, M. Roubens (Eds.), Interactive Fuzzy Optimization. VII, 216 pages. 1991.

Vol. 369: R. Koblo, The Visible Hand. VIII, 131 pages.1991.

Vol. 370: M. J. Beckmann, M. N. Gopalan, R. Subramanian (Eds.), Stochastic Processes and their Applications. Proceedings, 1990. XLI, 292 pages. 1991.

Vol. 371: A. Schmutzler, Flexibility and Adjustment to Information in Sequential Decision Problems. VIII, 198 pages. 1991.

Vol. 372: J. Esteban, The Social Viability of Money. X, 202 pages. 1991.

Vol. 373: A. Billot, Economic Theory of Fuzzy Equilibria. XIII, 164 pages. 1992.

Vol. 374: G. Pflug, U. Dieter (Eds.), Simulation and Optimization. Proceedings, 1990. X, 162 pages. 1992.

Vol. 375: S.-J. Chen, Ch.-L. Hwang, Fuzzy Multiple Attribute Decision Making. XII, 536 pages. 1992.

Vol. 376: K.-H. Jöckel, G. Rothe, W. Sendler (Eds.), Bootstrapping and Related Techniques. Proceedings, 1990. VIII, 247 pages. 1992.

Vol. 377: A. Villar, Operator Theorems with Applications to Distributive Problems and Equilibrium Models. XVI, 160 pages. 1992.

Vol. 378: W. Krabs, J. Zowe (Eds.), Modern Methods of Optimization. Proceedings, 1990. VIII, 348 pages. 1992.

Vol. 379: K. Marti (Ed.), Stochastic Optimization. Proceedings, 1990. VII, 182 pages. 1992.

Vol. 380: J. Odelstad, Invariance and Structural Dependence. XII, 245 pages. 1992.

Vol. 381: C. Giannini, Topics in Structural VAR Econometrics. XI, 131 pages. 1992.

Vol. 382: W. Oettli, D. Pallaschke (Eds.), Advances in Optimization. Proceedings, 1991. X, 527 pages. 1992.

Vol. 383: J. Vartiainen, Capital Accumulation in a Corporatist Economy. VII, 177 pages. 1992.

Vol. 384: A. Martina, Lectures on the Economic Theory of Taxation. XII, 313 pages. 1992.

Vol. 385: J. Gardeazabal, M. Regúlez, The Monetary Model of Exchange Rates and Cointegration. X, 194 pages. 1992.

Vol. 386: M. Desrochers, J.-M. Rousseau (Eds.), Computer-Aided Transit Scheduling. Proceedings, 1990. XIII, 432 pages. 1992.

Vol. 387: W. Gaertner, M. Klemisch-Ahlert, Social Choice and Bargaining Perspectives on Distributive Justice. VIII, 131 pages. 1992.

Vol. 388: D. Bartmann, M. J. Beckmann, Inventory Control. XV, 252 pages. 1992.

Vol. 389: B. Dutta, D. Mookherjee, T. Parthasarathy, T. Raghavan, D. Ray, S. Tijs (Eds.), Game Theory and Economic Applications. Proceedings, 1990. IX, 454 pages. 1992.

Vol. 390: G. Sorger, Minimum Impatience Theorem for Recursive Economic Models. X, 162 pages. 1992.

Vol. 391: C. Keser, Experimental Duopoly Markets with Demand Inertia. X, 150 pages. 1992.

Vol. 392: K. Frauendorfer, Stochastic Two-Stage Programming. VIII, 228 pages. 1992.

Vol. 393: B. Lucke, Price Stabilization on World Agricultural Markets. XI, 274 pages. 1992.

Vol. 394: Y.-J. Lai, C.-L. Hwang, Fuzzy Mathematical Programming. XIII, 301 pages. 1992.

Vol. 395: G. Haag, U. Mueller, K. G. Troitzsch (Eds.), Economic Evolution and Demographic Change. XVI, 409 pages. 1992.

Vol. 396: R. V. V. Vidal (Ed.), Applied Simulated Annealing. VIII, 358 pages. 1992.

Vol. 397: J. Wessels, A. P. Wierzbicki (Eds.), User-Oriented Methodology and Techniques of Decision Analysis and Support. Proceedings, 1991. XII, 295 pages. 1993.

Vol. 398: J.-P. Urbain, Exogeneity in Error Correction Models. XI, 189 pages. 1993.

Vol. 399: F. Gori, L. Geronazzo, M. Galeotti (Eds.), Nonlinear Dynamics in Economics and Social Sciences. Proceedings, 1991. VIII, 367 pages. 1993.

Vol. 400: H. Tanizaki, Nonlinear Filters. XII, 203 pages. 1993.

Vol. 401: K. Mosler, M. Scarsini, Stochastic Orders and Applications. V, 379 pages. 1993.

Vol. 402: A. van den Elzen, Adjustment Processes for Exchange Economies and Noncooperative Games. VII, 146 pages. 1993.

Vol. 403: G. Brennscheidt, Predictive Behavior. VI, 227 pages. 1993.

Vol. 404: Y.-J. Lai, Ch.-L. Hwang, Fuzzy Multiple Objective Decision Making. XIV, 475 pages. 1994.

Vol. 405: S. Komlósi, T. Rapcsák, S. Schaible (Eds.), Generalized Convexity. Proceedings, 1992. VIII, 404 pages. 1994.

Vol. 406: N. M. Hung, N. V. Quyen, Dynamic Timing Decisions Under Uncertainty. X, 194 pages. 1994.

Vol. 407: M. Ooms, Empirical Vector Autoregressive Modeling. XIII, 380 pages. 1994.

Vol. 408: K. Haase, Lotsizing and Scheduling for Production Planning. VIII, 118 pages. 1994.

Vol. 409: A. Sprecher, Resource-Constrained Project Scheduling. XII, 142 pages. 1994.

Vol. 410: R. Winkelmann, Count Data Models. XI, 213 pages. 1994.

Vol. 411: S. Dauzère-Péres, J.-B. Lasserre, An Integrated Approach in Production Planning and Scheduling. XVI, 137 pages. 1994.

Vol. 412: B. Kuon, Two-Person Bargaining Experiments with Incomplete Information. IX, 293 pages. 1994.

Vol. 413: R. Fiorito (Ed.), Inventory, Business Cycles and Monetary Transmission. VI, 287 pages. 1994.

Vol. 414: Y. Crama, A. Oerlemans, F. Spieksma, Production Planning in Automated Manufacturing. X, 210 pages. 1994.

Vol. 415: P. C. Nicola, Imperfect General Equilibrium. XI, 167 pages. 1994.

Vol. 416: H. S. J. Cesar, Control and Game Models of the Greenhouse Effect. XI, 225 pages. 1994.

Vol. 417: B. Ran, D. E. Boyce, Dynamic Urban Transportation Network Models. XV, 391 pages. 1994.

Vol. 418: P. Bogetoft, Non-Cooperative Planning Theory. XI, 309 pages. 1994.

Vol. 419: T. Maruyama, W. Takahashi (Eds.), Nonlinear and Convex Analysis in Economic Theory. VIII, 306 pages. 1995.

Vol. 420: M. Peeters, Time-To-Build. Interrelated Investment and Labour Demand Modelling. With Applications to Six OECD Countries. IX, 204 pages. 1995.

Vol. 421: C. Dang, Triangulations and Simplicial Methods. IX, 196 pages. 1995.

Vol. 422: D. S. Bridges, G. B. Mehta, Representations of Preference Orderings. X, 165 pages. 1995.

Vol. 423: K. Marti, P. Kall (Eds.), Stochastic Programming. Numerical Techniques and Engineering Applications. VIII, 351 pages. 1995.